I0019666

Studies in
Logic
Logic and Cognitive Systems
Volume 9

Handbook of
Paraconsistency

Studies in Logic Series Editor
Dov Gabbay   dov.gabbay@kcl.ac.uk

Logic and Cogntive Systems editors
                    D. Gabbay, J. Siekmann, J. van Benthem and J. Woods

Handbook of Paraconsistency

Edited by

Jean-Yves Béziau
Walter Carnielli
Dov Gabbay

© Individual author and College Publications, 2007. All rights reserved.

ISBN 978-1-904987-73-4
College Publications
Scientific Director: Dov Gabbay
Managing Director: Jane Spurr
Department of Computer Science
King's College London
Strand, London WC2R 2LS, UK

Original cover design by orchid creative   www.orchidcreative.co.uk

All rights reserved. No part of this publication may be reproduced, stored in a retrieval system
or transmitted, in any form, or by any means, electronic, mechanical, photocopying, recording
or otherwise, without prior permission, in writing, from the publisher.

CONTENTS

III. Concepts and Tools for Paraconsistent Logics

IV. Results about Paraconsistent Logics

V. Philosophical Aspects of Paraconsistent Logic

Preface to the "Handbook of Paraconsistency"

The project of this *Handbook of Paraconsistency* was developed following the III World Congress on Paraconsistency held in 2003 in Toulouse, France.

It became apparent that Paraconsistent Logic was a fast growing field, interesting more and more people and that a reference book that could be useful for all people having interest in the subject, from those knowing almost nothing about the subject up to the more advanced researchers in the field, was still missing. This book is conceived in this spirit, with a first part dedicated to historical aspects of paraconsistent logic, a second part dealing with a variety of systems of paraconsistent logic, a third part with some basic concepts and tools useful for the study, construction, and development of paraconsistent systems, a fourth part with new results about already existing systems, and finally a fifth part discussing philosophical questions related to paraconsistency. The reader will find along hundreds of bibliographical references for further readings.

Let us briefly recall that paraconsistent logics are logics which allow to reason deductively under contradictions, or to put it in a more technical manner, to deal with contradictory yet non-trivial theories. The subject is a mathematical field surrounded by philosophical debate, with many applications in technology.

The field may be traced back to the beginning of modern logic with the work of Nicolai Vasiliev at the end of the XXth century proposing the idea of a non-Aristotelean logic by analogy to the work on non-Euclidean geometry of his fellow citizen from Kazan, Lobatchevski, and also with the work of the famous Polish logician, Jan Łukasiewicz, developing a critical analysis of the way that Aristotle is supporting the principle of non-contradiction.

But the expression *paraconsistent logic* was only coined in 1976 by the Peruvian philosopher Francisco Miró Quesada, later on President of the highest non-governmental world organization for philosophy, the FISP (International Federation of Philosophical Societies). His friend, the Brazilian logician Newton da Costa, had been working since the 1950s on systems allowing contradictory yet non-trivial theories and was asking him to find a good name for such systems. There was a discussion, Quesada proposed him several names and da Costa chose the term *paraconsistency*. It was indeed a good choice: after the official divulgation of this name (during the meeting of the Association for Symbolic Logic in Campinas, Brazil, 1976), it quickly spread around the world, strongly contributing to the popularity and development of this field of research. And in 1991 was created a special section for paraconsistency in the *Mathematical Reviews* of the *American Mathematical Society*, giving mathematical recognition to paraconsistent logic, a privilege shared by few other non-classical logics.

The 1st World Congress on Paraconsistency was organized in Ghent, Belgium, in 1997. and a 2nd World Congress on Paraconsistency in Juquehy, Brazil, in 2000. Two books resulted from these events, essential for anyone doing research in the field:

Frontiers of Paraconsistent Logic, D. Batens, C. Mortensen, G. Priest, and J.-P. van Bendegem, eds, Research Studies Press, Baldock, 2000.
Paraconsistency: The Logical Way to the Inconsistent, W. A. Carnielli, M. E. Coniglio, and I. M. L. D'Ottaviano, eds, Marcel Dekker, New York, 2002.

For the publication following the 3rd World Congress on Paraconsistency (WCP3), we decided, instead of publishing another proceedings volume, to work on a more ambitious project. This gave rise to the present *Handbook of Paraconsistency*. It contains some papers based on talks presented at the WCP3, but seriously reworked, and other papers not presented there but in the spirit of a Handbook on the field.

Since we think that events such as workshops and congresses are fundamental to develop scientific research by giving people the possibility to maintain discussions and more direct exchanges in a friendly human atmosphere beyond the cyberspace, we will say a few words about the WCP3.

The 3rd World Congress on Paraconsistency was held at the premises of the Institut de Recherche en Informatique de Toulouse (IRIT) in Toulouse, France, from 28-31 July, 2003. It was sponsored by the IRIT and co-sponsored by the Centre for Logic, Epistemology and History of Science (CLE) at Campinas, SP, Brazil and by the Stanford University at Stanford, USA. The proposal of organizing WCP3 in France intended to celebrate the collaboration between a French mathematician, Marcel Guillaume, and the main promotor of paraconsistent logic, the Brazilian logician Newton da Costa. Guillaume was a key figure in helping to disseminate da Costa's work, introducing the ideas of paraconsistency as a formal science to the prestigious *Comptes Rendus de l'Académie des Sciences de Paris* at the beginning of the sixties. During more than ten years da Costa and his collaborators published numerous notes on the advancement of their work in this prestigious scientific journal. Nowadays a new generation of researchers in France and in several other locations is interested in paraconsistent logic, working in particular on its applications to computer science and artificial intelligence. The IRIT itself was the scenery of the first developments of application of paraconsistent logic in the nineties: In 1992 a Ph.D. thesis by M. Lima-Marques supervised by L. Farinas del Cerro and co-supervised by W. A. Carnielli marked the beginning application of paraconsistent design to solve conflicts in air traffic control. Not any surprising research line, considering that Toulouse is one of the most important centers of aeronautic in the world.

Though an important language tool, English was not taken, as it so often happens, as the official language of the Conference: To avoid monolinguistic dominance and encourage linguistic democracy, participants were welcomed and encouraged to communicate in several different languages including *Interlingua*, a modern form of the *Latino sine Flexione*, an international language proposed by the famous Italian logician and mathematician Giuseppe Peano.

Almost one hundred participants from more than 20 countries came to Toulouse to share ideas, discuss not-so-easily conciliable philosophical views on paraconsistency, and compare new theories and applications of the paraconsistent paradigm in different areas as intelligent agents, belief change and logic programming. There was also a parallel poster session which permitted students and researchers not able to come in person or willing to present additional material to express themselves in poster format.

To complement the scientific program several social activities were planned by the local organizers, Andreas Herzig, Philippe Besnard and Luis Fariñas del Cerro (head of the IRIT). A cocktail party was offered by the mayor of Toulouse, welcoming the participants at the Town Hall, and an informal but savorous on board banquet (the "Captain Buffet") was organized during a cruise on the Garonne river and the Canal du Midi. Excellent local food and fine wine was served. *La crème de la crème* of such a live and productive gathering is the heart of this volume.

We would like to thank the numerous referees who help to us to prepare this book, to Jane Spurr for her competent work on file processing and to the contributors who patiently remained waiting for their publication in this *Handbook of Paraconsistency.*

Jean-Yves Béziau[1], Walter Carnielli[2] and Dov Gabbay
June 29th, 2007.

---

[1]Work supported by a grant of the Swiss National Science Foundation.
[2]Work supported by a CNPq (National Council for Scientific and Technological Development) grant and by FAPESP ConsRel Project, Brazil.

Part I

Survey Papers

da Costa 1964 Logical Seminar: Revisited Memories

MARCEL GUILLAUME

ABSTRACT.   A new insight on the structuring of some kinds of
paraconsistent logics is sketched by using quite elementary technical
means, after a recollection of the methods which, during the summer
of 1964, permitted us to discover, while working on the systems C_n of
da Costa, some new results (at that time). Some of these are echoed
by some of the ideas of Waragai and Shidori in their contribution to
the Congress. Strongly inspired by Carnielli and Marcos' *Taxonomy*,
this insight differs nevertheless from it in its structuring, and rests on
a prerequisite study of the links between the acceptance of some rules
pertaining to the method of semantic tableaux and the acceptance of
some corresponding schemes in Hilbert's style axiomatics of logics
which are intermediate between positive intuitionistic logic and full
classical logic.

Warnings.

All technical topics in the sequel are devoted to systems of *propositional*
calculuses, to their axioms, their formal proofs and theorems, which will always be given in the form of *schemes*: schemes of axioms, of formal proofs
and theorems, which, for the sake of brevity, will be respectively misnamed,
almost everywhere afterwards, axioms, formal proofs and theorems; similarly, for the same reason, the encountered formal systems will be improp-
erly termed *logics*. Among them, all those which are of the Hilbert type
obey *modus ponens* as their *unique inference rule*. Moreover, each of them
extends (properly or not) positive intuitionistic logic; this entails that the
so-called "deduction (meta)theorem" *holds* for all of them.

1   Recollections of Da Costa 1964 logic seminar at Curitiba.

I came to Curitiba as an invited professor with all my family for three
months during the summer of 1964. The initiator of that invitation was

my colleague of the University of Paraná, Newton da Costa, with whom I had got in contact since the beginning of 1962 through the intermediary of a mutual colleague and friend, Artibano Micali, a brasilian algebraist who came to Clermont-Ferrand a few years earlier in order to prepare a doctoral thesis under my thesis' advisor, Pierre Samuel, who had been an invited Professor at the University of São-Paulo during the year 1958. Micali, while staying in Brazil on vacation, had heard of plans concerning Newton's work, which was then in progress. At that time, I had just defended my doctoral thesis, and Artibano thought that there might be an opportunity to start a cooperation.

Finally such a cooperation was established, as I have related elsewhere [Guillaume, 1996]. I do not intend to repeat that paper, but merely to recall how the way was paved for the beginning of that cooperation and also to give more details on the manner in which my results were obtained. Some formulas on whose possible roles we were led to reflect on were reintroduced by Waragai and Shidori in their contribution to this Congress in a way that revived my interest in these axiomatic questions and led me to a reflection which I see throwing some light on a number of concepts and of new ways which are open to investigation.

To get back to my earliest contacts with Newton, I remember that a first exchange of letters began early in 1962. During 1963 he defended his own doctoral Thesis and undertook publishing his own ideas. Through Artibano he knew that at the time I had been hired as a lecturer at the University of Clermont-Ferrand, Samuel, who was leaving to teach at the Sorbonne, had introduced me in a kind of informal reviewing system around the *Académie des Sciences de Paris*. The point of such a system was that, according to the rules of the *Académie*, to present a *Note*, which is afterwards published in the *Comptes-Rendus des Séances hebdomadaires de l'Académie des Sciences de Paris*, is a privilege limited to the members of the *Académie*. So, at that time, to promote research by finding valuable works to present and to publish, the academicians trusted the scholars who had been presented to one of them by some earlier reviewer they knew[1]. The publication following the presentation to the *Académie* was quick – a few weeks. I had been thus presented by Samuel to the academician René Garnier.

Then in 1963 Newton sent me a planned *Note*, in which he explained his systems C_n, asking for its transmission to the *Académie*. He announced that some schemes were provable, others not, but many of these propositions were put forward without proofs – *e.g.* only the simplest matrices were

---

[1]Sometimes the work, of which it was required that it was written in french at that time, was not accepted without debates; so René Garnier asked me to justify the use of the french term "inconsistant", which I did in a way that he judged satisfactory.

produced. In fact, Newton's results were obtained by the use of matrices for disproofs and for proofs often by proof-theoretic methods with which I was not very familiar. My reaction was to verify all the details before sending the text. I did send it indeed, after having established my own proofs, often by semantic methods with which Newton was not very familiar, and René Garnier presented the *Note* to the *Académie*.

Amongst the methods that I brought into operation was Beth's "semantic tableaux" method, that I applied quite heuristically[2] to the case of the C_n ; some rules appeared clearly sound, if not complete, and gave me proofs.

Incidentally, I should say that today it seems to me that my fate in those times was to be in beginnings: during the academic year 1956-1957, I had been sent to Amsterdam with a grant from the French government to study logic under the firm rule of E.W. Beth. He had recently published his works on his "semantic tableaux" method, first in a classical version, and the previous year in an intuitionistic version[3]. So I became one of the first to learn his theory of "semantic tableaux", and further I had an opportunity to expose it to a session of the Bourbaki seminar, in 1958. He assigned to me the task to extend it to some systems of modal logic – in fact, at this time, for such systems, there were neither characteristic matrices nor Gentzen's type systems. I chose to work on Von Wright's systems M, M' and M"; the equivalence of the last two with Lewis' S4 and S5 respectively was known. It was not too difficult to find the good rules to apply – this was settled in the midspring of 1957 – but a completeness theorem was needed, and when, during the summer of 1957, I went to Beth's lounge to explain the complicated proof-theoretic method of proof I had conceived some days before (the point was to prove the cut rule), Beth, who was just coming home from a stay in the States was opening its mail in which he found the booklet of Stig Kanger [Kanger, 1957] in which the same Gentzen's type systems were introduced and proved sound and complete, by using semantics very different from my proof-theoretical arguments. Furthermore I used the fact that the axioms of M' are equivalent to Kuratowski's axioms for the general topology for providing sound and complete semantics for the same three systems; equivalently, Mac Kinsey and Tarski's closure algebras can be used instead.

There was some analogy between the heuristics with which I attacked the search for an extension of the method to some modal systems and those with which I used to verify Newton's assertions. In fact, the proofs and

---

[2] As Beth himself when he began to use it, before theorizing it.

[3] Some years after, facing with bitter criticisms against the completeness proof for the second, he changed his mind, succeeded in justifying the same rules by proof-theoretic methods, and presented them anew under the name of "deductive tableaux"; see [Beth, 1962].

theses I discovered at Curitiba during the summer of 1964 were all obtained this way, and the vision which I intend to explain now hereafter rests on the use of similar heuristics.

For disproofs, I had coined matrices. In the case of C_ω, even an infinite one, regular but not characteristic, built on the power set of the set of natural numbers, in order to prove that in this logic, no formula resulting from prefixing h negation signs before a propositional variable can imply any another formula obtained from the first by continuing to prefix an even number of supplementary negation signs: that C_ω is not finitely trivialisable follows[4].

After I had sent Newton's *Note*, I wrote to him to tell him what means I had employed to verify its assertions. At that time, he had no semantical insight on his systems, and he receveid my letter with enthusiasm. This was the origin of my invitation to Curitiba, and of the running of the summer logic seminar led by Newton in 1964. In fact, we were rarely more than three – Newton, young Ayda Arruda, and I.

The study of the C_n made substantial progress during this time. Our results were reported in two papers written in common: [Da Costa and Guillaume, 1964], [Da Costa and Guillaume, 1965], too forgotten afterwards, above all that of the *Portugaliae Mathematicae* [Da Costa and Guillaume, 1965] – it is true that as it is written, it is difficult to read. However it seems to me that the spirit in which it was meant has been revived in recent writings.

I attempted to build closed tableaux for detecting theses, Newton and Ayda used to build matrices as they were in the habit of doing. One day, haphazardly, I discovered a proof of Newton's axiom for the transmission of consistency by prefixing a negation sign, which made recourse to the other axioms only; the tableau was more or less the one I reproduced from memory on the figure 1 – the tableau of figure 1 is for C_1, but the proof which will be drawn from it can easily be adapted to all the C_n's.

---

[4]This matrix is described in [Da Costa and Guillaume, 1964].

	Valid ?	Not valid ?
A°		$(NA)^\circ$
		$\cdots\cdots$ abbrev.
		$N(\underbrace{NA \wedge NNA}_{B})$
$[B \vee NB]$		
$NA \wedge NNA$	NB	
NA	$===$	$===$
NNA		
$[NNA \to A]$		
A		(A)
$==============$		$==============$

Fig. 1

Let me explain according to which reasoning this tableau is built. To check the sequent $A^\circ \vdash (NA)^\circ$ – which holds iff the formula $A^\circ \to (NA)^\circ$ is provable – we start by proposing the formula A° as valid, and the formula $(NA)^\circ$ as not valid. So we started two lists of formulas: the first one is put in a column at the left of the tableau, and lists formulas going to be be proposed as valid; the second one is put in a column at the right of the tableau, and lists formulas going to be proposed as not valid. These two columns are said to be conjugate (in the tableau) to express the connexions between the two collected lists, above all those which follow from the relations which are accepted between the validity, or non-validity, of a formula, and the validity, or non-validity, of its immediate subformulas (or of its penultimate subformula, in the case of a formula dominated by a double negation). Now, recall that Newton's formula A° is just a shorthand for the formula $N(A \wedge NA)$; so we replace the proposal of $(NA)^\circ$ as not valid, written above the dotted line indicating that it is an abbreviation which is to be replaced by the shorten formula, by the represented proposal of $N(NA \wedge NNA)$ as not valid. For saving space, I propose also the use of B as a shorthand for the formula $NA \wedge NNA$, which we *could* propose as valid, if we reason on an usual logic. However, at best, we reason on positive logic, since all axioms of positive logic are assumed to be axioms of C_1. This permits us to apply all the rules for the proposals of formulas as valid or as not valid when their dominating connective is not a negation. Precisely, NB is dominated by a negation. What to do ?

Remember that C_1 accepts all instances of the schema $A \vee NA$ as axioms. So, the next step of our reasoning consists in recalling that the instance $B \vee NB$ can be proposed as valid. It is written in square brackets because we intend to forget this proposal after having recognized its effects. Indeed, to this formula, which is dominated by a disjunction sign, we can apply the rule which allows to split our attempt into two cases, represented by the splitting of the columns of the tableau into two pairs of conjugate columns, the furtherst left column being associated with the third from the left for forming a "left" subtableau, the furtherst right being associated with the third from the right for forming a "right" subtableau.

Let us start this right subtableau by the proposal of NB as valid. But, remember that NB was proposed as not valid: contradiction. Then, we have to abandon our proposal of NB as valid. Thus we stop the development of the right subtableau and indicate this outcome by closing its two columns by a double horizontal line (here just dashed) and by saying that it is closed. This means that the sole information provided by it is that the set of proposals which is written in it is untenable.

Let us start the left subtableau by the proposal of B as valid. But B is a shorthand, and I write the shorten formula $NA \wedge NNA$ in the left column of the subtableau. As this formula is dominated by a conjunction, we can apply the usual rule which allows to add to the list of formulas proposed as valid both formulas NA and NNA. Both are dominated by a negation, but we can use again of the previously used trick: it can be remembered that the convenient instance of the schema $NNA \rightarrow A$ (here this is that which is obtained without substitution whatsoever) can be proposed as valid, since we postulated it. This permits us to apply the rule *modus ponens*, which allows to propose the formula A as valid.

Thus the three formulas $A°$, NA, A are proposed as valid, and it is known that the sequent $A°$, NA, $A \vdash$ holds in C_1 (*cf. e.g.* [Carnielli and Marcos, 2002]). For some time, I called this argument the *supplementary closure rule 1*, but I just changed my mind: it seems better to put in evidence the role of *control* played by the formula $A°$ for applying the usual rule from the proposal of the formula NA as valid, and to add the proposal of the formula A as not valid (I put it in brackets), which leads to the closure of the left subtableau as usual, since A already occurs in the list of formulas proposed as valid. Having all its subtableaux closed, the tableau itself is said to be closed; this means that the set of proposals at the start of the reasoning is untenable: it leads to unacceptable contradictions. Intuitively, the checked sequence holds, since our attempt to *invalidate* it *failed*.

Better, and irrefutable: this tableau, which resumes a tree of sequents, can be transformed in an Hilbert-style *formal proof*. This may be done as follows: taking in account the recourses to the axioms hidden in the rules, let us start by writing Newton's axiom (12) $A° \rightarrow ((B \rightarrow A) \rightarrow ((B \rightarrow NA) \rightarrow NB))$;   but B was introduced in the tableau as a shorthand for $NA \wedge NNA$; so $B \rightarrow NA$ becomes a provable formula, and (being positively intuitionistically exchangeable with the other hypotheses, A° and $B \to A$), is going to be detached. The same is true for $B \to A$ because B, which shortens $NA \wedge NNA$, provably implies NNA, which in its turn provably implies A, since $NNA \to A$ is an axiom. And what remains after the removal of these provable hypotheses is precisely the desired $A^\circ \to (NA)^\circ$.

I don't remember in which order the three theses discovered by using such a method have been obtained. What I have a clear recollection of is that I became quickly conscious of the need to be cautious in applying the rules, since *only* the axioms of positive *intuitionistic* logic were accepted: so, in the case of implicative formulas, the *intuitionistic* rule *only* can be applied.

Well, to explain the differences between the intuitionistic and the classical rules, Beth had the habit to compare the results of applying the ones and the others to a formula serving as a criterium, the so-called Peirce's law, to wit, the schema $((A \to B) \to A) \to A$. So, one nice day, I was able to build, much to my surprise, a tableau more or less like that of figure 2.

Valid ?		Not valid ?
$[(A \land NA) \lor A^\circ]$		$((A \to B) \to A) \to A$
$[(B \land NB) \lor B^\circ]$		

$$A^\circ \qquad A \land NA$$
$$A$$
$$NA$$
$$(A \to B) \to A \qquad\qquad A$$
$$============ \qquad ============$$

$$B^\circ \qquad B \land NB$$
$$B \searrow$$
$$(A \to B) \to A \qquad NB \qquad \diagup A \qquad\qquad A$$
$$(A \to B) \to A$$

$$A \qquad\qquad A \to B$$
$$A$$
$$==== \qquad ==== \qquad A \qquad ==== \qquad B \qquad ==== \quad A \to B$$
$$A \diagup \qquad\qquad\qquad\qquad\qquad\qquad\qquad \searrow B$$
$$========== \quad ========== \qquad ========== \qquad ============$$

classical	intuitionistic
logic	logic:
applicable	delete all
to formulas	formulas
built out	above B at
A and B	the right side

Fig. 2

To apply the rules intuitionistically means that when we draw the consequences of the proposal of a formula of the form $A \to B$ as not valid (for example, here, that of the Peirce's law, by which the tableau is started), we have to reduce the list of the formulas proposed as not valid to the formula which comes just behind the dominating implication sign; in Peirce's law as it was written above, this is A. As for the formula before this sign, it can be added to the list of the formulas proposed as valid. In the tableau of figure 2, this rule has not been applied at first, but belatedly; meanwhile, splittings in subtableaux not yet closed were performed; so, we have to apply the rule one time per not closed subtableau (note that, in the right subtableau stemming from the first splitting, the occurrence of A proposed as not valid does not come from the application of a rule to the occurrence of NA proposed as valid – the application of this rule is blocked, because in this subtableau, the formula A° is *not* proposed as valid – it comes from the application of the rule just explained to Peirce's law in this subtableau).

In view of the explanations furnished in the case of the tableau of figure 1, the tableau of figure 2 calls for a little number of supplementary comments.

All closures arise with the last formula proposed as not valid in the sub-tableau where they happen, except two. A typical case is that of the two occurrences of B which are marked with slanted arrows, of which each points in the direction of that from which the closure happens; at the final step (before closure) B is proposed as not valid, but it was already proposed before as valid. Another typical case (the exceptional one) is that of the two occurrences of A similarly indicated; this needs further explanation.

Note that in the long furthest left subtableau, the formulas A° and B° are proposed as valid. Then, according to a theorem proved by da Costa in his doctoral Thesis [Da Costa, 1963], all formulas built with A and B satisfy all theses of *classical* logics; hence, *after* the proposals of both A° and B° as valid, the consequences to be drawn from the proposals of formulas built out A and B *solely*, as valid or as not valid, are to be drawn according to the *classical* rules. The two occurrences of A which are indicated are precisely in this case[5].

So, this tableau was perhaps the first mixed tableau produced in history, intuitionistic in the majority of its parts, but classical in a final section of one of its subtableaux !

There remains another little point: as previously, the square brackets surround formulas listed as valid for being known as such. In the case of figure 2, this seems to call for some explanation, because this seems not to be immediately obvious. Let us take the example of the second added formula, which might be written $(B \land NB) \lor N(B \land NB)$, an instance of an axiom. What was effectively written is just the result of replacing the subformula $N(B \land NB)$ by its shorthand B°.

This discovery entails that C_1 extends positive *classical* logic, what allows to apply the *classical* rules – except, of course, in the case of formulas dominated by a negation sign and proposed as valid. I remember clearly that this was made well *before* Ayda Arruda discovered the strong negation, which was coming as an explanation: meanwhile we were so astonished!

The third instructive tableau was that which permitted us to put in evidence certain properties of three formulas of similar forms[6], to wit

---

[5]The marked occurrence of A at the bottom of the left column of the subtableau can be obtained only by handling the proposal of $A \to B$ as not valid according to the corresponding rule. But if this last is the *intuitionistic* one, the marked proposal of A as not valid in the right column is to be dropped, and the closure vanish. This is not the case when the *classical* rule is applied, since that implies that all previous proposals of formulas as not valid are maintained.

[6]For which I kept the labels that were assigned to them in [Da Costa and Guillaume, 1965].

$$(m_1) \qquad N(A \wedge B) \to (NA \vee NB),$$
$$(m_2) \qquad N(A \vee B) \to (NA \wedge NB),$$
$$(m_3) \qquad N(A \to B) \to (A \wedge NB),$$

to which I add, for reasons which will appear later,

$$(m_0) \qquad NNA \to A.$$

By building a tableau similar to that of the figure 3 below, I discovered that (m_1) is a thesis of the C_n's.

Valid ?			Not valid ?	
			$N(A \wedge B) \to (NA \vee NB)$	
$N(A \wedge B)$			$NA \vee NB$	
			NA	
			NB	
A				
B				
$[(A \wedge NA) \vee A^\circ]$				
$A \wedge NA$	A°			
A				
NA				
======		======		
$[(B \wedge NB) \vee B^\circ]$				
$B \wedge NB$	B°			
B				
NB				
======		======		
$[A^\circ \to (B^\circ \to (A \wedge B)^\circ)]$				
$[B^\circ \to (A \wedge B)^\circ]$				
$[(A \wedge B)^\circ]$				
			$A \wedge B$	
		A	B	
======	==========	======	==========	

Fig. 3

After the explanations which have been given about the tableaux of figures 1 and 2, there remains almost no comment to be made about the tableau of figure 3. It need not to be "intuitionistic", because except the first formula, which plays no closure role, it is deprived of formulas dominated by an implication sign. In the cases of figure 1 and 2 we saw how the rules applied for deducing the proposals of A and B as valid in view of the previous proposals of NA and NB as not valid, and how the proposals of $(A \wedge NA) \vee A^\circ$ and of $(B \wedge NB) \vee B^\circ$ as valid are justified by classical axioms of C_1. Then, without being intuitionistic, this tableau is justified by the axiomatics of C_1.

2 A survey of some reflections on the present.

2.1 An architectural view on paraconsistent logics.

Turning my mind over Waragai and Shidori's contribution to this Congress led me to explore some architectural features of a certain type of paraconsistent systems which are in large part common to da Costa's systems C_n and to Waragai and Shidori's system PCL1, using to that end the *consistency connective* $°A$ introduced by Carnielli and Marcos [Carnielli and Marcos, 2002] as a new *primitive* of which the meaning is to be axiomatically determined[7].

Carnielli and Marcos [Carnielli and Marcos, 2002] also sketched a distinction, in paraconsistent logic, between three levels of axioms. At a more basic level, formulas dominated by the consistency connective are just present as supplementary hypotheses under which, according to the paraconsistent view, other hypotheses are restricted to entail the conclusions that they entail traditionally. Formulas without consistency connectives and which imply such supplementary hypotheses and so give to the consistency connective some content in terms of the usual connectives appear in a further somewhat less basic level. In a third level the formulas which occur require some properties of "propagation of consistency", some of which being said "structural", some other not.

I mean to label here as "abstract" the paraconsistent logics of which the axioms, either are such that the subformulas dominated by a consistency sign occuring in them are just present as "certificate of applicability" of some traditional rule(s) (in which the consistency connective, of course, do not appear), or express some *structural* property of "propagation of consistency"[8]. Also, I will say that such formulas are "abstract", and that the use of them comes under what I mean to call here the "abstract" point of view.

On another side, I mean to label as "concrete" the points of view under which $°A$ either is viewed as a shorthand and can always be freely replaced in all or in some of its occurrences by the abbreviated formula – this I will call the *strongly concrete* point of view – or is explained by axioms intending to transfer at the level of the object language such a meaning, first

---

[7] Carnielli and Marcos also introduced as primitive an "inconsistency connective" that, despite its interest, I will completely disregard here, since I mean to treat before all of the aforesaid features.

[8] No "non-structural" property of propagation of consistency will be taken in account here.

in isolation, second, in the scope of any connective escaping the replacement metatheorem, which in general is the case for a paraconsistent negation – this I will call the *weakly concrete* point of view.

Also, I mean to label as "concrete" the paraconsistent formulas and the paraconsistent logics falling under any concrete point of view, but, in practice, this will be almost always under the *weakly* concrete one. Such logics also always obeys axioms where the consistency connective appears abstractly, in the aforesaid sense, and so extend some abstract paraconsistent logic. So, their axiomatics can be seen as consisting of *two or three* levels of axioms, according as one remains at the abstract level or as one goes up to the concrete level.

Firstly, a *ground* level gathers together what I mean to call the *ground axioms*: schematic metaformulas *deprived of any explicit occurrence of the consistency connective*; this means that what is *specifically* required of the consistency connective *as such* is *not expressed* by these schemes; the formulas dominated by a consistency connective obeys them exactly in the same way as the formulas dominated or not by anyone of the other connectives. I mean to call the logic axiomatized by these schemes, the *ground logic* of any paraconsistent logic having all of them and no other of the same character accepted amongst its axioms.

Second, one or two *paraconsistent* level(s), according as one remains at the abstract level or as one goes up to the concrete level, gather(s) together what I mean to call *paraconsistent axioms*: schematic metaformulas in which the consistency connective appears *explicitly;* so these schemes express *what is specifically required of the consistency connective as such*, in its relations to the other connectives. These axioms will be said "abstract" or "concrete" according to the previously explained criteria.

For example, C_ω is the *ground logic* of the C_n 's as they have been axiomatized by da Costa[9], rather than their limit, and their *paraconsistent axioms* are those in which occurs at least one subformula under the abridged form $A^{(n)}$. These last axioms are all *abstract* [10], because the true C_n are *strongly concrete*, so that no concrete paraconsistent axiom was written for them.

The shorten formula, in the strong case, or the *definiens* of a definition of the consistency connective, in the weak case, are presented by da Costa, followed by Waragai and Shidori, as defining the "well-behavior" (a synonym for "possessing consistency") of the formula in the scope of a consistency sign[11]. This notion is *individual*, in the sense that it defines as such the

---

[9]Carnielli and Marcos [Carnielli and Marcos, 2002] noticed that their logic *min* can be taken as the ground logic for their axiomatization of the C_n 's.

[10]Despite the fact that, for bringing their abstract character to the fore, we have to replace in our mind each occurrence of $A^{(n)}$ by an occurrence of $\circ A$.

[11]So the formula obtained by prefixing a consistency sign appears as a "certificate of

formulas one by one, each independently of the other. Of course, such a defining formula is *not* provable in the ground logic; indeed, otherwise, all formulas would be well-behaved, which makes this notion useless. It can then be asked what are the extensions of the ground logic of which this defining formula is a thesis ? Clearly, there is a minimal such extension, which can be axiomatized by adding this formula to the ground logic.

Here the surprises begin, when it is asked what happens under this respect to the C_n and to PCL1 ? For the first ones and the second one, the answer is that, added to the ground logic, neither any $A^{(n)}$ nor $A \leftrightarrow NNA$ is sufficient for obtaining full classical logic.[12]

Nevertheless it is generally thought that in all the preceding cases, the behavior of the well-behaved formulas is classical. This is true, but watch out! A *classical* behavior *cannot be individual*[13]; it is *collective*; thus, if you term well-behavior the behavior of the well-behaved formulas, you are speaking of the *collective* well-behavior of the well-behaved formulas, not of their *individual* well-behavior: these two notions of well-behavior have to be carefully distinguished.

Then, the study of the collective well-behavior of formulas is the study of the structural properties of the set of the well-behaved formulas; that is to say, in the world of formulas, of the axioms and inference rules that the well-behaved formulas meet. These axioms and rules define, in their turn, a logic, that I will venture to call the *aimed logic* of the paraconsistent logic in question. This aimed logic necessarily *extends the ground logic*, just because the well-behaved formulas satisfy the ground axioms, as all formulas, including the bad-behaved, and *accepts as a thesis any formula defining the individual well-behavior* for the simple reason that this is said by a conjunct of the definition itself.

On another side, the *aimed* logic of a paraconsistent logic necessarily *extends* this logic. Indeed, just in virtue of its very definition, the aimed logic accepts as a thesis every formula A which is such that the formula B which says that A is implied by the condition that all its prime components are well-behaved (*i.e.* by the conjunction of their certificates of well-behavior) is provable in the paraconsistent logic, a condition which is satisfied by the theses of this last.

Axioms for the aimed logic are induced by the set of those of the abstract paraconsistent axioms which prescribe, under conditions of presentation of certificates of well-behavior, supplementary special relations between well-

---

consistency", or of "well-behavior", for the formula to which it is prefixed.

[12]What is obtained is only a *fragment* of C_n in the first cases, of PCL1 in the second. Notice moreover that the $A^{(n)}$'s are already provable in *intuitionistic* logic, as is $A \rightarrow NNA$, which is the conjunct of $A \leftrightarrow NNA$ which is not an axiom of PCL1.

[13]This is *not* a feature particular to classical logic; it holds *e.g.* for intuitionistic logic.

behaved formulas and formulas of which the behavior is not necessarily
specified. If, from any of these axioms, we detach the certificates of well-
behavior (of the form $^\circ A$, remember), we obtain formulas in which these
certificates no longer occur, and which are able to complete, together with
the ground axioms, a set of axioms for the aimed logic: the condition that
these formulas are implied, in the paraconsistent logic, by the conjunction of
all the certificates of their prime components, is satisfied, since we assumed
that this logic extends positive intuitionistic logic. However, the *inference
rules* the aimed logic obeys are not in evidence.

It was only by a detour that da Costa solved this problem in his Thesis
[Da Costa, 1963]. His proof can be reproduced word for word in all the
main cases of paraconsistent logics which I mean to somewhat study here;
indeed, in his proof, that he thought as aimed to classical logic, *no specific
axiom of classical logic* was playing any part whatever, so that this proof
bears in reality upon intuitionistic logic as well.

But the price to pay was to set the abstract paraconsistent *stability axioms* for the well-behaved formulas, which transcribe, at the object language
level, the prescription that the set of well-behaved formulas has to be stable under the actions of building formulas, in which Carnielli and Marcos
[Carnielli and Marcos, 2002] saw properties of "propagation of consistency".
No such other properties than da Costa's four stability axioms will be taken
in consideration here[14].

In all the same cases, the paraconsistent logic admits of some non-paraconsistent
extension accepting as a thesis the formula which defines the individual well-
behavior. From this it follows that this extension extends also the aimed
logic, because then, any certificate of well-behavior is a thesis of this extension, in particular those of the prime components of any thesis of the aimed
logic, by which this thesis is implied in the paraconsistent logic and in all
its extensions. Also, in these cases, to add the stability axioms allows to
establish that the rule *modus ponens* holds in the aimed logic, so that this
last is well determined by its axioms.

For example, in the C_n's, there is only one abstract paraconsistent axiom
not expressing a property of propagation of consistency, *i.e.* that which was
labelled (12') in da Costa's Thesis [Da Costa, 1963] and which, in abstract
form, is written
$$^\circ B \to ((A \to B) \to ((A \to NB) \to NA)).$$
By detaching the hypothesis $^\circ B$, which is a certificate of "consistency"
for B, we restore Kleene's axiom 7 (*cf.* [Kleene, 1952]), which is just the
axiom lacking to C_ω for completing an axiomatization of full classical logic.
Moreover, the formulas $A^{(n)}$ of da Costa are theses of this logic, which hence

---

[14]Which, however, are just *reinforced* by most of the proposed alternative axioms.

is well the aimed logic of the C_n 's.

However, the axioms presenting the said characters do not warrant that *there exist*, effectively, well-behaved formulas. A formula defining what is a certificate of individual well-behavior, resulting from an interpretation of $^\circ A$ as a shorthand, or from suitable concrete axioms, solve this question by allowing to *produce* well-behaved formulas entering in the picture.

2.2 A look onto the following sections.

The next section will be devoted to the study of a set of ground logics, all extensions of positive intuitionistic logic, all fragments (proper of not) of full classical logic. Each of them is meant to have a "semantic tableaux" *à la Beth* [Beth, 1959] version, for being axiomatized as extending positive intuitionistic logic, and moreover – except of course positive intuitionistic logic itself – by what I mean to call *rule-making* axioms for implication and negation, each of which but one intimately tied to the classical rule for implication or to some intuitionistic or classical rule for negation.

The rule-making axiom that is an exception is (m_0), which is so tied to a rule which is primitive neither in the intuitionistic nor in the classical sense. The role played by this axiom is twofold; it is also tied, in another way, to a paraconsistent stability axiom, as we saw in the first section; but I discovered by developing the present study that this link is stronger and more fundamental than I thought. As its rule-making role is well-known, I will content myself in the next section with evoking it and the two ground logics in which it is accepted as an axiom, to wit C_ω and Carnielli and Marcos' *min* [Carnielli and Marcos, 2002], which has the same theses as Da Costa and Guillaume's C'_ω [Da Costa and Guillaume, 1964][15], and then with focusing on its link with the first of Carnielli and Marcos "Guillaume's theses" [Carnielli and Marcos, 2002].

This also announces that further sections will afterwards be devoted to paraconsistent axioms evoked in what follows, first of all to those which are roughly tied to some of these rule-making axioms as da Costa's axiom (12) for C_1 is tied to Kleene's axiom 7, afterwards to paraconsistent axioms thus tied to other classical theses, then to definitions of concrete consistency connectives.

The (not linear) hierarchy of logics which emerges from these studies allows to locate the axiomatic levels where certain theses begin to appear.

---

[15]These last two logics extend C_ω by an axiom warranting that all theses of positive classical logic become provable; to this effect, the first makes use of the "Dummett's formula", the second, of Peirce's law.

3   A hierarchy of ground logics.

3.1   Positive intuitionistic logic.

The *really minimal* axiomatic system in what follows is the *positive intu-itionistic logic*[16] P^I1, of which we will not precise here the axioms[17].

The reader can think, for example, either of Kleene's axioms 1a, 1b, 3, 4a, 4b, 5a, 5b, 6 in its *Introduction to Metamathematics* [Kleene, 1952], or of Rasiowa and Sikorski's axioms in their *The Mathematics of the Meta-mathematics* [Rasiowa and Sikorski, 1970] to which Waragai and Shidori have given the labels from $A1$ to $A9$ in their contribution to this Congress: these two systems are equivalent, in the sense that each formal theorem of each of them is a formal theorem of the other.

The tableaux version $^tP^I1$ of this logic obeys all *intuitionistic* rules con-cerning the cases of formulas of the forms $A \wedge B$, $A \vee B$, and $A \rightarrow B$; these rules are also *classical,* in the sense that they are applicable to the tableaux version of the positive classical logic, **except** the one where it is to be inferred, as a consequence of the proposal of a formula of the form $A \rightarrow B$ as *not valid,* that the formula A can be added to the list of formulas proposed as *valid*, and the formula B, proposed as *not valid*: the *classical* rule allows also to add B to the list of formulas proposed as *not valid*, while following the *intuitionistic* rule, there is a price to pay for adding A to the list of formulas proposed as *valid* – to cancel the previous list of formulas proposed as *not valid*, including $A \rightarrow B$, so that the new list of formulas proposed as not valid is reduced to B.[18]

---

[16]I will even so much consider it as basic that I will frequently take as being the same two formulas which are only positively intuitionistically equivalent.

[17]However, the paraconsistent negation and the abstract connective of consistency will be considered as belonging also to the language of P^I1. As they are concerned by none of the axioms of this logic, the occurences of formulas dominated by the corresponding signs have to be taken as functioning as prime components in its axioms, proofs and theses.

[18]This introduces in practice an element of strategy in the order in which the conclu-sions of the proposals of formulas as valid or as not valid are drawn according to the rules for leading to a closed tableau as a test of correctness for a given sequent.

Beth devised his algorithm so that, when it is rigorously applied, the corresponding guesses can entirely be eliminated, by allowing to keep all possibilities open; this is obtained by introducing a *disjunctive* splitting in subtableaux ("disjunctive" means that the closure of just one of the new subtableaux entails the closure of the splitted one), one for checking the new proposals as they have been just explained, the other for repeating the previous proposals in the hope that paying attention to some other of them will lead to closure.

All intuitionistic or hybrid (semi-classical) closed tableaux which will be exhibited or evoked further on will be sufficiently short and simple for allowing to spare the use of such disjunctive splittings, leaving them implicit, but remember that correctly checking these tableaux supposes to check with particular care that the rule to be applied to formulas proposed as not valid and dominated by an implication (or by a negation) sign has not been violated.

The system of rules of this tableaux version is complete in the following sense: on one side, a closed tableau, started for checking a given sequent, can be algorithmically converted in a proof of this sequent in a Gentzen-type system of positive intuitionistic logic, which in its turn can be algorithmically converted, when the first member of the checked sequent is void and its second member is reduced to one formula, in an Hilbert-style proof of this formula; on the other side, a saturated[19] tableau which is not closed allows to build a matrix[20] which invalidates the checked sequent.

However, only closed tableaux will be exhibited or evoked further on as proofs of sequents of the form $\Gamma \vdash A$, where Γ is a finite set of formulas taken as formal hypotheses[21], A is a formula, and $\vdash$ represents the idea that there exists, in the logic in question in the context, a formal deduction of A under the hypotheses.

The matrices used in order to invalidate certain sequents - in fact in order to provide independence proofs for axioms - are all very simple, and will be described directly[22].

Each of the logics in question in what follows will extend P^I1: it will even be understood, without any further mention, that all axioms of P^I1 are included in its own axioms, so that this logic will be characterized just by listing those of its axioms which are not axioms of P^I1.

3.2   Positive classical logic.

Only one of the schemes here taken in consideration has no explicit occurrence of a negation sign: to wit, the so-called "Dummett's formula"

DF $A \vee (A \to B)$,

of which the adjunction to P^I1 gives the *positive **classical** logic* P^C1.[23]

This logic also has a tableaux version $^tP^C1$, which obeys all *classical* rules

---

[19]A tableau is said to be saturated when to apply the relevant rule to any of the occurrences of a compound formula in it does not add to it new information whatever. More precisely, this new application produces correlative lists of proposals of formulas as valid and as not valid which can already be found in one of the concerned branches of the tableau.

[20]In substance, an upper bounded distributive lattice with its top element as only distinguished value.

[21]To speak of a set rather than of a list exempts to handle structural rules.

[22]The completeness of similar tableaux versions of the logics in which the consistency connective is treated under the *abstract* point of view can be deduced from the completeness of the versions for positive, either intuitionistic or classical, logics. I am not sure whether the same holds when consistency is treated under a concrete point of view. But, if I correctly understood the contribution of Itala d'Ottaviano and Milton de Castro to this Congress, this question can be recursively decided by applying their method.

[23]Beth (*e.g.* in [Beth, 1959]), and, following him, Da Costa and Guillaume [Da Costa and Guillaume, 1965], made use, with the same aim, of the more complicated Peirce's law.

concerning the cases of formulas of the forms $A \wedge B$, $A \vee B$, and $A \rightarrow B$.

Correlatively, what allows us to pass from $^tP^I1$ to $^tP^C1$ is just the replacement of the intuitionistic rule according to which the conclusions to be inferred from the proposal of a formula of the form $A \rightarrow B$ as not valid are drawn, by the corresponding classical rule. It is of interest to observe that one can *justify* that this change corresponds to the adjunction of DF to P^I1.

Indeed, first, it can immediately be verified that by using the classical rule for drawing the consequences of the proposal of a formula of the form $A \rightarrow B$ as not valid (so, in $^tP^C1$), every instance of DF can be proven (To be sure, this seems quite obvious, but that is because we pass to a very well known logic; the corresponding property for passing to other logics, which are not so well-known, can by no means seem evident).

Second, conversely, if DF is accepted as valid, every application of the classical rule can be justified in *positive intuitionistic logic*.

The clearest way to proceed is by local modifications of tableaux in $^tP^I1$. Let us first explain how we will do it, in the cases of three auxiliary *derived* rules, which thus are *dispensable* for *not primitive* in $^tP^I1$, but which allow to build simplified tableaux, which nevertheless are as reliable and concluding as the tableaux which are built just by applying primitive rules. These rules are the following:

Auxiliary derived rule 1: If two formulas of the forms A and $A \rightarrow B$ are proposed as *valid*, then the formula B can be added to the list of the formulas proposed as *valid*;

Auxiliary derived rule 2: If a formula of the form $A \rightarrow B$ is proposed as *valid*, and if moreover the formula B is proposed as *not valid*, then the formula A can be added to the list of formulas proposed as *not valid*;

auxiliary derived rule 3: If a formula of the form $A \rightarrow B$ is proposed as *not valid*, then the formula B can be proposed as *not valid*.

To be sure, it is easy to establish the rectitude of these rules directly, without resorting to the primitive rules for correctly building tableaux[24]; but this does nothing but confirm that the primitive rules have been accurately devised, since these auxiliary rules become derivable. The point here is to show *how* they can be derived.

The figures 4 and 5 are devoted to explaining the reasoning following which the auxiliary derived rules 1 and 2, respectively, can be justified by resorting to the primitive rule according to which the conclusions to be inferred from the proposal of a formula of the form $A \rightarrow B$ as valid

---

[24]In particular, the auxiliary derived rule 1 clearly does just repeat what the inference rule *modus ponens* says.

are drawn. This primitive rule is common to the intuitionistic and to the classical sets of rules, and we will see that all we will do applies to each of the extensions of $^t\mathrm{P}^\mathrm{I}\mathrm{l}$ considered here and that thus our auxiliary derived rules are at our disposal in all these extensions.

Valid ?	Not valid ?		Valid ?	Not valid ?
•	•		•	•
•	•		•	•
$A \to B$	•		•	B
•	•		•	•
•	•		•	•
A	•		$A \to B$	•
•	•		•	•
•	•		•	•

Fig. 4: *Auxiliary derived rule 1*   Fig. 5:*Auxiliary derived rule 2*

Each of these figures represents a pair of conjugate columns of some tableau, as if they were next. The situation from which the reasoning is assumed to begin is represented by the part of the figure which lies above the first horizontal dashed line. The vertical dotted lines represent the lists of formulas taken into account in these conjugate columns, according to the rules governing the way to do so. Only the formulas which will be considered in the reasoning are explicited, replacing a dot in one of the two conjugate columns. The choice of the order in which the explicited formulas appear is arbitrary; it is to be understood that the reasoning would be the same if the places of these formulas are permuted, provided they are not moved from one column to the other.

What is under the first horizontal dashed line is firstly the result of the application to the formula $A \to B$, proposed as valid, of the rule which allows to split the conjugate mother columns in two pairs of conjugate extensions (as explained in section 1 by commenting figure 1), and then to propose, in the left pair of conjugate daughter columns, the formula B as valid, and in the right pair of conjugate daughter columns, the formula A as not valid; and second, the result of the later simplification which leads to what boils down to an application of the auxiliary rule 1 in figure 4, and of the auxiliary rule 2 the figure 5.

Indeed, in the case of figure 4, the right pair of conjugate daughter colums is closed (indeed, the formula A occurs in the two lists of formulas corresponding to this pair of conjugate columns, so is proposed both as valid and as not valid); in the case of figure 5, the left pair of conjugate daughter columns is closed, because in this case the formula B is proposed both as valid and as not valid. In each of the corresponding tableaux, to pruning the nascent necessarily immediately closed pair of conjugate columns spares a splitting and avoid various drawings, and can be made by erasing all that enters in the frame limited by the two horizontal dashed lines and by the two columns of ‡'s. What results just extends the mother conjugate columns as required by the application of the auxiliary rule.

Observe that only positively intuitionistically holding rules were applied. Then, these auxiliary rules are good in the tableaux version of every one of the logics here taken in consideration, provided that it admits of such a version, and one can say that a tableau built by applying also on occasion these auxiliary rules is an *abridged* tableau in its tableaux version. The same is true for the auxiliary derived rule 3, as we will now see.

The figures 6 and 7 are similar to the figures 3 and 4, except in what concerns the formulas placed in square brackets under the horizontal dotted line above which is represented the situation before the start of the reasoning. The square brackets here indicate that the formula inside is meant to be written at the place where it occurs for the time of the reasoning and then to be in principle erased.

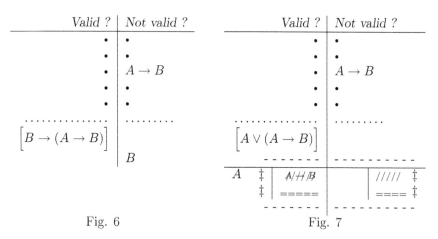

Fig. 6   Fig. 7

The reasoning corresponding to figure 6 goes as follows. In $^{t}\mathrm{P}^{I}1$, we can add to the list of the formulas proposed as valid the formula $B \rightarrow (A \rightarrow B)$, since this formula is an axiom of $\mathrm{P}^{I}1$! Then, we can apply the auxiliary derived rule 2, add the formula B to the list of the formulas proposed as not valid, and forget the proposal of an axiom as valid: the result is an application of the auxiliary derived rule 3, since no rule was applied to the occurrence of the formula $A \to B$ explicited in the list of formulas proposed as not valid; thus *from the list of formulas proposed as not valid, no formula was removed.*

The reasoning corresponding to figure 7 goes similarly, with an intermediate step analogous to the step carried out in the case of auxiliary rule 1, except if we run this reasoning in $^{t}P^{I}1$; in this case, the formula $A \vee (A \to B)$ in square brackets must be kept as an additional hypothesis introduced after the start of the building of the tableau; if this one was started in order to check a given sequent $\Gamma \vdash \Delta$, the sequent that will actually be checked will be the sequent $A \vee (A \to B)$, $\Gamma \vdash \Delta$.

The primitive rule which was applied to the formula $A \vee (A \to B)$, proposed as valid, is the one which allows to extend the list of formulas proposed as valid following two different ways generating a splitting into two extensions, to wit one in which the formula A is added to this list, the other in which it is the formula $A \to B$ which is so added – which in this second case leads to the immediate closure of the conjugate daughter columns so introduced, since then this last formula remains at the same time proposed as not valid.

After the pruning of the immediately closed branch, what remains is the adjunction of the formula A to the list of formulas proposed as valid. Again, no rule was applied to the occurrence of the formula $A \to B$ explicited in the list of formulas proposed as not valid, and *no formula proposed as not valid was removed.*

But the tableau of figure 7 can also be read otherwise ! By erasing the formula in square brackets, which is a thesis, hence valid, in *classical* logic, we transform this tableau into a tableau which can *locally* be read as a *classical* one. Indeed, by reading, as we did before, the tableau as intuitionistic, we justified the application of the *classical* **half**-rule which says that if a formula of the form $A \to B$ is proposed as *not valid*, then the formula A can be added to the list of formulas proposed as *valid*.

Putting together the *intuitionistic* auxiliary derived rule 3 and the previous *classical* "half-rule", we obtain the *classical* rule according to which the conclusions to be inferred from the proposal of a formula of the form $A \to B$ as not valid are drawn. So, to build a classical tableau by applying this rule has been shown to be reducible to building an intuitionistic tableau, recursively produced from the classical one by replacing each application of this rule by the adjunction of a suitable instance of DF to the corresponding list of formulas proposed as valid and by subsequent applications of intuitionistic rules *without effect on the lists of the formulas proposed as not valid* [25].

3.3   The two "bounded" logics.

For a reason to be explained later, I will call *bounded intuitionistic logic,* resp. *bounded classical logic,* the logic $B^I 1$, resp. $B^C 1$, which is obtained by adjoining to $P^I 1$, resp. $P^C 1$, the schema

EF $NA \rightarrow (A \rightarrow B)$ [26],

which is common to the axiomatics given by Kleene [Kleene 1952] and Rasiowa and Sikorski [Rasiowa and Sikorski, 1970] for (full) classical logic $F^C 1$.

The logic $B^I 1$, resp. $B^C 1$, admits of a tableaux version $^t B^I 1$, resp. $^t B^C 1$, of which the rules concerning the cases of formulas of the forms $A \wedge B$, $A \vee B$ and $A \rightarrow B$ are the same as the corresponding rules of $^t P^I 1$, resp. $^t P^C 1$ – hereafter called the *basic* intuitionistic, resp. classical, rules – and are complemented by the rule, common to $^t B^I 1$ and to $^t B^C 1$, which allows to infer, from a proposal of a formula of the form NA as *valid,* that the formula A can be added to the list of formulas proposed as *not valid.*

The manner in which the adjunction of this rule to $^t P^I 1$, resp. to $^t P^C 1$, corresponds to the adjunction of EF to $P^I 1$, resp. to $P^C 1$, is very similar to the manner in which the passage from an intuitionistic rule to the corresponding classical one corresponds to the adjunction of DF to $P^I 1$.

Indeed, on one side, one can immediately verify that by using the rule to draw the consequences of the proposal of a formula of the form NA as valid, every instance of EF can be proven in $^t B^I 1$. The same is true for $^t B^C 1$; indeed, a closed tableau of $^t B^I 1$ is *a fortiori* also a closed tableau of $^t B^C 1$.[27]

On the other side, one can *justify* that the rule is entailed by the adjunction of EF to $P^I 1$, then also to $P^C 1$.

We intend to proceed as we did for DF, referring to the figure 8 below, where the tableau can be classical as well as intuitionistic.

---

[25] To be sure, the full rigour would call for developing this reasoning in a formal recursive proof along the building of a tableau. However, the pieces of such a proof just provided should be enough to allow the reader to have a clear idea of this proof.

[26] "*Ex falso*" (*sequitur quodlibet*), a label suggested in [Carnielli and Marcos, 2002].

[27] Just as a closed tableau of $^t P^I 1$ is also a closed tableau of $^t P^C 1$.

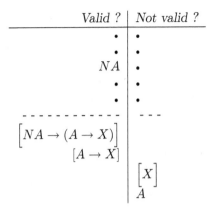

Fig. 8

So it is assumed that the explicited formula NA occurs in the list of formulas proposed as valid before the beginning of the reasoning, which begins by adding the formula $NA \to (A \to X)$ to this list. Then, according to the auxiliary derived rule 1, the formula $A \to X$ is also added to this list.

The step which follows consists in adding the formula X to the list of the formulas proposed as not valid. Just after can be applied the auxiliary derived rule 2, which allows to add the formula A to the list of the formulas proposed as not valid. Then, after the erasure of the formulas in square brackets (*i.e.* of all formulas added, except the last), what remains is the adjunction of A to the list of formulas proposed as not valid, as required by the rule to be justified.

However, this cannot be done without proviso. Our choice of the instance of EF to introduce was partly justified by our will to set X as not valid; but what gave to us the right to decree that the formula X *can* be added to the not valid formulas ? How to know whether such a decree might not be guilty of some irrelevant closure, not really resulting from the assumed failure of validity of the sequence checked at the start, and arising from the possible further appearance, in the tableau to be transformed, of an occurrence of X proposed as valid ?

A first response is to specify that X, at the time of its introduction, must be chosen among the list of the formulas already proposed as not valid. But, what to do if this list is void ?

In this case, any subsequent disruptive closure can be avoided by choosing as X a variable which *do not occur in the sequent to be checked at the start of the tableau's building* [28]. Moreover, so chosen, X can be introduced, not each

---

[28]This solution is also convenient if there are formulas proposed as not valid.

time that an instance of EF is introduced, but as the *first* formula proposed
as not valid in the transformed tableau. No occurrence of X is possible in
the tableau to be transformed; and in the transformed tableau, simplified
by the use of auxiliary rules, no other occurrences of the subformula reduced
to X can appear explicitly. Finally, after the erasure of all formulas added
for the time of the reasoning, in order to reduce every application of the
rule to a step in the building of a tableau in $^t\mathrm{P}^\mathrm{I}1$, this occurrence of X will
also be removed.

So, finally, to build a tableau in $^t\mathrm{B}^\mathrm{I}1$, resp. in $^t\mathrm{B}^\mathrm{C}1$, by applying the
rule according to which the conclusion to be inferred from the proposal of
a formula of the form NA as valid are drawn, is recursively reducible to
building a tableau in $^t\mathrm{P}^\mathrm{I}1$, resp. in $^t\mathrm{P}^\mathrm{C}1$.

In another respect, it is clear that $\mathrm{B}^\mathrm{C}1$ is an extension of $\mathrm{B}^\mathrm{I}1$. This last
one, and then every one of its extensions, presents a special feature: the
Lindenbaum algebra associated to its set of formulas (where the prefixation of a negation sign does not necessarily have to preserve equivalence
classes modulo provable reciprocal implication, when it has not to obey a
replacement theorem – as this is just the case for $\mathrm{B}^\mathrm{I}1$ and $\mathrm{B}^\mathrm{C}1$) is also *lower*
bounded, in virtue of the fact which follows – this is my reason for christen-
ing "bounded" the two extensions, respectively of $\mathrm{P}^\mathrm{I}1$ and of $\mathrm{P}^\mathrm{C}1$, obtained
by adjoining EF.

Indeed, in $\mathrm{P}^\mathrm{I}1$, the schema $(A \wedge NA) \to B$ is provably equivalent to
EF, and in particular, we can take NT for B, where T is provable in $\mathrm{B}^\mathrm{I}1$;
then, we can cancel the occurrence of T from the antecedent of the sequent
$T, NT \vdash B$, in particular when $A \wedge NA$ is taken for B; hence holds the

FACT 1. In every extension of $\mathrm{B}^\mathrm{I}1$, the formulas $(A \wedge NA) \leftrightarrow (B \wedge NB)$
and $(A \wedge NA) \leftrightarrow NT$, where T is provable, are formally provable.

CAUTION. Neither $\mathrm{B}^\mathrm{C}1$ nor *a fortiori* $\mathrm{B}^\mathrm{I}1$ have $N(A \wedge NA)$ amongst their
formal theorems. This can be seen by endowing any boolean algebra with
the constant taking the value 0 as interpreting operation for the prefixation
of a negation sign[29].

---

[29]Going farther is possible only when the theorem of replacement can be applied to
formulas dominated by a negation sign; then holds, in particular, the well-known fact
that all formulas of the form $N(A \wedge NA)$ are provable in every extension of full intuitionistic logic $\mathrm{F}^\mathrm{I}1$, from which follows that two distinct formulas of that form are provably
equivalent.

3.4   The three "unbounded" logics.

A walk between intuitionistic and classical fragments.

Waragai and Shidori, by their clever reference to Rasiowa and Sikorski's axiomatisation of P^C1, guide us to a complete analogous illumination on the rules governing the way of drawing the consequences of the proposal of a formula of the form NA as *not valid.*   Indeed, besides axioms for P^I1 supplemented by EF, Rasiowa and Sikorski propound, as additional axioms,

RA* $(A \rightarrow (A \wedge NA)) \rightarrow NA$ [30],

and the so-called *principle of the Excluded Third,*

ET $A \vee NA$.

What follows will soon bring to the fore the interest of RA*; however, it says no more than a simpler, somewhat implicit form of the *reductio ad absurdum,*

RA- $(A \rightarrow NA) \rightarrow NA$,

which is *equivalent* to RA* in P^I1, of which $(A \rightarrow (A \wedge B)) \leftrightarrow (A \rightarrow B)$ is a thesis.

On another side, the following (meta)theorem has interesting philosophical incidences:

THEOREM 2.   *The sequent $A \vee B \vdash (A \rightarrow B) \rightarrow B$ holds in P^I1,*

as can be verified by building in $^tP^I1$ a closed tableau, started by checking it, which is straightforward.

Indeed, substituting NA for B, we find that, in P^I1, the principle of Excluded Third implies RA-, and then also RA*! [31]   So, as early in the axiomatic building of P^C1 as in P^I1, the principle of Excluded Third entails the rejection of some particular form of contradiction !

Now, let me recall the following

FACT 3.   The schema RA- is a formal theorem of *intuitionistic logic*

– contrary to ET.

Proof. In Kleene's axiomatization of intuitionistic logic is included the full schema of *reductio ad absurdum* $(A \rightarrow B) \rightarrow ((A \rightarrow NB) \rightarrow NA)$, labelled 7 by him – I will prefer in what follows the positively intuitionistically equivalent

RA $(A \rightarrow NB) \rightarrow ((A \rightarrow B) \rightarrow NA)$.

Now, let us substitute A for B in this Kleene's schema 7, and detach the provable in P^I1 subformula $A \rightarrow A$: so, RA- is obtained intuitionistically.

---

[30]This is an *explicit* restricted form of the *reductio ad absurdum.*

[31]This was my reason for mentioning explicitly RA* here.

Observe that RA- *cannot* entail ET in *positive* intuitionistic logic, since otherwise, the same would hold in *full* intuitionistic logic, where ET would then be provable, contrary to what is well-known.

The situation is not the same in the classical case, where holds the

FACT 4. The sequent $(A \to B) \to B \vdash A \lor B$ holds in classical logic –

as is quasi-instantaneously shown by building a closed tableau in $^t\mathrm{P}^\mathrm{C}1$, started by checking this sequent.

Hence, to add RA- to $\mathrm{P}^\mathrm{C}1$ boils down to adding ET to $\mathrm{P}^\mathrm{C}1$, which gives a fragment $\mathrm{U}^\mathrm{C}1$ of classical logic, while to add ET to $\mathrm{P}^\mathrm{I}1$ produces a strict extension $\mathrm{U}^\mathrm{H}1$ of the fragment $\mathrm{U}^\mathrm{I}1$ of intuitionistic logic obtained by adding RA- to $\mathrm{P}^\mathrm{I}1$. These three logics, that I label as "unbounded" by opposition to the "bounded" ones where EF is accepted (the "H" is put for "hybrid"), deserve a few words to come.

To complete the picture concerning the axioms linked to the rules which apply, in order to build tableaux, in presence of proposals of formulas dominated by an implication sign or by a negation sign as valid or as not valid – as we will soon have done by proving that such are the three cases remaining to be looked into – remains to be established the

THEOREM 5. *(Full) intuitionistic logic* $\mathrm{F}^\mathrm{I}1$ *can be axiomatized by adding* EF *and* RA- *to axioms for* $\mathrm{P}^\mathrm{I}1$.

Proof. Kleene [Kleene, 1952] axiomatizes $\mathrm{F}^\mathrm{I}1$ by adding, to the aforementioned axioms for $\mathrm{P}^\mathrm{I}1$, his axioms 7 and 8^I. First, the last one is nothing else than the one I labelled EF. Second, it will be proved that Kleene's axiom 7 (or RA which is equivalent to it relatively to $\mathrm{P}^\mathrm{I}1$) is entailed in $\mathrm{P}^\mathrm{I}1$ by (suitable instances of) EF and RA-.

Indeed, it is clear that in $\mathrm{P}^\mathrm{I}1$, the sequent
$NB \to (B \to NA)\,, (A \to NA) \to NA\,, A\,, A \to NB\,, A \to B \vdash NA\,.$
holds. So, applying the deduction theorem, the sequent
$NB \to (B \to NA)\,, (A \to NA) \to NA\,, A \to NB\,, A \to B \vdash A \to NA$
also holds, as then the sequent
$NB \to (B \to NA)\,, (A \to NA) \to NA\,, A \to NB\,, A \to B \vdash NA$
and, applying again the deduction theorem, the announced sequent[32]
$NB \to (B \to NA)\,, (A \to NA) \to NA \vdash (A \to B) \to ((A \to NB) \to NA)\,.$

COROLLARY 6. *(Full) classical logic* $\mathrm{F}^\mathrm{C}1$ *can be axiomatized by adding* EF *and* ET *to axioms for* $\mathrm{P}^\mathrm{I}1$.

Indeed, adding ET to the axioms listed in the theorem 5 axiomatizes $\mathrm{F}^\mathrm{C}1$

---

[32]Note and remember for later use which instance of EF the argument resorted to.

(as proved *e.g.* by Kleene in [Kleene, 1952])[33] and moreover we know that RA⁻ can be deduced from ET.

On "unbounded" intuitionistic and classical fragments.

The logic U^I1, resp. U^C1, admits of a tableau version $^tU^I1$, resp. $^tU^C1$, in which the basic intuitionistic, resp. classical, rules are complemented by the intuitionistic, resp. classical, rule saying that it can be infered, as a consequence of the proposal of a formula of the form NA as *not valid,* that the formula A can be added to the list of the formulas proposed as *valid,* the list of formulas proposed as *not valid* being correlatively, in the intuitionistic, resp. classical, case, reduced to the formula NA, resp. maintained.

The manner in which to adjoin this rule to $^tP^I1$, resp. to $^tP^C1$, corresponds to adjoining RA⁻ to P^I1, resp. to P^C1, is the same as that in which to adjoin to $^tP^I1$ the rule for drawing the consequences of the proposal of a formula of the form NA as valid corresponds to adjoining EF to P^I1.

Indeed, on one side, one can immediately verify that by using the rule, every instance of RA⁻ can be proven in $^tU^I1$, resp. in $^tU^C1$.

On the other side, one can *justify* that the rule is entailed by the adjunction of RA⁻ to P^I1, resp. to P^C1.

Valid ?	Not valid ?		Valid ?	Not valid ?
•	•		•	•
•	•		•	•
•	NA		•	NA
•	•		•	•
•	•		•	•
$\big[(A \to NA) \to NA\big]$		‡ $\big[A \vee NA\big]$		‡
	$[A \to NA]$	A ‡	$//N\!/\!A$	$///$ ‡
		‡	$======$	$===$ ‡
A	NA			

| Fig. 9 | | Fig. 10 |

We will do as we did previously, referring to the figure 9, where the tableau can be classical as well as intuitionistic. Thus, it is assumed that the explicited formula NA occurs in the list of formulas proposed as not valid before the beginning of the reasoning, which begins by adding the

---

[33]This implies that DF can be deduced from EF and ET. Indeed, it is well known that the formula $(C \to D) \to ((A \vee C) \to (A \vee D))$ is a thesis of P^I1. Now, substitute NA for C and $A \to B$ for D.

formula $(A \rightarrow NA) \rightarrow NA$ to the list of formulas proposed as valid. Then, according to the auxiliary derived rule 2, the formula $A \rightarrow NA$ is added to the list of formulas proposed as not valid.

The step which follows consists in applying the rule according to which the conclusions to be inferred from this last proposal are drawn. Then, in the *intuitionistic* case, all formulas above the dotted line are removed from the list of formulas proposed as not valid, the formula A is proposed as valid, and the proposal of the formula NA as not valid is regained, while in the classical case, all formulas proposed as not valid remain such. Thus, after the erasure of the formulas in square brackets, what remains is indeed, in both intuitionistic and classical cases, the result of an application of the rule to justify.

Then, to build a tableau in $^tU^I1$, resp. in $^tU^C1$, by applying the intuitionistic, resp. classical, rule according to which the conclusions to be infered from the proposal of a formula of the form NA as valid are drawn is recursively reducible to building a tableau in $^tP^I1$, resp. in $^tP^C1$.

The "unbounded" hybrid fragment, besides the "unbounded" classical one.

The logic U^H1, resp. U^C1, admits of a tableaux version $^tU^H1$, resp. $^tU^C1$, in which the basic intuitionistic, resp. classical, rules are complemented by the *classical* rule saying that it can be inferred, as a consequence of the proposal of a formula of the form NA as *not valid,* that the formula A can be added to the list of the formulas proposed as *valid.*

The manner in which to adjoin this rule to $^tP^I1$, resp. to $^tP^C1$, corresponds to adjoining ET to P^I1, resp. to P^C1, is the same as that in which to adjoin to $^tP^I1$ the rule for drawing the consequences of the proposal of a formula of the form NA as valid corresponds to adjoining EF to P^I1 [34].

Indeed, on one side, one can immediately verify that by using the rule, every instance of ET can be proven in $^tU^H1$, resp. in $^tU^C1$.

On the other side, one can *justify* that the rule is entailed by the adjunction of ET to P^I1, resp. to P^C1, by proceeding as we did previously, referring to figure 10 above, where the tableau can be classical as well as intuitionistic. Thus, it is assumed that the explicited formula NA occurs in the list of formulas proposed as not valid before the beginning of the reasoning, which begins by adding the formula $A \vee NA$ to the list of formulas proposed as valid.

The next step is the same as that which was accomplished in the case of

---

[34]We are encountering here two equivalent axiomatizations of U^C1, resulting from the adjunction, to P^C1, in the first of RA-, in the second of ET, and we are led to the same rule, but with two different justifications.

the adjunction of DF to P^I1, after the inscription of the formula $A \lor (A \to B)$ in the list of formulas proposed as valid (see figure 7); what remains after the pruning of the immediately closed branch, followed by the erasure of the formula $A \lor NA$, is the adjunction of A to the list of formulas proposed as valid.

Then, to build a tableau in $^t U^H 1$, resp. in $^t U^C 1$, by applying the classical rule according to which the conclusions to be inferred from the proposal of a formula of the form NA as valid are drawn, is recursively reducible to building a tableau in $^t P^I 1$, resp. in $^t P^C 1$.

Notice that thus, in $^t U^H 1$, the intuitionistic rule prescribing what to do after the proposal of a formula of the form $A \to B$ as not valid, and the classical rule prescribing what to do after the proposal of a formula of the form NA as not valid, coexist. Observe that, nevertheless, $U^C 1$ is a *proper extension* of $U^H 1$; indeed, otherwise, DF would be provable in $U^H 1$, hence in C_ω which is an extension of it, and we know since a long time that this is not true.

3.5 Links of extension in a set of ground logics.

The figure 11 below shows the Hasse diagram, ordered by the relation of covering by strict extension, of the set of the ground logics which have been evoked until now.

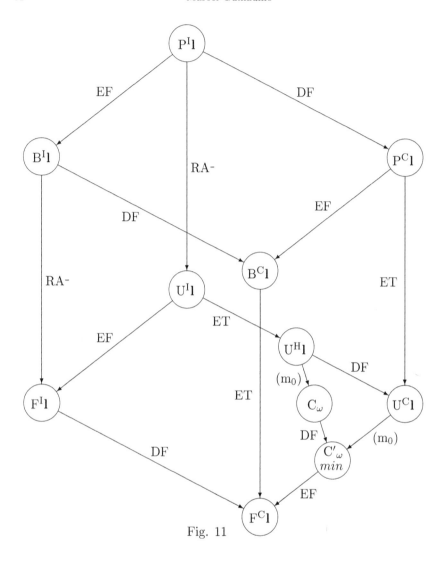

Fig. 11

Each arrow means that the logic at its target is obtained by adding the axiom at its side to the logic at its source and strictly extends the last one.

We could, of course, add many other ground logics to the logics so listed; but amongst them appear already all those in which we are more specially interested here and those which complete a clear vision of the structuring of the logics governed by the rule-making axioms which are tied to the primitive intuitionistically or classically recognized rules[35].

The extensions represented hold clearly in virtue of the above produced (or well-known) axiomatizations. From the remarks and lemmas which follow it can easily be deduced that all are strict, and that no other extension than those resulting from them by transitivity hold between the represented logics.

Observe first that the following extensions are well known as strict: those of P^I1 by P^C1, of F^I1 and of min by F^C1, and of C_ω by min. Also, we already noted that of U^H1 by U^C1.

In the lemmas which follow, the matrices which are described are either three- or four-element linear Heyting algebras or four-elements boolean algebras, of which we retain the three binary operations by which conjunction, disjunction, and implication are interpreted; so in the boolean case, DF is satisfied; it is not in the heytingian case, for then, if a is an element which is neither 0 nor 1, then $a \vee (a \rightarrow 0) = a \vee 0 = a \neq 1$ hold. On the contrary, the (pseudo-)complement will be left aside and replaced by an unary operation $\boldsymbol{n}$ of which the table will be provided, and by which we will interpret the prefixation of the negation sign N. Finally, recall that if RA- is not satisfied, then ET cannot any more be satisfied.

LEMMA 7. P^I1 *does not prove any of* DF, EF, RA-, ET, (m_0).

Proof. Consider the four elements linear Heyting algebra in which a covers 0, and b covers a (and is covered by 1); let $\boldsymbol{n}$ be the unary operation which sends 1 onto 0 and each other element onto the one by which it is covered. Then a invalidates RA-, (m_0), and EF, with 0 for interpreting B.

LEMMA 8. P^C1 *does not prove any of* EF, RA-, ET, (m_0).

Proof. Consider the four elements boolean algebra of which the atoms are a and b, taking for $\boldsymbol{n}$ the unary operation such that $\boldsymbol{n}0 = 1, \boldsymbol{n}a = 1, \boldsymbol{n}b = b$ and $\boldsymbol{n}1 = a$. Then b invalidates RA- and EF, and 0 invalidates (m_0).

LEMMA 9. B^I1 *does not prove any of* DF, RA-, ET, (m_0).

Proof. Going back to the algebra used for proving lemma 7, take this time for $\boldsymbol{n}$ the unary operation which sends 0 onto b and all other elements onto 0. Then 0 invalidates RA- and a invalidates (m_0).

LEMMA 10. U^I1 *does not prove any of* DF, EF, ET, (m_0).

---

[35] It is specially simple to prove, along the way followed three times since the subsection 2.3, that to add (m_0) as an axiom corresponds, for building tableaux, to adding the rule allowing, after a proposal of a formula of the form NNA as valid, to propose the formula A as valid.

Proof. Consider the three elements linear Heyting algebra of which the atom is a, taking for $\boldsymbol{n}$ the unary operation which sends 0 onto 1 and each other element onto its predecessor. Then a invalidates ET and (m_0), and 1 and 0 invalidate EF.

LEMMA 11. $U^H 1$ *does not prove any of* DF, EF, (m_0).

Proof. Consider the three elements linear Heyting algebra of the preceding lemma, taking for $\boldsymbol{n}$ the unary operation which sends 1 onto a and each other element onto 1. Then 1 invalidates EF, and 0 invalidates (m_0).

LEMMA 12. $B^C 1$ *does not prove any of* RA⁻, ET, (m_0).

Proof. Going back to the algebra used for proving lemma 8, take this time for $\boldsymbol{n}$ the unary operation such that $\boldsymbol{n}0 = b, \boldsymbol{n}b = a, \boldsymbol{n}a = 0$ and $\boldsymbol{n}1 = 0$. Then 0 invalidates both RA⁻ and (m_0).

LEMMA 13. *3.5.7.* $U^C 1$ *proves neither* EF *nor* (m_0).

Proof. Going back to the algebra used for proving lemma 8, take this time for $\boldsymbol{n}$ the unary operation such that $\boldsymbol{n}0 = \boldsymbol{n}a = 1, \boldsymbol{n}b = a$ and $\boldsymbol{n}1 = b$. Then a invalidates both EF and (m_0).

4 A hierarchy of abstract paraconsistent logics.

4.1 Paraconsistent controlled rule-making axioms.

Carnielli and Marcos [Carnielli and Marcos, 2002] obtain their rule (bc1) from their schema (tPS) – my EF – just as da Costa's axiom (12) is obtained from Kleene's axiom 7, but this time, the operation is performed on a rule-making axiom, and produces another rule-making axiom. In tableaux versions, the rule which is induced by the adjunction of EF is the one that says that if a formula of the form NA is proposed as valid, then the formula A can be added to the list of formulas proposed as not valid. Now, along the lines that I follow here, the same operation gives the axiom

cEF $\qquad °A \rightarrow (NA \rightarrow (A \rightarrow B))$.

Until now, I explained the rule induced by the adjunction of cEF as a *supplementary closure rule* saying that a subtableau, in which the formulas NA, A, $°A$ are simultaneously proposed as valid, is closed; but when writing this contribution, I changed my mind, and came to a view according to which to check out if the formula $°A$ was proposed as valid earlier is to exercise a *control* over the application of the rule saying that if a formula of the form NA is proposed as valid, then the formula A can be added to the list of formulas proposed as not valid, a positive answer being mandatory to perform this instruction[36]; of course, if the permission is given, the closure

---

[36]Whence the "c" in cEF, meaning "controlled".

follows. This allows us to extend this treatment to analogous rules, derived
from those which correspond to other rule-making axioms by adding such a
control, and which would be laborious to state in the form of closure rules[37].

Now, recall da Costa's axiom (12)[38], which I prefer to label cRA and
spell

cRA $°B \rightarrow ((A \rightarrow NB) \rightarrow ((A \rightarrow B) \rightarrow NA))$.

Using (m_0), we can deduce cEF from it, just as EF can be deduced from
RA. Now, what happens if we proceed similarly with the deduction we made
of RA- from RA ? We obtain

cRA- $°A \rightarrow ((A \rightarrow NA) \rightarrow NA)$,

of which the adjunction will induce a *controlled* intuitionistic rule applicable
when a formula of the form NA has been proposed as not valid. This rule
says the same thing as the rule induced by the adjunction of RA- , but *adds*
to it a *restriction* forbidding to apply it for lack of *permission* given by an
earlier proposal of the formula $°A$ as valid.

I will say that cRA- is a *controlled rule-making* axiom *associated* with the
ground rule-making axiom RA- (or that the first is a *controlled version* of
the second). Indeed, the manner in which the adjunction of this controlled
rule is induced by the adjunction of cRA- as an axiom is very similar to
the manner in which the adjunction of the ordinary, not controlled, rule is
induced by the adjunction of its associated ground axiom RA- : first, using
the controlled rule, every instance of cRA- can be instantaneously recognized as provable by checking it by way of building a closed tableau; second,
we can reason as we did in the subsubsection devoted to the "unbounded"
fragments, in order to establish that, in presence of a previous proposal of
$°A$ as valid and of a previous proposal of NA as not valid, an occurrence
of A can be added to the list of formulas proposed as valid (the fate of the
other previous proposals of formulas as not valid depending on the nature,
intuitionistic or classical, of the rule applied to the proposals of the formulas of the form $A \rightarrow B$ as not valid, by adding transiently the successive
proposals of the formulas $°A \rightarrow ((A \rightarrow NA) \rightarrow NA)$ and $(A \rightarrow NA) \rightarrow NA$
as valid (note in passing that this last formula is an instance of RA-), then
the proposal of $A \rightarrow NA$ as not valid, followed by the proposals of A as
valid and of NA as not valid, both to be kept only in order to take the now
justified rule into account.

In what follows, controlled rule-making axioms associated with the rule-
making axioms ET and EF will be similarly introduced; I will not state

---

[37]This allows us also to imagine a world of logics derived from the idea of subjecting
the applications of some rules to certifications which can be of natures differing totally
from consistency.

[38]In its abstract form.

explicitly the rules induced by the adjunction of one of these paraconsistent rule-making axioms; each of them just adds to the rule induced by the adjunction of the associated ground rule-making axiom, a restriction forbidding to apply the rule when one (or two) suitable certificate(s) of well behavior are missing; the required certificates are those which are added to the associated ground rule-making axiom as hypotheses (in front of a newly introduced implication sign).

Now, recall that from EF and RA⁻ taken as hypotheses, RA is formally provable. If we redo the proof[39] by taking cEF and cRA⁻ as hypotheses, the mandatory controls that are introduced for using these hypotheses during the proof entail that the formula at the end of the obtained formal proof is *weaker* that cRA; this formula is

fcRA $\qquad\qquad {}^{\circ}A \to \Big({}^{\circ}B \to ((A \to NB) \to ((A \to B) \to NA))\Big)$,

in which the control is *full* in the sense that it bears upon the formula A which is to be rejected for implying a contradiction as well as upon the formula B which is a term of this contradiction.

Put your mind at rest, fcRA is still strong enough for entailing cRA⁻! But if we try to deduce cEF from it, what we obtain a formal proof of is restricted, by the certificates required for applying this hypothesis, to

fcEF $\qquad\qquad {}^{\circ}B \to \Big(({}^{\circ}A \to (NA \to (A \to B)))\Big)$ [40]

which gives rise to many thoughts, the first of which is the questioning about the now conceivable "conclusion-controlled"

ccEF $\qquad\qquad {}^{\circ}B \to (NA \to (A \to B))$.

Is this relevant to paraconsistency ? In order to control explosions, to allow all explosives but by confining their use to the room of the formulas possessing consistency, as ccEF is doing, is it worse than to limit the allowed explosive but by leaving their effects unlimited, as cEF is doing ?

No such doubt about fcEF, but the study of logics in which it holds remains to be done, at the risk that some may turn out not to be rewarding (the same for ccEF, whether relevant to paraconsistency or not; the same also for a number of logics including cEF).

Many doors open out similarly by considering cRA⁻. Why could a formula A be, at the same time: not necessarily rejected, according to da Costa's cRA, when it implies a contradiction $B \wedge NB$ where the formula B is not known as well-behaved, and necessarily rejected, according to RA⁻, when it implies the contradiction $A \wedge NA$, even if A is not known as well-behaved

---

[39] At the end of the proof of theorem 2.

[40] This time the cycle is fully described: fcRA is formally deducible from the conjunction of fcEF and cRA⁻.

? Then, it seems to me that it could be thought to paraconsistent logics accepting cRA-.

On another side, in view of theorem 2, it is useless to accept cRA- if ET is accepted. If one think that cRA- has to be accepted, then one *must* at most accept

cET $\qquad\qquad °A \to (A \vee NA)$,

which contributes more to pave the way to a large amount of combinations of ground and abstract paraconsistent schemes corresponding to controlled or not controlled rules to apply in presence of some formula of the form NA proposed as valid or as not valid in the tableaux versions of a plenty of paraconsistent logics.

In what follows, by a *widest* paraconsistent logic, I will refer to a paraconsistent logic, extending P^l1 and in which, for every proposal of a compound formula as valid or as not valid, there is a proof of a rule-making axiom inducing a rule allowing to propose the immediate subformula(s) of that formula as valid or as not valid, possibly at the price of a splitting into two cases, but also possibly only under conditions on a required context of previous proposals, mainly of certificates of consistency for one or two subformulas – in the cases of fcEF and of ccEF, a previous proposal of a subformula as not valid is also required; this renders technically difficult thorough general studies about the logics accepting one of these axioms; so, as far as what regards them, I will content myself with simple and easy remarks, or with indirect conclusions.

Among these logics, in what concerns those of which the tableaux version admits of rules to apply when a formula of the form NA is proposed as not valid, a difficulty comes from the fact that we are not accustomed to working with these rules when they are *controlled*. In fact, further on, I will only skim over such logics; this is because the emphasis is put here on da Costa's C_n and on Waragai and Shidori's PCL1, which both allow axiomatizations in which cEF is the sole controlled abstract paraconsistent axiom accepted[41].

I will say somewhat more on another way which also opens out, and which is to combine controlled or not controlled versions of EF and RA- for obtaining *intuitionistic* paraconsistent logics of which the *aimed* logic is *intuitionistic* and in which the *individual well-behavior* of formulas is defined by one of the $A^{(n)}$ or by $A \to NNA$!

REMARK 14. I did not introduce schemes such that

---

[41] Of course, in the case of PCL1, I do not allude here to the own axiomatization of Waragai and Shidori.

cDF	$°A \rightarrow (A \vee (A \rightarrow B))$,
ccDF	$°B \rightarrow (A \vee (A \rightarrow B))$, or
fcDF	$°B \rightarrow \left(°A \rightarrow (A \vee (A \rightarrow B))\right)$,

but we can deduce each of them from suitable versions of EF and ET by a formal proof reproducing with evident modifications the formal deduction of DF from EF and ET (*cf* footnote 33, p. 22): among the hypotheses, one at least has to be submitted to (one of) the same control(s) as the wanted version of DF.

4.2   A hierarchy of basic logics of consistency.

The figure 12 at the top of the next page shows, as the figure 11 did, the Hasse diagram, ordered by the relation of covering by extension, of a set of paraconsistent "basic logics of consistency", as I am going to explain soon that I will call them, and of their ground logics. Labels framed in dashed lines are those of these ground logics, and labels framed in circles, those of the basic logics of consistency. As in figure 11, each arrow means that the logic at its target is obtained by adding the axiom(s) at its side, to the logic at its source, and extends the latter[42].

The idea of these widest paraconsistent logics stems from Carnielli and Marcos [Carnielli and Marcos, 2002], where they introduced, as a model of a more general kind of logics similarly obtained from other ground logics, that logic they called the (abstract) "basic logic of consistency" **bC** of which the ground logic is *min* and which can be formalized by adding cEF as unique abstract paraconsistent axiom[43].

---

[42]It can reasonably be assumed that all these extensions are strict, what is well known to hold in some of these cases.

[43]Carnielli and Marcos adopt instead their "rule (bc1)", which gives an equivalent formalization, using the deduction theorem and *modus ponens*.

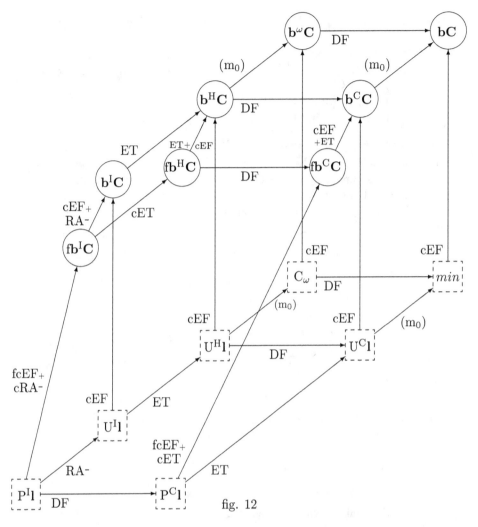

fig. 12

This insight is more explicitly expanded here by dubbing "basic logics of consistency" a few neighbouring logics obtained first by weakening the ground logic, by reinforcing the controls on one or several of the rule-making axioms, sometimes after having replaced ET by RA-. However, in what follows, I will make allusions only to **bC** and to the following other such logics, named in the figure:

fb^I**C**, of which the ground logic is P^{I}l, to which are added fcEF and cRA-;

b^I**C**, of which the ground logic is U^{I}l, to which is added cEF;

$\mathbf{fb^H C}$, of which the ground logic is $P^I 1$, to which are added fcEF and cET;

$\mathbf{b^H C}$, of which the ground logic is $U^H 1$, to which is added cEF;
$\mathbf{fb^C C}$, of which the ground logic is $P^C 1$, to which are added fcEF and cET;

$\mathbf{b^C C}$, of which the ground logic is $U^C 1$, to which is added cEF;
$\mathbf{b^\omega C}$, of which the ground logic is C_ω, to which is added cEF.

4.3 Some very basic theses of PCL1 and bC.

Some theses of PCL1 and $\mathbf{bC}$ become formally provable at a lower level than that of $\mathbf{bC}$, as can easily be verified by building simple closed tableaux. I will merely list them hereafter, keeping the labels given to each of them in the corresponding work (some are common to both; in this case, I keep the label given in PCL1).

FACTS 15. The following schemes are formally provable from $\mathbf{b^I C}$ on:
3.16 (i) [Or (ii)] $\quad NA \to (A \to N^\circ A)$,
T5 $\qquad \qquad \ ^\circ B \to ((A \to B) \to (NB \to NA))$,
3.17 (ii) $\qquad \ \ ^\circ B \to ((A \to NB) \to (B \to NA))$,
T8 $\qquad \qquad \ ^\circ A \to (NA \to ((A \vee B) \to B))$,
T9 $\qquad \qquad \ ^\circ A \to (^\circ B \to ((NA \vee NB) \to N(A \wedge B)))$,
T10 $\qquad \qquad \ ^\circ A \to (^\circ B \to ((NA \wedge NB) \to N(A \vee B)))$,
to which I add
U1 $\qquad \qquad \ ^\circ B \to ((A \wedge NB) \to N(A \to B))$.

FACT 16. The following schemes are formally provable from $\mathbf{b^H C}$ on:
T6 $\qquad \qquad \ ^\circ B \to ((NA \to NB) \to (B \to A))$,
3.17 (iii) $\qquad \ ^\circ B \to ((NA \to B) \to (NB \to A))$.

4.4 Abstract paraconsistent stability axioms.

These are the well known four following:
(n) $\qquad \qquad \qquad \ ^\circ A \to {}^\circ NA$,
(c) $\qquad \qquad \qquad \ ^\circ A \to (^\circ B \to {}^\circ (A \wedge B))$,
(d) $\qquad \qquad \qquad \ ^\circ A \to (^\circ B \to {}^\circ (A \vee B))$,
(i) $\qquad \qquad \qquad \ ^\circ A \to (^\circ B \to {}^\circ (A \to B))$.

First links of these axioms with the schemes (m_0), (m_1), (m_2), (m_3) were discovered by da Costa and Guillaume [Da Costa and Guillaume, 1965]; some new ones have recently been discovered by Waragai and Shidori; but more can still be said on this topic. We get down now to its abstract side.[44]

---

[44] I leave it to the reader to check the facts that follow, for example by building almost

FACTS 17. The schemes

(an)	$^\circ NA \to (\mathrm{m_0})$,
(ac)	$^\circ(A \land B) \to (\mathrm{m_1})$,
(ai)	$^\circ(A \to B) \to (^\circ A \to (\mathrm{m_3}))$

are formally provable[45] from $\mathbf{b^H C}$ on.

FACT 18. The schema

(ad)	$^\circ(A \lor B) \to (\mathrm{m_2})$

is formally provable[46] from $\mathbf{b^I C}$ on.

4.5 Abstract logics of consistency.

In what follows, I will substitute the letter $\mathbf{a}$ for the letter $\mathbf{b}$ in the label of one of the basic logics of the penultimate subsection in order to form a label of the logic obtained by adding *all* the last axioms to those of the basic logic. I mean to call the logics resulting from these adjunctions "abstract full logics of consistency".

For example, $\mathbf{a^I C}$ will denote the abstract full intuitionistic logic of consistency, which results from the adjunction of all stability axioms to the basic intuitionistic logic of consistency $\mathbf{b^I C}$, and $\mathbf{a^\omega C}$ can similarly denote the logic which has as axioms all axioms, under abstract form, of the C_n, as they were axiomatized by da Costa in his doctoral Thesis [Da Costa, 1963][47].

Note, in passing, that this last logic, in which the consistency connective remains *uninterpreted*, is *neither* PCL1 (which moreover satisfy an abstract paraconsistent axiom which has not been mentioned until now) nor any of the C_n. At most, its theses are theses which are *common to all the* C_n.

Clearly, each of the abstract full logic of consistency extends the basic logic from which it stems, and the relations of extension which hold between

---

immediately closed tableaux in the tableaux versions of the logics indicated. In the formulas which occur in these facts, and in some analogous ones to come later, $(\mathrm{m_0})$ is conceived as written using the metavariable A, and $(\mathrm{m_1})$, $(\mathrm{m_2})$, $(\mathrm{m_3})$, by using metavariables A and B as in section 1.

[45] Regarding (an), $(A \lor NA) \to (^\circ NA \to (NNA \to A))$ is in fact a thesis of $\mathbf{b^I C}$.

On another hand, the good strategy for building the hinted tableau in the case of (ai) is to use the derived rule 3 in the branch which begins by the proposal of NB as not valid, and as for the branch which begins by the proposal of A as not valid, to *delay to the utmost* the application of the intuitionistic rule taking in account the proposal of $A \to B$ as not valid.

[46] This is due to the fact that $(\mathrm{m_2})$ is an *intuitionistic* thesis.

[47] Up to equivalence, since, in presence of ET and $(\mathrm{m_0})$, each of cEF and cRA can be deduced from the other.

two basic logics hold between the abstract full logics which stem from them.

It will sometimes happen that we mention other abstract logics of consistency, which result from the adjunction of only *some* of the stability axioms. In order to indicate that such a logic has so been axiomatized, the letters which occur into the brackets in the labels of the added stability axioms will be placed before the letter $\mathbf{C}$ for forming a label of this logic, in the order in which the stability axioms occur in the list of the preceding subsection.

For example, $\mathbf{a}^H nc\mathbf{C}$ will denote the logic of which an axiomatization is obtained by adding, to that of $\mathbf{b}^H\mathbf{C}$, the axioms (n) and (c).

I leave it to the reader to get an idea of the relations of extension which hold between all these logics and between them and those which have been previously mentioned.

5   Concrete paraconsistent logics.

5.1   Low (weakly) concrete paraconsistent logics.

I will now examine the adjunction of a type of axioms, to which I assign the label (Ia)[48], and which are of the form $\circ A \leftrightarrow D$, where D is a schema of which A is the unique prime component. Note that if we wish to regard D as a shorthand for $\circ A$, this axiom becomes provable, being of the form $D \leftrightarrow D$.

The paraconsistent logic obtained by this adjunction will here be labelled by transforming the label of the logic to which it is added, replacing the "$\mathbf{C}$" by what is effectively in the brackets of the axiom's label.

So $\mathbf{b}^I\mathbf{C}$ gives rise to the extension labelled $\mathbf{b}^I$Ia which might be called the "low (weak) concrete basic intuitionistic Ia paraconsistent logic" (and so forth for other basic logics of consistency), and $\mathbf{a}^\omega\mathbf{C}$ gives rise to $\mathbf{a}^\omega$Ia which might be called the "low (weak) concrete full Ia paraconsistent logic" (and so forth for other full abstract logics of consistency)[49].

I already noticed that if (Ia) is accepted as an axiom, then, $\circ A \leftrightarrow D$ must be a thesis of the paraconsistent logic, so that D is a thesis of the aimed logic, which cannot be a thesis of the ground logic without making the paraconsistency disappear.

On the borderline of concrete considerations are the following abstract

FACTS 19. The following schemes are formally provable from $\mathbf{b}^I\mathbf{C}$ on:

---

[48] "Initials of the author".

[49] I do not whish to rack my brain over the names to give to the low concrete Ia paraconsistent logics which are not full.

U2	$^\circ A \to N(A \wedge NA)$,
U3	$^\circ A \to (A \to NNA)$,
U4	$^\circ A \to (A \leftrightarrow NA)$.

For the first two cases, proofs are brought by the tableau of the figure 13 at the top of the next page, which condenses, in one only, four ordinary tableaux, and on which we can read some remarkable peculiarities it will be interesting to make use of farther.

On the abstract side, the tableau condenses in one two tableaux of $^t\mathbf{b}^I\mathbf{C}$, which have the same left column except for the formula in *simple* square brackets, and also the same part of the right column above the first horizontal dashed lines, but with two different continuations under that line, which are distributed into two right columns separated by a vertical line made of ‡'s. In the tableau, the notation $A^\star$, which is the only explicited expression above the first horizontal dashed line, is meant to represent a shorthand for one of the two formulas in simple square brackets by which each the two separated continuations of the right column begins. All in all, the inscription of the formula located just under the first horizontal dashed line corresponds, in each of the separated right continuations, merely to the replacement of the shorthand by the shortened formula.

	Valid ?	Not valid ?	
	$\bullet$	$\bullet$	
	$[\![^\circ A]\!]$	$\bullet$	
	$\bullet$	$\bullet$	
	$\bullet$	$A^\star$	
	$\bullet$	$\bullet$	
	$\bullet$	$\bullet$	

$[A \wedge NA]$ $[N(A \wedge NA)]$ ‡ $[A \to NNA]$

A ‡

NA ‡ $[NNA]$

    ‡

$[\![A]\!]$

Fig. 13

Then, on the side of $N(A \wedge NA)$, the formula $A \wedge NA$ is proposed as valid, and so are next the formulas A and NA; on the side of $A \to NNA$, the formula A is proposed as valid and the formula NNA, as not valid: then, the formula NA is proposed as valid. As $^\circ A$ is in the list of formulas proposed as valid, then, in both cases, the formula A can be added to the list of formulas proposed as not valid, which entails the closure of both the tableaux we have condensed in one.

In the case where $A^\star$ would be meant to shorten the formula $N(A \leftrightarrow NA)$, a splitting in two subtableaux results from the inscription of the formula $NA \to A$ in the list of formulas proposed as valid, after which the formula A is proposed as valid in both the newly introduced subtableaux, and then NA is proposed as valid in both also, by applying the derived rule 1 in virtue of the earlier introduction of the proposal of the formula $A \to NA$ as valid. Then, the closure intervenes as previously, in both the subtableaux.

Now, let us delete from the condensed tableau the two formulas in *double* square brackets; what is obtained is a tableau which condenses, in one only, two tableaux of $^t\mathrm{U}^\mathrm{I}1$. Then, the later erasure of the formulas in *simple* square brackets (and the same operations in the evoked but not exhibited analogous tableau for the case where the formula $N(A \leftrightarrow NA)$ is shortened into $A^\star$) enters in a reasoning justifying the application of the following

Auxiliary special concrete rule 1: In any extension of $^t\mathbf{b}^\mathrm{I}\mathbf{C}$, if a formula of the form $A^\star$ is proposed as *not valid*, where $A^\star$ is a shorthand of one of the formulas $N(A \land NA)$, $A \to NNA$, $N(A \leftrightarrow NA)$, then both formulas A and NA can *simultaneously* be proposed as *valid*; correlatively, the list of the formulas proposed as not valid is reduced to one unique formula of the form NB (B is respectively the formula $A \land NA$, NA, $A \leftrightarrow NA$), except in the case of an extension of $^t\mathbf{b}^\mathrm{H}\mathbf{C}$, in which the list of formulas previously proposed as not valid remains unaltered.

Now, each of these formulas represents one of the meanings attributed to the notion of consistency of a formula by various authors: $N(A \land NA)$ is the shorthand of $^\circ A$ in da Costa C_1; in PCL1, this shorthand is equivalent to $A \to NNA$ [50]; $N(A \leftrightarrow NA)$ is another possible interpretation that I add in order to extend the set of terms to be compared.[51]

Owing to the facts 19, it is equivalent to adjoin, to any extension of $\mathbf{b}^\mathrm{I}\mathbf{C}$:

(a) the axiom	(DC)	$^\circ A \leftrightarrow N(A \land NA)$
or the axiom	(DC)	$N(A \land NA) \to {}^\circ A$ [52];
(b) the axiom	(WS)	$^\circ A \leftrightarrow (A \to NNA)$
or the axiom	(WS)	$(A \to NNA) \to {}^\circ A$;
(c) the axiom	(GM)	$^\circ A \leftrightarrow N(A \leftrightarrow NA)$

---

[50] More precisely, this shorthand is $A \leftrightarrow NNA$, but $NNA \to A$ is an axiom of the ground logic of PCL1 and takes no part in selecting the well-behaved formulas from the not necessarily well-behaved ones.

[51] Before all in the technicalities to come.

[52] Thus, my $\mathbf{b}$DC is Carnielli and Marcos' **Cil** (*cf* [Carnielli and Marcos, 2002]).

or the axiom   (GM) $N(A \leftrightarrow NA) \to {}^{\circ}A$.

The usefulness of the auxiliary special concrete rule 1 consists in opening the way towards a unified treatment of some properties which are common to concrete logics in which one of the formulas (DC), (WS), (GM) is accepted as a thesis. In contexts which are related to such properties, the logics concerned will be indicated by labels in which "Ia" will be replaced by "$T^{\star}$".

Besides, there exist other meanings attributed to the possession of consistency which are not in the situation of the facts 19:

FACT 20. For $n > 0$ the formula ${}^{\circ}A \to A^{(n)}$ is a thesis of $\mathbf{b}^{\omega}\mathbf{C}$ and of $\mathbf{a}^{\mathrm{I}}\mathrm{nc}\mathbf{C}$.

Indeed, to build a tableau in order to check this for $n > 1$ introduces proposals of formulas of the form NNB as valid (for $n = 1$, the formula to prove is nothing else than U2, which is already provable, according to the facts 19, in the fragment $\mathbf{b}^{\mathrm{I}}\mathbf{C}$ of $\mathbf{b}^{\omega}\mathbf{C}$). Now, as for finding a rule to apply for going farther, the one which will appear below as slightly the most efficient is that which is associated to (m_0) [53] and allows to propose formula B as valid – in the case of $\mathbf{a}^{\mathrm{I}}\mathrm{nc}\mathbf{C}$, the same outcome requires the proposal of the formula ${}^{\circ}B$ as valid, which can be deduced from the proposal of ${}^{\circ}A$ as valid when B is built by using negation and conjunction, from prime components of which all are occurrences of A. This leads to a cascade of proposals of formulas of the form $NN(A^i \wedge NA^i)$ as valid, which ends when i becomes equal to 0, with the proposals of the formulas A and NA as valid, leading to closure, in reason of the initial proposal of ${}^{\circ}A$ as valid.

Hence, for $n > 1$, I will take the formula

(DC_n) $A^{(n)} \to {}^{\circ}A$

as an axiom for the weakly concrete fragments of $\mathbf{C}_n$ extending $\mathbf{b}^{\omega}\mathbf{C}$ or $\mathbf{a}^{\mathrm{I}}\mathrm{nc}\mathbf{C}$.

An *auxiliary special concrete rule* 1_n is also *at our disposal* in the case where $A^{\star}$ is a shorthand for $A^{(n)}$ with $n > 1$; it says the same thing as the other special concrete rule, *but only* for the extensions of ${}^{\mathrm{t}}\mathbf{b}^{\omega}\mathbf{C}$ or of ${}^{\mathrm{t}}\mathbf{a}^{\mathrm{I}}\mathrm{nc}\mathbf{C}$.

The use as an axiom of (DC), (WS) or (GM), resp. of (DC_n) for $n > 1$, allows, when ${}^{\circ}A$ is proposed as not valid, to employ the derived rule 2, after an imagined proposal as valid of that of these axioms corresponding to the context, in order to perform the first step which was explained on the tableau of figure 13 after the supposed proposal of $A^{\star}$ as not valid; to imagine to perform later the next steps boils down to applying the auxiliary

---

[53] The reader can also consult the proof given in 1964 by da Costa and Guillaume [Da Costa and Guillaume, 1965].

special concrete rule 1, resp. 1_n, directly in virtue of the earlier proposal of $°A$ as not valid in place of that of $A^\star$.

In such contexts, in which some properties, common to the logics the label of which is one of those that the label $\text{T}^\star$ symbolizes, are shared also by the logics accepting as a thesis one of the formulas (DC_n) for $n > 1$, the label $\text{T}_n^\star$ will symbolize one of the labels symbolized by $\text{T}^\star$ or by DC_n for $n > 1$.

As first applications of the preceding remarks, it is quite easy to build closed tableaux in order to prove the propositions which follow[54].

THEOREM 21.   *The following schemes are formally provable from* $\mathbf{b}^\mathrm{I}\text{T}^\star$ *on:*

(na)	$(\mathrm{m}_0) \to (°A \to °NA)$,
(ca)	$(\mathrm{m}_1) \to \left(°A \to (°B \to °(A \wedge B))\right)$,
(da)	$(\mathrm{m}_2) \to \left(°A \to (°B \to °(A \vee B))\right)$,
(ia)	$(\mathrm{m}_3) \to (°B \to °(A \to B))$.

COROLLARY 22.   *The following schemes are formally provable from* $\mathbf{b}^\mathrm{H}\text{T}^\star$ *on:*

(ne)	$(°A \to °NA) \leftrightarrow (°A \to (\mathrm{m}_0))$,
(ce)	$\left(°A \to (°B \to °(A \wedge B))\right) \leftrightarrow \left(°A \to (°B \to (\mathrm{m}_1))\right)$,
(ie)	$\left(°A \to (°B \to °(A \to B))\right) \leftrightarrow \left(°A \to (°B \to (\mathrm{m}_3))\right)$.

COROLLARY 23.   *The schema*
(de) $\qquad \left(°A \to (°B \to °(A \vee B))\right) \leftrightarrow \left(°A \to (°B \to (\mathrm{m}_2))\right)$
is formally provable from $\mathbf{b}^\mathrm{I}\text{T}^\star$ *on.*

COROLLARY 24.   *To any extension of* $\mathbf{b}^\mathrm{H}\text{T}^\star$, *it amounts to the same to adjoin independently as axiom:*

(i)	(n) *or*	(a$_0$)	$°A \to (\mathrm{m}_0)$,
(ii)	(c) *or*	(a$_1$)	$°A \to (°B \to (\mathrm{m}_1))$,
(iii)	(i) *or*	(a$_3$)	$°A \to (°B \to (\mathrm{m}_3))$.

COROLLARY 25.   *To any extension of* $\mathbf{b}^\mathrm{I}\text{T}^\star$, *it amounts to the same to adjoin independently as axiom:*

---

[54]The theorem 21 dates from [Da Costa and Guillaume, 1965]. Waragai and Shidori recently discovered the implications from the right to the left in the schemes (ce) of corollary 22 and (de) of corollary 23 below.

(d) *or* (a_2) $°A \to (°B \to (m_2))$.

To the preceding results are to be added those which follow, which can be established inductively by reasoning through the building of more complex tableaux.

THEOREM 26. *"Guillaume's thesis"*[55] $°A \to °NA$ *is formally provable in all extensions of* $\mathbf{b}^\omega \mathrm{T}_n^\star$ [56].

THEOREM 27. *For* $n > 1$, *the schemes* (ca), (da), (ia), (ce), (de), (ie) *are formally provable in* $\mathbf{b}^\omega \mathrm{DC}_n$.

COROLLARY 28. *To any extension of* $\mathbf{b}^\omega \mathrm{DC}_n$ *with* $n > 1$, *it amounts to the same to adjoin* independently *as axiom:*
(i)   (c) *or* (a_1),   (ii)   (d) *or* (a_2),   (iii)   (i) *or* (a_3).

Henceforth any logic resulting by adjunction of one or several of the axioms (a_0), (a_1), (a_2), (a_3) to an extension of $\mathbf{b}^H \mathrm{T}^\star$ or of $\mathbf{b}^\omega \mathrm{DC}_n$ will be labeled as the logic resulting of this extension by adjunction of the corresponding axiom(s) (n), (c), (d), (i); the same for the extensions of $\mathbf{b}^I \mathrm{T}^\star$ *not satisfying* ET is allowed *only* in the case of (a_2) and (d); this is due to the fact that each of (a_0), (a_1) and (a_3) requires from the well-behaved formulas the satisfaction, respectively, of (m_0), (m_1) or (m_3), among which each imposes a touch of *classical* well-behavior, which is not the case of (m_2). Then, it can be *conjectured* that the aimed logics which are not quite classical are in general more than intuitionistic: *intermediate*.

5.2 The strong negation in concrete paraconsistent logics.

Intuitionistic characteristics.

The strong negation $\sim$ is defined so that $\sim A$ is a shorthand for $°A \wedge NA$ for every formula A.

This definition has two abstract immediate consequences, stated at once.

FACT 29. $°A \to (\sim A \leftrightarrow NA)$ holds even in $\mathrm{P}^I 1$, so in all the logics considered here (in other words, the restrictions of $\sim$ and of N to the well-behaved formulas coincide).

THEOREM 30. *In* $\mathbf{b}^I \mathbf{C}$, *so in all its extensions, the strong negation satisfies EF, i.e.* $\sim A \to (A \to B)$ *holds.*

---

[55] According to Carnielli and Marcos [Carnielli and M rcos, 2002].

[56] It will be seen below that $\mathbf{a}^H \mathrm{n}\mathrm{T}^\star$ coincide with $\mathbf{b}^\omega \mathrm{T}^\star$.

Thus, the properties of the strong negation are the same as those of the paraconsistent negation *in what concerns the well-behaved* formulas, but, in what concerns *all formulas, whether well-behaved or not*, they appear as being closer to those of the negation of the aimed logic than to the properties of the paraconsistent negation. In almost all the cases of the logics here under consideration, this closeness is maximal. Indeed, the theorems which follow are then at our disposal.

THEOREM 31. *From* $\mathbf{b}^{\mathrm{I}}\mathrm{T}^{\star}$ *on and, for* $n > 1$, *from* $\mathbf{a}^{\mathrm{I}}\mathrm{ncDC}_n$ *on, the strong negation is* intuitionistic.

Proof. This can be seen by building a suitable closed tableau showing that the strong negation satisfies RA-, *i.e.* that $(A \to \sim A) \to \sim A$ holds. The good strategy is to take in account forthwith the proposal of the formula $\sim A$ as not valid by splitting the tableau into two subtableaux. The one which begins by the proposal of NA as not valid can be continued by the proposal of A as valid. The other, which begins by the proposal of $°A$ as not valid, can also be continued by the proposal of A as valid by applying an auxiliary special concrete rule. Then, applying the auxiliary derived rule 1 to the earlier proposal of $A \to \sim A$ as valid, we are led to closure in both subtableaux by the further proposals of NA and of $°A$ as valid.

REMARK 32. A queer point about the just contemplated concrete paraconsistent logics is the following: it is perilous for paraconsistency that the replacement theorem holds for the paraconsistent negation, *i.e.* that $NA \leftrightarrow NB$ would have to be provable if $A \leftrightarrow B$ is provable. On another side, one does not see why $°A \leftrightarrow °B$ would have to be provable in the same circumstance. And yet, if $A \leftrightarrow B$ is provable, $(NA \wedge °A) \leftrightarrow (NB \wedge °B)$ has to be provable.

Some paraconsistent logics with classical strong negation.

THEOREM 33. *From* $\mathbf{b}^{\mathrm{H}}\mathrm{DC}$ *on, from* $\mathbf{b}^{\mathrm{H}}\mathrm{GM}$ *on, and for* $n > 1$, *from* $\mathbf{b}^{\omega}\mathrm{DC}_n$ *on, the strong negation is* classical.

Proof. We will proceed by a series of lemmas.

FACT 34. The schema $((A \vee (NA \wedge °A)) \leftrightarrow (A \vee NA) \wedge (A \vee °A))$ is formally provable in $\mathrm{P}^{\mathrm{I}}\mathrm{1}$.

COROLLARY 35. *The schema* $(A \vee \sim A) \leftrightarrow (A \vee °A)$ *is formally provable in* $\mathrm{U}^{\mathrm{H}}\mathrm{1}$ *and so in its extension* $\mathbf{b}^{\mathrm{H}}\mathbf{C}$.

Now, the theorem follows at once from the

LEMMA 36. *The schemes* $NA \vee {}^{\circ}A$ *and* $A \vee {}^{\circ}A$ *are formally provable from* $\mathbf{b}^H\mathrm{DC}$ *on, from* $\mathbf{b}^H\mathrm{GM}$ *on, and for* $n > 1$, *from* $\mathbf{b}^{\omega}\mathrm{DC}_n$ *on.*

Proof. We check this by building a suitable closed tableau, which begins by the proposal, say, of $A \vee {}^{\circ}A$ as not valid, and continue by adding successively the separate proposals of A and ${}^{\circ}A$ as not valid. Note here that if a further proposal of a formula of the form $B \to C$ as not valid is exploited, then the proposal of the formula A as not valid will be removed, while it is needed for obtaining closure[57]. On the other hand, a proposal of a formula of the form NB as not valid can be exploited, provided this is done by using the *classical* rule. Now, according to an auxiliary special concrete rule, the proposal as valid of the formula A follows, leading to closure. In this reasoning, the initial proposal of A as not valid can be replaced by a proposal of NA as not valid, since the special rule entails also the proposal of that formula as valid.

COROLLARY 37. $\sim\sim A \to A$ *and* DF *are provable in all extensions of* $\mathbf{b}^{\omega}\mathrm{DC}_n$, *for* $n > 1$, *of* $\mathbf{b}^H\mathrm{DC}$ *and of* $\mathbf{b}^H\mathrm{GM}$.

Indeed, in order to see that, me have merely to apply the transcription of the corollary 6 to the last two lemmas.

COROLLARY 38. *For every* $n > 0$, *the logics* $\mathbf{b}^{\omega}\mathrm{DC}_n$ *and* $\mathbf{b}\mathrm{DC}_n$ *coincide, as coincide the logics* $\mathbf{b}^H\mathrm{GM}$ *and* $\mathbf{b}\mathrm{GM}$.

The power of the transmission of consistency by passing to negation.

In fact, we can say more:

THEOREM 39. *The schema* (m_0) *is provable in all extensions of* $\mathbf{a}^H\mathrm{nT}^{\star}$.

Proof[58]. Beginning by the cases of the logics to which the lemma 36 apply and which do not count (m_0) amongst their axioms, we notice that $A \to (\mathrm{m}_0)$ is a thesis of any such logic, as being an instance of an axiom of $\mathrm{P}^I 1$, and recall that ${}^{\circ}A \to (\mathrm{m}_0)$ also is a thesis, in virtue of corollary 24 *(i)*; whence (m_0) also is provable.

Now we turn towards the case of $\mathbf{b}^H\mathrm{WS}$, and proceed by a series of lemmas of independent interest.

FACT 40. *The schemes* $NNA \to {}^{\circ}A$ *and* $NA \vee {}^{\circ}A$ *are formally provable in* $\mathbf{b}^H\mathrm{WS}$.

---

[57]Here is the blocking point of the reasoning, in case where (WS) is used.

[58]In a manuscript for the first chapter of a projected *Introduction to Paraconsistent Logic*, J.Y. Béziau made use of this result in order to give a very simple Hilbert's style axiomatization of C_1.

Proof. The formula $NNA \to (A \to NNA)$ is an instance of an axiom of $\mathrm{P^I1}$, and so the formula $NNA \to {}^\circ A$ follows in this logic from the hypothesis WS, *i.e.* $(A \to NNA) \to {}^\circ A$. Now, a formal proof of $NA \lor {}^\circ A$ can be obtained by substituting NA for A and ${}^\circ A$ for B in the following

FACT 41. The schema $(A \lor NA) \to ((NA \to B) \to (A \lor B))$ is formally provable in $\mathrm{P^I1}$.

The following step consists in demonstrating the strong peculiarity of fragments of PCL1 which follows.

LEMMA 42. *The schema* ${}^\circ NA$, *i.e. the PCL1's axiom T11, is formally provable in* $\mathbf{a}^H$nws.

Proof. By using now (n), *i.e.* ${}^\circ A \to {}^\circ NA$, we obtain, by inserting a supplementary step in the preceding proof, formal proofs first of $NNA \to {}^\circ NA$ and then of $(NNA \lor {}^\circ NA) \to {}^\circ NA$. But $NNA \lor {}^\circ NA$ is an instance of a schema which is formally provable according to the fact 40; then, ${}^\circ NA$ is formally provable.

Now, we reach the asserted theorem by using the schema (an) from the facts 17, which allow us to obtain a formal proof of (m_0).

COROLLARY 43. *The logics* $\mathbf{a}^H$nT* *and* $\mathbf{b}^\omega$T* *coincide*[59].

The power of the transmission of consistency by passing to conjunction.

In sum, taking into account that the formula (na) of theorem 21 is a thesis of all concerned systems, what we proved is that (n) $\leftrightarrow$ (m_0) is a thesis of $\mathbf{b}^H$T*.

In this respect, the situation of (m_1) is similar: (c) $\leftrightarrow$ (m_1) is a thesis of $\mathbf{b}^H$T*, this time taking into account the formula (ca) and the following

LEMMA 44. *The schema* (m_1) *is provable in all extensions of* $\mathbf{a}^H$cT*.

Proof[60]. Assume (c). Then ${}^\circ A \to ({}^\circ B \to (m_1))$ becomes provable, owing to the provability of the formula (ac) of facts 17. Now note that $NA \to (m_1)$ and $NB \to (m_1)$ are provable in $\mathrm{P^I1}$, as thus such is $NA \to ({}^\circ B \to (m_1))$; then from the provability of $NA \lor {}^\circ A$ follows that ${}^\circ B \to (m_1)$ is provable; finally, repeat that argumentation.

---

[59] Also with $\mathbf{a}^\omega$nT*, but the last coincides with the preceding ones, by using the formula (na) of theorem 21.
[60] Da Costa and Guillaume [Da Costa end Guillaume, 1965].

More on the aimed logics.

We are now in a position to say more on the links between any concrete full paraconsistent logic and its aimed logic, which obey the following

THEOREM 45. *Let* $\mathbf{p}^*$Ia *be a concrete* full *paraconsistent logic admitting of a tableaux version* $^t\mathbf{p}^*$Ia, *and let* D *be the formula defining the meaning of the consistency connective in* $\mathbf{p}^*$Ia *in terms of occurrences of some metavariable* A *as prime components and of the usual (not paraconsistent) connectives.*

Assume moreover that the logic $\mathbf{p}^*$Ia *is a fragment of a not paraconsistent logic* N^*1 *which proves* D *and admits of a* complete *tableaux version* $^tN^*1$.

Then, N^*1 *is the* aimed *logic of* $\mathbf{p}^*$Ia *iff any ground rule-making axiom of which every instance is provable in* N^*1 *admits of some version of which every instance is provable in* $\mathbf{p}^*$Ia.

CAUTION. The consistency connective, in principle, does *not belong to the language which is proper to not paraconsistent logics* as P^I1, N^*1 and the aimed logic A^*1 of $\mathbf{p}^*$Ia. Hence, any possible occurrence of a (sub)formula *dominated by a consistency connective* functions, and must be counted, as a *prime component* of any formula of one of these logics in which it appears.

Having said this, I can offer here, for the last theorem, the following *sketch of a proof*.

First of all, notice too that to say that a formula F is a thesis of the aimed logic does *not* mean the same as to say that a formula is a thesis of the other logics which are here involved: we lack any notion of formal proof for the aimed logic. Let us prefer to say that F is *accepted* in the aimed logic for expressing that F fulfils the criterion which is required for that, to wit, that $\wedge°[F] \to F$ is a thesis of $\mathbf{p}^*$Ia, where $\wedge°[F]$ is meant to denote the conjunction of the certificates of consistency of the prime components of F.

As N^*1 is assumed to extend $\mathbf{p}^*$Ia, the formula $\wedge°[F] \to F$ will also be a thesis in it, as are also the instances of (Ia) corresponding to the prime components of F and the corresponding instances of D, so that, by *modus ponens*, F itself is a thesis of N^*1: then, *every formula accepted by* A^*1 *is a thesis of* N^*1.

Then, to prove the theorem 45 is reduced to prove the converse, which ensues, by taking Γ void and Δ reduced to F, as an easy corollary from the following

THEOREM 46. *Assume that the logics* $\mathbf{p}^*$Ia *and* N^*1 *fulfil the assumptions of theorem 45. For every sequent* $\Gamma \vdash \Delta$ *of* N^*1, *let* $°[\Gamma\Delta]$ *be the set of the certificates of consistency of all the prime components of the formulas* *occurring in this sequent.*

Then, the sequent $°[\Gamma\Delta]$, $\Gamma \vdash \Delta$ *holds in* $\mathbf{p}^*$Ia *iff the sequent* $\Gamma \vdash \Delta$ *holds in* N^*1.

First, let us *prove* that the condition is *necessary*[61]. Indeed, as N^*1 extends $\mathbf{p}^*$Ia, if the sequent $°[\Gamma\Delta]$, $\Gamma \vdash \Delta$ holds in the second, then it holds in the first, in which all formulas of $°[\Gamma\Delta]$ are provable.

Now, I claim that the condition is *sufficient*. I sketch a *proof*.

Suppose that the sequent $\Gamma \vdash \Delta$ holds in N^*1. We proceed step by step.

First, since the tableaux version of N^*1 is complete, we can build a closed tableau of $^tN^*1$, started by checking this sequent.

Second, by applying systematically the methods "of justification" of the *not* basic intuitionistic rules described in section 3, modified just by agreeing to put the added proposals of instances of rule-making axioms as valid in the first places of the left column at the root of the tableau (and if needed the proposal of a variable X not occurring in $\Gamma\cup\Delta$ as valid, in the first place of the right column at the root), we can transform this first tableau into a kept closed tableau of $^t\mathrm{P}^I1$, which could have been built in conformity with a correct application of the rules as started by checking the sequent $R[\Sigma(\Gamma\Delta)]$, $\Gamma \vdash (X)$, Δ, where $R[\Sigma(\Gamma\Delta)]$ denotes a set of instances of not basic intuitionistic rule-making axioms, built with subformulas of formulas of $\Gamma \cup \Delta$, and where (X) denotes the set reduced to X if we had resorted to an instance of EF, or the void set otherwise.

Third, we form a set $cR[\Sigma(\Gamma\Delta)]$ from the set $R[\Sigma(\Gamma\Delta)]$ by applying, to each formula of this last, the transformation which consists in restablishing the prefixation(s) of the implication(s) by a certificate of consistency, required by the version, which holds in $\mathbf{p}^*$Ia, of the ground rule-making axiom of which this formula is an instance. In the same time, we collect the reintroduced certificates, which at the end form the set we denote $°\Sigma(\Gamma\Delta)$, and add them, as well as the formulas of $cR[\Sigma(\Gamma\Delta)]$, at the beginning of the list of the proposals of formulas as valid, in the left column at the root of the tableau.

We cannot come against an obstruction here: in virtue of the assumptions made in the statement of the theorem 45, we can always find, amongst the rule-making axioms holding in $\mathbf{p}^*$Ia, some more or less controlled version of each ground rule-making axiom holding in N^*1.

Thus we have transformed the second tableau in a kept closed tableau of $^t\mathrm{P}^I1$, which could have been built in conformity with a correct application of the rules as started by checking the sequent $°\Sigma(\Gamma\Delta)$, $cR[\Sigma(\Gamma\Delta)]$, $\Gamma \vdash (X)$, Δ. Indeed, the proposal as valid of every formula belonging to $R[\Sigma(\Gamma\Delta)]$

---

[61] After Newton da Costa [Da Costa, 1963].

can now be seen as resulting from an application of the auxiliary derived rule 1 in virtue of the proposals as valid of the corresponding formulas of $cR[\Sigma(\Gamma\Delta)]$ and $°\Sigma(\Gamma\Delta)$.

Fourth, we can now transform the just obtained third tableau in a still closed tableau, this time of $^t\mathbf{p}^*$Ia, and which could have been built, in conformity with a correct application of the rules, as started by checking the sequent $°\Sigma(\Gamma\Delta)$, $\Gamma \vdash \Delta$. This can be made by erasing the formulas of $cR[\Sigma(\Gamma\Delta)]$, those of $R[\Sigma(\Gamma\Delta)]$ and those of their subformulas viewed, in conformity with the explanations of section 3, as transiently written in order to justify the applications of the corresponding rules by their imagined presence.

As was seen in section 3, none of these processes did affect the pairs of occurrences of formulas proposed both as valid and as not valid in one and the same branch of the initial tableau, which so transmits its closure to its successive transforms, until the last, of which the closure proves that the sequent $°\Sigma(\Gamma\Delta)$, $\Gamma \vdash \Delta$ holds in $\mathbf{p}^*$Ia.

Fifth, as $\mathbf{p}^*$Ia is assumed to be *full*, the sequent $°[\Gamma\Delta] \vdash °\Sigma(\Gamma\Delta)$ holds in it, and from the two lastly established facts, the desired conclusion follows, Q.E.D.

Now, suppose that $\mathbf{p}^*$Ia is *widest*. Then, the same holds for N*1, and the only open possibilities are that N*1 is either $\mathrm{F^I1}$ or $\mathrm{F^C1}$. We are led this way to the two following results.

THEOREM 47. $\mathrm{F^I1}$ *is the aimed logic of* $\mathbf{a}^\mathrm{I}\mathrm{T}_n^\star$ *(or more generally of* fa$^\mathrm{I}$Ia *if the formula D in* (Ia) *is a thesis of* $\mathrm{F^I1}$).

THEOREM 48. $\mathrm{F^C1}$ *is the aimed logic of* $\mathbf{a}^\mathrm{H}\mathrm{T}^\star$ *and of* $\mathbf{a}^\omega\mathrm{DC}_n$ *for* $n > 1$.

The main original features of PCL1.

As was already shown by the lemma 42, by comparison with da Costa's initial views, what happens through the development of PCL1 makes a *big* difference: all formulas of the form NA are well-behaved – and this is inescapable; moreover, as we will see, the strong negation is *not classical*, which is not in itself a sin, but also not what happens to da Costa's systems C_n, even just to C_1. Indeed:

THEOREM 49. *Neither of the formulas* DF, $A \vee °A$, $N(A \wedge NA) \to °A$ *is formally provable in* $\mathbf{a}^\omega$ws [62].

Proof. Let us consider the linear Heyting algebra with three elements 0, a, 1, ordered as in this enumeration, with 1 as only distinguished ele-

---

[62]This results extends to the logic $\mathbf{a}^\omega$ws' introduced in the next subsection.

ment. Keep the heytingian interpretations of the conjunction, disjunction, implication: so, all the theses of $\mathrm{P^I1}$ are satisfied by the matrix so obtained. As interpretation of the paraconsistent negation, let us take the unary operation $\boldsymbol{n}$ defined by the equalities $\boldsymbol{n}0 = 1$, $\boldsymbol{n}a = 1$, $\boldsymbol{n}1 = 0$. According to WS, compute $°0 = 0 \rightarrow \boldsymbol{nn}0 = 1$; $°1 = 1 \rightarrow \boldsymbol{nn}1 = 1$; $°a = a \rightarrow \boldsymbol{nn}a = 0$: then, the equation $°x \wedge \boldsymbol{n}x \wedge x = 0$ holds, from what it results that cEF is satisfied; and ET also, since the equation $x \vee \boldsymbol{n}x = 1$ holds. But $a \vee (a \rightarrow 0) = a \vee °a = a \vee 0 = a$ hold; hence DF and $A \vee °A$ are not satisfied. Also, $\boldsymbol{n}(a \wedge \boldsymbol{n}a) = 1$ holds; so $\boldsymbol{n}(a \wedge \boldsymbol{n}a) \rightarrow °a = 0$ holds, and thus $N(A \wedge NA) \rightarrow °A$ is not satisfied[63].

Moreover, it is very easy to verify that all stability axioms are satisfied by this matrix, so that it shows that the logic $\mathbf{a}$WS is a *strict* extension of the logic $\mathbf{a}^\omega$WS[64]. Hence, the fragments of the first *strictly* extend the corresponding fragments of the other; so, for example, $\mathbf{b}^\omega\mathbf{C}$ is a *strict* fragment of $\mathbf{bC}$ (and also of $\mathbf{b}^C\mathbf{C}$, since the first does not satisfy DF which is an axiom of the second).

The following theorem, which will be useful farther, is as striking as the previous one:

THEOREM 50. *The schema*   nEF $NNA \rightarrow (NA \rightarrow B)$ *is formally provable in all extensions of* $\mathbf{b}^\omega$WS.

Proof. From cEF it follows that the sequent $°A, NA, A \vdash B$ holds; from the fact 40, that the sequent $NNA \vdash °A$ holds; and from (m_0), that the sequent $NNA \vdash A$ holds. Hence, the sequent $NNA, NA \vdash B$ holds; this entails the theorem by repeated applications of the deduction theorem.

COROLLARY 51. *Da Costa's formula* A^{n+1} *is a thesis of every extension of* $\mathbf{b}^\omega$WS *for every* $n > 1$.

Proof. A^{n+1} is defined as a shorthand for $N(A^n \wedge NA^n)$, for every $n > 0$. Try to propose it as not valid; then, both A^n and NA^n have to be proposed as valid. But by assuming n greater than 1, one assumes that A^n itself abbreviates a formula of the form NB, where B is $A^{n-1} \wedge NA^{n-1}$; thus, the formulas just proposed as valid are NB and NNB, and according to nEF, the formula A^{n+1} would have to be proposed as valid, a contradiction.

---

[63]Note that $\sim 0 = \boldsymbol{n}0 \wedge °0 = 1$, $\sim 1 = \boldsymbol{n}1 \wedge °1 = 0$, $\sim a = \boldsymbol{n}a \wedge °a = 0$ hold, so that on this matrix (the idea of which I owe to some attempt of Waragai and Shidori's), one can see how the interpretation of the strong negation coincides with the heytingian negation, and how its restriction coincides with the restriction of the interpretation of the paraconsistent negation on the two well-behaved elements 0 and 1.

[64]This is also true of the axiom (WS') of the next subsection, and of its logics $\mathbf{a}$WS' and $\mathbf{a}^\omega$WS'.

5.3   High (weakly) concrete paraconsistent logics.

What the low level of weak concrete paraconsistency allows to do.

Let us return to the generic axiom (Ia) of 5.1, assuming now that the implication $°A \to D$ is provable in $\mathbf{b^I C}$ as it happened in three of the cases effectively encounterered until now – recall that in the other cases, $°A \to D$ is provable in $\mathbf{a^I ncC}$; this boils down to assuming that to add (Ia) is equivalent to adding the implication in the opposite direction $D \to °A$, to which I will attribute the generic label (IA).

Now, what will happen, if it becomes necessary to use the meaning attributed to $°A$ in a context where this formula occurs as a subformula of a formula of the form $N°A$? We have to contrapose the two conjuncts of (Ia), so to use instances of T5 (*cf.* facts 15), and in doing this, we have to find, on one hand, a proof of $°D$, and on the other hand, a proof of $°°A$.

The last formula means that a certificate of well-behavior should be itself well-behaved, and this appears by no means as something expected. So, let us first explore another way: to begin with, to assess what can be proved; then, if something is lacking, to postulate it.

It can quickly be realized that in the previously treated cases, one can prove the formula $ND \to N°A$ at the price set by the construction of D, which requires the use of some stability axiom at each of its steps; that amounts to saying that some differences intervene at this level between formulas of which the behaviors at the lower levels were similar. By building tableaux and researching to reduce the requirements for building them[65] one is led to the following *abstract*

FACTS 52. The formulas $NA^{(n)} \to N°A$ for $n > 0$, $N(A \to NNA) \to N°A$, $N(A \leftrightarrow NA) \to N°A$ are formally provable in $\mathbf{a^I ncC}_n$, $\mathbf{a^I niC}$, $\mathbf{a^I nciC}$ respectively[66].

Fortunately, the proposal of $N°A$ as not valid has to appear at the beginning of the construction, followed immediately by the proposal of $°A$ as valid; in the present cases, where D is $A^{(n)}$, $N(A \to NNA)$, $N(A \leftrightarrow NA)$ respectively, the stability axioms which are added are those which allow, starting from this last proposal, to end with the proposal of $°D$ as valid, so

---

[65] This leads to favour the use of a certificate of well-behavior towards the recourse to (m_0), in order to pass from the proposal of a formula of the form NNA as valid to the proposal of the formula A as valid. However, in virtue of what was established in the preceding subsection, the savings so realized are totally illusory when ET is a thesis, since the passage to the negation has to be taken into account in all systems so extending the systems to which the facts to establish at present apply.

[66] The formula $(A \wedge NA) \to N°A$ is even provable in $\mathbf{b^I C}$.

that, owing to the earlier proposal of ND as valid, it becomes permitted to propose D as not valid, and we are then in the situation described by figure 13, with D at the place indicated by $A^\star$. The closure follows.

Consistent certificates of consistency.

Now, as it seems that, to prove the reverse implications, we need to prove $^{\circ\circ}A$, we will try to take them as axioms. Then, to the axiom (IA), I will make a new axiom (IA') correspond, which will be $N^\circ A \to ND$, and start the study of what results from the *simultaneous* adjunction of both axioms (IA) and (IA'). The paraconsistent logic so obtained will here be labelled by appending the sign "prime" to the label of its fragment obtained by removing just the axiom (IA').

By the described simultaneous adjunctions, $\mathbf{b^I C}$ gives rise to the extension labelled $\mathbf{b^I}$IA' which might be called the "high (weak) concrete basic intuitionistic IA' paraconsistent logic" (and so forth for other basic logics), and $\mathbf{aC}$ gives rise to $\mathbf{a}$IA' which might be called the "high (weak) concrete full IA' paraconsistent logic" (and so forth for other full abstract logics)[67].

Alas, as can be seen by building suitable tableaux (Recipe: begin by applying the auxiliary special concrete rule to the proposal of $^{\circ\circ}A$ as not valid; continue by using as soon as possible the derived rule 1 applied to (IA'), meant as added to the list of formulas proposed as valid, in order to introduce the formula ND corresponding to the suitable formula D from the very beginning of the present section), the trick does not work:

FACTS 53. The formula $^{\circ\circ}A$ is provable in $\mathbf{a^I}$ncDC'$_n$ for $n > 0$, in $\mathbf{a^I}$niws', and in $\mathbf{a^I}$nciGM'.

That is to say, exactly in the logics in which the implications in the sense which is opposite to (DC'$_n$), (WS'), (GM'), respectively, are provable. But the link is more narrow than what precedes:

THEOREM 54. *The equivalences* (DC'$_n$) $\leftrightarrow$ $^{\circ\circ}A$ *for* $n > 0$, (WS') $\leftrightarrow$ $^{\circ\circ}A$, (GM') $\leftrightarrow$ $^{\circ\circ}A$ *are provable from* $\mathbf{a^I}$ncDC$_n$, $\mathbf{a^I}$niwS, $\mathbf{a^I}$nciGM *on, respectively.*

Proof. The implications from the left say in other words just what the facts 53 are saying. The implications from the right are just instances of T5, in which $^\circ A$ is substituted for B and the suitable formula D for A. Then (IA') comes as substituted for $NB \to NA$, and (IA), as substituted for $A \to B$, using the suitable value of (IA), and so, can be detached.

---

[67]When IA is DC, the last logic is that which is labelled **Cila** by Carnielli and Marcos [Carnielli and Marcos, 2002].

A glance at some high weak concrete full paraconsistent logics.

Let us recapitulate what we have learned until now on the high weak concrete full paraconsistent logics of the kind in which we have been interested here.

For each $n > 0$ the following equalities have been established and the following inequality can be conjectured with good reasons:

$$\mathbf{a}^I \text{DC'}_n \neq \mathbf{a}^H \text{DC'}_n = \mathbf{a}^\omega \text{DC'}_n = \mathbf{a}\text{DC'}_n .$$

The same holds for the equalities and the inequality which follow:

$$\mathbf{a}^I \text{GM'} \neq \mathbf{a}^H \text{GM'} = \mathbf{a}^\omega \text{GM'} = \mathbf{a}\text{GM'} .$$

We also established the equality and the last of the inequalities which follow, the first one of the latter remaining conjectural:

$$\mathbf{a}^I \text{WS'} \neq \mathbf{a}^H \text{WS'} = \mathbf{a}^\omega \text{WS'} \neq \mathbf{a}\text{WS'} .$$

For each $n > 0$, the system $\mathbf{a}\text{DC'}_n$, of which Carnielli and Marcos [Carnielli and Marcos, 2002] showed that it is an extension of $\mathbf{bC}$, can be seen as a weakly concrete system which expresses what the strongly concrete "true" C_n expresses. It might be called the "*weak* C_n" and labelled $\text{w}C_n$. Its collective well-behavior and its strong negation are classical.

On another side, for each $n > 0$, the system $\mathbf{a}^I \text{DC'}_n$ has *intuitionistic* collective well-behavior and strong negation. It might be called the "*intuitionistic weak* C_n" and labelled $\text{w}C_n^I$. I did only skim its study in the preceding lines. To be sure, this study should be much more difficult than that of the corresponding C_n, but not uninteresting owing to the close links of computer science and intuitionistic logic.

The systems $\mathbf{a}^I \text{GM'}$ and $\mathbf{a}\text{GM'}$ were here introduced rather in order to learn a lesson from their comparison with the two other based on (DC) and (DC'), or on (WS) and (WS'), respectively, than owing to their own interest, of which I have no clear idea. Technically, they are clearly more intricate to handle. On another side, classically, $A \wedge NA$ and $A \leftrightarrow NA$ are equivalent formulas, which thus express equivalent assertions; it can be meant that the same holds for their negations, but what paraconsistently ? Perhaps is it permitted to conceive that the equivalence is maintained; in this case, this point would underline that da Costa instinctively chose the simpler way for expressing a condition of consistency. However, intuitionistically, $A \wedge NA$ is strictly stronger than $A \leftrightarrow NA$, but until now I have not had sufficient time to evaluate the consequences of this remark.

I am not in a position to draw more conclusions on the system $\mathbf{a}^I \text{WS'}$, which Waragai and Shidori clearly had intended neither to invent nor to study. I have no more to say about the system $\mathbf{a}\text{WS'}$; in some way, certain features presented by it could as well as PCL1 deceive some of the hopes they put on the latter.

The case of $\mathbf{a}^\omega \text{WS'}$ is different, since it is a fragment of PCL1, which is not apparent at first. Then, this remains to be cleared up, and is what I will do in the next subsubsection.

More on PCL1 and da Costa's hierarchy.

THEOREM 55. PCL1 *extends* $\mathbf{a}^\omega\mathrm{WS}$.

COROLLARY 56. $\mathbf{a}^\omega\mathbf{C}$ *is shared by* PCL1 *and the* wC_n *for all* $n > 0$ *as a common fragment*[68].

Theorem's proof. Leaving aside in a first time all stability axioms (of which one is not of the same type as those which have been considered until now), let us start from the fragment of PCL1 which is axiomatized by the axioms of $\mathrm{P}^{\mathrm{I}}1$ and the PCL1 axioms *A10*, this one to which the label ($\mathrm{m_0}$) was here allocated, *A11* which is $NA \rightarrow NNNA$ and *A12* which is $^\circ A \rightarrow ((NA \vee B) \leftrightarrow (A \rightarrow B))$ in its abstract version, to which I will add, in view to obtain a weakly concrete system, ($\mathrm{WS}^\star$) which is $A^{\mathrm{I}} \rightarrow {}^\circ A$, where A^{I} is Waragai and Shidori's shorthand for $A \leftrightarrow NNA$.

Note that a weakly concrete version of PCL1 requires the presence of ($\mathrm{WS}^\star$), which is $A^{\mathrm{I}} \leftrightarrow {}^\circ A$, as an axiom, which implies ($\mathrm{WS}^\star$). Note furthermore that, after replacing A^{I} by the formula it shortens, in ($\mathrm{WS}^\star$), it becomes apparent that this last is equivalent to $(NNA \rightarrow A) \rightarrow ((A \rightarrow NNA) \rightarrow {}^\circ A)$ in $\mathrm{P}^{\mathrm{I}}1$, so that, in presence of ($\mathrm{m_0}$), we have (WS) at our disposal. Conversely, (WS) implies ($\mathrm{WS}^\star$) in $\mathrm{P}^{\mathrm{I}}1$.

We turn now to *A12* which, taking into account the stability paraconsistent axioms, says, in virtue of what was explained in the subsubsection devoted in the present section to the aimed logics, that the collective *well-behavior is classical.*

Indeed, in $\mathrm{P}^{\mathrm{I}}1$: first, *A12* implies $^\circ A \rightarrow ((NA \vee B) \rightarrow (A \rightarrow B))$, from which cEF follows by resorting to the instance $NA \rightarrow (NA \vee B)$ of one of the axioms. Conversely $^\circ A \rightarrow ((NA \vee B) \rightarrow (A \rightarrow B))$ can be deduced from cEF which is $^\circ A \rightarrow (NA \rightarrow (A \rightarrow B))$ by taking into account the axiom $B \rightarrow (A \rightarrow B)$.

Second, *A12* implies $^\circ A \rightarrow ((A \rightarrow B) \rightarrow (NA \vee B))$, of which the instance obtained by substituting A for B allows at once to deduce cET. Conversely, the implication of $^\circ A \rightarrow ((A \rightarrow B) \rightarrow (NA \vee B))$ by cET as a thesis of $\mathrm{P}^{\mathrm{I}}1$ can be obtained by substituting NA for C in the thesis $(A \vee C) \rightarrow ((A \rightarrow B) \rightarrow (C \vee B))$ [69].

---

[68] I do not know whether this fragment is the *greatest* common to these logics.

[69] In the same way, one proves that to adjoin the sole $(A \rightarrow B) \leftrightarrow (NA \vee B)$ to $\mathrm{P}^{\mathrm{I}}1$ suffices for axiomatizing $\mathrm{F}^{\mathrm{C}}1$. The same is true for $(A \vee B) \leftrightarrow (NA \rightarrow B)$; the proof results from the previous one by permuting the roles of A and NA.

Hence, the systems obtained by adding to $P^I 1$, on one hand, $A12$, and on the other hand, both cEF and cET, are equivalent. Then, our first step will be to replace $A12$ by the set of cEF and cET, in the set of axioms for our fragment.

The second step[70] will be to replace cET by ET, by making use of $A11$, of the instances $(NA \to NNNA) \to {}^\circ NA$ of (ws) and ${}^\circ NA \to (NA \lor NNNA)$ of cET, and of (m_0). This step is reversible relatively to $P^I 1$ since the implication of cET by ET is an instance of one of its axioms.

Now it is made clear that our fragment extends $\mathbf{b^I C}$. Hence we can prove U3 in it (according to the facts 19), and as (m_0) is taken as an axiom, we can prove ${}^\circ A \to A^I$, from what it follows that (ws*) implies (ws*); moreover, since (ws) implies (ws*), as noticed above, we can replace, in the axiomatization of our fragment, (ws*) by (ws) without making the desired (ws*) cease being a thesis of our fragment.

The fourth step consists in noticing that the fragment extracted from our fragment by omitting $A11$ is $\mathbf{b^\omega}$ws. Hence, according to corollary 43 and lemma 42, ${}^\circ NA$ is a thesis of it. But ${}^\circ NA \to (NA \to NNNA)$ is an instance of U3, hence provable; then $A11$ is superfluous and can be suppressed, what reduces our fragment to just $\mathbf{b^\omega}$ws.

Now, PCL1 axioms which are labeled $A13$ and $A14$ are those which were here labeled (a_1) and (a_2) respectively, and according to the corollaries 24 and 25, it comes to the same to add them or to add (c) and (d). Hence, in order to prove that PCL1 extends $\mathbf{a^\omega}$ws, hence also $\mathbf{a^\omega C}$, as the wC_n are doing for $n > 0$, it remains just to examine what follows from the axiom of PCL1 labeled $A15$.

This axiom is equivalent to the conjunction of the formulas ${}^\circ A \to (m_3)$ and ${}^\circ A \to ((A \land NB) \to N(A \to B))$, whether it is given in the beautiful form ${}^\circ A \to (N(A \to B) \leftrightarrow (A \land NB))$ or by separately postulating on a side that ${}^\circ A \to ((A \land NB) \to N(A \to B))$ is an axiom and on another side that the sequent ${}^\circ A \vdash N(A \to B) \to (A \land NB)$ holds.

Now, it follows easily from the provability of the formula (ia) of theorem 21 that $({}^\circ A \to (m_3)) \to (i)$ is provable, so that the stability axiom (i) is provable in PCL1, what ends a proof of my claim that $\mathbf{a^\omega}$ws is a fragment of PCL1.

This time, I do *not* see how we could prove more; neither ${}^\circ A \to (m_3)$ nor ${}^\circ A \to ((A \land NB) \to N(A \to B))$ seem deducible from the axiom (i).

SCHOLIUM. I have still to complete a proof that I was right by replacing the formula $A \leftrightarrow NNA$ shortened by A^I by the formula $A \to NNA$ in my

---

[70]Inspired by what Waragai and Shidori are doing.

weakly concrete axiomatization of PCL1[71].

Now, previously, (ws*) was shown provable in $\mathbf{b}^H$ws; by now it remains to show that the formula $NA^I \leftrightarrow N°A$, that I will label (ws'*), is provable in the weakly concrete version of PCL1. I will content myself by showing that it is provable in $\mathbf{a}^H$ws'.

First, from (ws*) we can deduce the conjunct (ws'*), *i.e.* $N°A \to NA^I$, by having recourse to the instance of T5 obtained by substituting A^I for A and $°A$ for B, noticing that the permission required is then the provable $°°A$ (here is where we use (ws').)

Second, for proving the reverse implication, we can start a tableau in $^t\mathbf{a}^H\mathbf{C}$ with the intention to check the sequent $N(A \leftrightarrow NNA) \vdash N°A$. Then the formula $°A$ is quickly proposed as valid; as the only prime component occurring in the checked sequence is A, we can, by iterated uses of stability axioms, assume that for all formulas of the form NB proposed as valid, the formula $°B$ is proposed as valid. Then, the tableau can be built just by applying the classical rules for drawing the consequences of the proposals of formulas of the form NB as valid or as not valid. As the formula $N(A \leftrightarrow NNA)$ was proposed as valid, there is a splitting in two subtableaux after the proposal of the formula $A \leftrightarrow NNA$ as not valid; in one of them, the formula $A \to NNA$ is proposed as not valid, now just under the conditions under which we checked the formula U3; so this subtableau is closed, as was the case in figure 13. In the other subtableau, the formula $NNA \to A$ is proposed as not valid; after application of the rule for drawing the consequences of such a proposal, there remains no new application of this rule to do, and we are led to closure as if we were building a classical tableau for checking this last classical thesis. This ends the proof.

It is known that fot every $n > 0$, the formula $A^{(n)} \to A^{(n+1)}$ is provable in wC_n [72], but *not* the formula $A^{(n+1)} \to A^{(n)}$. I will show that a similar phenomenon arises between $\mathbf{a}^\omega$ws' and wC_n^I : $(A \to NNA) \to A^{(n)}$, where $A^{(n)}$ is da Costa's formula, is provable in $\mathbf{b}^\omega$ws, since $°A \to A^{(n)}$ is so provable according to the fact 20, and since (ws) is precisely $(A \to NNA) \to °A$; but on the other side:

THEOREM 57. *For no $n > 0$ is the formula $A^{(n)} \to (A \to NNA)$ provable in $\mathbf{a}^\omega$ws'.*

Proof. We begin with the case $n = 1$:

LEMMA 58. *The formula*   U5 $N(A \wedge NA) \to (A \to NNA)$ *is*

---

[71]Note that, in the strongly concrete version of PCL1, such a proof is by no means necessary, since the formula $A \leftrightarrow NNA$ becomes equivalent *in meaning* to its conjunct $A \to NNA$ after the acceptance of its other conjunct $NNA \to A$ as a postulate.

[72]In fact, the proof rests only on wC_n^I's axioms.

not provable *in* $\mathbf{a}^\omega$ws'

– just because a proof of $N(A \wedge NA) \to {}^\circ A$ could immediately be built by extending a proof of U5, contrary to the theorem 49. Now,

LEMMA 59. *The formula* $N(A \wedge NA) \to A^{(n)}$ *is provable in* $\mathbf{b}^\omega$ws

– just because $N(A \wedge NA)$, *alias* A^1, is precisely just the conjunct which lacks to the conjunction of the A^i for i going from 2 to n, all provable in $\mathbf{b}^\omega$ws according to the corollary 51, for forming $A^{(n)}$.

Hence, in $\mathbf{a}^\omega$ws', a proof of $A^{(n)} \to (A \to NNA)$ would allow immediately to build a proof of the forbidden $N(A \wedge NA) \to (A \to NNA)$.

THEOREM 60. *There is no* $n > 0$ *such that the formula* $(A \to NNA) \to A^{(n)}$ *is provable in* wC_n, *nor a fortiori in* wC_n^I.

Proof. We begin with the case $n = 1$:

LEMMA 61. *The formula*   U6 $(A \to NNA) \to N(A \wedge NA)$ *is not provable in* wC_1, *nor, a fortiori, in* wC_1^I.

Proof of the lemma. I will have recourse to the matrix $\mathcal{M}_2$ long ago devised by da Costa (*cf.* [Da Costa and Guillaume, 1964]), which has three elements 0, β and 1, from which the two last are distinguished. The values indicated in its table for $a \to b$ are 1 everywhere, *except* for the cases $b = 0$ and $a = 1$ or $a = \beta$, where it is 0. In its table for $a \wedge b$ the values indicated are all 0 when $a = 0$ or $b = 0$, and 1 in all other cases. The operation $\mathbf{n}$ intended to interpret the negation is defined by the equalities $\mathbf{n}1 = 0, \mathbf{n}0 = 1, \mathbf{n}\beta = \beta$. Then all axioms of wC_1 are satisfied, but $\mathbf{n}(\beta \wedge \mathbf{n}\beta) = \mathbf{n}(\beta \wedge \beta) = \mathbf{n}1 = 0$ and $\beta \to \mathbf{nn}\beta = 1$ hold, so that $(\beta \to \mathbf{nn}\beta) \to \mathbf{n}(\beta \wedge \mathbf{n}\beta) = 0$ holds, and thus U6 is not satisfied[73].

Proof of the theorem. Now, as noticed above, $A^{(n)} \to N(A \wedge NA)$ is provable in P^I1, and it is known for a long time that wC_1 *extends* wC_n, for every $n > 0$. Hence, in wC_n, a proof of $(A \to NNA) \to A^{(n)}$ would allow immediately to build a proof of U6, which in its turn allows to build a proof in C_1 of the forbidden U6.

REMARK 62. The matrix used in the proof of theorem 49 provides also a proof of non provability of the formula $N(A \leftrightarrow NA) \to (A \to NNA)$ in $\mathbf{a}^\omega$ws', and the just described matrix $\mathcal{M}_2$ can be used to prove the non provability of the formula $(A \to NNA) \to N(A \leftrightarrow NA)$ in $\mathbf{a}^\omega$GM'. Yet these matrices assign the same operation for interpreting the schemes $N(A \wedge NA)$ and $N(A \leftrightarrow NA)$, which blocks the above reasonings in what regards the links of $\mathbf{a}^\omega$GM' with wC_1, and beyond with the wC_n's. I am compelled to leave this question open.

---

[73]By luck, since this matrix was by no means devised with the aim to prove the lemma !

Acknowledgements

I thank the organizers of the Congress for their invitation, as well as Professor Kevin Clarke and my childhood friend Jean Grissolange, who helped me, in record time, to improve my inaccurate English wording.

BIBLIOGRAPHY

[Kleene, 1952] S.C. Kleene, **Introduction to Metamathematics.** Van Nostrand, New-York, 1952.

[Kanger, 1957] S. Kanger, **Provability in Logic.** Acta Universitatis Stockholmiensis, Stockholm Studies in Philosophy 1. Almqvist and Wiksell, Stockholm, 1957.

[Beth, 1959] E.W. Beth, **The Foundations of Mathematics. A Study in the Philosophy of Science.** North Holland, Amsterdam, 1959.

[Beth, 1962] E.W. Beth, **Formal Methods. An Introduction to Symbolic Logic and to the Effective Operations in Arithmetic and Logic.** Reidel, Dordrecht, 1962.

[Da Costa, 1963] N.C.A. da Costa, **Sistemas formais inconsistentes.** Tese, Universidade Federal do Paraná. Núcleo de Estudos e Pesquisas Científicas, Rio de Janeiro, 1963.

[da Costa and Guillaume, 1964] N.C.A. da Costa et M. Guillaume, *Sur les calculs C_n* , **Anais da Academia Brasileira de Ciências 36** (1964), 379-382.

[da Costa and Guillaume, 1965] N.C.A. da Costa et M. Guillaume, *Négations composées et loi de Peirce dans les systèmes C_n* , **Portugaliae Mathematicae 24** (1965), 201-210.

[Rasiowa and Sikorski, 1970] H. Rasiowa and R. Sikorski, **The Mathematics of Metamathematics.** Polish Scientific Publishers, Warszawa, 1970.

[Guillaume, 1996] M. Guillaume, *Regard en arrière sur quinze années de coopération douce avec l'Ecole brésilienne de logique paraconsistante,* **Logique et Analyse 153-154** (1996), 5-14.

[Carnielli and Marcos, 2002] W.A. Carnielli and J. Marcos, *A taxonomy of C-systems,* *in* W.A. Carnielli, M.E. Coniglio, I.M Loffredo d'Ottaviano, **Paraconsistency, the logical way to the inconsistent (Lecture Notes in Pure and Applied Mathematics Series 228).** Dekker, New-York, 2002.

Marcel Guillaume

LLAIC, Université de Clermont-Ferrant I, France

guill@llaic.u-clermont1.fr

Adventures in the Paraconsistent Jungle

JEAN-YVES BEZIAU

ABSTRACT. This is a survey of the main contributions of the author to the field of paraconsistent logic. After a brief introduction explaining how the author entered this field, Section 2 describes his work on C-systems: reformulation of the semantics of C1, creation of a sequent systems for C1, proof of cut-elimintaion for this system, extension of C1 into a stronger logic C1+. Section 3 is about his investigations on the definition of paraconsistent logic and a general theory of negation. Section 4 relates what he considers as its main contribution: the discovering that classical first-order logic and the modal logic S5 are paraconsistent logics and how this led him to a new theory of opposition, where a polyhedron replaces the traditional square of opposition. In Section 5 he explains that he conceives the philosophical aspects of paraconsistency mainly in relation with applications and he says a word of what kind of situations paraconsitent logic may apply to. Section 6 describes his work as paraconsistent promoter, editor, translator, and organizer. Finally in Section 7, he indicates his future lines of paraconsistent investigations and frothcoming works. The paper includes a complete bibliography of his works in paraconsitent logic. This paper can be read by anyone interested in logic even with few or no knowledge of paraconsistent logic.

1 Introduction

> *Faire de la vie un rêve et d'un rêve une réalité.*
>
> Pierre Curie

In this paper,[1] I tell my adventures in the Paraconsistent Jungle, from my discovery of paraconsistent logic in the Spring 1989 through the meeting with a donkey, to the encounter of strange creatures in Brazil, to the frequentation of the mysterious O-corner of the square of oppositon in California, to contructions of polyhedra on the bank of the lake of Neuchâtel.

---

[1]Work supported by a grant of the Swiss National Science Foundation

As I already recalled in another paper [17], these adventures started in the Spring 1989 in France when I read a paper about Newton da Costa and paraconsistent logic in the French Lacanian psychoanalysis journal *L'Âne*. I was not a Lacanian and/or mentally sick, but I like to read any kind of magazines: from *The Notices of the American Mathematical Society* to *Les Cahiers du Cinéma*, from *National Geographic* to *Penthouse*, from *Cosmopolitan* to *The Economist*. I never had the ambition to be a one dimensional dull man. I like to watch was is going around and basically that's how I discovered Paraconsistent Logic.

In reading the donkey paper I was fascinated by this logic violating the principle of non contradiction, presented as a possible logic of the incouscious, as much as by da Costa's personality. I remember in particular that da Costa was quoting the above words of Pierre Curie and also there was a photo of him with a subtitle saying "I am not considering myself as a simple technician of logic."

At the beginning of 1991 da Costa was in Paris to bungee jump from the Eiffel Tower. I had the opportunity to meet him and to show him works I have been doing on his systems (I didn't spend all my time riding donkeys). He liked it and he invited me to come to Brazil to work with him. Few months later I was in the Urban Jungle of São Paulo.

2   C-systems

Relação contraditória, portanto relação lógica,
Cadela que me prende, para me libertar ...
Lars Eriksen

2.1   Reformulation of the semantics of C1

I started to work on paraconsistent logic in Paris by studying the C-systems of da Costa, one of the first and most known systems of paraconsistent logic. My aim was to understand properly how C1 and its semantics work. This tentative led me to a reformulation of the axiomatization and semantics of C1. I tried to reformulate C1 in the most intuitive way.

C1 was originally conceived by da Costa as an Hilbert-type proof system. In the orginal version, the only rule is *modus ponens*, and the axioms for C1 are the axioms of positive intiutionistic logic plus the following ones:

[10] $a \vee \neg a$
[11] $\neg\neg a \to a$
[12] $(b^o \to ((a \to b) \to ((a \to \neg b)) \to \neg a))$
[13-15] $a^o \wedge b^o \to (a\copyright b)^o$, ($\copyright \in \{\vee, \wedge, \to\}$)
[16] $a^o \to (\neg a)^o$

a^o means $\neg(a \wedge \neg a)$

A sound and complete semantics bivalent for this systen was given by da Costa and Alves in 1976. This semantics is non truth-functional and the bivaluations are defined directly on the whole set of formulas. The conditions for conjunction, disjunction and implication are the same as the classical ones. Then we have the following set of conditions for negation.

[1] if $\beta(a) = 0$, then $\beta(\neg a) = 1$
[2] if $\beta(\neg\neg a) = 1$, then $\beta(a) = 1$
[3] if $\beta(b^o) = \beta(a \rightarrow b) = (a \rightarrow \neg b) = 1$, then $\beta(\neg a) = 0$
[7] if $\beta(a^o) = \beta(b^o) = 1$, then $\beta(\neg(a \copyright b)) = 1$, $(\copyright \in \{\vee, \wedge, \rightarrow\})$

The first of these conditions is half of the standard one for classical negation, and generates the excluded middle given by axiom [10]. The meaning of the other conditions is not necessarily clear. They look like *ad hoc* conditions to get completeness. Obviously, condition [2] is a semantical description of axiom [11], condition [3] of axioms [13-15] and condition [7] of [16]. I succeeded to show that we can replace these artificial conditions by a set of intuitive conditions $[C1 - JYB]$, which are the following:

$[\neg_{lo}]$ if $\beta(a \wedge \neg a) = 1$, then $\beta(\neg(a \wedge \neg a)) = 0$
$[\neg_{l}\copyright]$ if $\beta(a \copyright b) = 1$ and $\beta(a) = 0$ or $\beta(\neg a) = 0$ and $\beta(b) = 0$ or $\beta(\neg b) = 0$, then $\beta(\neg(a \copyright b)) = 0$, $(\copyright \in \{\vee, \wedge, \rightarrow\})$
$[\neg_{l\neg}]$ if $\beta(\neg a) = 1$ and $\beta(a) = 0$ or $\beta(\neg a) = 0$ then $\beta(\neg\neg a) = 0$

These conditions are all restricted forms of the other half of the condition which defines classical negation, i.e. if $\beta(a) = 1$, then $\beta(\neg a) = 0$. This condition is a semantical formulation of the principle of non contradiction, it says that a formula and its negation cannot both be true. The principle of non contradiction does not hold in general in C1, but the idea of C1 is not just to withdraw half of the semantical condition which defines classical negation, which leads to a quite weak logic, but to replace it with a restricted version expressed by the conditions $[C1 - JYB]$.

The condition $[\neg_{l}\copyright]$ says that if the principle of non contradiction holds for two formulas a and b it also holds for the conjunction, disjunction and implication. The condition $[\neg_{l\neg}]$ says that if the principle of non contradiction holds for a formula a it also holds for its negation. These two conditions can be unified by some notational device and can be expressed inutuitively as meaning: the principle of non contradiction is preserved by complexification. This is one of the two main ideas of the semantics of C1.

The other main idea is given by condition $[\neg_{lo}]$ which can be described as: a contradiction obeys the principle of non contradiction. This condition allows to have a classical negation within C1 and therefore to translate classical logic into C1.

In my reconstruction, I introduced Ci which is the logic semantically defined with classical conditions for conjunction, disjunction, implication plus [[1]], plus [[$\neg_{lo}$]]. Ci is a very weak logic, it only allows to define a classical negation with the paraconsistent negation. If we add the preservation principle, we get much more: double negation and some de Morgan laws.

Having reformulated the semantics of C1, I was also able to give another more intuitive Hilbert system for C1. In this system we have *modus ponens*, plus the axioms for *classical* positive logic, plus the following ones:

$a \vee \neg a$

$(a \wedge \neg a) \wedge \neg(a \wedge \neg a) \to b)$

$a^o \wedge b^0 \to (a \copyright b)^0$

$a^o \to (\neg a)^0$

For Ci, we just have to take the axioms for classical positive logic plus the first two axioms above.

With this new formulation of C1 I was able on the one hand to describe in a more intuitive way the method of truth-table for C1 and to give some set-theoretical diagrammatic representations of how it works , on the other hand to built a sequent systems for C1. I will not talk about the first point here. The second point is the subject of the next section.

2.2   Sequent systems for C1 and cut-elimination

Andrès Raggio, an Argentinian logician former student of Paul Bernays, had tried to formulate a sequent calculus for C1 in the 1960s, but he didn't quite succeed. The formulation I gave in 1990 is based on my reformulation of the semantics of C1 presented in the preceding section and on the discovery of an intimate connection between sequents and bivaluations described in details in My PhD and in .

The sequent system S1 for C1 has the same rules as the sequent system for classical propositional logic, except the rules for negations which are the following:

RULES FOR NEGATION IN $\mathcal{S}_1$

$$[\neg_r] \quad \frac{a \Rightarrow}{\Rightarrow \neg a} \qquad\qquad \frac{\Rightarrow a \wedge \neg a}{\neg(a \wedge \neg a) \Rightarrow} \quad [\neg_{lo}]$$

$$\frac{\Rightarrow a \copyright b \qquad a, \neg a \Rightarrow \qquad b, \neg b \Rightarrow}{\neg(a \copyright b) \Rightarrow} \quad [\neg_l \copyright]$$

$$\frac{\Rightarrow \neg a \qquad a, \neg a \Rightarrow}{\neg\neg a \Rightarrow} \quad [\neg l \neg]$$

In S1 we have all the structural rules, in particular the cut rule. But this rule can be elimated. Cut-elimination of S1 is not a big deal. You just have to check that these monstruous rules do not interfer with the standard process. By studying cut-elimination for S1 I was able to give a general formulation of the cut-elimination theorem in my PhD.

S1 rules don't have the subformula property, but they have something analogeous: the subnegformula. A subnegformula of a formula is a subformula or a negation of a proper subformula. From this property and cut-elimination, we have another method of decidability for C1 using S1.

All this work about C1 appears in my Master's thesis [3] and was subsequently published in [5].

2.3 C1+

There is an obvious and intuitive way to strengthen C1 in a logic C1+. I had this idea in 1990 and it appears in several of my works, in particluar [4]. The idea is to replace the axiom

$a^o \wedge b^0 \rightarrow (a \copyright b)^0$

by the axiom

$a^o \vee b^0 \rightarrow (a \copyright b)^0$

This corresponds to a new preservation principle: in order for the principle of non contradiction to be preserved by complexification, we just need that one of the proper subformulas of a complex formula obeys the principle of non contradiction.

With this strengthening we get 4 more de Morgan laws. In my paper I presented the following comparative table of the logics C1, C1+ and their paraconsistent duals. In this paper I used the notation Cx, C+ instead of C1 and C1+, to emphasize the conjonctive (x) or disjunctive (+) formulation of the preservation principle.

Cx	x	+	C+
$\neg(a \wedge b) \vdash \neg a \vee \neg b$			$\neg(a \vee b) \vdash \neg a \wedge \neg b$
$\neg(\neg a \wedge \neg b) \vdash a \vee b$			$\neg(\neg a \vee \neg b) \vdash a \wedge b$
$\neg(a \wedge \neg b) \vdash \neg a \vee b$			$\neg(a \vee \neg b) \vdash \neg a \wedge b$
$\neg(\neg a \wedge b) \vdash a \vee \neg b$			$\neg(a \vee b) \vdash a \wedge \neg b$
C			C
K			K
$\neg a \wedge \neg b \vdash \neg(a \vee b)$			$\neg a \vee \neg b \vdash \neg(a \wedge b)$
$a \wedge b \vdash \neg(\neg a \vee \neg b)$			$a \vee b \vdash \neg(\neg a \wedge \neg b)$
$\neg a \wedge b \vdash \neg(a \vee \neg b)$			$\neg a \vee b \vdash \neg(a \wedge \neg b)$
$a \wedge \neg b \vdash \neg(\neg a \vee b)$			$a \vee \neg b \vdash \neg(\neg a \wedge b)$
Kx	x	+	K+

The table has to be read as follows: what hold for the x-systems, hold for the +-systemes (but not the contrary). What hold for the K-systems do not hold for the C-systems and vice-versa.

In [4] I also introduced the notion of non truth-functional many-valuedness (subsequently developed in [22]). I present non truth-functional semantics for all these systems. This allows to get back the subformula property.

3   Negation theory

> *There are no paraconsistent negations.*
> Barry Slater

One thing that really interests me in paraconsistent logic is that it gives a better idea of what negation is, and it gives in particular a better idea of what classical negation itself is and is not. From the point of view of someone who is trying to understand negation, there is no opposition between classical logic and paraconsisten logic. My work in paraconsistent logic naturally led me to a general study of negation.

3.1   Definition of paraconsitent negation

This started with the question of definition of paraconsistent logic. I was interested in this problem since the beginning and in particular I had read with interest the excellent paper of Igor Urbas, where he gave an improved version of the standard definition of paraconsistent negation. The standard definition is based on the rejection of the *ex-contradictione sequitur quodlibet* (EC hereafter): $a, \neg a \vdash b$. But Urbas remarked that we want to reject more quodlibet than the usual quodlibet, for example we want also to reject: $a, \neg a \vdash \neg b$.

The definition of Urbas is certainly a great improvement of the standard definition, but it is still not a very satisfactory definition, since what we

want is not only a (reinforced) negative criterium for paraconsistent negation, in order to have a *paraconsistent* negation, but we want also a positive
criterium, in order to have a paraconsistent *negation*. Many unary operators not obeying Urbas' version of EC, have absolutely nothing to do with
negation, for example the identity operator which transforms any formula
into itself. From this viewpoint, any logic is paraconsistent(since a paraconsistent logic is defined as a logic in which there is at least one paraconsistent
negation).

I think there is a great deal of confusion, hypocrisy and crookedeness in
defining paraconsistent logic only negatively and talking as there were no
problems with paraconsistent negation. I discussed this question mainly in
two papers, one presented at the First World Congress on Paraconsisteny
in Ghent in 1997, entitled "What is paraconsistent logic?" and the other
presented at the Second World Congress of Paraconsistency in Juquehy in
2000, entitled "Are paraconsistent negations negations?". Both are related
to a paper written by Barry Slater, "Paraconsistent logics?" published in
the Journal of Philosophical Logic in which he doubts that there are any
paraconsistent negations. Although Slater is wrong, as I explained in a
paper which is a reply to his, "Paraconsistent logics!", he found the Achilles'
heel of paraconsistentists and rightly tickled it. In my Juquehy's paper I
argue that the existence of paraconsistent logic has not been proved so far
(a position which is neither Slaterian nor anti-Slaterian). I also developed a
classification of properties for negation and of the so-called paraconsistent
logics.

3.2   Pure negation and incompatibility results

Classical negation can be formulated with only one axiom, the *reductio ad
absurdum*, but this axiom can be decomposed into a multiplicity of weaker
axioms. In my paper "Théorie législative de la négation pure" [6], I made
a study of negation alone, independently of other connectives (implication,
conjunction, etc.). I showed that with very few assumptions on the consequence relation, basic properties such as *contrapositio, weak reductio*, etc.
have to be rejected if one rejects EC. This study gave me a rather pes-
simistic vision of paraconsistent logic. If we want a paraconsistent negation
defined by Urbas' version of EC, then nearly no properties of negation are
compatible with this rejection. Positive properties have to be found within
the interaction with other connectives, in particular with conjunction and
disjunction, such as de Morgan's laws.

In a paper published in *Notre Dame Journal of Formal Logic* [12], I
showed a basic incompatibility result which applies both to Asenjo's logic
A3, Priest's logic LP and the logic J3 of da Costa and D'Ottaviano. In a

paraconsistent logic, If $\neg(a \wedge \neg a)$ is valid and also the double negation is valid, then the replacement theorem cannot hold. In fact the replacement does not hold in A3, LP and J3, it seems I was the first to realize that, a typical counter-example is that $\neg(p \wedge \neg p) \dashv\vdash \neg(q \wedge \neg q)$ but not $\neg\neg(p \wedge \neg p) \dashv\vdash \neg\neg(q \wedge \neg q)$. So from the point of view A3, LP and J3 are no better systems than the C-Systems (the failure of the replacement theorem in C1 and related systems is traditionally considered as fatal defect, see for an analysis of this problem).

4 Modal logic, paraconsistent logic and the square of oppositions

Ça crevait les yeux,
mais la poufiasse n'y avait vu que du feu.
Arthur Oursipan, *Panique à Rio*

4.1 Classical logic is paraconsistent

I can say that my main contribution to paraconsistent logic, up to now, is the discovering that classical logic is paraconsistent. I know that this sounds terribly paradoxical, and that even the most crazy paraconsistent logicians, those living in the antipodes, are not ready to consider this as true. But it is. In the next subsections I tell the story of my discovery that the modal logic S5 and classical first logic are paraconsistent logics, and different consequences of this discovery.

My impression is that this discovery was not very welcomed and since the time I discovered this, about five years ago, few people have talked about it, although I have presented lectures on this topic in many places around the world: Moscow, Amsterdam, Melbourne, Rio de Janeiro, Naples, Vancouver, etc. I think the reason is that such fact is disgusting both for paraconsistentists and anti-paraconsistentists.

On the one hand, people believing in contradictions or wanting to built logics in which from a contradiction not everything is derivable, have been working hard during many years to built new logics, sometimes quite complicated, not realizing that there were very simple logics already at hand able to do that, namely good old classical first-order logic and the most well-known modal propositional logic, S5.

On the other hand, for a classicist, like the late Quine, believing that the rejection of the law of non-contradiction is a "popular extravaganza" (*Philosophy of logic*, p.81), realizing that there is a paraconsistent negation within classical logic is a shocking obscene reality, similar e.g. as realizing that his wife is a hermaphrodite. Quine wrote "the classical logic of truth-functions and quantification is free of paradox, and incidentally it is a paradigm of clarity" (*Ibidem*, p.85). He could never have imagined that the beautiful classical fruit was infected by a logic of paradoxes.

4.2   The origin of the discovery of this amazing fact: Fortaleza, Copacabana and Torun

I discovered this fact several years ago. It is connected with things I had been thinking of since more than fifteen years. But things took shape in 1997 in Fortaleza. At the Beira Mar, I was discussing with my friend Arthur de Vallauris about the notion of contradiction, he was repeating a Sufi story with had inspired him in his research in paraconsistent logic. What I was saying to him, is that a thing may appear black to someone, and white to someone else, that doesn't mean that it is black and white. In the "logic of appearances" (expression coined by our friend Tarcisio Pequeno), an object may have contradictory properties, which correspond to different viewpoints. I derived such conceptions from a study I had made on the work of the physicist David Bohm at the end of the eighties. In his book *Wholeness and the implicate order*, he gave several metaphors supporting such a view.

So my idea was to develop a logic that can be used to reason with different contradictory viewpoints. The idea was that the negation of a proposition is false only if the proposition is true from all viewpoints. If from all evidences it is true that light is a wave, then it is false that light is not a wave. On the hand, if according to some experiments it is true that light is a wave and according to some other experiments, it is false that light is a wave, then what we can say from the point of view of all experiments is that it is true that light is a wave and that it is not a wave. What light really is, only God knows. After my discussion with Arthur, when I was back to my room at the Hôtel Olympo, I started to develop a possible worlds semantics based on this idea.

When I was back to Rio, were I was living at this time, I developed a full study of the related logic, that I called Z. I saw a connection between S5 and Z. The paraconsistent negation of Z was equivalent to $\neg\Box$. From this I infered that it was possible to define Z into S5, that Z was in S5, that therefore S5 was paraconsistent and that also, via Wasjberg's theorem, classical first-order logic was paraconsistent. I discussed my discovery with the Italian logician Claudio Pizzi, who has a little castle in Copacabana and uses to come there frequently. Pizzi told me two important things: one right and one wrong. The wrong was that the operator $\neg\Box$ corresponds to contingency, the right was that S5 and Z are equivalent since it is possible in S5 to define classical negation and necessity with $\neg\Box$, conjunction and implication.

The intuition underlying Z is connected with the basic intuition of Jaśkowski's discussive logic. Jaśkowski wanted to formalize the logic of a discussion group where people may have contradictory opinions. To formalize this he switched the universal-like definition of truth in a Kripke Structure by a existential-like one: a proposition is true in a Kripke Structure iff it is true in some worlds. This leads to a logic which has some quite strange properties (non adjunction). It seemed to me that my formalization was a fomalization of Jaśkowski's intuition, which led to a more interesting paraconsistent logic that the standard formulation of Jaśkowski's logic. I emphasized this fact in a lecture I gave on the logic Z in July 1998 in Torun, Poland, at the conference commemorating the 50 years of Jaśkowski's paper.

4.3   The mysterious 0-corner of the square of oppositions

I made further important discoveries about the paraconsistent negation $\neg\Box$ when I was in California (I had moved to Stanford at the beginning of 2000). In 2001 I was attending Johan van Benthem's seminar on modal logic and he draw the square of modalities, with the four corners $\Box$, $\Diamond$, $\neg\Box$ and $\neg\Diamond$ saying that there was no name for the modality $\neg\Box$, and that was the same for the square of quantifiers, there was no name for $\neg\forall$. Few days later I attended a lunch lecture at the CSLI by another Dutch, Pieter Seuren, about quantifiers, and he was talking about the non-lexicalization of this O-corner of the square: linguists had been studying many languages around the world and never found a primitive word for the quantifier $\neg\forall$. The conclusion of Seuren was that the square of opposition was a wrong theory of quantification.

It was very funny to see that paraconsistent negation was just showing its nose at this mysterious O-corner. I decided to explore the paraconsistent negation within this square problematic and I made quick advances, due to the fact that I was able to put several different information together and complete the jigsaw. I had read some years ago something interesting on the square of opposition by Robert Blanché and I decided to look back at Blanché. The analysis of the square of opposition by Blanché is something very nice and profound. He suggests to replace the square of the quantifiers by a hexagon made of two triangles, one of contrariety, one of subcontrariety. The six vertices of the hexagon represents six quantifiers. If we made a similar construction for modalities, we have an hexagon with six modalities, and we see clearly where stands contingency, it is the vertice $\Diamond \wedge \neg\Diamond$ of the triangle of subcontrariety. This modality and its negation (non contingency) have been studied by Montgomery, Routley at the end of the 1960s and more recently by Humbertsone and Kuhn. It is clearly different from $\neg\Box$. $\neg\Box$ is the dual of $\neg\Diamond$, the impossible, which corresponds to intuitionistic negation in S4, and which is in general a paracomplete negation.

So in the hexagon of modalities appear clearly two negations, but what do not appear are the relations between these negations and a proposition *tout court*. My idea was then to introduce the proposition *tout court* in the hexagon, and by symmetry, the classical negation of the proposition. Then I transformed the hexagon of Blanché into an octagon. In this octagon, the relation between a proposition and its paraconsistent negation ¬□ is a relation of subcontrariety and the relation between a proposition and its paracomplete negation ¬◇ is a relation of contrariety. All this shows very clearly that the three notions of oppositions met in the square of Aristotle correspond to three negations: contradictory to classical negation, contrariety to paracomplete negation, subcontrariety to paraconsistent negation. My impression was that square of opposition could be used as a philosophical basis for a theory of negation admitting various type of negations, including paraconsistent negation.

I made some further progress in this square approach to paraconsistent negation at the beginning of 2003 in Switzerland (where I moved in august 2002). I had never been very satisfied by the octagon because it was not so simple and nice as Blanché's hexagon. I was trying to put everything together in a more beautiful geometrical objet. I discovered then that the octagon can be decomposed in three hexagons having the same structure as Blanché's hexagon, i.e. being the composition of two triangles. Among these three hexagons, there is Blanché's hexagon itself, a paraconsistent hexagon in which appears the paraconsistent negation and a paracomplete hexagon in which appears the paracomplete negation. Then my idea was to put these three hexagons together by constructing a three dimensional object, a polyhedron. One possible polyhedron is the first stellation of the rombic dodecahedron. But my friend Alessio Moretti has contructed a more interesting polyhedron. In this polyhedron appears a fourth hexagon that he and Hans Smessaert discovered. This polyhedraic constructions are a great achievement which gave an exact idea of the place of the paraconsistent negation ¬□ and its relation with other negations and modalities. Moreover these relations can be understood easily through geometrical intuitons, you just have to look at the objets. Due to the ridiculous limitations of Latex, I will not present pictures here, but they can be found on my website at www.unine.ch/unilog.

5   Philosophy and applications of paraconsistent logic

Endive aumenta o colesterol. Ovo, manteiga, bacon fazem baixar o
colesterol,
é a última descoberta dos pesquisadores de uma universidade sueca.
Rubem Fonseca *Idiotas que falam outra lingua*

For me, philosophical aspects of paraconsistent logic are directly related
with the question of its applications. I consider philosophy as something
related with reality not with Nephelokokkugia (The city of cuckoos in the
clouds). As it is known, it is not necessary the case of philosophers in general
and not either the case of logical philosophers. Paraconsistent logicians are
most of the time not interested in a world where there is an Eiffel Tower in
Paris, black birds in the sky of Hiroshima, nice girls on Copacabana beach.
They prefer to live in a world full of "funny" paradoxes or, instead of looking
at the world at it is, they try to reduce it to a terrible machine ruled by the
God of Contradiction, where there are contradictions everywhere: between
night and day, cats and dogs, cheese and wine. Many people are interested
in paraconsistent logic because they think that contradiction is the "thing".
They don't want to eliminate contradictions, they want to play with them.
And if you say that there are no contradictions, if you take out their toys,
their daily bred, they become hungry like spoiled children.

In different papers I wrote, I tried to provide interpretation of paraconsistent logic, by giving some examples. It is striking to see that most of the
time, formal logicians give only very few examples or trivial ones (Snow is
white, Bush is a pacifist, Bachelors love married women). Unfortunately
paraconsistent logicians are no exceptions. So we have the impression that
what they are doing is some silly formal games, without meaning, with no
relations with everyday life. In the best case, when they are related to
something, it is to some "funny" paradoxes that nobody care about except
those leaving in Nephelokokkugia.

It is a pity because one of the strongest features of paraconsistent logic is
its potential application, not to metaphysics, but to technology. Inconsistent
databases are a "reality". We need paraconsistent logic to construct robots
able to behave in a surrounding inconsistent world of information. Such
robots in fact have already been conceived, the first of them is a female
called Sophia, born in Brazil in 1999. She had now a little sister, Emmy.

In the different examples I have been developed, I was especially inspired
by Flaubert. In his last unfinished book *Bouvard et Pécuchet*, Gustave
Flaubert describes the adventures of two sympathetic old Parisians who
decided to retire in the countryside and study all sort of things: physics,
agriculture, how to cure a hysterical woman, etc. In each case, they try

to get some information, either by reading books or asking some people, and in each case they always find contradictory information, so at the end they don't know what to to, what to think. This is expressed by Flaubert in a very funny style, characterized by Maupassant as "juxtaposition antithétique".

In [25] I gave some examples of contradictory information given by two fictitious Physicians, Dr. Bouvard and Dr. Pécuchet. Medecine is a rich source of examples for paraconsistent logic, we find contradictions everywhere, contradictions between diagnoses, contradictions between theories (homeopathy vs. allopathy), etc. Medecine was also a favourite target of Flaubert's irony (cf. the couple physician-pharmacian Bovary-Homais in *Madame Bovary.*)

Imagine that Rintintin is sick. On the one hand Dr.Bouvard tells him he has a genuine disease, like cancer, and that he will irremediably die before Easter, on the other hand Dr. Pécuchet tells him it is a benign disease, that he has just to put his feet on hot water everyday while smoking a special herbal mix. So what will Rintintin do? Put his feet on hot whater while smoking everyday this special herbal mix until Easter and see if he will die? At this point, it may be useful for him to use paraconsistent logic...

Other applications I work with are related with justice and law. Imagine that someone has been killed and that Pécuchard has been seen close to the place of the crime at the time it happened. Imagine also that another witness has seen Pécuchard at this same time in the zoo giving bananas to the monkeys. Here again paraconsistent logic can be useful and can avoid to believe a witness rather another one on the basis of subjective criteria like, for example: the guy who has seen Pécuchard at the zoo is a "crazy" guy doing research on iterative forcing.

6   My contribution to the development of paraconsistent logic

When I started my researches in paraconsistent logic at the end of the eighties, it was an exotic topic known only by the happy few. It is amazing how the situation has changed 15 years.

I have myself contributed to the development of paraconsistent logic, not only by writing papers, but also by discussing with people, organizing events and doing some editorial work.

In France, when I started to work on paraconsistent logic, very few people knew what it was. Most of the people never had heard this curious name. This was the case of Daniel Andler, who nevertheless thought that the subject was interesting and accepted to directed my Master on the subject. By studying da Costa's papers, I discovered that in the early 1960s he had

collaborated with a French mathematician from Clermont-Ferrand called
Marcel Guillaume. I asked Andler if he knew something about Guillaume,
if he was still alive, etc. He knew him and told me that he was still in
Clermont-Ferrand. I wrote to Guillaume and he was kind enough to answer
me and sent me some papers he wrote with da Costa I was not able to find
in Paris. He also told me that he was not working in the subject since many
years.

When I met da Costa in Paris in 1991, he was also not really working
anymore on paraconsistent logic since several years. His main interest was
on the applications of logic to physics, in particular versions of Gödel's
theorem for physics.[2] And when I went to work with him in São Paulo, there
were very few people working with him on paraconsistent logic. But slowly
people started to be interested again in paraconsistent logic, in particular da
Costa himself. In some sense I contributed to the revival of paraconsistent
logic in Brazil and also to its spreading all over the world: after spending
one year in Brazil I did some researches in Poland and California and gave
lectures on paraconsistent logic in many countries. I gave also several mini-
courses on paraconsistent logic to advanced students: at the ENS in Lyon in
2000, at the ESSLLI in Birmingham in 2000, at the XXI SBC in Fortaleza
in 2001, etc.

I translated into French, or something alike, the book of da Costa *Ensaio sobre os fundamentos da logic*. The book in French is called *Logiques
classiques et non classiques* and was published by Masson in 1997. This
book is not only about paraconsistent logic, but as you guess, paraconsistent logic is widely presents in it. There is in particular a discussion about
Lukasiewicz's analysis of Aristotle's analysis the principle of contraciction.
There is also a discussion about paradoxes and what da Costa calls Hegel's
thesis, the thesis according to which there are true contradictions. In this
book da Costa writes: "On the one hand, there are some people for whom
contradictions play a quasi mystical rôle, being used as explaining nearly
all the universe; on the other hand, there are some excellent specialists who
believe that contradiction is something ununderstable. Paraconsistent logic
not only contributes to demystify contradiction but also to calm down those
who are afraid of it."

The French edition is a revised version of the original Portuguese edition
with furthermore two appendices I wrote myself. One of them is on paraconsistent logic. I tried to present the logic C1 in a intuitive way, with a
natural deduction system and providing some examples.

---

[2]He got some positive results in this direction together with the Brazilian physicist,
F.A.Doria, who never wrote papers on paraconsistent logic, but is one of the most paraconsistent being in the world.

I edited two special numbers of *Logique et Analyse* dedicated to Contemporary Brazilian Research in Logic containing several papers on paraconsistent logic. The first volume contains a paper of Guillaume recalling his adventures in snowy Curitiba at the beginning of the 1960s. The second volume is quite unique because you can find in it a paper by Quine standing by some papers on paraconsistent logic. Quine's paper itself is not, as you guess, on paraconsistent logic, but it is called "Mission to Brazil" and is related with the year Quine spent in Brazil in 1942-43 and the book which was the fruit of this stay, *O sentido da nova logica*, a book Quine wrote in Portuguese or something alike, and which has never been translated in English.

I participated in a significant way to three paraconsistent events. The first of them is the WCP2 (Second World Congress on Paraconsistency) which was organized in Brazil in 2000, for the commemoration of da Costa 70th birthday, following the WCP1 (the 1st World Congress on Paraconsisteny) organized in Ghent in 1991. I was part of the scientific committee of the WCP2, but my main contribution was the finding of a very nice location for the realization of the congress. I did that with my friend João Marcos, we spent about two days driving on the littoral of the State of São Paulo to find a beautiful place and we choosed Juquehy Beach. It seems that everybody liked the place. The cousin of da Costa, Lars Eriksen, told me it was the best congress of his life and that he particularly enjoyed the food.

Together with Walter Carnielli, I was the main organizer of the WCP3 which happened in Toulouse, France, 2003. I organized also a small paraconsistent event in Las Vegas, USA in 2001.

7   My future work in paraconsistent logic

> *Eu que nao me sento no trono de um apartamento,*
> *com a boca escancarada, cheia de dentes, esperando a morte chegar.*
> Raul Seixas *Ouro de tolo*

I think I will work on paraconsistent logic until the end of my life, and maybe after, if there are any possibilities to do logical researches in paradise. Here are my projects for the next few years.

I am presently writing a book on paraconsistent logic, with my PhD student Alexandre Costa Leite. The title of the book is *A Panoramic introduction to paraconsistent logic*.

At the present time, there is no introductory book of paraconsistent logic, where someone can learn the basic techniques of paraconsistent logic. I think this is a very bad thing which reflects the confuse situation prevailing at the present time in paraconsistent logic: paraconsistentists talk as if paraconsistency was obvious, as if there were a lot of satisfactory and well-known systems of paraconsistent logic. But it is not at all the case. A typical example is that, as pointed by João Marcos, most of time when someone is talking about Jaśkowski he doesn't know what he is speaking about. There is no clear definition of what is Jaśkowski's logic. So if someone wants really to learn what is paraconsistent logic, how it works and not just to read a lot of pseudo-philosophical blah blah blah, or some very technical papers, he has nothing at hand.

Paraconsistent logic is not only one logic as is intuitionistic logic, neither a multiplicity of logics constructed with one technique like modal logics. This is why paraconsistent logic is not something easy to study. Our book will reflect this state of affairs. In *A Panoramic introduction to paraconsistent logic*, each chapter will present a different paraconsistent logic based on a different technique. Through the book many examples will be given. And in a final chapter we will present a general theory of paraconsistent negation.

A line of investigation I will develop is related to the square of opposition and paraconsistent negation as $\neg\Box$. My work can be developed in several directions, that can be explored with the geometrical intuitions given by the polyhedra. One direction is the study of $\neg\Box$ and the related polyhedra in the cases of some specific modal logics. Alessio Moretti has been studying the case of S4. In S4, and in each modal logic, $\neg\Box$ behaves as a paraconsistent negation. So there are a lot of paraconsistent negation to be (re)discovered. I am also studying the behaviour of $\neg\Box$ in modal logics based on many-valued semantics.

Another direction is the generalization of the Star-of-David that can be found in Blanché's hexagon into a star with eight vertices constructed by the combination of two squares (This symbol is quite popular in the Muslim world, I saw it for example in the great mosque Hagia Sophia in Istambul). One square is a contrary square, and the other one is a subcontrary square, they generalize the contrary triangle and subcontrary triangle of Blanché's hexagon. In the same way that Blanché's Star-of-David is based on tri-chotomy, the Muslim-Star is based on quatritomy and we can go on and constructed stars based on any politomy. This allows to develop a general theory of polytomy and shows that Kant was totally wrong when saying that only dichotomy is non empirical and that polytomy cannot be teached in logic (*Logik, 113.*)

I have also many other different projects, working on some specific para-consistent logics, like De Morgan logic, writing on the philosophical aspects of paraconsistency, developing connexion between paraconsistent logic and universal logic, etc.

BIBLIOGRAPHY

[1] J.-Y.Béziau, 1987, *L'holomouvement chez David Bohm*, Master of Philosophy, Department of Philosophy, University Panthéon-Sorbonne, Paris.

[2] J.-Y.Béziau, 1989, "Calcul des séquents pour logique non althétiques", *Logique et Analyse*, **30**, pp.143-155.

[3] J.-Y.Béziau, 1990, *La logique paraconsitante C1 de Newton C.A. da Costa*, MD of Logic, Department of Mathematics, University Denis Diderot, Paris.

[4] J.-Y.Béziau, 1990, "Logiques construites suivant les méthodes de da Costa", *Logique et Analyse*, **33**, 259-272.

[5] J.-Y.Béziau, 1993, "Nouveaux résultats et nouveau regard sur la logique paraconsistante C_1", *Logique et Analyse*, **36**, 45-58.

[6] Jean-Yves Béziau, 1994, "Théorie législative de la négation pure", *Logique et Analyse*, **147-148**, 209–225.

[7] Jean-Yves Béziau, 1994, "Universal Logic", in *Logica '94 - Proceedings of the 8th International Symposium*, T.Childers and O.Majers (eds), Czech Academy of Science, Prague, 73–93.

[8] J.-Y.Béziau, 1995, "Negation : what it is and what it is not", *Boletim da Sociedade Paranaense de Matemática*, **15**, 37-43.

[9] J.-Y.Béziau, 1995, *Recherches sur la logique universelle*. PhD, Department of Mathematics, University of Paris 7, 1995.

[10] J.-Y.Béziau, 1995, *Sur la vérité logique* PhD, Department of Philosophy, University of São Paulo, 1996.

[11] J.-Y.Béziau, 1997, "Logic may be simple", *Logic and Logical Philosophy*, **5**, 129–147.

[12] J.-Y. Béziau, 1998, "Idempotent full paraconsistent negations are not algebraizable", *Notre Dame Journal of Formal Logic*, **39**, pp.135-139.

[13] J.-Y. Béziau, 1998, "De Morgan lattices, paraconsistency and the excluded middle", *Boletim da Sociedade Paranaense de Matemática*, **18**, pp.169-172.

[14] J.-Y.Béziau, 1999, "Classical negation can be expressed by one of its halves", *Logic Journal of the Interest Group on Pure and Applied Logics*, **7**, 145–151.

[15] J.-Y. Béziau, 1999, "The future of paraconsistent logic", *Logical Studies*, **7**, 1–20, http://www.logic.ru.

[16] J.-Y. Béziau, 1999, "What is paraconsistent logic ?", in *Frontiers of paraconsistent logic*, Research Studies Press, Baldock, 2000, pp.95-112.

[17] J.-Y. Béziau, 2000, "From paraconsistent logic to universal logic", *Sorites*, **12**, 1–33, http://www.ifs.csic.es/sorites.

[18] Jean-Yves Béziau, 2001, "Universal logic: Towards a general theory of logics", *Bulletin of Symbolic Logic*, **7**, 101.

[19] J.-Y. Béziau, 2001, "The logic of confusion", in *Proceedings of the International Conference on Artificial Intelligence IC-AI 2001*, CSREA Press, Las Vegas, pp.821-826.

[20] J.-Y. Béziau, 2002, "S5 is a a paraconsistent logic and so is first-order logic", *Logical Investigations*, **9**, 301–309.

[21] J.-Y. Béziau, 2003, "The square of opposition, modal logic and paraconsistent logic", *Logical Investigations*, **10**, 220–237.

[22] J.-Y. Béziau, 2004, "Non truth-functional many-valuedness", in *Aspects of Universal Logic*, J.-Y. Béziau, A.Costa Leite and A.Facchini (eds), University of Neuchâtel, Neuchâtel, 2004, pp.199-218.

[23] J.-Y. Béziau, 2006, "The paraconsistent logic Z", Paper presented at the Jaśkwoski Memorial Symposium, Torun, July 15-18, 1998, to appear in *Logic and Logical Philosophy*.

[24] J.-Y. Béziau, 2006, "Paraconsistent logic!", *Sorites*, **13**, 1–11, http://www.ifs.csic.es/sorites.

[25] N.C.A. da Costa, 1997 *Logiques Classiques et non classiques*, Masson, Paris.

[26] N.C.A. da Costa and J.-Y.Béziau, 1994, "Théorie de la valuation", *Logique et Analyse*, **37**, 95–117.

[27] N.C.A. da Costa and J.-Y.Béziau, 1996, "Théories paraconsistantes des ensembles", *Logique et Analyse*, **39**, 51–67.

[28] N.C.A. da Costa and J.-Y.Béziau, 1998, "Définitions, théories des objets et paraconsistance", *Theoria*, **32**, 367–379.

[29] N.C.A. da Costa, J.-Y.Béziau and O.A.Bueno, 1995, "Aspects of paraconsistent logic", *Logic Journal of the IGLP*, **4**, 597-614.

[30] N.C.A. da Costa, J.-Y.Béziau and O.A.Bueno, 1995, "Paraconsistent logic in a historical perspective", *Logique et Analyse*, **38**, 150-152.

[31] N.C.A. da Costa, J.-Y.Béziau and O.A.Bueno, 1998, *Elementos de teoria paraconsistente de conjuntos*, State University of Campinas, Campinas.

Jean-Yves Beziau

Institute of Logic of the University of Neuchâtel, Espace Louis-Agassiz 1, 2000 Neuchtel, Switzerland.

jean-yves.beziau@unine.ch

Recent Results by the Inconsistency-Adaptive Labourers[1]

Diderik Batens and Joke Meheus

ABSTRACT. This paper offers an incomplete survey of recent results on inconsistency-adaptive logics (disregarding results on other adaptive logics). Much attention is paid to the so-called standard format, because it provides most of the required metatheory for adaptive logics that are phrased in this format. Combined adaptive logics are also briefly discussed. Other results briefly reported on concern (i) rethinking some philosophical theories from a paraconsistent viewpoint, (ii) the characterization of some (further) inconsistency-handling mechanisms in terms of adaptive logics, and (iii) the problem of defining criteria for final derivability. The final section concerns a further step towards eliminating (undesired) inconsistencies from a theory.

1 Introduction

As the title suggests, this paper briefly reports on recent results in inconsistency-adaptive logics. Completeness was not sought. The aim is rather to report on some technical results, and on the way in which they shape the programme from a philosophical point of view.

Adaptive logics are meant to characterize consequence relations for which there is no positive test (that are not even partially recursive)—see [30, 29]. All but not only non-monotonic consequence relations lack a positive test. Adaptive logics require dynamic proofs, viz. proofs in which formulas that are considered as derived at some stage are considered as not derived at a later stage, and *vice versa*. During the last five years, most results on adaptive logics concern ampliative adaptive logics—logics that extend **CL** (Classical Logic). These will not be reported on in the present paper. However, we shall briefly mention some ampliative results that apply generally, including in inconsistent situations—in principle all ampliative logics have

---

[1] Research for this paper was supported by subventions from Ghent University and from the Fund for Scientific Research – Flanders.

variants that apply in inconsistent situations, but unfortunately not all of
these have been studied.

For readers that are not familiar with inconsistency-adaptive logics, it
may be useful to recall their effect in intuitive terms. Consider the premise
set $\Gamma = \{p, \sim p, p \vee q, \sim r, r \vee s\}$. On nearly all (monotonic) paraconsistent
logics,[1] Disjunctive Syllogism is invalid and neither q nor s is a consequence
of Γ. An inconsistency-adaptive logic **AL** interprets a premise set *as consistently as possible*. While Γ requires $p \wedge \sim p$ to be true, it does not require $r \wedge \sim r$ to be true. So, on a minimally inconsistent interpretation
of Γ, r comes out false and hence s comes out true. Put differently, s is
an **AL**-consequence of Γ, whereas q is not. Monotonic paraconsistent logics invalidate certain **CL**-rules, for example Disjunctive Syllogism, whereas
inconsistency-adaptive logics invalidate certain *applications* of some **CL**-
rules, and it depends on the premises which applications are invalidated.

The adaptive programme is neutral with respect to the question whether
there are true inconsistencies. Dialetheists, who answer the question in the
positive, will argue that most true statements are consistent, and hence
that consistency can be presupposed unless and until proven otherwise—
this is precisely what inconsistency-adaptive logics do. Graham Priest has
spelled this out in [54]. People who deny that there are true inconsistencies, for example classical logicians and intuitionists, cannot get around
the fact that inconsistencies occurred in the history of the sciences—see
[31, 36, 37, 41, 49, 50, 51, 59] for some case studies—and that, in the presence of inconsistent theories, one should *reason from them* in order to find
consistent replacements for them. As was argued already in [3], the first step
to be taken in such circumstances is to interpret the inconsistent theory as
consistent as possible, in other words to apply an inconsistency-adaptive
logic. The two aforementioned positions do not exhaust all possibilities.
Thus our own position, which is outlined and argued for in [9], is that there
is no warrant that there is a true consistent description of the world (in a
conceptual system that humans can handle), but that there are good reasons
to adopt the methodological maxim that one should try to eliminate inconsistencies from our knowledge—this entails that inconsistencies are seen as
a problem, although not necessarily as the most urgent one.

2   The Standard Format

The growing multiplicity of adaptive logics called for systematization. The
idea was not to restrict the variety of applications that the logics capture,
but rather to find a common formal characterization. This characterization

---

[1]Meheus' **AN∅** from [39] is an exception, but we disregard it for the present intuitive
discussion.

is provided by the so-called standard format. The basic mechanism behind all adaptive logics is the same. So the standard format, if well designed, should do most if not all of the work. We shall see that it does.

In many situations one needs to combine adaptive logics. In this respect, the aim was to rely on the standard format in order to design general stratagems for combining adaptive logics. Again, the combination stratagems, rather than specific properties of the combined logics, should warrant that the combination does the desired job. Combined adaptive logics are discussed in Section 3.

The standard format, first presented in [8], is both simple and perspicuous. An adaptive logic **AL** is in standard format if it is characterized as a triple consisting of the following elements:

(i) **LLL**, a lower limit logic,
(ii) Ω, a set of abnormalities that all have the same logical form,
(iii) an adaptive strategy.

The lower limit logic **LLL** determines the part of the adaptive logic that is not subject to adaptation. From a proof theoretic point of view, the lower limit logic delineates the rules of inference that hold unexceptionally. From a semantic point of view, the adaptive models of Γ are a selection of the lower limit models of Γ. It follows that $Cn_{\mathbf{LLL}}(\Gamma) \subseteq Cn_{\mathbf{AL}}(\Gamma)$. In principle, the lower limit logic is a monotonic and compact logic—see Section 3 on combined adaptive logics.

Abnormalities are formulas that are presupposed to be false, unless and until proven otherwise. Ω comprises all formulas of a certain logical form, which may be restricted—see below.

For many inconsistency-adaptive logics, Ω is the set of formulas of the form $\exists(A \wedge {\sim}A)$, in which $\exists A$ abbreviates the existential closure of A. For other inconsistency-adaptive logics, the set is restricted, for example, to formulas in which A is a primitive formula—a formula that contains no logical symbols except for identity.[2] Where introduced, the restriction is justifiable or even desirable, as we shall explain in the paragraph on flip-flop logics. Examples of such inconsistency-adaptive logics are those that have as their lower limit logic, for example, Schütte's $\mathbf{\Phi}_v$ from [58],[3] Priest's **LP** from [53], or Meheus' **AN∅** from [39], and modal adaptive logics that characterize paraconsistent inference relations under a translation—several examples follow in Section 4. Incidentally, in all these logics, every formula

---

[2]Similar restrictions are imposed on many ampliative adaptive logics. See, for example, [19] and [18].

[3]The Ghent name for the predicative extension of this logic is **CLuNs**—see [17]. Schütte, who was not a paraconsistent logician, introduced $\mathbf{\Phi}_v$ for a special purpose. Many paraconsistent logicians rediscovered this logic or its predicative extension.

$\exists(A \land \sim A)$ in which A is not primitive entails a disjunction of formulas of the form $\exists(B \land \sim B)$ in which B is primitive.

If the lower limit logic is extended with the requirement that no abnormality is logically possible, one obtains the *upper limit logic* **ULL**. Syntactically, **ULL** is obtained by extending **LLL** with an axiom stating that members of Ω entail triviality. The lower limit logic of inconsistency-adaptive logics is a paraconsistent logic, and the axiom schema $(A \land \sim A) \supset B$ is the most popular candidate for obtaining the upper limit logic. Even if Ω is characterized by a restricted logical form, there is no need to impose a restriction on the axiom schema if, for any C of the unrestricted form, there is a $\Delta \subseteq \Omega$ for which $C \vdash_{\textbf{LLL}} \bigvee(\Delta)$. There also is no need to restate the axiom schema as $\exists(A \land \sim A) \supset B$, provided quantification behaves normally. Semantically, the upper limit logic is characterized by the lower limit models that verify no abnormality. **ULL** requires premise sets to be normal, and 'explodes' abnormal premise sets (assigns the trivial consequence set to them).

If, as is the case for many inconsistency-adaptive logics, the lower limit logic is a paraconsistent logic that contains full positive **CL** as well as excluded middle, for example expressed as $(\sim A \supset A) \supset A$,[4] and the set of abnormalities comprises the formulas of the form $\exists(A \land \sim A)$, possibly restricted as described before, then the upper limit logic is **CL**.[5]

If the premise set does not require any abnormality to obtain, the adaptive logic will deliver the same consequences as the upper limit logic. If the premise set requires some abnormalities to obtain, the adaptive logic will still deliver more consequences than the lower limit logic, viz. all upper limit consequences that are not 'blocked' by those abnormalities. In sum, the adaptive logic interprets the set of premises 'as normally as possible'; it takes abnormalities to be false 'in as far as' the premises permit.

The lower limit logic and the upper limit logic do not determine the set of abnormalities Ω. A few paragraphs ago, we considered the case where Ω contains all formulas of the form $\exists(A \land \sim A)$ in which A is a primitive formula. If the restriction is removed, the upper limit logic is still **CL**, but the adaptive logic is different—see the paragraph on flip-flop logics.

An *adaptive strategy* is required because many premise sets **LLL**-entail

---

[4]The weakest such logic is **CLuN**, **CL** allowing for gluts with respect to negation, which consists exactly of full positive **CL** together with $(\sim A \supset A) \supset A$—see [35] for a proof of the propositional case. Remark that Replacement of Identicals does not hold in **CLuN**, viz. does not hold in the scope of a negation. From a technical point of view, it is easier to handle **CLuN** if it is extended with a classical negation $\neg$ (or with $\bot$, characterized by $\bot \supset A$). This greatly simplifies the metatheory. In our preferred application contexts, classical negation does not occur in the premises.

[5]Priest's **LP** does not contain a detachable implication, and hence the requirement that all abnormalities are false should be introduced by a rule, viz. $A \land \sim A / B$, in which case the upper limit logic is **CL**.

a disjunction of abnormalities (members of Ω) without entailing any of its disjuncts. Disjunctions of abnormalities will be called *Dab-formulas*. In the sequel, any expression of the form $Dab(\Delta)$ will refer to the (classical) disjunction of the members of a finite $\Delta \subseteq \Omega$. *Dab*-formulas that are derivable by the lower limit logic from the premise set Γ will be called *Dab-consequences* of Γ. If $Dab(\Delta)$ is a *Dab*-consequence of Γ, then so is $Dab(\Delta \cup \Theta)$ for any finite $\Theta \subset \Omega$. For this reason, it is important to concentrate on the minimal *Dab*-consequences of the premise set: $Dab(\Delta)$ is a *minimal Dab*-consequence of Γ iff $\Gamma \vdash_{\mathbf{LLL}} Dab(\Delta)$ and there is no $\Theta \subset \Delta$ such that $\Gamma \vdash_{\mathbf{LLL}} Dab(\Theta)$. If $Dab(\Delta)$ is a minimal *Dab*-consequence of Γ, then Γ determines that some member of Δ behaves abnormally, but fails to determine which member of Δ behaves abnormally. We have seen that adaptive logics interpret a premise set 'as normally as possible'. As some minimal *Dab*-consequences of Γ may contain more than one disjunct, this phrase is not unambiguous. It is disambiguated by choosing a specific adaptive strategy.

The oldest known strategy is *Reliability* from [3]. The *Minimal Abnormality* strategy was first presented in [2]. It delivers at least the same consequences as the Reliability strategy, and for some premise sets it delivers more consequences.

Some lower limit logics and sets of abnormalities are such that Δ is a singleton whenever $Dab(\Delta)$ is a minimal *Dab*-consequence of a premise set. If this is the case, the Reliability strategy and the Minimal Abnormality strategy lead to the same result and coincide with what is called the *Simple* strategy: a formula behaves abnormally just in case the abnormality is derivable from the premise set—see [19], [39] and [40] for examples. Several other strategies have been studied. Most of them are needed to characterize an existing consequence relation by an adaptive logic. Examples may be found in [7], [11], [22], [32] and [64].

Every line of an annotated dynamic proof consists of a line number, a formula, a justification, and a *condition*. The proofs are governed by three (generic) rules and a marking definition. Let

$$A \qquad \Delta$$

abbreviate that A occurs in the proof on the condition Δ, the rules may then be phrased as follows:

PREM   If $A \in \Gamma$:

$$\frac{\begin{array}{cc} \cdots & \cdots \end{array}}{\begin{array}{cc} A & \emptyset \end{array}}$$

RU   If $A_1, \ldots, A_n \vdash_{\mathbf{LLL}} B$:

$$\begin{array}{cc} A_1 & \Delta_1 \\ \vdots & \vdots \\ A_n & \Delta_n \\ \hline B & \Delta_1 \cup \ldots \cup \Delta_n \end{array}$$

RC   If $A_1, \ldots, A_n \vdash_{\mathbf{LLL}} B \vee Dab(\Theta)$

$$\begin{array}{cc} A_1 & \Delta_1 \\ \vdots & \vdots \\ A_n & \Delta_n \\ \hline B & \Delta_1 \cup \ldots \cup \Delta_n \cup \Theta \end{array}$$

Given a dynamic proof, we shall say that $Dab(\Delta)$ is a *minimal Dab-formula* at stage s of the proof if, at stage s, $Dab(\Delta)$ occurs in the proof on the empty condition and, for any $\Delta' \subset \Delta$, $Dab(\Delta')$ does not occur in the proof on the empty condition. Where $Dab(\Delta_1)$, ..., $Dab(\Delta_n)$ are the minimal Dab-formulas at stage s of the proof, $U_s(\Gamma) = \Delta_1 \cup \ldots \cup \Delta_n$ is the set of unreliable formulas at stage s.

DEFINITION 1. Marking for Reliability: Line i is marked at stage s iff, where Θ is the condition of line i, $\Theta \cap U_s(\Gamma) \neq \emptyset$.

We refer to [8] (and elsewhere) for the marking definition of the Minimal Abnormality strategy and for that of the Simple strategy.

Remark that the rules depend on the lower limit logic and on the set of abnormalities, whereas the marking definition depends on the strategy.

A formula is *derived* from Γ *at a stage* of the proof iff it is the formula of a line that is unmarked at that stage. As the proof proceeds, unmarked lines may be marked and vice versa. So, it is important that one defines a different, stable, kind of derivability:

DEFINITION 2. A is *finally derived* from Γ on line i of a proof at stage s iff (i) A is the formula of line i, (ii) line i is not marked at stage s, and (iii) any extension of the proof in which line i is marked may be further extended in such a way that line i is unmarked.

This means that there is a (possibly infinite) proof in which line i is unmarked and that is *stable* with respect to line i (line i is unmarked in all extensions of the proof). The previous definition is more appealing. The only way to establish the existence of an infinite proof is by a metalinguistic reasoning anyway. Moreover, the definition has a nice game-theoretic interpretation: whenever an opponent is able to extend the proof in such a way that line i is marked, the proponent is able to extend it further in such a way that line i is unmarked.

DEFINITION 3. $\Gamma \vdash_{\textbf{AL}} A$ (A is *finally* **AL**-*derivable* from Γ) iff A is finally derived on a line of a proof from Γ.

The *semantics* of all adaptive logics is defined in the same way. $M \models A$ will denote that M assigns a designated value to A, in other words that M verifies A. If the semantics is two-valued—and it is shown in [60] that any semantic system may be rephrased in two-valued terms—then $M \models A$ comes to $v_M(A) = 1$. $M \models \Gamma$ will denote that M verifies all members of Γ.

The abnormal part of a **LLL**-model M is defined as follows:

DEFINITION 4. $Ab(M) = \{A \in \Omega \mid M \models A\}$

Where $Dab(\Delta_1)$, $Dab(\Delta_2)$, ... are the minimal Dab-consequences of a premise set Γ,[6] $U(\Gamma) = \Delta_1 \cup \Delta_2 \cup \ldots$ is the set of formulas that are unreliable with respect to Γ. Let $\textbf{AL}^r$ and $\textbf{AL}^m$ be the adaptive logics defined from **LLL** and Ω by the Reliability strategy and the Minimal Abnormality strategy respectively.

DEFINITION 5. A **LLL**-model M of Γ is *reliable* iff $Ab(M) \subseteq U(\Gamma)$.

In other words, M is a reliable model of Γ iff it verifies no abnormalities outside of members of $U(\Gamma)$, the set of formulas that are unreliable with respect to Γ.

DEFINITION 6. $\Gamma \models_{\textbf{AL}^r} A$ iff A is verified by all reliable models of Γ.

DEFINITION 7. A **LLL**-model M of Γ is *minimally abnormal* iff there is no **LLL**-model M' of Γ such that $Ab(M') \subset Ab(M)$.

DEFINITION 8. $\Gamma \models_{\textbf{AL}^m} A$ iff A is verified by all minimally abnormal models of Γ.

So the proof theory and the semantics of an adaptive logic are fixed by the standard format. But there is more. Many metatheoretic properties of adaptive logics can be proved from the format itself, rather than from the specific properties of the logic. We mention only some examples. First and foremost, there are the *Soundness* and *Completeness* proof (given that the lower limit logic is sound and complete with respect to its semantics). The motor for the dynamic proofs is the *Derivability Adjustment* Theorem: $\Gamma \vdash_{\textbf{ULL}} A$ iff there is a finite $\Delta \subset \Omega$ such that $\Gamma \vdash_{\textbf{LLL}} A \vee Dab(\Delta)$. An important further property is *Proof Invariance*, which states that it does not depend on the way in which a proof sets out whether a conclusion is finally derivable from a premise set: If $\Gamma \vdash_{\textbf{AL}} A$, then any proof from

---

[6]The minimal Dab-consequences of Γ may be semantically defined in view of the soundness and completeness of **LLL** with respect to its semantics.

Γ can be extended into a proof in which A is finally derived from Γ. A semantically essential property is *Strong Reassurance*: If M is a **LLL**-model of Γ but not an **AL**-model of Γ, then there is an **AL**-model M' such that $Ab(M') \subset Ab(M)$. For proofs of these and more, see [12] and [16].

Nearly all known inconsistency-adaptive logics have a characterization in standard format.[7] In some cases, forging a consequence relation into standard format may require a translation, for example to a modal language—see Section 5 for an illustration. Even where it is useful to provide 'direct dynamic proofs' in untranslated terms—see [25] or [45]—the formulation in standard format has the advantage to provide the proof theory, semantics and metatheoretic properties, and to warrant (by an easy demonstration) that the direct proof theory is correct.

Flip-flops Some adaptive logics are called flip-flops because they have the following weird property: if Γ is normal (has upper limit models), then $Cn_{\mathbf{AL}}(\Gamma) = Cn_{\mathbf{ULL}}(\Gamma)$, if Γ is abnormal, then $Cn_{\mathbf{AL}}(\Gamma) = Cn_{\mathbf{LLL}}(\Gamma)$. Flip-flops typically result from an unsuitable choice of the set of abnormalities Ω. If, for some adaptive logic, this set is characterized by a restricted logical form, for example $\exists(A \wedge \sim A)$ in which A is a primitive formula, and is replaced by the set characterized by the unrestricted form, in the example all formulas of the form $\exists(A \wedge \sim A)$, a flip-flop results.[8]

Formula-preferential systems Some adaptive logics were characterized by formula-preferential systems in [35]—see also [1]. The idea is close to that behind adaptive logics: the consequence set is determined by the premises, a regular logic (this actually is the lower limit logic **LLL**) and a set of preferred formulas. A subset of the preferred formulas is added to the premises, and the subset should be maximal with respect to the premises and **LLL**. It is not clear whether all adaptive logics can be characterized by a formula-preferential system, but it is easily provable that all formula-preferential systems are characterized by an adaptive logic; the characterization requires that the preferences are expressed in the object language—see Section 3 for means to do so.

The best-studied (and very simple) inconsistency-adaptive logic is **ACLuN1**—see [3] for its propositional proof theory and [5] for both the syntax and semantics at the predicative level. Its lower limit logic is **CLuN**, full positive **CL** plus Excluded Middle—see footnote 4. The set of abnormalities, Ω, is

---

[7]An exception is Priest's $\mathbf{LP}^m$ from [54] and emended in [55]. This adaptive logic proceeds in terms of properties of the model, rather than in terms of the formulas verified by the model.

[8]Flip-flops were considered as utterly uninteresting, until some interesting prioritized adaptive logics turned out to be flip-flops—see [18].

the set of all formulas of the form $\exists(A \wedge \sim A)$, and the strategy is Reliability. The logic has many nice properties, studied in [5] but also (in a more general framework) in [8] and [12]. Other well-studied inconsistency-adaptive logics have another strategy or another lower limit logic.

Some new inconsistency-adaptive logics were developed recently, mainly as a result of a Ghent–Torun cooperation. A first example is the discussive logic $\mathbf{DL}^r$ that constitutes a (non-monotonic) alternative for Jaśkowski's paraconsistent system $\mathbf{D}_2$ (see [44]). Like $\mathbf{D}_2$, $\mathbf{DL}^r$ validates all single-premise rules of $\mathbf{CL}$. However, for formulas that behave consistently, $\mathbf{DL}^r$ moreover validates all multiple-premise rules of $\mathbf{CL}$—this is realized without the introduction of the discussive connectives. It is stipulated that A is a $\mathbf{DL}^r$-consequence of Γ iff $\Diamond A$ is an $\mathbf{AJ}^r$-consequence of $\Gamma^\Diamond = \{\Diamond A \mid A \in \Gamma\}$. $\mathbf{AJ}^r$ is in standard format, its elements being $\mathbf{S5}$, $\Omega = \{\exists(\Diamond A \wedge \Diamond \sim A) \mid A \in \mathcal{F}^p\}$ ($\mathcal{F}^p$ are the sentential letters and primitive predicative formulas), and Reliability.

A second example is the logic $\mathbf{AD}_2$, developed by Marek Nasieniewski in [47] and partly presented in [48]. $\mathbf{AD}_2$ is an adaptive extension of $\mathbf{D}_2$, including the discussive connectives. In [48], several important properties of $\mathbf{AD}_2$ are studied and $\mathbf{AD}_2$ is compared to $\mathbf{DL}^r$.

3   Combining Adaptive Logics

Several ways to combine adaptive logics have been studied. Basically three methods were obtained, but there is no reason why these should be exhaustive—we shall not discuss all variants of the combination methods.

If two adaptive logics have the same lower limit logic and the same strategy, the obvious way to combine them is by taking the union of their sets of abnormalities as the set of abnormalities of the combined logic. The premise set is then interpreted as normally as possible with respect to both kinds of abnormalities.

If two or more adaptive logics share their upper limit logic and their strategy, a combined adaptive logic is obtained by taking as the lower limit logic the intersection of the lower limit logics and as the set of abnormalities the union of the sets of abnormalities (and retaining the strategy). The specific example we want to present deserves a philosophical comment.

Paraconsistent logics allow for negation gluts—they have models that verify formulas of the form $\exists(A \wedge \sim A)$. However, some inconsistent premise sets also have a non-trivial consequence set if they are handled by a logic that allows for conjunction gluts—these have models that verify formulas of the form $\exists((A \wedge B) \wedge \sim A)$ and of the form $\exists((A \wedge B) \wedge \sim B)$. The same is true for logics that have gluts with respect to some other logical symbol, including the quantifiers and identity. The same even holds for logics that allow for gaps with respect to a logical symbol. Consider, for example, the case of negation gaps. If a model falsifies both p and $\sim p$, it verifies the premise set $\{p \supset q, p \supset \sim q, \sim p \supset q, \sim p \supset \sim q\}$. If this premise set is closed by **CL**, its consequence set is trivial. It follows that paraconsistency need not be the best road to handling an inconsistent premise set. In specific cases, a logic allowing for other gluts, or allowing for gaps, may be more suitable. Let us call such logics paralogics.

Every paralogic gives rise to adaptive logics that interpret the premises as normally as possible. In other words, the adaptive logics presuppose that formulas describing an allowed gap or glut are false unless and until proven otherwise.[9]

All such logics may be combined by the method discussed two paragraphs ago. In [6] they are even combined with (a variant of) the ambiguity-adaptive logic from [61]. The resulting lower limit logic is zero-logic, according to which no formula (not even a premise) is derivable from a premise set. The adaptive logic still interprets the premises as normally as possible. In the special case where the premise set has **CL**-models, the adaptive logic delivers all **CL**-consequences of the premises.[10]

A last way to combine adaptive logics is by superimposing adaptive logics on top of each other. From a definitional point of view, the consequence set of the combined logic is $Cn_{\mathbf{AL2}}(Cn_{\mathbf{AL1}}(\Gamma))$. So, Γ is first interpreted as normally as possible with respect to the abnormalities of **AL1**, and the resulting consequence set is then interpreted as normally as possible with respect to the abnormalities of **AL2**. Two problems seem to arise here. First, it appears that the lower limit logic of **AL2** is the adaptive logic **AL1**, which puts **AL2** outside of the standard format. This is rather easily solved. One can take as the lower limit logic of **AL2** any monotonic logic that does not trivialize $Cn_{\mathbf{AL1}}(\Gamma)$—the lower limit logic of **AL1** is a good candidate, as $Cn_{\mathbf{AL1}}(\Gamma)$ is closed under this logic, but there are other suitable candidates as well.

The second problem is that the above construction seems to require that all **AL1**-consequences from the premises are obtained before the result is closed by **AL2**, which makes the construction problematic from a computational point of view. Fortunately, the combined adaptive logic avoids this problem in a way that seems impressively elegant to us. Let us restrict attention to the special case where the lower logics of **AL1** and **AL2** are

---

[9]Formulas describing gaps may require that the language is extended by the classical logical symbol that corresponds to the symbol displaying gluts or gaps. This is an easily solvable technical problem.

[10]The basic idea behind the adaptive logic is obviously that all properties of all logical symbols are defeasible in the sense that one can interpret sensibly a 'text' that requires gluts or gaps with respect to any logical symbol.

the same, which makes their unconditional rules identical. One then simply applies the unconditional rule and both conditional rules in the usual way. The only difference is that the marking definition has to be changed into a two-step definition. It comes to this: first lines are marked with a mark of type 1 according to the marking definition of **AL1**, next lines are marked with a mark of type 2 according to the marking definition of **AL2**, except that the later definition takes only into account the lines that are type 1 unmarked. We refer to [8] for details.

A nice example enables one to build combined adaptive logics that handle preferred consequence sets. Let $\mathcal{W}$ be the set of closed formulas of the usual (non-modal) predicative language, and consider a modal language in which the diamond is interpreted as "it is plausible that", more diamonds indicating a lower plausibility. Moreover, let $\diamond^n$ abbreviate a sequence of n diamonds. We now define an infinity of adaptive logics, $\mathbf{AML}_n$ having as their lower limit logic some standard modal logic $\mathbf{ML}$, $\{\diamond^n A \wedge \sim A \mid A \in \mathcal{W}\}$ (in which the n is the same as the one occurring in the name of the logic) as their set of abnormalities, Ω, and either Reliability or minimal Abnormality as their common strategy.[11] The upper limit logic of each of these adaptive logics is **Triv**—the system in which A is logically equivalent to $\Box A$ as well as to $\diamond A$. All these adaptive logics can be easily combined by the method under discussion in order to handle preferred statements. The premises will be expressed by a single set of formulas of the form $\diamond^n A$, in which $n = 0$ for statements that are taken to be certain, $n = 1$ for statements of the next highest priority, etc. The deductive closure of a premise set under the combined adaptive logic will contain the following members of $\mathcal{W}$: (i) all **CL**-consequences of premises that belong to $\mathcal{W}$, (ii) all **CL**-consequences of the previous set together with as many as can be consistently added from the following set: the $A \in \mathcal{W}$ for which $\diamond^1 A$ is a **ML**-consequence of Γ, etc.

This construction has been applied (for specific purposes) in [18], [23], [62] and other papers. As it is described here, it functions only as a way to prevent inconsistencies from arising. However, it can also be applied to stepwise extend an inconsistent set with (consequences of) statements that have (a higher or lower) priority and are compatible with the inconsistent premise set or with its previous extension—this problem was solved by the results from [43].

4   Rethinking in Terms of Paraconsistency

There is a kind of consistency-laziness in philosophy. Even after being convinced that many theories are inconsistent, either in themselves or with

---

[11]If the purpose is different, the set of abnormalities will be taken to be $\{\Box^n A \mid A \in \mathcal{W};$ $\nvdash_{\mathbf{ML}} \Box A\}$—see [19] for an example.

respect to data or to other theories, many philosophers will stay content with solving a problem for the consistent case. Some will argue that it only is the task of philosophers to offer an explication for or idealization of actual thinking, and then invoke some fallacious reasoning to conclude that the idealized situation is consistent. Of course they are free to define whatever they like as the idealized situation. The point is that they should spell out the relation between their ideal (their theory) and reality (actual reasoning), and precisely this they usually fail to do. Others might admit that their theory should handle the inconsistent case to be correct, but may still argue that they solved an important problem and that it is now someone else's task to consider the inconsistent case. Sometimes they even think the remaining problem to be easily solvable by some paraconsistent trick. Given this situation, we consider it essential that paraconsistent logicians put the rethinking of philosophical concepts and theories very high on their agenda. In this section we consider a few examples where the challenge was met (in terms of the Ghent approach).

Compatibility is a concept that is invoked in numerous situations. To give just one example, any extension (not revision) of our knowledge should be compatible with our present knowledge. So, if the task is to extend our knowledge with a new hypothesis or theory, one has to check whether it is compatible with our present knowledge. There obviously is no positive test for compatibility (at the predicative level), whence, even in the consistent case, compatibility is characterized by a adaptive logic, which was described in [19].

As our present knowledge may be inconsistent, this does not fully solve the problem. Moreover, even defining compatibility in inconsistent environments raises an interesting problem in that most definitions for consistent compatibility classify either nothing or everything as compatible with an inconsistent set of statements. A proposal (possibly the only sensible one) was put forward in [43]. There A is defined as compatible with Γ iff $\Gamma \cup \{A\}$ is not more inconsistent than Γ—this vague criterion obtains an exact articulation. Relying on it, the adaptive logic of compatibility for the inconsistent case is spelled out. This logic is a true generalization of the logic for the consistent case in the sense that both lead to the same consequence set if Γ is consistent.

In [14] a similar job was done for a very different theory, viz. Hintikka's theory of the process of explanation from [34] and elsewhere. Hintikka's criteria for an explanation are first adjusted in such a way that they lead to exactly the same result as the original criteria in the consistent case, but allow for explanation from inconsistent theories. Next, it is shown that these criteria can be met by proceeding exactly as Hintikka, except that

CL is replaced by the inconsistency-adaptive **ACLuN1**. It is shown that the result is a proper generalization of Hintikka's theory in that it leads to precisely the same results in the consistent case, and adjusts this theory in a natural and minimal way in the inconsistent case.

The criteria from [14] are applied in [46] to design an adaptive logic for abduction that enables one to generate explanations for novel facts as well as for anomalous facts (facts not consistent with one's background assumptions), and that moreover can handle possibly inconsistent theories.

An interesting theory on question evocation (by a set of declarative premises) and question implication (by a set of declarative premises together with a question) was proposed by Andrzej Wiśniewski, for example in [65] and [66]. Although Wiśniewski intends his theory to be general and relatively independent of the underlying logic (for the declarative premises), it cannot be applied as it stands in inconsistent situations. This situation was repaired for question evocation by Kristof De Clercq in a section in [33] and by Joke Meheus in [38]. Both contributions contain a system that allows for sensible question evocation in the presence of inconsistent declarative premises and in both cases answers to questions will repair the undesired incompleteness of the premises—this triggered the questions. However, the first approach (relying on **ACLuN1**) is more suitable if the premises are considered as true, whereas the second is more suitable if one sets out to obtain a consistent replacement (on the basis of new evidence) for the premises—it presupposes that the answers obtained have a higher preference than the original premises. Both approaches generalize Wiśniewski's theory in the sense that they lead to the same results as this theory in the consistent case.

It is well-known that there has hardly been any attention for inconsistent beliefs in the literature on belief revision. A major contribution to repair this is contained in [33]. In the second part of this book, Kristof De Clercq presents a full-fledged replacement for most of the standard theory. Taking inconsistent belief sets into account, he stepwise adjusts the postulates, the (tiresome) definitions, and the construction. He goes on to present and prove an impressive set of metatheorems, often developing new proof techniques, to show that the whole framework is adequate. He also shows that his framework leads to the same changes as the standard one if it is applied to consistent belief sets.

5 Further Work on Integration

Some of the Rescher–Manor mechanisms from [57]—for an overview see [26] for the flat ones and [27] for the prioritized ones—were among the first inconsistency-handling mechanisms. The flat ones were characterized by an adaptive logic in [7] (in terms of inconsistency-adaptive logics close to **ACLuN1**). Insights provided by [44] led to new results in [11]: the flat mechanisms are characterized in terms of modal adaptive logics, and sensible extensions of them are presented (mainly with discussive application contexts in mind—see also [63]). The prioritized mechanisms were characterized by adaptive logics in [64].

Another approach to paraconsistent reasoning that has been gaining popularity are the signed systems from [28] and other papers. These systems are characterized by adaptive logics in [22].

6   Criteria for Final Derivability

The proofs of adaptive logics are dynamic: a line that was introduced may later be marked, still later unmarked, etc. This is not a disadvantage of adaptive logics, but a result of the properties of the consequence relations that adaptive logics explicate. As there is no positive test for these consequence relations (at the predicative level), a non-dynamic proof theory cannot possibly be provided.

Even in the absence of a positive test, there may be criteria that enable one to decide that a finite proof warrants that one of its lines will not be marked in any extension of the proof—in other words that the extensions of the proof are stable with respect to this line. The search for criteria led, first to [4], which provided interesting but weak criteria, and next to [20] and [21], which provided criteria in terms of tableau methods. The disadvantage of the latter is that they are rather remote from proofs and moreover require many steps that seem intuitively useless.

Work on a completely different problem provided a way out. In [24], a special proof format was devised (for propositional **CL**) in which part of the proof heuristics is pushed into the proof—in [56] the result is extended to full (predicative) **CL**. In [10] and [13]—the latter paper is in the most recent format and contains the metatheory—the approach is applied to propositional **CLuN** and next is turned into a procedural criterion for final derivability in **ACLuN1**.[12] It can easily be seen that the approach can be generalized to the predicative case (where it really pays), and next can be generalized to any adaptive logic—although further work is clearly required to incorporate other strategies. The advantage of the procedural criterion is that it sets off from a clear goal (the formula one tries to derive), that every step in the prospective proof is demonstrably sensible (in view of the stage of the proof) for reaching this goal, and that, if the procedure stops, which obviously cannot be warranted at the predicative level, one knows whether

---

[12]A computer program that implements the procedure can be downloaded from `http://logica.ugent.be/centrum/writings/programs.php`.

the goal is or is not derivable from the premises. If desired, the prospective proof can be algorithmically turned into a standard adaptive logic proof.

7   Moving towards Consistency

We saved a philosophical problem for the last section. Suppose that a theory was intended as consistent but turns out to be inconsistent, which is the preferred application context of inconsistency-adaptive logics. The adaptive logic locates the inconsistencies and isolates them (prevents them from spreading). As we have seen, the **CL**-consequences that cannot be drawn from the theory (or premise set) will be those that are prevented by the minimal disjunctions of abnormalities that are derivable from the premises by the lower limit logic. Suppose that the following formula is a minimal disjunction:

$$(p \wedge \sim p) \vee (q \wedge \sim q) \vee (r \wedge \sim r).$$

So the theory states that either p or q or r behaves inconsistently, but fails to specify which of the three does.

One clearly will like to go beyond isolating the inconsistencies: one will want to obtain a consistent replacement for the inconsistent theory, provided the latter is an improvement of the former. Given this qualification the task is not one for logic—it may require gathering new evidence, forging a new conceptual system, etc. However, the task will require guidance by logic.

In agreement with Wiśniewski's erotetic logic, the minimal *Dab*-consequence of the premises will evoke the question $?\{p \wedge \sim p, q \wedge \sim q, r \wedge \sim r\}$. Depending on the epistemic situation, it may be possible to obtain a (full or partial) answer to this question. Nearly every answer will free one of the three formulas from the suspicion of inconsistency. As an effect, more **CL**-consequences of the premise set will be finally derivable and the remaining problems are better located. The explication of such processes proceeds in terms of the method from [38].

In some cases, no answer to the aforementioned question can be obtained. Even then, a researcher may discover reasons to narrow down the suspicion. For example, q may be well entrenched in other theories that are considered unproblematic, whence the researcher may decide to consider q as behaving consistently. Or the researcher might have reasons to suspect p, and decide that the inconsistency lies there. Clearly such decisions should themselves be defeasible. New information may put the blame elsewhere.

These problems are considered in [15] and the required adaptive logics are spelled out and shown to be adequate. Actually two approaches turn out sensible for expressing suspicion or freedom of suspicion in a defeasible way, one relying on the modal approach mentioned at the end of Section 3, the other in terms of a hierarchy of negations of inconsistencies.

There are many open problems, even for logic, concerning the road from inconsistent theories to consistent replacements for them (in cases where there are such replacements). Still, the previous results clarify a further bit of that road. We'll get there, step by step.[13]

BIBLIOGRAPHY

[1] Arnon Avron and Iddo Lev. A formula-preferential base for paraconsistent and plausible non-monotonic reasoning. In *Proceedings of the Workshop on Inconsistency in Data and Knowledge (KRR-4) Int. Joint Conf. on AI (Ijcai 2001)*, pages 60–70, 2001.

[2] Diderik Batens. Dialectical dynamics within formal logics. *Logique et Analyse*, 114:161–173, 1986.

[3] Diderik Batens. Dynamic dialectical logics. In Graham Priest, Richard Routley, and Jean Norman, editors, *Paraconsistent Logic. Essays on the Inconsistent*, pages 187–217. Philosophia Verlag, München, 1989.

[4] Diderik Batens. Blocks. The clue to dynamic aspects of logic. *Logique et Analyse*, 150–152:285–328, 1995. Appeared 1997.

[5] Diderik Batens. Inconsistency-adaptive logics. In Orłowska [52], pages 445–472.

[6] Diderik Batens. Zero logic adding up to classical logic. *Logical Studies*, 2:15, 1999. (Electronic Journal: `http://www.logic.ru/LogStud/02/LS2.html`).

[7] Diderik Batens. Towards the unification of inconsistency handling mechanisms. *Logic and Logical Philosophy*, 8:5–31, 2000. Appeared 2002.

[8] Diderik Batens. A general characterization of adaptive logics. *Logique et Analyse*, 173–175:45–68, 2001. Appeared 2003.

[9] Diderik Batens. In defence of a programme for handling inconsistencies. In Meheus [42], pages 129–150.

[10] Diderik Batens. On a partial decision method for dynamic proofs. In Hendrik Decker, Jørgen Villadsen, and Toshiharu Waragai, editors, *PCL 2002. Paraconsistent Computational Logic*, pages 91–108. (= *Datalogiske Skrifter* vol. 95), 2002. Also available as cs.LO/0207090 at `http://arxiv.org/archive/cs/intro.html`.

[11] Diderik Batens. A strengthening of the Rescher–Manor consequence relations. *Logique et Analyse*, 183–184:289–313, 2003. Appeared 2005.

[12] Diderik Batens. The need for adaptive logics in epistemology. In Dov Gabbay, S. Rahman, J. Symons, and J. P. Van Bendegem, editors, *Logic, Epistemology and the Unity of Science*, pages 459–485. Dordrecht, Kluwer, 2004.

[13] Diderik Batens. A procedural criterion for final derivability in inconsistency-adaptive logics. *Journal of Applied Logic*, 3:221–250, 2005.

[14] Diderik Batens. The theory of the process of explanation generalized to include the inconsistent case. *Synthese*, 143:36–88, 2005.

[15] Diderik Batens. Narrowing down suspicion in inconsistent premise sets. In Jacek Malinowski and Andrzej Pietruszczak, editors, *Essays in Logic and Ontology: dedicated to Jerzy Perzanowski*, Poznań Studies in the Philosophy of the Sciences and the Humanities. Rodopi, Amsterdam, in print.

[16] Diderik Batens. A universal logic approach to adaptive logics. In Jean-Yves Béziau, editor, *Directions in Universal Logic*. Birkhauser, To appear.

[17] Diderik Batens and Kristof De Clercq. A rich paraconsistent extension of full positive logic. *Logique et Analyse*, 185–188:227–257, 2004. Appeared 2005.

---

[13]Unpublished papers in the reference section (and many others) are available from the internet address `http://logica.ugent.be/centrum/writings/`.

[18] Diderik Batens and Lieven Haesaert. On classical adaptive logics of induction. *Logique et Analyse*, 173–175:255–290, 2001. Appeared 2003.

[19] Diderik Batens and Joke Meheus. The adaptive logic of compatibility. *Studia Logica*, 66:327–348, 2000.

[20] Diderik Batens and Joke Meheus. A tableau method for inconsistency-adaptive logics. In Roy Dyckhoff, editor, *Automated Reasoning with Analytic Tableaux and Related Methods*, volume 1847 of *Lecture Notes in Artificial Intelligence*, pages 127–142. Springer, 2000.

[21] Diderik Batens and Joke Meheus. Shortcuts and dynamic marking in the tableau method for adaptive logics. *Studia Logica*, 69:221–248, 2001.

[22] Diderik Batens, Joke Meheus, and Dagmar Provijn. An adaptive characterization of signed systems for paraconsistent reasoning. To appear.

[23] Diderik Batens, Joke Meheus, Dagmar Provijn, and Liza Verhoeven. Some adaptive logics for diagnosis. *Logic and Logical Philosophy*, 11/12:39–65, 2003.

[24] Diderik Batens and Dagmar Provijn. Pushing the search paths in the proofs. A study in proof heuristics. *Logique et Analyse*, 173–175:113–134, 2001. Appeared 2003.

[25] Diderik Batens and Timothy Vermeir. Direct dynamic proofs for the Rescher–Manor consequence relations: The flat case. *Journal of Applied Non-Classical Logics*, 12:63–84, 2002.

[26] Salem Benferhat, Didier Dubois, and Henri Prade. Some syntactic approaches to the handling of inconsistent knowledge bases: A comparative study. Part 1: The flat case. *Studia Logica*, 58:17–45, 1997.

[27] Salem Benferhat, Didier Dubois, and Henri Prade. Some syntactic approaches to the handling of inconsistent knowledge bases: A comparative study. Part 2: The prioritized case. In Orłowska [52], pages 473–511.

[28] Philippe Besnard and Torsten Schaub. Signed systems for paraconsistent reasoning. *Journal of Automated Reasoning*, 20:191–213, 1998.

[29] George S. Boolos, John P. Burgess, and Richard J. Jeffrey. *Computability and Logic*. Cambridge University Press, 2002. (Fourth edition).

[30] George S. Boolos and Richard J. Jeffrey. *Computability and Logic*. Cambridge University Press, 1989. (Third edition).

[31] Bryson Brown. How to be realistic about inconsistency in science. *Studies in History and Philosophy of Science*, 21:281–294, 1990.

[32] Kristof De Clercq. Two new strategies for inconsistency-adaptive logics. *Logic and Logical Philosophy*, 8:65–80, 2000. Appeared 2002.

[33] Kristof De Clercq. *Logica in communicatie*, volume 14 of *Cahiers du Centre de Logique*. Academia-Bruylant, Louvain-la-Neuve, 2005.

[34] Jaakko Hintikka. *Inquiry as Inquiry: A Logic of Scientific Discovery*. Kluwer, Dordrecht, 1999.

[35] Iddo Lev. Preferential systems for plausible non-classical reasoning. Master's thesis, Department of Computer Science, Tel-Aviv University, 2000. Unpublished M.A. dissertation.

[36] Joke Meheus. Adaptive logic in scientific discovery: the case of Clausius. *Logique et Analyse*, 143–144:359–389, 1993. Appeared 1996.

[37] Joke Meheus. Clausius' discovery of the first two laws of thermodynamics. A paradigm of reasoning from inconsistencies. *Philosophica*, 63:89–117, 1999. Appeared 2001.

[38] Joke Meheus. Erotetic arguments from inconsistent premises. *Logique et Analyse*, 165–166:49–80, 1999. Appeared 2002.

[39] Joke Meheus. An extremely rich paraconsistent logic and the adaptive logic based on it. In Diderik Batens, Chris Mortensen, Graham Priest, and Jean Paul Van Bendegem, editors, *Frontiers of Paraconsistent Logic*, pages 189–201. Research Studies Press, Baldock, UK, 2000.

[40] Joke Meheus. Adaptive logics for question evocation. *Logique et Analyse*, 173–175:135–164, 2001. Appeared 2003.

[41] Joke Meheus. Inconsistencies in scientific discovery. Clausius's remarkable derivation of Carnot's theorem. In Helge Krach, Geert Vanpaemel, and Pierre Marage, editors, *History of Modern Physics. Acta of the XXth International Congress of History of Science*, pages 143–154. Brepols, Turnhout (Belgium), 2002.

[42] Joke Meheus, editor. *Inconsistency in Science*. Kluwer, Dordrecht, 2002.

[43] Joke Meheus. Paraconsistent compatibility. *Logique et Analyse*, 183–184:251–287, 2003. Appeared 2005.

[44] Joke Meheus. An adaptive logic based on Jaśkowski's approach to paraconsistency. *Journal of Philosophical Logic*, in print.

[45] Joke Meheus and Dagmar Provijn. Direct dynamic proofs for classical compatibility. *Logique et Analyse*, 185–188:305–317, 2004. Appeared 2005.

[46] Joke Meheus, Liza Verhoeven, Maarten Van Dyck, and Dagmar Provijn. Ampliative adaptive logics and the foundation of logic-based approaches to abduction. In Lorenzo Magnani, Nancy J. Nersessian, and Claudio Pizzi, editors, *Logical and Computational Aspects of Model-Based Reasoning*, pages 39–71. Kluwer, Dordrecht, 2002.

[47] Marek Nasieniewski. *Logiki adaptujące sprzeczność (Inconsistency adapting logics)*. Phd thesis, Chair of Logic, N. Copernicus University, Toruń, Poland, 2002.

[48] Marek Nasieniewski. An adaptive logic based on Jaśkowski's logic D_2. *Logique et Analyse*, 185–188:287–304, 2004. Appeared 2005.

[49] Nancy Nersessian. Inconsistency, generic modeling, and conceptual change in science. In Meheus [42], pages 197–211.

[50] John Norton. The logical inconsistency of the old quantum theory of black body radiation. *Philosophy of Science*, 54:327–350, 1987.

[51] John Norton. A paradox in Newtonian gravitation theory. *PSA 1992*, 2:421–420, 1993.

[52] Ewa Orłowska, editor. *Logic at Work. Essays Dedicated to the Memory of Helena Rasiowa*. Physica Verlag (Springer), Heidelberg, New York, 1999.

[53] Graham Priest. *In Contradiction. A Study of the Transconsistent*. Nijhoff, Dordrecht, 1987.

[54] Graham Priest. Minimally inconsistent **LP**. *Studia Logica*, 50:321–331, 1991.

[55] Graham Priest. Paraconsistent logic. In D. Gabbay and F. Guenthner, editors, *Handbook of Philosophical Logic*, volume 2. Kluwer, Dordrecht, 2 edition, in print.

[56] Dagmar Provijn. *Prospectieve dynamiek. Filosofische en technische onderbouwing van doelgerichte bewijzen en bewijsheuristieken*. PhD thesis, Universiteit Gent (Belgium), 2005. Unpublished PhD thesis.

[57] Nicholas Rescher and Ruth Manor. On inference from inconsistent premises. *Theory and Decision*, 1:179–217, 1970.

[58] Kurt Schütte. *Beweistheorie*. Springer, Berlin, 1960.

[59] Joel Smith. Inconsistency and scientific reasoning. *Studies in History and Philosophy of Science*, 19:429–445, 1988.

[60] Roman Suszko. The Fregean axiom and Polish mathematical logic in the 1920s. *Studia Logica*, 36:377–380, 1977.

[61] Guido Vanackere. Ambiguity-adaptive logic. *Logique et Analyse*, 159:261–280, 1997. Appeared 1999.

[62] Liza Verhoeven. All premises are equal, but some are more equal than others. *Logique et Analyse*, 173–174–175:165–188, 2001. Appeared 2003.

[63] Liza Verhoeven. Changing one's position in discussions. Some adaptive approaches. *Logic and Logical Philosophy*, 11–12:277–197, 2003.

[64] Liza Verhoeven. Proof theories for some prioritized consequence relations. *Logique et Analyse*, 183–184:325–344, 2003. appeared 2005.

[65] Andrzej Wiśniewski. *The Posing of Questions. Logical Foundations of Erotetic Inferences*. Kluwer, Dordrecht, 1995.

[66] Andrzej Wiśniewski. The logic of questions as a theory of erotetic arguments. *Synthese*, 109:1–25, 1996.

Diderik Batens and Joke Meheus
Centre for Logic and Philosophy of Science, Universiteit Gent, Blandijnberg
2, 9000 Gent, Belgium.
{Diderik.Batens, Joke.Meheus}@UGent.be

Part II
Systems of Paraconsistent Logic

Semi-Paraconsistent Deontic Logic[1]

CASEY MCGINNIS

ABSTRACT. This paper explores the question of what one's deontic logic ought to look like if one believes that normative conflicts (situations in which something is obligatory and forbidden) are logically possible, but does not believe that dialetheias (true contradictions) are. After dismissing the standard, (fully) paraconsistent, and "non-aggregative" approaches to deontic logic as inadequate under the present assumption, I propose a "semi-paraconsistent" approach which results in a deontic logic that is paraconsistent within the scope of the deontic operators (and thus conflict-tolerant) but fully classical otherwise (and thus explosive). I highlight some interesting features of this logic, and address the question of whether it is philosophically well-motivated.

1 Starting Point

The starting point for this paper is the assumption that normative conflicts[1] (situations in which something is both obligatory and forbidden) are logically possible, but dialetheias (situations in which something is both true and false) are not. I advance this assumption not as a rigid doctrine, but rather as a somewhat plausible thesis whose logical and philosophical implications may be interesting to investigate. Why might one find this thesis plausible? Because one might hold that what is obligatory and forbidden can, at least in some contexts, be determined by *convention*; and, as everyone knows, conventions can conflict. However, while we may be able to bring it about that something is both obligatory and forbidden, we cannot bring it about that something is both true and false.

[1]This work was supported in part by a University of Minnesota Graduate School Doctoral Dissertation Fellowship (2002-2003). An early version of the paper was presented at the 2002 meeting of the Australasian Association for Logic in Canberra, Australia (see [23]). I would like to thank Katalin Bimbó, Manuel Bremer, William Hanson, Geoffrey Hellman, Graham Priest, Jack Stecher, two anonymous referees, and especially Lou Goble for helpful comments. Any mistakes or other infelicities are mine alone.
[1]What I am calling 'normative conflicts' are also variously known as 'quandaries', '(deontic) dilemmas', 'conflicts of obligation', etc.

Given this starting point, what should our deontic[2] logic look like? Most importantly we want it to reject *deontic explosion* $(\mathcal{O}\varphi, \mathcal{O}\neg\varphi/\mathcal{O}\psi)$[3]—an inference that is valid in every extension of K, and *a fortiori* in every standard deontic logic.[4] We want to allow for *non-catastrophic* normative conflicts—situations in which $\mathcal{O}\varphi$ and $\mathcal{O}\neg\varphi$ hold for *some* φ without holding for *all* φ. This is easily achieved by constructing our deontic logic on the basis of a paraconsistent logic, i.e. one that rejects *explosion* $(\varphi, \neg\varphi/\psi)$. It is interesting to note that G. H. von Wright, the founder of modern deontic logic,[5] has come titillatingly close to endorsing this approach:

> Most systems of deontic logic are built within a classical frame. But there are alternative ways of building them. Within, for example, a paraconsistent deontic logic the notion of doability is seen under a different aspect. And seeing it thus may actually answer to a way in which we sometimes see it in 'real life'. An order of the form $O(p \ \& \sim p)$ is thus not necessarily and unconditionally spurious. [36, p. 386]

To my knowledge, von Wright never explored this approach to deontic logic in any detail. Other logicians have.[6] There are significant costs associated with "going paraconsistent," however. It is well known that anyone who rejects explosion and accepts transitivity of entailment is committed to rejecting either disjunctive addition or disjunctive syllogism, as shown by the following Mates-style [22] derivation:

$\{1\}$	1.	φ	assumption
$\{2\}$	2.	$\neg\varphi$	assumption
$\{1\}$	3.	$\varphi \vee \psi$	from $\{1\}$ by disjunctive addition
$\{1,2\}$	4.	ψ	from $\{2,3\}$ by disjunctive syllogism

Thus, given that we do not believe in the logical possibility of dialetheias (and are not particularly concerned with "relevance"), the straightforward paraconsistent approach will not do.

---

[2]The term 'deontic' is derived from the Greek δεοντως ('as it should be' or 'of that which is binding').

[3]$\mathcal{O}\varphi$ can be read as 'it is obligatory that φ' or 'it ought to be (the case) that φ'. There is undoubtedly a subtle difference between these two locutions (whether semantic or merely pragmatic, I will not venture to say). For simplicity's sake, however, I will treat them as interchangeable.

[4]The widely-known system K is defined in, e.g., [4], [9], [19], [26] It is easy to see that deontic explosion is valid in K, and thus in every extension thereof, if one observes that, in a K model, $\mathcal{O}\varphi$ and $\mathcal{O}\neg\varphi$ can be jointly true only at "dead-end" worlds, i.e. worlds that do not "see" any worlds and thus render $\mathcal{O}\psi$ true for all ψ.

By 'standard deontic logic' I mean any normal modal logic that validates either $\mathcal{O}\varphi \rightarrow \mathcal{P}\varphi$ or $\mathcal{O}(\mathcal{O}\varphi \rightarrow \varphi)$, and invalidates $\mathcal{O}\varphi \rightarrow \varphi$. This includes, e.g., the systems D, DM, DB, DS4, and DS5 (in the nomenclature of [17]).

[5]See his seminal [35].

[6]See e.g. [1], [3], [10], [14], [16], [21], [24], [25], [30], [32].

2   The Non-Aggregative Approach

There is at least one approach in the literature—the "non-aggregative" approach developed by van Fraassen [34] and others[7]—which, at least superficially, satisfies our desiderata. In particular, non-aggregative deontic logics reject deontic explosion while retaining all of classical propositional logic (including, of course, disjunctive addition, disjunctive syllogism, and explosion). These systems are called 'non-aggregative' because they invalidate deontic aggregation: $\mathcal{O}\varphi, \mathcal{O}\psi / \mathcal{O}(\varphi \wedge \psi)$. For example, they deem inferences like the following invalid:

> It ought to be that you are honest.
> It ought to be that you are kind.
> ∴ It ought to be that: you are honest and you are kind.

One way of constructing a semantics for a non-aggregative deontic logic—not the only way[8]—is to replace the relation $\mathcal{R}$ of standard deontic logic with a *set* of deontic accessibility relations $\Re = \{\mathcal{R}_1, \mathcal{R}_2, \cdots\}$. (We can think of the elements of $\Re$ as representing various binding normative systems.) The truth condition for obligation is then formulated as follows: $\mathcal{O}\varphi$ holds at a world w iff there is an $\mathcal{R} \in \Re$ such that for every world w', if $w\mathcal{R}w'$ then φ holds at w'. (That is, φ is obligatory just in case some binding normative system prescribes it.) This construction invalidates deontic explosion, as the following counterexample shows:

$$\overset{\curvearrowright 0}{\boxed{w_0}} \longrightarrow_1 \boxed{w_1}$$
$$\begin{array}{cc} p & \neg p \\ \neg q & \neg q \end{array}$$

The arrows represent the elements of $\Re$, our deontic accessibility relations:

$$\boxed{w} \longrightarrow_n \boxed{w'} \quad \text{means that} \quad w\mathcal{R}_n w'.$$

$$\overset{\curvearrowright n}{\boxed{w}} \quad \text{means that} \quad w\mathcal{R}_n w.$$

A formula φ beneath a world w indicates that φ is true at w. It is routine to verify that $\mathcal{O}p$ and $\mathcal{O}\neg p$ are true at w_0, while $\mathcal{O}q$ is not. "Regular" explosion remains valid in this construction, since φ and $\neg\varphi$ can never be jointly true at a world.

---

[7]See e.g. [15], [31].

[8]Another common approach involves the use of "preference-based models," as in e.g. [15], [20], [33].

The main problem with the non-aggregative deontic approach—aside from the fact that it lives up to its name![9]—is that, while it invalidates deontic explosion as desired, it nevertheless validates what I take to be an equally undesirable schema, namely, *conjunctive* deontic explosion: $\mathcal{O}(\varphi \wedge \neg\varphi)/\mathcal{O}\psi$. I see no reason to rule out the possibility of obligations that are logically impossible to fulfill.[10] Thus the non-aggregative approach will not do.

3   The Semi-Paraconsistent Approach

Let us return to the straightforward paraconsistent approach to deontic logic. As I have suggested, the main—really the only—downside to this approach is that it forces us to throw out the baby (classical propositional logic) with the bathwater (deontic explosion). What we would like is for our deontic logic to be paraconsistent within, but fully classical without, the scope of the deontic operators—"semi-paraconsistent," as it were.

This can be achieved as follows. First, we designate a "home world," w_0, and stipulate that it is always consistent and complete. (Intuitively, w_0 can be thought of as the *actual* or *real* world—the world we call home.) All other worlds are permitted to be inconsistent and/or incomplete. We then construct our deontic logic on the basis of our favorite paraconsistent propositional logic (say, BN4)[11] in the standard way, except that we define semantic consequence as truth preservation not at every world, but just at the home world, w_0. We will (tentatively) call this deontic logic DX.

Now for some details:

Our object language, $\mathcal{L}$, is the set of atomic formulas $\mathcal{A} = \{p, q, r, p', \cdots\}$ closed under negation ($\neg$), conjunction ($\wedge$), implication ($\rightarrow$), and obligation ($\mathcal{O}$). I will use φ, ψ, and χ to range over elements of $\mathcal{L}$, and Σ to range over subsets of $\mathcal{L}$. Disjunction ($\vee$) and equivalence ($\leftrightarrow$) are defined in the usual ways. $\mathcal{F}\varphi$ ('it is forbidden that φ') and $\mathcal{P}\varphi$ ('it is permissible that φ') are defined as $\mathcal{O}\neg\varphi$ and $\neg\mathcal{F}\varphi$, respectively. I will omit outer parentheses and let $\wedge$ and $\vee$ bind more tightly than $\rightarrow$ and $\leftrightarrow$.

Our set of truth values is $\mathcal{TV} = \wp\{1, 0\} = \{\{1\}, \{0\}, \{1, 0\}, \varnothing\}$. Intuitively, $\{1\}$ represents *exclusively true* (i.e. true and not false), $\{0\}$ represents *exclusively false* (i.e. false and not true), $\{1, 0\}$ represents *both true*

---

[9]I take aggregation to be an intuitively correct principle of deontic logic. Intuitions differ widely on this issue, however.

[10]Suppose, for example, that a congressional body deems the *conjunction* of a long list of (individually consistent) propositions, some of which happen to contradict each other, legally obligatory.

[11]BN4 is presented and discussed in e.g. [5], [6], [28]. I have chosen it here primarily for its simplicity. One can imagine constructing a more sophisticated version of the semi-paraconsistent system presented here on the basis of, e.g., a relevance logic such as Anderson and Belnap's R [2].

and false, and $\varnothing$ represents *neither true nor false*. (Cf. [11].) I will refer to the elements of $\mathcal{CTV} = \{\{1\}, \{0\}\}$ as "classical" truth values, and to $\{1, 0\}$ and $\varnothing$ as "non-classical" truth values. Our *designated* truth values—i.e., the ones that are preserved in a valid inference—are $\{1\}$ and $\{1, 0\}$.

A DX model is a quadruple $\langle \mathcal{W}, w_0, \mathcal{R}, v \rangle$ such that $\mathcal{W}$ is a set (of worlds); $w_0 \in \mathcal{W}$; $\mathcal{R} \subseteq \mathcal{W} \times \mathcal{W}$ such that $(\forall w)(\forall w')(w\mathcal{R}w' \Rightarrow w'\mathcal{R}w')$;[12] and $v : \mathcal{A} \times \mathcal{W} \mapsto \mathcal{TV}$ such that for all $\alpha \in \mathcal{A}$, $v(\alpha, w_0) \in \mathcal{CTV}$. v is extended to $\bar{v} : \mathcal{L} \times \mathcal{W} \mapsto \mathcal{TV}$ as follows. (I will write $\bar{v}_w(\varphi)$ to abbreviate $\bar{v}(\varphi, w)$.) For all $\alpha \in \mathcal{A}$, $\varphi, \psi \in \mathcal{L}$ and $w \in \mathcal{W}$:

$$\bar{v}_w(\alpha) \quad = \quad v(\alpha, w)$$

$$1 \in \bar{v}_w(\neg\varphi) \quad \Leftrightarrow \quad 0 \in \bar{v}_w(\varphi)$$
$$0 \in \bar{v}_w(\neg\varphi) \quad \Leftrightarrow \quad 1 \in \bar{v}_w(\varphi)$$

$$1 \in \bar{v}_w(\varphi \wedge \psi) \quad \Leftrightarrow \quad 1 \in \bar{v}_w(\varphi) \ \& \ 1 \in \bar{v}_w(\psi)$$
$$0 \in \bar{v}_w(\varphi \wedge \psi) \quad \Leftrightarrow \quad 0 \in \bar{v}_w(\varphi) \text{ or } 0 \in \bar{v}_w(\psi)$$

$$1 \in \bar{v}_w(\varphi \rightarrow \psi) \quad \Leftrightarrow \quad (1 \in \bar{v}_w(\varphi) \Rightarrow 1 \in \bar{v}_w(\psi)) \ \& \ (0 \in \bar{v}_w(\psi) \Rightarrow 0 \in \bar{v}_w(\varphi))$$
$$0 \in \bar{v}_w(\varphi \rightarrow \psi) \quad \Leftrightarrow \quad 1 \in \bar{v}_w(\varphi) \ \& \ 0 \in \bar{v}_w(\psi)$$

$$1 \in \bar{v}_w(\mathcal{O}\varphi) \quad \Leftrightarrow \quad (\forall w' \in \mathcal{W})(w\mathcal{R}w' \Rightarrow 1 \in \bar{v}_{w'}(\varphi))$$
$$0 \in \bar{v}_w(\mathcal{O}\varphi) \quad \Leftrightarrow \quad (\exists w' \in \mathcal{W})(w\mathcal{R}w' \ \& \ 0 \in \bar{v}_{w'}(\varphi))$$

φ is a DX-semantic-consequence of Σ (in symbols, $\Sigma \models_{\mathsf{DX}} \varphi$) iff for all DX models $\langle \mathcal{W}, w_0, \mathcal{R}, v \rangle$, if $1 \in \bar{v}_{w_0}(\psi)$ for all $\psi \in \Sigma$, then $1 \in \bar{v}_{w_0}(\varphi)$. (I write $\models_{\mathsf{DX}} \varphi$ to abbreviate $\varnothing \models_{\mathsf{DX}} \varphi$.)

There is a problem with this construction: it guarantees neither consistency nor completeness at the home world. While we have ensured (by stipulation) that no *atomic* formula takes a non-classical truth value at the home world, this guarantee does not extend to all formulas. The clauses for the extensional connectives ($\neg, \wedge, \rightarrow$) are not the problem, since they agree with their classical counterparts with respect to the classical truth values (and thus never generate a non-classical output from purely classical inputs). Clearly the problem is with the clauses for $\mathcal{O}$. We need to ensure that no formula of the form $\mathcal{O}\varphi$ takes a non-classical truth value at the home world. The easiest way to do this is to simply revise the falsity condition for $\mathcal{O}$ as follows:

$$0 \in \bar{v}_w(\mathcal{O}\varphi) \quad \Leftrightarrow \quad 1 \notin \bar{v}_w(\mathcal{O}\varphi)$$

This guarantees that $\mathcal{O}\varphi$ will never receive a non-classical truth value at any world, and *a fortiori* never at the home world. However, this has (at least) two unfortunate side effects. First, it invalidates the (U) schema,

---

[12]This condition is commonly referred to as 'secondary reflexivity' or 'shift reflexivity'.

$\mathcal{O}(\mathcal{O}\varphi \to \varphi)$, which I take to be a fundamental axiom of deontic logic.[13] Second, it validates a special case of deontic explosion, $\mathcal{O}\mathcal{O}\varphi, \mathcal{O}\neg\mathcal{O}\varphi/\mathcal{O}\psi$.[14]

A better approach is to compromise by reformulating $\mathcal{O}$'s falsity condition as follows:

$$0 \in \bar{v}_w(\mathcal{O}\varphi) \quad \Leftrightarrow \quad \begin{array}{l} (w = w_0 \ \& \ 1 \notin \bar{v}_w(\mathcal{O}\varphi)) \text{ or} \\ (w \neq w_0 \ \& \ (\exists w' \in \mathcal{W})(w\mathcal{R}w' \ \& \ 0 \in \bar{v}_{w'}(\varphi))) \end{array}$$

This falsity condition practices a "double standard" of sorts, treating the home world differently from all other worlds. (Note that these differently formulated falsity conditions are actually equivalent in standard deontic logic.) This revision validates $\mathcal{O}(\mathcal{O}\varphi \to \varphi)$ and invalidates $\mathcal{O}\mathcal{O}\varphi, \mathcal{O}\neg\mathcal{O}\varphi/\mathcal{O}\psi$, as desired. Let us call the resulting system SPDL, for "semi-paraconsistent deontic logic."

HOME WORLD LEMMA. The home world is consistent and complete with respect to all formulas. That is, for all SPDL models $\langle \mathcal{W}, w_0, \mathcal{R}, v \rangle$ and all $\varphi \in \mathcal{L}$, $\bar{v}_{w_0}(\varphi) \in \mathcal{CTV}$.

Proof. A simple induction on the length of formulas. ⊠

I now specify a tableau-style proof theory for SPDL. (I assume some familiarity with tableau proof theories for multi-valued and modal logics, as presented in e.g. [4], [26].) The initial list for an inference Σ/φ, where $\Sigma = \{\psi_1, \cdots, \psi_n\}$,[15] is $\mathcal{I}(\Sigma/\varphi) =$

$$\psi_1 \ 0+$$
$$\vdots$$
$$\psi_n \ 0+$$
$$\varphi \ 0-$$

---

[13]Counterexample: $\mathcal{W} = \{w_0, w_1\}$; $\mathcal{R} = \{\langle w_0, w_1 \rangle, \langle w_1, w_1 \rangle\}$; $v(p, w_1) = \{1, 0\}$. On this model $\bar{v}_{w_0}(O(Op \to p)) = \{0\}$.

[14]Indeed, it validates the stronger inference $\mathcal{O}\mathcal{O}\varphi, \mathcal{O}\neg\mathcal{O}\varphi/\psi$, a special case of of *deontic hyperexplosion* ($\mathcal{O}\varphi, \mathcal{O}\neg\varphi/\psi$). It also validates $\mathcal{O}(\mathcal{O}\varphi \vee \neg\mathcal{O}\varphi)$, a special case of deontic excluded middle ($\mathcal{O}(\varphi \vee \neg\varphi)$).

[15]Note that 'initial list' is undefined for inferences with infinite premise sets. We could have defined the *set* of initial lists for Σ/φ as the set of all lists of the form

$$\psi_1 \ 0+$$
$$\vdots$$
$$\psi_n \ 0+$$
$$\varphi \ 0-$$

where $\{\psi_1, \cdots, \psi_n\} \subseteq \Sigma$. This approach introduces complications, however. Thus, since we have no particular interest in inferences with infinite premise sets, we restrict our proof theory to finite premise sets.

SPDL has ten non-branching tableau rules:

$$\frac{[\neg\neg]}{\neg\neg\varphi\ x\pm}{\varphi\ x\pm}$$

$$\frac{[\wedge+]}{\varphi\wedge\psi\ x+}{\begin{array}{c}\varphi\ x+\\ \psi\ x+\end{array}}$$

$$\frac{[\neg\wedge-]}{\neg(\varphi\wedge\psi)\ x-}{\begin{array}{c}\neg\varphi\ x-\\ \neg\psi\ x-\end{array}}$$

$$\frac{[\neg\rightarrow]}{\neg(\varphi\rightarrow\psi)\ x\pm}{\varphi\wedge\neg\psi\ x\pm}$$

$$\frac{[\mathcal{O}+]}{\begin{array}{c}\mathcal{O}\varphi\ x+\\ xry\end{array}}{\varphi\ y+}$$

$$\frac{[\mathcal{O}-]}{\begin{array}{c}\mathcal{O}\varphi\ x-\\ xri\end{array}}{\varphi\ i-}$$

$$\frac{[\neg\mathcal{O}_0]}{\neg\mathcal{O}\varphi\ 0\pm}{\mathcal{O}\varphi\ 0\mp}$$

$$\frac{[\neg\mathcal{O}+]}{\begin{array}{c}\neg\mathcal{O}\varphi\ z+\\ zri\end{array}}{\neg\varphi\ i+}$$

$$\frac{[\neg\mathcal{O}-]}{\begin{array}{c}\neg\mathcal{O}\varphi\ z-\\ zry\end{array}}{\neg\varphi\ y-}$$

$$\frac{[\text{s-r}]}{xry}{yry}$$

And four branching rules:

$$\frac{[\wedge-]}{\varphi\wedge\psi\ x-}{\varphi\ x-\ |\ \psi\ x-}$$

$$\frac{[\neg\wedge+]}{\neg(\varphi\wedge\psi)\ x+}{\neg\varphi\ x+\ |\ \neg\psi\ x+}$$

$$\frac{[\rightarrow+]}{\varphi\rightarrow\psi\ x+}{\begin{array}{c|c|c|c}\varphi\ x- & \varphi\ x- & \psi\ x+ & \psi\ x+\\ \neg\psi\ x- & \neg\varphi\ x+ & \neg\psi\ x- & \neg\varphi\ x+\end{array}}$$

$$\frac{[\rightarrow-]}{\varphi\rightarrow\psi\ x-}{\begin{array}{c|c}\varphi\ x+ & \neg\psi\ x+\\ \psi\ x- & \neg\varphi\ x-\end{array}}$$

x and y are natural numbers already on the branch. z is a *positive* natural number already on the branch. i is a natural number new to the branch. We use $\pm$ and $\mp$ to condense multiple rules. For example, $[\neg\mathcal{O}_0]$ is a condensation of the following two rules:

$$\frac{[\neg\mathcal{O}_0+]}{\neg\mathcal{O}\varphi\ 0+}{\mathcal{O}\varphi\ 0-}\qquad\frac{[\neg\mathcal{O}_0-]}{\neg\mathcal{O}\varphi\ 0-}{\mathcal{O}\varphi\ 0+}$$

A branch is *closed* iff nodes of the forms $[\varphi\ x+$ and $\varphi\ x-]$, $[\varphi\ 0+$ and $\neg\varphi\ 0+]$, or $[\varphi\ 0-$ and $\neg\varphi\ 0-]$ occur on it. (Here x can be any natural number.) The set $\mathcal{T}(\Sigma/\varphi)$ of tableaux for Σ/φ is the smallest superset of $\{\mathcal{I}(\Sigma/\varphi)\}$ such that if $\tau \in \mathcal{T}(\Sigma/\varphi)$ and τ' is the result of applying a tableau rule to an open (i.e. unclosed) branch of τ, then $\tau' \in \mathcal{T}(\Sigma/\varphi)$. A tableau is closed iff all of its branches are closed. A branch of a tableau is *complete* iff no further rules can be applied to it. A tableau is complete iff all of its branches are complete. φ is an SPDL-proof-theoretic-consequence of Σ (in symbols, $\Sigma \vdash_{\mathsf{SPDL}} \varphi$) iff there is a closed tableau for Σ/φ. (I write $\vdash_{\mathsf{SPDL}} \varphi$ to abbreviate $\varnothing \vdash_{\mathsf{SPDL}} \varphi$.)

Each of the following is proved in the Appendix:

SOUNDNESS THEOREM. $\Sigma \vdash_{\mathsf{SPDL}} \varphi$ only if $\Sigma \models_{\mathsf{SPDL}} \varphi$.

COMPLETENESS THEOREM. For finite Σ, if $\Sigma \models_{\mathsf{SPDL}} \varphi$ then $\Sigma \vdash_{\mathsf{SPDL}} \varphi$.

ALL-OR-NOTHING COROLLARY. If there is an open, complete tableau for Σ/φ, then there is no closed tableau for Σ/φ.

I now provide some examples of the proof theory in use. (I will use '∗' to indicate that a branch is closed, and '↑' to indicate that a branch is open and complete.)

EXAMPLE 1. Here is a proof that SPDL is explosive, i.e. $\varphi, \neg\varphi \vdash_{\mathsf{SPDL}} \psi$:[16]

$$
\begin{array}{l}
1.\ \varphi\ 0+ \\
2.\ \neg\varphi\ 0+ \\
3.\ \psi\ 0- \\
\quad *
\end{array}
$$

Explanation. Nodes (1)-(3) constitute the initial list for the inference. The branch can already be closed, since φ 0+ and $\neg\varphi$ 0+ occur on it. The tableau is closed, since its only branch is closed. Thus $\varphi, \neg\varphi \vdash_{\mathsf{SPDL}} \psi$.

EXAMPLE 2. Here is a proof that SPDL validates the (U) schema, i.e. $\vdash_{\mathsf{SPDL}} \mathcal{O}(\mathcal{O}\varphi \to \varphi)$:

$$
\begin{array}{c}
1.\ \mathcal{O}(\mathcal{O}\varphi \to \varphi)\ 0- \\
2.\ 0r1 \\
3.\ \mathcal{O}\varphi \to \varphi\ 1- \\
4.\ 1r1 \\
\swarrow \qquad \searrow \\
5.\ \mathcal{O}\varphi\ 1+ \qquad 6.\ \neg\varphi\ 1+ \\
7.\ \varphi\ 1- \qquad 8.\ \neg\mathcal{O}\varphi\ 1- \\
9.\ \varphi\ 1+ \qquad 10.\ \neg\varphi\ 1- \\
\quad * \qquad\qquad\qquad *
\end{array}
$$

Explanation. Node (1) constitutes the initial list for $\mathcal{O}(\mathcal{O}\varphi \to \varphi)$, which we construe as an inference with no premises: $\varnothing/\mathcal{O}(\mathcal{O}\varphi \to \varphi)$. Applying $[\mathcal{O}-]$ to (1) yields (2) and (3). Applying [s-r] to (2) yields (4). Applying $[\to -]$ to (3) yields (5)-(8). Applying $[\mathcal{O}+]$ to (4) and (5) yields (9). The

---

[16] Note that the node numbering and the '∗' are not official parts of the tableau; they are included for explanatory purposes only.

left branch is closed, as φ 1+ and φ 1− occur on it. Applying $[\neg\mathcal{O}-]$ to (4) and (8) yields (10). The right branch is closed, as $\neg\varphi$ 1+ and $\neg\varphi$ 1− occur on it. Both branches of the tableau are closed, so the tableau is closed. Hence $\vdash_{\text{SPDL}} \mathcal{O}(\mathcal{O}\varphi \to \varphi)$.

EXAMPLE 3. Here is a proof that $\vdash_{\text{SPDL}} \mathcal{O}(\mathcal{O}\varphi \to \mathcal{P}\varphi)$:

$$\mathcal{O}(\mathcal{O}\varphi \to \neg\mathcal{O}\neg\varphi)\ 0-$$
$$0r1$$
$$\mathcal{O}\varphi \to \neg\mathcal{O}\neg\varphi\ 1-$$
$$1r1$$

$\mathcal{O}\varphi$ 1+	$\neg\neg\mathcal{O}\neg\varphi$ 1+
$\neg\mathcal{O}\neg\varphi$ 1−	$\neg\mathcal{O}\varphi$ 1−
φ 1+	$\mathcal{O}\neg\varphi$ 1+
$\neg\neg\varphi$ 1−	$\neg\varphi$ 1+
φ 1−	$\neg\varphi$ 1−
*	*

EXAMPLE 4. Here is a proof that SPDL is not deontically explosive, i.e. $\mathcal{O}\varphi, \mathcal{F}\varphi \nvdash_{\text{SPDL}} \mathcal{O}\psi$:

1. $\mathcal{O}\varphi$ 0+
2. $\mathcal{O}\neg\varphi$ 0+
3. $\mathcal{O}\psi$ 0−
4. 0r1
5. ψ 1−
6. φ 1+
7. $\neg\varphi$ 1+
8. 1r1
↑

Explanation. Nodes (1)-(3) are the initial list. Applying $[\mathcal{O}-]$ to (3) yields (4) and (5). Applying $[\mathcal{O}+]$ to (1) and (4) yields (6). Applying $[\mathcal{O}+]$ to (2) and (4) yields (7). Applying [s-r] to (4) yields (8). No further rules can be applied, and the branch is open. We know then, by the all-or-nothing corollary, that there is no closed tableau for $\mathcal{O}\varphi, \mathcal{F}\varphi/\mathcal{O}\psi$. Thus $\mathcal{O}\varphi, \mathcal{F}\varphi \nvdash_{\text{SPDL}} \mathcal{O}\psi$.

We can read a counterexample off the open branch as follows. First, we uniformly convert all small Greek letters to atomic formulas:

$$\mathcal{O}p\ 0+$$
$$\mathcal{O}\neg p\ 0+$$
$$\mathcal{O}q\ 0-$$
$$0r1$$
$$q\ 1-$$
$$p\ 1+$$
$$\neg p\ 1+$$
$$1r1$$

We now define our SPDL model as follows.

- $\mathcal{W} = \{w_i : i \text{ is a natural number appearing on the branch}\}$;

- $\mathcal{R} = \{\langle w_i, w_j \rangle : irj \text{ appears on the branch}\}$;

- if $\alpha\ i+$ appears on the branch, $1 \in v(\alpha, w_i)$;

- if $\alpha\ i-$ appears on the branch, $1 \notin v(\alpha, w_i)$;

- if $\neg\alpha\ i+$ appears on the branch, $0 \in v(\alpha, w_i)$;

- if $\neg\alpha\ i-$ appears on the branch, $0 \notin v(\alpha, w_i)$.

(Any parameters that remain undetermined may be arbitrarily specified.) When applied to the open branch above, this procedure induces the following SPDL model:

$$
\begin{array}{llllll}
\mathcal{W} & = & \{w_0, w_1\} & v(p, w_1) & = & \{1, 0\} \\
\mathcal{R} & = & \{\langle w_0, w_1 \rangle, \langle w_1, w_1 \rangle\} & 1 & \notin & v(q, w_1)
\end{array}
$$

It is routine to check that on this SPDL model, $\mathcal{O}p$ and $\mathcal{O}\neg p$ are (at least) true at the home world but $\mathcal{O}q$ is not.

EXAMPLE 5. Here is a proof that SPDL invalidates the (D) schema, i.e. $\nvdash_{\mathsf{SPDL}} \mathcal{O}\varphi \to \mathcal{P}\varphi$:

$$\mathcal{O}\varphi \to \neg\mathcal{O}\neg\varphi\ 0-$$

$$\mathcal{O}\varphi\ 0+ \qquad \neg\neg\mathcal{O}\neg\varphi\ 0+$$
$$\neg\mathcal{O}\neg\varphi\ 0- \qquad \neg\mathcal{O}\varphi\ 0-$$

$$\mathcal{O}\neg\varphi\ 0+ \qquad \vdots$$
$$0r1$$
$$1r1$$
$$\varphi\ 1+$$
$$\neg\varphi\ 1+$$
$$\uparrow$$

The SPDL model specified above as a counterexample to deontic explosion works in this case as well.

4 Some Noteworthy Features of SPDL

The first thing I observe is that, unlike standard and non-aggregative deontic logics, SPDL invalidates both deontic explosion $(\mathcal{O}\varphi, \mathcal{F}\varphi/\mathcal{O}\psi)$ and conjunctive deontic explosion $(\mathcal{O}(\varphi \wedge \neg\varphi)/\mathcal{O}\psi)$. For related reasons, SPDL invalidates deontic disjunctive syllogism $(\mathcal{O}(\varphi \vee \psi), \mathcal{F}\varphi/\mathcal{O}\psi)$ and "Prior's paradox," $\mathcal{F}\varphi \rightarrow \mathcal{O}(\varphi \rightarrow \psi)$ [27]. I have already explained why deontic explosion and conjunctive deontic explosion are undesirable. If deontic explosion is bad, then so is deontic disjunctive syllogism; for since (*contra* those who find "Ross's paradox"[17] troubling) $\mathcal{O}\varphi$ entails $\mathcal{O}(\varphi \vee \psi)$, any situation in which $\mathcal{O}\varphi$ and $\mathcal{F}\varphi$ hold but $\mathcal{O}\psi$ does not is also one in which $\mathcal{O}(\varphi \vee \psi)$ and $\mathcal{F}\varphi$ hold but $\mathcal{O}\psi$ does not. As for Prior's paradox, it seems rather harsh to suppose that 'Bob ought not to jaywalk' entails 'It ought to be that if Bob jaywalks, he is given the death penalty.'[18] (Even the most ardent defenders of capital punishment are likely to cringe at such a cruel and unusual inference.)

The second thing I will point out is that, unlike (fully) paraconsistent deontic logics, SPDL is fully classical at the non-deontic level, in the following sense. Let $\models_{\mathsf{CPL}}$ be the semantic consequence relation on $\mathcal{L}$ that is defined by treating formulas of the form $\mathcal{O}\varphi$ as atomic, and using the usual classical clauses for $\neg$, $\wedge$, and $\rightarrow$.

CLASSICALITY THEOREM. If $\Sigma \models_{\mathsf{CPL}} \varphi$ then $\Sigma \models_{\mathsf{SPDL}} \varphi$.

Proof. Suppose $\Sigma \not\models_{\mathsf{SPDL}} \varphi$. Then there is an SPDL model $\langle \mathcal{W}, w_0, \mathcal{R}, v \rangle$ such that $1 \in \bar{v}_{w_0}(\psi)$ for all $\psi \in \Sigma$ but $1 \notin \bar{v}_{w_0}(\varphi)$. Let $\mathcal{A}_{\mathsf{CPL}} = \mathcal{A} \cup \{\mathcal{O}\varphi : \varphi \in \mathcal{L}\}$. Define a classical valuation $v_{\mathsf{CPL}} : \mathcal{A}_{\mathsf{CPL}} \mapsto \{1, 0\}$ as follows. For all $\theta \in \mathcal{A}_{\mathsf{CPL}}$, let $v_{\mathsf{CPL}}(\theta) = 1$ if $\bar{v}_{w_0}(\theta) = \{1\}$, and let $v_{\mathsf{CPL}}(\theta) = 0$ if $\bar{v}_{w_0}(\theta) = \{0\}$. (The home world lemma assures us that these two cases are exhaustive.) Now extend v_{CPL} to $\bar{v}_{\mathsf{CPL}}$ using the usual classical clauses for $\neg$, $\wedge$, and $\rightarrow$. Since the clauses for $\neg$, $\wedge$, and $\rightarrow$ in SPDL agree with their classical counterparts with respect to the classical truth values, it is easy to show by induction that for all $\chi \in \mathcal{L}$, $\bar{v}_{\mathsf{CPL}}(\chi) = 1$ iff $1 \in \bar{v}_{w_0}(\chi)$. Thus $\bar{v}_{\mathsf{CPL}}(\psi) = 1$ for all $\psi \in \Sigma$ but $\bar{v}_{\mathsf{CPL}}(\varphi) \neq 1$. Thus $\Sigma \not\models_{\mathsf{CPL}} \varphi$. $\boxtimes$

I now highlight some less important, but still noteworthy, features of the system.

---

[17]'Ross's paradox' (so named due to [29]) denotes the alleged counterintuitiveness of the inference from, e.g., 'You ought to mail the letter' to 'You ought to either mail the letter or burn it'. For a nice explanation of why this "paradox" shouldn't bother us, see [8, pp. 63-5].

[18]Cf. [21, p. 5].

It is easy to show that SPDL satisfies each of the following principles discussed in [7]:

Non-Contradiction: $(\exists\Sigma)(\forall\varphi)(\Sigma \not\models \varphi$ or $\Sigma \not\models \neg\varphi)$
Non-Triviality: $(\exists\Sigma)(\exists\varphi)(\Sigma \not\models \varphi)$
(Generalized) Explosion: $(\forall\Sigma)(\forall\varphi)(\forall\psi)(\Sigma, \varphi, \neg\varphi \models \psi)$
Ex Falso Sequitur Quodlibet: $(\exists\varphi)(\forall\Sigma)(\forall\psi)(\Sigma, \varphi \models \psi)$

SPDL satisfies each of the following basic adequacy conditions for a deontic logic:

CONDITION 1.   Any adequate deontic logic must invalidate all of the basic naturalistic-fallacy-like schemas: that is, both $\varphi \to \odot\varphi$ and $\odot\varphi \to \varphi$ must fail for all $\odot \in \{\mathcal{O}, \mathcal{F}, \mathcal{P}\}$.

CONDITION 2.   Any adequate deontic logic must invalidate the (4) schema, $\mathcal{O}\varphi \to \mathcal{O}\mathcal{O}\varphi$, and the (5) schema, $\mathcal{P}\varphi \to \mathcal{O}\mathcal{P}\varphi$. (These schemas are as implausible on a deontic reading as they are plausible on an alethic one.)

CONDITION 3.   Any adequate deontic logic must validate each of the good aggregation-/distribution-like schemas on the left and invalidate each of the bad ones on the right:

$$\mathcal{O}\varphi \vee \mathcal{O}\psi \to \mathcal{O}(\varphi \vee \psi) \qquad \mathcal{O}(\varphi \vee \psi) \to \mathcal{O}\varphi \vee \mathcal{O}\psi$$
$$\mathcal{P}(\varphi \wedge \psi) \to \mathcal{P}\varphi \wedge \mathcal{P}\psi \qquad \mathcal{P}\varphi \wedge \mathcal{P}\psi \to \mathcal{P}(\varphi \wedge \psi)$$
$$\mathcal{F}\varphi \vee \mathcal{F}\psi \to \mathcal{F}(\varphi \wedge \psi) \qquad \mathcal{F}(\varphi \wedge \psi) \to \mathcal{F}\varphi \vee \mathcal{F}\psi$$
$$\mathcal{O}(\varphi \wedge \psi) \leftrightarrow \mathcal{O}\varphi \wedge \mathcal{O}\psi$$
$$\mathcal{P}(\varphi \vee \psi) \leftrightarrow \mathcal{P}\varphi \vee \mathcal{P}\psi$$
$$\mathcal{F}\varphi \wedge \mathcal{F}\psi \leftrightarrow \mathcal{F}(\varphi \vee \psi)$$

CONDITION 4.   Any adequate deontic logic must validate each of the following "(K)-like" schemas:

$$\mathcal{O}(\varphi \to \psi) \to (\mathcal{O}\varphi \to \mathcal{O}\psi)$$
$$\mathcal{O}(\varphi \to \psi) \to (\mathcal{F}\psi \to \mathcal{F}\varphi)$$
$$\mathcal{O}(\varphi \to \psi) \to (\mathcal{P}\varphi \to \mathcal{P}\psi)$$
$$\mathcal{O}(\varphi \to \psi) \wedge \mathcal{O}(\psi \to \chi) \to \mathcal{O}(\varphi \to \chi)$$

CONDITION 5.   Any adequate deontic logic must validate the (U) schema $\mathcal{O}(\mathcal{O}\varphi \to \varphi)$ as well as its fraternal twin, $\mathcal{O}(\varphi \to \mathcal{P}\varphi)$.

I take Condition 1 to be uncontroversial. Conditions 2-4 are not beyond reasonable doubt, but I will not defend them here. Condition 5 deserves comment. The intuition behind $\mathcal{O}(\mathcal{O}\varphi \to \varphi)$ is the apparent truism that obligations ought to be fulfilled: $\mathcal{O}\varphi \to \varphi$ is not in general true; but it *ought* to be, and *would* be in a deontically perfect world.[19] The second schema, $\mathcal{O}(\varphi \to \mathcal{P}\varphi)$, though equivalent to $\mathcal{O}(\mathcal{O}\varphi \to \varphi)$, is less intuitive to some, for it seems to say that whatever one does ought to be permitted.   ("It ought to be that if I kill my mother, I am permitted to do so.") This is a misinterpretation of the schema, however. $\mathcal{O}(\varphi \to \mathcal{P}\varphi)$ says that it ought to be that φ holds only if φ is permissible; i.e., it ought to be that only permissible states of affairs obtain.   Thus, it ought to be that I kill my mother only if I am permitted to do so—which (presumably) I am not!

An important respect in which SPDL differs from standard (and many non-standard) deontic logics is that it invalidates the (D) schema, $\mathcal{O}\varphi \to \mathcal{P}\varphi$, as well as the "no-conflicts" schema, $\neg(\mathcal{O}\varphi \wedge \mathcal{F}\varphi)$.[20] $\mathcal{O}\varphi \to \mathcal{P}\varphi$ and $\neg(\mathcal{O}\varphi \wedge \mathcal{F}\varphi)$ are equivalent in SPDL, as they are in standard deontic logic. Clearly $\neg(\mathcal{O}\varphi \wedge \mathcal{F}\varphi)$ will not be missed. $\mathcal{O}\varphi \to \mathcal{P}\varphi$, on the other hand, is undesirable only if 'it is permissible that' is understood/defined as 'it is not obligatory that not'.   But we need not understand or define the locution this way.   We could, if so inclined, make $\mathcal{P}$ primitive, specifying its truth and falsity conditions as follows:

$$1 \in \bar{v}_w(\mathcal{P}\varphi) \quad \Leftrightarrow \quad (\exists w' \in \mathcal{W})(w\mathcal{R}w' \ \& \ 1 \in \bar{v}_{w'}(\varphi))$$
$$0 \in \bar{v}_w(\mathcal{P}\varphi) \quad \Leftrightarrow \quad (w = w_0 \ \& \ 1 \notin \bar{v}_w(\mathcal{P}\varphi)) \text{ or}$$
$$(w \neq w_0 \ \& \ (\forall w' \in \mathcal{W})(w\mathcal{R}w' \Rightarrow 0 \in \bar{v}_{w'}(\varphi)))$$

When combined with the seriality condition, $(\forall w)(\exists w')w\mathcal{R}w'$, this revision validates the (D) schema, but leaves the "no-conflicts" schema invalid. Not surprisingly, it also invalidates $\mathcal{P}\varphi \leftrightarrow \neg\mathcal{F}\varphi$ in both directions. The biggest problem with this revision is that it raises the difficult question of why $\mathcal{F}\varphi$ should be defined as $\mathcal{O}\neg\varphi$ rather than $\neg\mathcal{P}\varphi$.

SPDL also invalidates each of the following schemas:

$$\mathcal{O}(\varphi \vee \neg\varphi) \qquad \mathcal{F}(\varphi \wedge \neg\varphi)$$
$$\mathcal{P}(\varphi \vee \neg\varphi) \qquad \mathcal{P}\varphi \vee \mathcal{P}\neg\varphi$$
$$\mathcal{O}(\varphi \wedge (\varphi \to \psi) \to \psi)$$
$$\mathcal{O}((\varphi \to (\varphi \to \psi)) \to (\varphi \to \psi))$$

Incidentally, all of these but $\mathcal{P}(\varphi \vee \neg\varphi)$ and $\mathcal{P}\varphi \vee \mathcal{P}\neg\varphi$ can be regained by simply eliminating $\varnothing$ from our collection of truth values, i.e., requiring that all formulas be at least true or false.   (In that case, our underlying paraconsistent logic would be RM3 rather than BN4.)[21]   We can make

---

[19]Cf. [1, p. 356].

[20]As demonstrated in Example 3, SPDL *does* validate the "obligification" of the (D) schema, $\mathcal{O}(\mathcal{O}\varphi \to \mathcal{P}\varphi)$. Even $\mathcal{O}\neg(\mathcal{O}\varphi\wedge\mathcal{F}\varphi)$, the obligification of "no conflicts," is invalid in SPDL, however.

[21]RM3 is presented in, e.g., [4], [5], [6], [26].

$\mathcal{P}(\varphi \vee \neg\varphi)$ and $\mathcal{P}\varphi \vee \mathcal{P}\neg\varphi$ valid by eliminating $\varnothing$ and making $\mathcal{P}$ primitive in the way suggested above.

It is easy to verify that the semantic deduction theorem holds for SPDL:

$$\Sigma \cup \{\varphi\} \models_{\mathsf{SPDL}} \psi \text{ iff } \Sigma \models_{\mathsf{SPDL}} \varphi \rightarrow \psi.$$

That SPDL enjoys the deduction theorem may not seem like a big deal. But it is noteworthy in light of the fact that its underlying paraconsistent logic, BN4, does not. (Counterexample: $p, q \models_{\mathsf{BN4}} p$ but $p \not\models_{\mathsf{BN4}} q \rightarrow p$.)

Unlike standard (and many non-standard) deontic logics, SPDL does not satisfy the rule of "obligification" (the deontic analog of "necessitation"),

(1) $\Sigma \models \varphi \Rightarrow \mathcal{O}\Sigma \models \mathcal{O}\varphi,$

where $\mathcal{O}\Sigma = \{\mathcal{O}\psi : \psi \in \Sigma\}$. (Counterexample: $p, \neg p \models_{\mathsf{SPDL}} q$ but $\mathcal{O}p, \mathcal{O}\neg p \not\models_{\mathsf{SPDL}} \mathcal{O}q$.) The rough idea behind (Obl)—that obligation is closed under entailment—often manifests itself in a slightly different form, as

(2) $\models \varphi \rightarrow \psi \Rightarrow \ \models \mathcal{O}\varphi \rightarrow \mathcal{O}\psi.$

Not surprisingly, (RM) too fails in SPDL. (Counterexample: $\models_{\mathsf{SPDL}} p \wedge \neg p \rightarrow q$ but $\not\models_{\mathsf{SPDL}} \mathcal{O}(p \wedge \neg p) \rightarrow \mathcal{O}q$.) Prima facie, the failure of (Obl) and (RM) in SPDL might seem like a benefit, since many people find these principles counterintuitive. Indeed, principles like (Obl) and (RM) have been fingered as the culprits in some of the nastier paradoxes of deontic logic, such as the "adverbial samaritan."[22] I cannot in good conscience tout the failure of these principles as a happy feature of SPDL, however, since the following variations on (Obl) and (RM) *do* hold:

(3) $\Sigma \models_{\mathsf{BN4}} \varphi \Rightarrow \mathcal{O}\Sigma \models_{\mathsf{SPDL}} \mathcal{O}\varphi$

(4) $\models_{\mathsf{BN4}} \varphi \rightarrow \psi \Rightarrow \ \models_{\mathsf{SPDL}} \mathcal{O}\varphi \rightarrow \mathcal{O}\psi$

If (Obl) and (RM) are counterintuitive, then surely so are (Obl*) and (RM*).

Finally, I note that unlike standard deontic logics, SPDL is not *self-extensional*, i.e., it does not enjoy the replacement theorem,

(5) $\models \varphi \leftrightarrow \varphi' \Rightarrow \ \models \psi \leftrightarrow \psi[\varphi/\varphi'],$

---

[22]See [12], [13].

where $\psi[\varphi/\varphi']$ is the result of replacing one or more occurrences of φ in ψ with φ'. (Counterexample: $\models_{\mathsf{SPDL}} p \wedge \neg p \leftrightarrow q \wedge \neg q$ but $\not\models_{\mathsf{SPDL}} \mathcal{O}(p \wedge \neg p) \leftrightarrow \mathcal{O}(q \wedge \neg q)$.) Of course, SPDL's underlying paraconsistent logic, BN4, does not satisfy (Rep) either.[23] It is quite obvious, however, that even a semi-paraconsistent deontic logic constructed on the basis of a self-extensional paraconsistent logic will not itself be self-extensional. Some might view this as an undesirable feature.

5 Is SPDL Well-Motivated?

A likely objection to SPDL is that certain features of its semantics lack an intuitive or philosophical foundation. In particular, the following features may strike one as *ad hoc* or undermotivated:

- the stipulation that only the home world must be consistent and complete.

- the "double standard" inherent in the falsity condition for $\mathcal{O}$.

I now attempt to address these concerns.

Let's think about the intuitive significance of worlds in *alethic* modal logic (in particular, S5). $\Diamond\varphi$ ('it is possible that φ') is true iff φ is true at some some world' and $\Box\varphi$ ('it is necessary that φ') is true iff φ is true at every world. Here, worlds represent ways that the actual world might have been. The actual world is necessarily consistent and complete. Thus it makes sense to build into alethic modal logic the assumption that each world is consistent and complete, no matter how much it may otherwise differ from the actual world.

Contrast this with deontic logic. In deontic logic, the worlds in a model have an entirely different import. They represent not ways that the actual world might have been, but rather ways that a normative system could dictate that the actual world *ought to be*—or, what is ultimately equivalent, ways that a normative system could dictate that deontically perfect worlds *are*. Now, a normative system can be silent (i.e. incomplete) with respect to some matters and inconsistent with respect to others. Legal theorists, for example, often speak of gaps and conflicts (gluts) in the law; ethicists often speak of moral dilemmas. Thus it would be a mistake to build into our deontic logic the assumption that worlds are, *in general*, consistent and complete. It still makes sense, however, to assume that the *home world* is consistent and complete, since the home world intuitively represents the way things actually are.

---

[23]Counterexample: $\vdash_{\mathsf{BN4}} (p \rightarrow p) \leftrightarrow (q \rightarrow q)$ but $\not\vdash_{\mathsf{BN4}} \neg(p \rightarrow p) \leftrightarrow \neg(q \rightarrow q)$.

One may argue that if we can stipulate what is obligatory, we must also be able to stipulate what is *not* obligatory. (Indeed, if we can stipulate what is *permissible*, and 'permissible' just means 'not obligatory that not', it follows that we can stipulate what is *not obligatory*.) And if we can stipulate what is *not* obligatory, then we ought to allow for the possibility not just of normative conflicts, but of full-blown deontic dialetheias (of the form $\mathcal{O}\varphi \wedge \neg\mathcal{O}\varphi$) at the home world. We may still hold that *non*-deontic formulas (i.e. ones containing no deontic operators) are always consistent at the home world. If we limit our dialethism to deontic formulas in the way being suggested, we can preserve, in *qualified* forms, certain inferences that would otherwise be categorically invalid. For example, despite the fact that DX is paraconsistent we still have $\varphi, \neg\varphi \models_{DX} \psi$ and $\varphi \vee \psi, \neg\varphi \models_{DX} \psi$, *where φ is non-deontic.*

Let me respond to this argument with an analogy. Making it obligatory that φ is like placing φ into an "obligation box"—a big box of propositions that magically closes its contents under (some variety of) logical consequence. A proposition φ is obligatory just in case it is in the box. Clearly it is possible for both φ and $\neg\varphi$ to be in the box. (In the simplest case, we just put them directly in; in other cases, they appear in more complicated ways as a result of the box's magical closure property.) However, it is not possible for φ to be both in the box and not in the box. Thus it is not possible for φ to be both obligatory and non-obligatory.

This is all quite fanciful, of course, but I believe it might point the way to a more detailed philosophical defense of this paper's starting point.

As for the "double standard" inherent in the falsity condition for $\mathcal{O}$, this is motivated by the idea that negation necessarily behaves in a "Boolean"[24] manner at the home world, as the home world is necessarily consistent and complete. We could build the double standard into all of our semantic clauses, and it wouldn't make a difference. For example, we could reformulate the clauses for negation as:

$$1 \in \bar{v}_w(\neg\varphi) \quad \Leftrightarrow \quad 0 \in \bar{v}_w(\varphi)$$
$$0 \in \bar{v}_w(\neg\varphi) \quad \Leftrightarrow \quad (w = w_0 \ \& \ 1 \notin \bar{v}_w(\neg\varphi)) \text{ or}$$
$$(w \neq w_0 \ \& \ 1 \in \bar{v}_w(\varphi))$$

The only reason we don't do this is because the two conditions ($1 \notin \bar{v}_w(\neg\varphi)$ and $1 \in \bar{v}_w(\varphi)$) *collapse* at the home world (as $1 \notin \bar{v}_{w_0}(\neg\varphi) \Leftrightarrow 1 \in \bar{v}_{w_0}(\varphi)$), and so it is unnecessary to make the double standard explicit. The two conditions do not collapse in the clause for $\mathcal{O}$, however, since this clause "reaches out," as it were, to other worlds—worlds which, unlike the home

---

[24]A negation operator $\sim$ is "Boolean" iff it satisfies the following condition: $\sim\varphi$ is true (at a world) iff φ is not true (at that world).

world, are not necessarily consistent or complete. Thus, we have to make the double standard explicit in this case.

At this point one might object that we could just as well define the clauses for $\mathcal{O}$ as follows:

$$1 \in \bar{v}_w(\mathcal{O}\varphi) \quad \Leftrightarrow \quad (w = w_0 \ \& \ 0 \notin \bar{v}_w(\mathcal{O}\varphi)) \text{ or}$$
$$(w \neq w_0 \ \& \ (\forall w' \in \mathcal{W})(w\mathcal{R}w' \Rightarrow 1 \in \bar{v}_{w'}(\varphi)))$$
$$0 \in \bar{v}_w(\mathcal{O}\varphi) \quad \Leftrightarrow \quad (\exists w' \in \mathcal{W})(w\mathcal{R}w' \ \& \ 0 \in \bar{v}_{w'}(\varphi))$$

What possible justification could there be for having $\mathcal{O}$'s *falsity* condition, rather than its *truth* condition, practice the "double standard"? The decision seems completely arbitrary. And observe what happens if, for symmetry's sake, we make *both* conditions observe the double standard:

$$1 \in \bar{v}_w(\mathcal{O}\varphi) \quad \Leftrightarrow \quad (w = w_0 \ \& \ 0 \notin \bar{v}_w(\mathcal{O}\varphi)) \text{ or}$$
$$(w \neq w_0 \ \& \ (\forall w' \in \mathcal{W})(w\mathcal{R}w' \Rightarrow 1 \in \bar{v}_{w'}(\varphi)))$$
$$0 \in \bar{v}_w(\mathcal{O}\varphi) \quad \Leftrightarrow \quad (w = w_0 \ \& \ 1 \notin \bar{v}_w(\mathcal{O}\varphi)) \text{ or}$$
$$(w \neq w_0 \ \& \ (\exists w' \in \mathcal{W})(w\mathcal{R}w' \ \& \ 0 \in \bar{v}_{w'}(\varphi)))$$

Now each formula of the form $\mathcal{O}\varphi$ takes a *completely arbitrary* (albeit classical) truth value at the home world. This is equivalent to treating formulas of the form $\mathcal{O}\varphi$ as *atomic* (as in the proof of the classicality theorem), and essentially reduces SPDL to a mere "Rube Goldberg" version of classical propositional logic!

I don't have a powerful or convincing response to this last objection. The best I can do is to simply concede that this feature of SPDL is *ad hoc*, but at the same time point out that nearly *every* logical system has at least some *ad hoc* features. Why, for example, does standard first-order logic validate $(\exists x)x = x$ but not $(\exists x)(\exists y)x \neq y$? Why should it be a logical truth that at least *one* thing exists, but not that at least *two* things exist? I submit that the asymmetry of the clauses for $\mathcal{O}$ in SPDL is no more *ad hoc*, and no less defensible, than this curious feature of classical first-order logic.

6   Conclusion

In light of the worries addressed above, I cannot with any confidence conclude that SPDL is the *ideal* system for anyone who agrees with our starting point. However, I hope that I have shown it to be an *interesting* system, and that it may stimulate further thought about possible new directions in deontic and paraconsistent logic.

7   Appendix: Soundness and Completeness

DEFINITION (faithful).   Let b be a branch of a tableau. An SPDL model $\mathcal{M} = \langle \mathcal{W}, w_0, \mathcal{R}, v \rangle$ is *faithful* to b iff there is a function $f : \mathbb{N} \mapsto \mathcal{W}$ such that:

1. $f(i) = w_0$ iff $i = 0$;
2. if irj is on b, then $f(i)\mathcal{R}f(j)$;
3. if $\varphi\ i+$ is on b, then $1 \in \bar{v}_{f(i)}(\varphi)$;
4. if $\varphi\ i-$ is on b, then $1 \notin \bar{v}_{f(i)}(\varphi)$;

We say that f *shows* $\mathcal{M}$ to be faithful to b.

SOUNDNESS LEMMA. Let b be an open branch, and $\mathcal{M}$ be an SPDL model that is faithful to b. If a tableau rule is applied to b, then $\mathcal{M}$ is faithful to at least one of the branches thereby generated.

Proof. Let b be an open branch, and $\mathcal{M}$ an SPDL model that is faithful to b. Suppose that a tableau rule is applied to b. We need to show that $\mathcal{M}$ is faithful to at least one of the branches generated. The proof is by cases. There are fourteen cases to consider (one for each tableau rule). Here I prove only the cases for $[\mathcal{O}+]$, $[\mathcal{O}-]$, $[\neg\mathcal{O}_0]$, $[\neg\mathcal{O}+]$, and $[\neg\mathcal{O}-]$. The remaining cases are relatively straightforward, and are thus omitted.

Notation. Where b is a branch of a tableau, $b(N_1, \cdots, N_n)$ is the branch generated by adding the nodes $N_1, \cdots, N_n$ to the tip of b.

Case 1. $\mathcal{O}\varphi\ x+$ and xry are on b, and $[\mathcal{O}+]$ is applied. Then $b(\varphi\ y+)$ is generated. Since $\mathcal{M}$ is faithful to b, $1 \in \bar{v}_{f(x)}(\mathcal{O}\varphi)$ and $f(x)\mathcal{R}f(y)$. Thus, by the truth condition for $\mathcal{O}$, $(\forall w \in \mathcal{W})(f(x)\mathcal{R}w \Rightarrow 1 \in \bar{v}_w(\varphi))$. Thus, since $f(x)\mathcal{R}f(y)$, $1 \in \bar{v}_{f(y)}(\varphi)$. Thus $\mathcal{M}$ is faithful to $b(\varphi\ y+)$.

Case 2. $\mathcal{O}\varphi\ x-$ is on b, and $[\mathcal{O}-]$ is applied. Then $b(xri, \varphi\ i-)$ is generated. Since $\mathcal{M}$ is faithful to b, $1 \notin \bar{v}_{f(x)}(\mathcal{O}\varphi)$. Thus, by the truth condition for $\mathcal{O}$, it is not the case that $(\forall w \in \mathcal{W})(f(x)\mathcal{R}w \Rightarrow 1 \in \bar{v}_w(\varphi))$. Thus there is an element of $\mathcal{W}$ (call it w) such that $f(x)\mathcal{R}w$ and $1 \notin \bar{v}_w(\varphi)$. Let f' be just like f except that $f'(i) = w$. Since i is new to the branch, f' shows $\mathcal{M}$ to be faithful to b. Moreover, $f'(x)\mathcal{R}f'(i)$ and $1 \notin \bar{v}_{f'(i)}(\varphi)$. Thus, f' shows $\mathcal{M}$ to be faithful to $b(xri, \varphi\ i-)$.

Case 3. $\neg\mathcal{O}\varphi\ 0+$ or $\neg\mathcal{O}\varphi\ 0-$ is on b, and $[\neg\mathcal{O}_0]$ is applied. There are two subcases:

Subcase 3.1. $\neg\mathcal{O}\varphi\ 0+$ is on the branch. Then $b(\mathcal{O}\varphi\ 0-)$ is generated. Since $\mathcal{M}$ is faithful to b, $1 \in \bar{v}_{f(0)}(\neg\mathcal{O}\varphi)$, whence by the truth condition for negation $0 \in \bar{v}_{f(0)}(\mathcal{O}\varphi)$. By the definition of faithful, $f(0)$ is the home world, w_0. Thus, by the home world lemma, $1 \notin \bar{v}_{f(0)}(\mathcal{O}\varphi)$. Thus $\mathcal{M}$ is faithful to $b(\mathcal{O}\varphi\ 0-)$.

Subcase 3.2. $\neg\mathcal{O}\varphi\ 0-$ is on the branch. Then $b(\mathcal{O}\varphi\ 0+)$ is generated. Since $\mathcal{M}$ is faithful to b, $1 \notin \bar{v}_{f(0)}(\neg\mathcal{O}\varphi)$, whence $0 \notin \bar{v}_{f(0)}(\mathcal{O}\varphi)$. By the definition of faithful, $f(0)$ is the home world, w_0. Thus, by the home world lemma, $1 \in \bar{v}_{f(0)}(\mathcal{O}\varphi)$. Thus $\mathcal{M}$ is faithful to $b(\mathcal{O}\varphi\ 0+)$.

In each subcase, $\mathcal{M}$ is faithful to the branch generated.

Case 4. $\neg\mathcal{O}\varphi\ z+$ is on b, and $[\neg\mathcal{O}+]$ is applied. Then $b(zri, \neg\varphi\ i+)$ is generated. Since $\mathcal{M}$ is faithful to b, $1 \in \bar{v}_{f(z)}(\neg\mathcal{O}\varphi)$, whence $0 \in \bar{v}_{f(z)}(\mathcal{O}\varphi)$.

Since $z \neq 0$, by the definition of faithful, $f(z) \neq w_0$. Thus, by the falsity condition for $\mathcal{O}$, there is an element of $\mathcal{W}$ (call it w) such that $f(z)\mathcal{R}w$ and $0 \in \bar{v}_w(\varphi)$. Thus, by the truth condition for $\neg$, $1 \in \bar{v}_w(\neg\varphi)$. Let f' be just like f except that $f'(i) = w$. Since i is new to the branch, f' shows $\mathcal{M}$ to be faithful to b. Moreover, $f'(z)\mathcal{R}f'(i)$ and $1 \in \bar{v}_{f'(i)}(\neg\varphi)$. Thus f' shows $\mathcal{M}$ to be faithful to $b(zri, \neg\varphi \, i+)$.

Case 5. $\neg\mathcal{O}\varphi \, z-$ and zry are on b, and $[\neg\mathcal{O}-]$ is applied. Then $b(\neg\varphi \, y-)$ is generated. Since $\mathcal{M}$ is faithful to b, $1 \notin \bar{v}_{f(z)}(\neg\mathcal{O}\varphi)$, whence $0 \notin \bar{v}_{f(z)}(\mathcal{O}\varphi)$, and $f(z)\mathcal{R}f(y)$. Since $z \neq 0$, by the definition of faithful, $f(z) \neq w_0$. Thus, by the falsity condition for $\mathcal{O}$, it is not the case that $(\exists w \in \mathcal{W})(f(z)\mathcal{R}w$ and $0 \in \bar{v}_w(\varphi))$. Thus $(\forall w \in \mathcal{W})(f(z)\mathcal{R}w \Rightarrow 0 \notin \bar{v}_w(\varphi))$. Thus, since $f(z)\mathcal{R}f(y)$, $0 \notin \bar{v}_{f(y)}(\varphi)$, whence $1 \notin \bar{v}_{f(y)}(\neg\varphi)$. Thus $\mathcal{M}$ is faithful to $b(\neg\varphi \, y-)$.

In each of the above cases, $\mathcal{M}$ is faithful to at least one of the branches generated. $\boxtimes$

SOUNDNESS THEOREM. $\Sigma \vdash_{\mathsf{SPDL}} \varphi$ only if $\Sigma \models_{\mathsf{SPDL}} \varphi$.

Proof. Suppose $\Sigma \not\models_{\mathsf{SPDL}} \varphi$. Then for some SPDL model $\mathcal{M} = \langle \mathcal{W}, w_0, \mathcal{R}, v \rangle$, $1 \in \bar{v}_{w_0}(\psi)$ for all $\psi \in \Sigma$, but $1 \notin \bar{v}_{w_0}(\varphi)$. $\mathcal{M}$ is faithful to $\mathcal{I}(\Sigma/\varphi)$. Moreover, by the soundness lemma, every subsequent application of a tableau rule will yield at least one branch to which $\mathcal{M}$ is faithful. Thus every $\tau \in \mathcal{T}(\Sigma/\varphi)$ has at least one branch to which $\mathcal{M}$ is faithful. Let $\tau \in T(\Sigma/\varphi)$, and let b be a branch of τ to which $\mathcal{M}$ is faithful. Suppose, for reductio, that b is closed. Then there are three cases:

Case 1. Nodes of the forms $\varphi \, x+$ and $\varphi \, x-$ occur on b. Then, since $\mathcal{M}$ is faithful to b, $1 \in \bar{v}_{f(x)}(\varphi)$ and $1 \notin \bar{v}_{f(x)}(\varphi)$. (Contradiction.)

Case 2. Nodes of the forms $\varphi \, 0+$ and $\neg\varphi \, 0+$ occur on b. Then, since $\mathcal{M}$ is faithful to b, $1 \in \bar{v}_{f(0)}(\varphi)$ and $0 \in \bar{v}_{f(0)}(\varphi)$. By the definition of faithful, $f(0)$ is the home world, w_0. Thus $1 \in \bar{v}_{w_0}(\varphi)$ and $0 \in \bar{v}_{w_0}(\varphi)$, contradicting the home world lemma.

Case 3. Nodes of the forms $\varphi \, 0-$ and $\neg\varphi \, 0-$ occur on b. Then, since $\mathcal{M}$ is faithful to b, $1 \notin \bar{v}_{f(0)}(\varphi)$ and $0 \notin \bar{v}_{f(0)}(\varphi)$. By the definition of faithful, $f(0)$ is the home world, w_0. Thus $1 \notin \bar{v}_{w_0}(\varphi)$ and $0 \notin \bar{v}_{w_0}(\varphi)$, contradicting the home world lemma.

Each case leads to a contradiction. Thus, by reductio, b is open. Thus every tableau for Σ/φ has at least open branch. Thus there are no closed tableaux for Σ/φ. Thus $\Sigma \not\vdash_{\mathsf{SPDL}} \varphi$. $\boxtimes$

DEFINITION (induced model). Let b be an open, complete branch of a tableau. An SPDL model *induced* by b is any SPDL model $\mathcal{M} = \langle \mathcal{W}, w_0, \mathcal{R}, v \rangle$ such that:

1. $\mathcal{W} = \{w_i : i$ is a natural number occurring on $b\}$;
2. $\mathcal{R} = \{\langle w_i, w_j \rangle : irj$ occurs on $b\}$;

    3. if α $i+$ is on b, then $1 \in v_{w_i}(\alpha)$;

    4. if α $i-$ is on b, then $1 \notin v_{w_i}(\alpha)$;

    5. if $\neg\alpha$ $i+$ is on b, then $0 \in v_{w_i}(\alpha)$;

    6. if $\neg\alpha$ $i-$ is on b, then $0 \notin v_{w_i}(\alpha)$.

(It is straightforward to verify that every open, complete branch induces at least one SPDL model.)

COMPLETENESS LEMMA. Let b be any open, complete branch of a tableau, and let $\mathcal{M} = \langle \mathcal{W}, w_0, \mathcal{R}, v \rangle$ be an SPDL model induced by b. Then:

    1. if φ $i+$ is on b, then $1 \in v_{w_i}(\varphi)$;

    2. if φ $i-$ is on b, then $1 \notin v_{w_i}(\varphi)$;

    3. if $\neg\varphi$ $i+$ is on b, then $0 \in v_{w_i}(\varphi)$;

    4. if $\neg\varphi$ $i-$ is on b, then $0 \notin v_{w_i}(\varphi)$.

Proof. An induction on the length of φ. If $\varphi \in \mathcal{A}$, the result holds by definition. We now need to show that if the result holds for φ and ψ then it also holds for $\neg\varphi$, $\varphi \wedge \psi$, $\varphi \to \psi$, and $\mathcal{O}\varphi$. Here I prove the result only for $\mathcal{O}\varphi$, leaving the rest of the proof to the reader.

1. Suppose $\mathcal{O}\varphi$ $x+$ is on b. Since b is complete, $[\mathcal{O}+]$ has been applied for all y such that xry is on b. Thus for all y such that xry is on b, φ $y+$ is on b. Thus, by the inductive hypothesis, $1 \in \bar{v}_{w_y}(\varphi)$ for all y such that xry is on b. By the definition of induced, $w_x \mathcal{R} w_y$ iff xry is on b. Thus $1 \in \bar{v}_{w_y}(\varphi)$ for all y such that $w_x \mathcal{R} w_y$. Thus, by the truth condition for $\mathcal{O}$, $1 \in \bar{v}_{w_x}(\mathcal{O}\varphi)$.

2. Suppose $\mathcal{O}\varphi$ $x-$ is on b. Since b is complete, $[\mathcal{O}-]$ has been applied to it. Thus xri and φ $i-$ are on b. By the definition of induced, $w_x \mathcal{R} w_i$. By the inductive hypothesis, $1 \notin \bar{v}_{w_i}(\varphi)$. Thus it is not the case that $(\forall w \in \mathcal{W})(w_x \mathcal{R} w \Rightarrow 1 \in \bar{v}_w(\varphi))$. Thus, by the truth condition for $\mathcal{O}$, $1 \notin \bar{v}_{w_x}(\mathcal{O}\varphi)$.

3. Suppose $\neg\mathcal{O}\varphi$ $x+$ is on b. There are two cases:

Case 3.1. $x = 0$. Since b is complete, $[\neg\mathcal{O}_0]$ has been applied. Thus $\mathcal{O}\varphi$ $0-$ is on b. Thus, since b is complete, $[\mathcal{O}-]$ has been applied to it. Thus $0ri$ and φ $i-$ are on b. By the definition of induced, $w_0 \mathcal{R} w_i$. By the inductive hypothesis, $1 \notin \bar{v}_{w_i}(\varphi)$. Thus it is not the case that $(\forall w \in \mathcal{W})(w_0 \mathcal{R} w \Rightarrow 1 \in \bar{v}_w(\varphi))$. Thus, by the truth condition for $\mathcal{O}$, $1 \notin \bar{v}_{w_0}(\mathcal{O}\varphi)$. Thus, by the home world lemma, $0 \in \bar{v}_{w_0}(\mathcal{O}\varphi)$.

Case 3.2. $x \neq 0$. Since b is complete, $[\neg\mathcal{O}+]$ has been applied. Thus xri and $\neg\varphi$ $i+$ are on b. By the definition of induced, $w_x \mathcal{R} w_i$. By the inductive hypothesis, $0 \in \bar{v}_{w_i}(\varphi)$. Thus $(\exists w \in \mathcal{W})(w_x \mathcal{R} w$ and $0 \in \bar{v}_w(\varphi))$. Moreover, by the definition of induced and the fact that $x \neq 0$, $w_x \neq w_0$. Thus, by the falsity condition for $\mathcal{O}$, $0 \in \bar{v}_{w_x}(\mathcal{O}\varphi)$.

In both cases, $0 \in \bar{v}_{w_x}(\mathcal{O}\varphi)$.

4. Suppose $\neg\mathcal{O}\varphi$ $x-$ is on b. There are two cases:

Case 4.1. $x = 0$.   Since b is complete, $[\neg \mathcal{O}_0]$ has been applied.   Thus $\mathcal{O}\varphi$ 0+ is on b.   Thus, since b is complete, $[\mathcal{O}+]$ has been applied for all y such that 0ry is on b.   Thus for all y such that 0ry is on b, φ y+ is on b. Thus, by the inductive hypothesis, $1 \in \bar{v}_{w_y}(\varphi)$ for all y such that 0ry is on b.   By the definition of induced, $w_0 \mathcal{R} w_y$ iff 0ry is on b.   Thus $1 \in \bar{v}_{w_y}(\varphi)$ for all y such that $w_0 \mathcal{R} w_y$ Thus, by the truth condition for $\mathcal{O}$, $1 \in \bar{v}_{w_0}(\mathcal{O}\varphi)$. Thus, by the home world lemma, $0 \notin \bar{v}_{w_x}(\mathcal{O}\varphi)$.

Case 4.2. $x \neq 0$.   Since b is complete, $[\neg \mathcal{O}-]$ has been applied for all y such that xry is on b.   Thus for all y such that xry is on b, $\neg\varphi$ $y-$ is on b.   Thus, by the inductive hypothesis, $0 \notin \bar{v}_{w_y}(\varphi)$ for all y such that xry is on b.   By the definition of induced, $w_x \mathcal{R} w_y$ iff xry is on b. Thus $(\forall w \in \mathcal{W})(w_x \mathcal{R} w \Rightarrow 0 \notin \bar{v}_w(\varphi))$.   Thus it is not the case that $(\exists w \in \mathcal{W})(w_x \mathcal{R} w$ and $0 \in \bar{v}_w(\varphi))$.   Moreover, by the definition of induced and the fact that $x \neq 0$, $w_x \neq w_0$.   Thus, by the falsity condition for $\mathcal{O}$, $0 \notin \bar{v}_{w_x}(\mathcal{O}\varphi)$.

In both cases, $0 \notin \bar{v}_{w_x}(\mathcal{O}\varphi)$. ⊠

COMPLETENESS THEOREM.   For finite Σ, if $\Sigma \models_{\mathsf{SPDL}} \varphi$ then $\Sigma \vdash_{\mathsf{SPDL}} \varphi$.

Proof.   Suppose $\Sigma \nvdash_{\mathsf{SPDL}} \varphi$.   Then there is no closed tableau for Σ/φ. Let τ be any open, complete tableau for Σ/φ, and choose some open branch b of τ.   Let $\mathcal{M} = \langle \mathcal{W}, w_0, \mathcal{R}, v \rangle$ be a model induced by b.   (We know that there is at least one such model.)   By the completeness lemma (and the definition of tableaux for an inference), $1 \in \bar{v}_{w_0}(\psi)$ for all $\psi \in \Sigma$, but $1 \notin \bar{v}_{w_0}(\varphi)$.   Thus $\Sigma \nvDash_{\mathsf{SPDL}} \varphi$. ⊠

ALL-OR-NOTHING COROLLARY.   If there is an open, complete tableau for Σ/φ, then there is no closed tableau for Σ/φ.

Proof.   Suppose there is an open, complete tableau for Σ/φ.   Pick an open branch, b, of this tableau.   Let $\mathcal{M} = \langle \mathcal{W}, w_0, \mathcal{R}, v \rangle$ be a model induced by b.   (We know that there is at least one such model.)   By the completeness lemma and the definition of tableaux for an inference, $1 \in \bar{v}_{w_0}(\psi)$ for all $\psi \in \Sigma$, but $1 \notin \bar{v}_{w_0}(\varphi)$.   Thus $\Sigma \nvDash_{\mathsf{SPDL}} \varphi$.   Thus, by the soundness theorem, $\Sigma \nvdash_{\mathsf{SPDL}} \varphi$.   Thus, by the definition of proof-theoretic consequence, there is no closed tableau for Σ/φ. ⊠

BIBLIOGRAPHY

[1] Anderson, A. R. 1967. Some nasty problems in the formal logic of ethics. *Noûs* 1: 345-60.

[2] Anderson, A. R. and Belnap, N. D. 1975. *Entailment: The Logic of Relevance and Necessity.* Princeton: Princeton University Press.

[3] Ausín, F. J. and Peña, L. 2000. Paraconsistent deontic logic with enforceable rights. In D. Batens et al. (eds.) *Frontiers of Paraconsistent Logic.* Philadelphia: Research Studies Press, 2000.

[4] Beall, J.C. and van Fraassen, B. C. 2003. *Possibilities and Paradox: An Introduction to Modal and Many-Valued Logic.* Oxford: Oxford University Press.

[5] Brady, R. 1982. Completeness proofs for the systems RM3 and BN4. *Logique et Analyse* **25**: 9-32.

[6] Bremer, M. 2005. *An Introduction to Paraconsistent Logics*. Frankfurt: Peter Lang.

[7] Carnielli, W. A. and Marcos, J. A taxonomy of C-systems. In Carnielli, W. A., Coniglio, M. E., and D'Ottaviano, I. M. L. (eds.), *Paraconsistency: The Logical Way to the Inconsistent*. New York: Marcel Dekker.

[8] Castañeda, H.-N. 1981. The paradoxes of deontic logic: the simplest solution to all of them in one fell swoop. In [18].

[9] Chellas, B. F. *Modal Logic: An Introduction*. Cambridge: Cambridge University Press..

[10] da Costa, N.C.A. and Carnielli, W. A. 1986. On paraconsistent deontic logic. *Philosophia* **16**: 293-305.

[11] Dunn, J. M. 1976. Intuitive semantics for first-degree entailments and 'coupled trees'. *Philosophical Studies* **29**: 149-68.

[12] Forrester, J. W. 1984. Gentle murder, or the adverbial samaritan. *Journal of Philosophy* **81**: 193-7.

[13] Goble, L. 1991. Murder most gentle: the paradox deepens. *Philosophical Studies* **64**: 217-27.

[14] Goble, L. 1999. Deontic logic with relevance. In P. McNamara and H. Prakken (eds.), *Norms, Logics, and Information Systems*. IOS Press: 1999.

[15] Goble, L. 2000. Multiplex semantics for deontic logic. *Nordic Journal of Philosophical Logic* **5**: 115-34.

[16] Grana, N. 1990. *Logica Deontica Paraconsistente*. Napoli: Liguori Editore.

[17] Hanson, W. H. 1965. Semantics for deontic logic. *Logique et Analyse* **8**: 177-90.

[18] Hilpinen, R. (ed.). 1981. *New Studies in Deontic Logic*. Dordrecht: D. Reidel.

[19] Hughes, G. E. and Cresswell, M. J. 1996. *A New Introduction to Modal Logic*. London: Routledge.

[20] Lewis, D. 1973, *Counterfactuals*, Blackwell, Oxford.

[21] Mares, E. 1992. Andersonian deontic logic. *Theoria* **58**: 3-20.

[22] Mates, B. 1972. *Elementary Logic*. New York: Oxford University Press.

[23] McGinnis, C. 2003. Some of the best worlds are impossible: semi-paraconsistent deontic logic and its motivation. (Abstract.) *Bulletin of Symbolic Logic* **9**(2): 262.

[24] McGinnis, C. 200?. *Paraconsistency and Deontic Logic*. Ph.D. thesis, University of Minnesota.

[25] Priest, G. 1987. *In Contradiction: A Study of the Transconsistent*. The Hague: Martinus Nijhoff.

[26] Priest, G. 2001. *An Introduction to Non-Classical Logic*. Cambridge: Cambridge University Press.

[27] Prior, A. N. 1954. The paradoxes of derived obligation. *Mind* **63**: 64-5.

[28] Restall, G. 2000. *Introduction to Substructural Logics*. London: Routledge.

[29] Ross, A. 1941. Imperatives and logic. *Theoria* **7**: 53-71.

[30] Routley, R. and Plumwood, V. 1989. Moral dilemmas and the logic of deontic notions. In G. Priest et al. (eds.), *Paraconsistent Logic: Essays on the Inconsistent*. Munich: Philosophia, 1989.

[31] Schotch, P. K. and Jennings, R. E. 1981. Non-Kripkean deontic logic. In [18].

[32] Stelzner, W. 1992. Relevant deontic logic. *Journal of Philosophical Logic* **21**: 193-216.

[33] van Fraassen, B. C. 1972. The logic of conditional obligation. *Journal of Philosophical Logic* **1**: 417-38.

[34] van Fraassen, B. C. 1973. Values and the heart's command. *Journal of Philosophy* **70**: 417-38.

[35] von Wright, G. H. 1951. Deontic logic. *Mind* **60**: 1-15.

[36] von Wright, G. H. 1991. Is there a logic of norms? *Ratio Juris* **4**: 265-83. Reprinted in A. Aarnio and N. MacCormick (eds.), *Legal Reasoning. Vol. I*. New York: New York University Press: 1992.

Casey McGinnis
Department of Philosophy, University of Minnesota, Minneapolis, MN 55455-0310, USA.
caseymcginnis@gmail.com

Entailment Logic — A Blueprint

ROSS T. BRADY

We aim to establish a blueprint[1] for the development of logic based on the weak relevant logic **MC** of entailment. (**MC** was previously called **DJ**, in line with the names of neighbouring logics.) Recall that **MC**, amongst other neighbouring logics, enables the simple consistency of naive set theory and naive predicate theory to be shown, thus providing a solution to the set-theoretic and semantic paradoxes. (For this, see [9] and [13].) Recall also that **MC** is the logic of meaning containment (hence the name '**MC**'), where the content of the consequent is contained in the content of the antecedent. (For this, see [8] and [13].) However, we first need to collect together some ideas as to how to proceed with the development of this logic, once we remove classical logic from its exalted status as the logic upon which all else must be built.

To see how to do this, let us go back to the beginnings of logic, forget much of what we have learnt, and put ourselves in the position of an intelligent critical student who might find the subject difficult to comprehend. Indeed, there are a number of oddities, paradoxes, surprises, questions, and the like, not only in the beginnings of logic but also, I suggest, amongst its celebrated results. I give a list of what I think these various concerns would be, with some commentary on each, under 6 headings and 23 sub-headings. The headings, covering the broad sweep of logic, are: arguments, inferential concepts, conjunction and disjunction, negation, quantifiers, truth-functional semantics. Admittedly, they are of various degrees of concern, ranging from those of great concern to some of lesser consequence.

At the end, we gather together the threads from these commentaries and weave them into a whole to obtain a coherent picture of how entailment logic could be set up and applied. To do this, we use the recently established

---

[1]This "paper" is a blueprint in that it is a gathering together of ideas about a large range of concerns about classical logic, marshalled into an entailment logic and especially its normalized natural deduction characterization, leaving much detail to be filled in. It is anticipated that a major book on this topic will be eventually written, with a much fuller technical and philosophical treatment. On the philosophical side, I am also in the process of writing a paper called "Four Basic Logical Issues", discussing the why's and wherefore's of the logical road I have chosen to go down.

normalized natural deduction systems for **MC** and **MCQ**, which we set up in 2. below. We will argue and explain along the way in an attempt to show why our particular picture seems to be appropriate.

1   The Classical Treatment of Logic - Oddities, paradoxes, surprises, questions, and the like.

We start with the classical treatment of logic and, with a fairly open mind, pick out such concerns and anomalies in each of the main areas.

1.1   Arguments.

(1) The Role of Meaning in Logic.
Unlike inductive arguments, the conclusions of deductive arguments follow as a matter of certainty from the truth of their premises. Let us reflect on the source of this certainty. It is clear that this is achieved through the determination of the meanings of logical and other words in the premises and conclusion of such an argument. Thus, meaning should have a central role in validity determination, and deductive logic generally. So, this meaning connection between premises and conclusion is such that the conclusion *follows from* the premises, rather than it just be the case that the conclusion is certain, regardless of the premises.

In fact, to make this point clearer, we should differentiate these two types of valid deductive argument. We will use the term, *classical deduction*, to apply to the standard (textbook-type) definition of deductive validity, with the inference involved understood as a material implication:

> If the premises are true then the conclusion is also true as a matter of certainty.

This leaves it open for a conclusion to be true as a matter of certainty, be irrelevant to the premises, and still justify a valid argument. Although this is the convention, it seems unintuitive. We will use the term, *relevant deduction*, to apply to the following definition:

> The conclusion is derivable from the premises, using rules that apply as a matter of certainty, at least one of the premises being used in the derivation.

These rules, which apply to the steps of an argument, would each require some analysis of the meanings of words in their premises and conclusion to ensure the requisite certainty. In this process, truth must still be preserved from premises to conclusion, to give each sentence the correct orientation so that meanings can be related. This process also would involve meta-logical change, away from the standard material implication of classical deduction. (See §12.8 of [11] for an account of a relevant deduction, captured in a natural deduction system.)

(2) The Role of Truth in Logic.

Classical logic is chiefly concerned with the analysis of the truth and falsity of compound statements in terms of that of their atomic components. Thus, truth of atomic statements is generally determined by the respective discipline (but see (3)) and not by logic. What constitutes meaning of logical constants in classical semantics seems to be lost in the quantification over the assignments of truth and falsity to atomic statements. There should be more to the meaning of a statement than its truth or falsity, even with quantification over its atomic constituents. Let us contrast this with the usual meaning of the word 'semantics' applied to a statement, for which one would expect, in a logical context, there to be some logical content, which may then be evaluated as true or false as part of validity determination of an argument or formula. (Logical statements should not be any different from other statements in this matter.) This suits an algebraic style of semantics, rather than a truth-functional style. (See (22) and (23).) Such an algebraic-style of semantics is my content semantics (see [5], [8] and [13]), in which truth plays a suitably minor role. So, truth, though still used in conjunction with valid relevant deduction as set out in (1), is overused in current truth-functional semantics. For further discussion of this, see (3) below.

(3) Truth in the World.

Logic essentially concerns argumentation wherever one may find it, e.g. in fictional and theoretical contexts. When one sets up an applied logic, one axiomatizes concepts by meaning postulates, which can then be taken together with some contingent statements that may relate to these concepts. These should not have to relate to truths in the world. Thus, logic is conceptually based rather than ontologically based, in that it is based on the concepts it is developing rather than on truths of the world, thereby suiting the range of contexts to which logic can apply. So, truth to a conceptual context matters more than truth in the world. However, for invalidity shown by true premises and false conclusion, these can refer to the world but could just as well refer to a conceptual context in which the argument is framed.

1.2   Inferential Concepts.

(4) Material Implication.

The classical truth-tables, in particular the truth-table for the material implication '⊃' creates certain anomalous tautologies. Much classical reasoning, in mathematical contexts especially, is done with an oversupply of logical laws, but the truth of basic atomic statements is not generally affected as material implication acts as a truth-preserver and the anomalies occur with compound expressions. These anomalies are exemplified by the following Paradoxes of Material Implication:

> Any statement implies a true statement.
> A false statement implies any statement.

Others follow:

> For any two statements, one implies the other.
> If the conjunction of 2 antecedents imply a consequent then one of them alone implies the consequent.

As for relevant deduction in (1), one would at least expect the consequent to be relevant to the antecedent(s). A certain level of relevance for sentential logics can be guaranteed through imposition of the following Relevance Condition:

> If $A \to B$ is a theorem of logic L then A and B share a sentential variable.

This prevents the logic L from having $\to$-inferences with nothing in common. Such a logic L, satisfying the Relevance Condition, is a called a *relevant logic*. The Relevance Condition is only a necessary condition on a good logic and more conditions are needed to pin down a good logic. Relevant logics can contain so-called "relevant junk" such as $A \vee (A \to B)$, which is not really in the spirit of relevant logic.

(5) The Fundamental Inferential Concept of Logic.

First compare implication, understood as truth-preservation, with entailment, understood as meaning containment. These are the two major inferences based on the two major semantic concepts of truth and meaning. Here, we take truth-preservation to be distinct from material implication in that it is rule-like, i.e. it is not activated when the premise(s) are false.

Let us consider the classical entailment concept of necessitated (or strict) implication, which induces the Paradoxes of Strict Implication:

> Any statement entails (i.e. strictly implies) a necessarily true statement.
> A necessarily false statement entails (i.e. strictly implies) any statement.

However, Meyer in [16] produces a strong criticism of strict implication as an entailment.

Let us also compare linguistic conditionals with the normative conceptual approach, i.e. 'actual' vs 'ought'. Here, logicians are concerned with distinguishing good from bad arguments, which requires some background concepts to compare them with. We are not concerned with how people actually reason, which would involve an empirical linguistic study of arguments, good and bad.

Entailment, understood as meaning containment, provides the inference-driver for the rules in the definition of relevant deduction in (1). On this basis, entailment is the fundamental inferential concept of logic. (It is also a single positive concept, which puts it at an advantage over strong relevant logics which are generally thought of as being like classical logic, but minus certain irrelevant or irrelevance-producing laws.) A suitable logic can then be built around this entailment using a number of key meaning-containment properties of the connectives and quantifiers, ultimately yielding the logic **MC** of meaning containment, set out in 2. (See [8] and [13].)

Further, it is more appropriate to symbolize as a connective (using '$\rightarrow$') the concept which bears the closer relationship between antecedent and consequent, leaving the deductive rule-'$\Rightarrow$' for the concept that typically requires a number of steps to reach the conclusion from the premises. The rule-'$\Rightarrow$' is quite appropriate too as it should not activate when premises are false. Here, meaning containment is the closer concept and truth-preservation is the more distant one. This more-or-less reverses the common Quinean perception that material implication is the connective, whilst the semantic (i.e. meta-theoretic or rule-based) concept is entailment.

(6) Perspicuous Characterization of Entailment.

If people are, fundamentally at least, using the concept of entailment, it cannot be too complex and so we need a simple characterization that picks out the concept in a perspicuous way. We will set out in section 2. a normalized natural deduction system **NMC** capturing the entailment concept, appropriate for the meaning containment logic **MC** in (5).

1.3   Conjunction and Disjunction.

(7) Duality between Conjunction and Disjunction.

As with the De Morgan and distribution properties, one would expect there to be a duality between conjunction and disjunction. In a standard Hilbert-style axiomatization, there are conjunction introduction and elimination rules, but only a disjunction introduction rule. There is no disjunction elimination rule, $A \vee B \Rightarrow A$ or B, (i.e. a priming rule), because of the lack of specificity of disjunctive steps in a standard Hilbert proof. However, Gentzen and natural deduction systems are reasonably dual, as they do have a version of this rule, as well as the others. (See **NMC** in 2.) So, we still need a way of including Hilbert-style disjunction elimination, with primeness being seen as a good thing to include. This leads us to section (8) below.

Slaney's **M1** and **M2** metacomplete logics in his [20], whilst embracing primeness through the admissible rule, $\vdash A \vee B \Rightarrow \ \vdash A$ or $\vdash B$, provide additional properties concerning negated entailments and inductive theoremhood. Indeed, **MC** is an **M1** logic in Slaney's sense. This has the consequence that **MC** has no negated entailment theorems, whilst an **M2** metacomplete logic has the property that $\vdash \ \sim (A \rightarrow B)$ iff $\vdash A$ and $\vdash \ \sim B$. What is good about both kinds of metacomplete logic is that they have an entailment focus, in that their theorems are either entailments or inductively buildable up from entailments.

(8) Assertion and Rejection Systems.

It is worthwhile adding rejection systems to Hilbert-style axiomatizations. We can include disjunctive properties, such as primeness, $\dashv A, \dashv B \Rightarrow \ \dashv A \vee B$, and Disjunctive Syllogism (DS), $\vdash A \vee B, \dashv A \Rightarrow \ \vdash B$ (the real one). Note that rejection-soundness (for all formulae A, not both $\vdash A$ and $\dashv A$) and rejection-completeness (for all A, either $\vdash A$ or $\dashv A$) are understood in framing these and other rules involving rejection, and although rejection-soundness is usually straight-forward to show, the same cannot be said for rejection-completeness.

Note that the admissible rule: $A, \ \sim B \Rightarrow \ \sim (A \rightarrow B)$, characteristic of **M2** metacomplete logics, can be derived from the further rejection rule: $\vdash A, \dashv B \Rightarrow \ \dashv A \rightarrow B$, using the consistency of $\ \sim B$, in the form: $\vdash \ \sim B \Rightarrow \ \dashv B$, and the Law of Excluded Middle (LEM) for $A \rightarrow B$, in the form: $\dashv A \rightarrow B \Rightarrow \ \vdash \ \sim (A \rightarrow B)$. So, since this admissible rule is derivable from these realistic assumptions, this leaves **M1** metacomplete logics as a class of logics neighbouring **MC** that are worthwhile studying.

1.4 Negation.

(9) Rule of Explosion.

Classical logic has the derived rule: $A, \ \sim A \Rightarrow B$, which enables everything to be derived from a contradiction. This has been variously called the Rule of Explosion, Ex Falso Quodlibet, and the Spread Law. It is easily establishable from the Disjunctive Syllogism or Modus Ponens for '$\supset$', using basic properties. Using it, conclusions can be drawn which are irrelevant to the premises. What is derivable should be so using the usual logical processes applied to A and to $\sim A$ and not by a special rule for contradictions. Thus, A, $\sim A \Rightarrow B$ should not be a derived rule of a good logic. Such a logic is called *paraconsistent*, providing a necessary condition on a logic. As with relevance, we still need further conditions to pin down a good logic. So, paraconsistency is one of a number of concerns about classical logic.

Let us pursue the further issue of *dialetheism*, i.e. whether there are true contradictions. Such true contradictions would at least be manifested in the base world of a suitable semantics. The invalidity of the Spread Law would allow this to happen in that a contradiction could be true in the base world without the world being trivial, i.e. by not having all formulae true. However, my understanding of dialetheism from p.228 of [18] is that such a base world would have to represent a theory with some sense of reality to it, i.e. that the base world is a real world or the real world. But, this takes us a step further into a metaphysical position, which is independent of our concerns about choice of logic, requiring only that the logic be paraconsistent. As stated in (3), logics can apply to arguments wherever we find them; they need not just apply to reality.

(10) Consistency and Completeness Assumptions.

Of course, classicality of negation is assumed through the usage of classical logic, but it still gets metatheoretically proved via $\sim$-consistency and $\sim$-completeness proofs. This is especially needed for classically-based applied logics, as classicality is not guaranteed in applications, which in turn raises some doubts about our assuming it in the first place. Indeed, there are instances where $\sim$-consistency and $\sim$-completeness fail. (See respectively (11) and (12) below.) Why do we then assume classicality within the logic and then feel the need to go and prove it meta-theoretically through $\sim$-consistency and $\sim$-completeness results? Obviously, classicality cannot be ensured from the basic logic through to its applications, and thus, classicality does not appear to be based on logical laws. Clearly, this anomaly does not apply to our basic logical concept of meaning containment, whose key properties persist through logical laws and derived rules. (See 2. on this.)

(11) Semantic and Set-Theoretic Paradoxes.

Classical logic, or indeed a basic relevant logic with LEM [$A \vee \sim A$] and MPA [$A \& (A \rightarrow B) \rightarrow B$], will yield the semantic paradoxes of the Liar and Heterologicality, and the set-theoretic paradoxes due to Russell and Curry, in all of which $\sim$-consistency fails. The point here is that the theories in which these paradoxes are encapsulated are quite reasonable and, despite the inherent self-reference, are based on seemingly intuitive concepts.

We show in logical terms how these paradoxes are generated using LEM and MPA. With basic logical assumptions, the LEM and MPA are deductively equivalent to the rules, $A \rightarrow \ \sim A \Rightarrow \ \sim A$ and $A \rightarrow .A \rightarrow B \Rightarrow A \rightarrow B$, respectively. The initial form leading to the Liar, Heterologicality and Russell Paradoxes is: $A \leftrightarrow \ \sim A$. Using the above LEM rule, once in negated form, $A\& \sim A$ follows. The initial form leading to the Curry Paradox is: $A \leftrightarrow .A \rightarrow B$. Using the above MPA rule, $A \rightarrow B$ follows and, by the converse implication, A, from which B is immediately derived. Since B can be an arbitrary formula, triviality results.

Further, the LEM and MPA are *logical villains* in that either or both cause other problems as well, i.e. to metacompleteness, normalization, and decidability. We take these three in turn. The soundness half of the metacompleteness proof in [19] fails for LEM and MPA. If the LEM is needed, it would have to be taken as an outright assumption in the main proof in a normalized natural deduction system for the logic **MC**, given in 2. According to Urquhart (in [11]), the MPA is central to the proof of undecidability of sentential relevant logics containing $\mathbf{TW}_+$.

(12) Incompleteness of Arithmetic.

Subject to the consistency of Peano Arithmetic, we know by Gœdel's Incompleteness Theorem that the Gœdel sentence G and its negation $\sim G$ are both unprovable. This is a failure of $\sim$-completeness for Peano Arithmetic, and indeed for any theory strong enough to express arithmetic in it. Using his semantic conception of truth, Tarski then assumed a classical meta-theory, which enabled him to re-assert the LEM in that theory, thus requiring a separation of proof from truth in order to carry this out.

(13) Meaning of Negation.

As can be seen from (10)-(12), classicality of negation and hence the truth-functional meaning of negation cannot be guaranteed in applications and hence the truth-functional meaning of negation is not generally correct. It is better that we take the meaning of negation from its interaction with entailment, similarly to that of conjunction and disjunction. (See the axioms in 2.1.) Indeed, as in [8] and [13], we obtain a mirror-image negation concept with contraposition, double negation and De Morgan properties, which yields a very neat symmetric view of negation, in combination with the conjunction and disjunction properties.

There remains the issue of the extent to which classicality holds, given that it does not generally apply. Note that there are no entailments in the two classicality properties of (10), which would have ensured some meaning relationships. As seen in [8] and [13], the classicality properties of $\sim$-

consistency and $\sim$-completeness do not follow from the meaning of negation, determined from the mirror-image concept. So, when we try to justify the (negation) classicality of a particular sentence, we cannot appeal to a general logical law. In (10)-(12), we saw that both of these classicality properties need proving for particular applied theories, and this also applies to single sentences which are being tested for classicality. Moreover, as presented in [8] and [13], we can ensure that there is a wide range of common classical sentences by using a criterion for classicality based on abstractions and idealizations of the physical world.

(14) The Classical Recapture in the Logic of Entailment.

As argued in (13), a wide variety of sentences are classical and so we need to find a way for them to re-enter the system. Thus, there is a need to consider the so-called "classical recapture". Classicality can be assumed for the appropriate atomic sentences using rejection: $\vdash A$ and $\dashv \sim A$ for classical truth, and $\vdash \sim A$ and $\dashv A$ for classical falsity. It can then be proved within the system, by proving the 2 formal classicality properties, viz. the LEM and Modus Ponens (MP) for '$\supset$'. The LEM is provable from one of its disjuncts by a disjunction introduction rule, as one would expect for a metacomplete system; MP is provable from the metacomplete rejection rule: $\dashv \sim A, \vdash A \supset B \Rightarrow \vdash B$ (easily seen to be the real DS), using the consistency of A in the form: $\vdash A \Rightarrow \dashv \sim A$. (Rejection soundness is assumed, i.e. for all A, not both $\vdash A$ and $\dashv A$.) This is "the metacomplete way of classical recapture", i.e. as opposed to dumping in a whole pre-justified classical theory into one sort of a 2-sorted theory, as in [13]. (There, I had 2 sorts of sentences: classical and general.) Whether classicality needs to be proved or not will depend on the circumstances within the theory being developed. Mostly, the two formal classicality properties would be bypassed in the process of deriving one's concluding theorem.

There are a number of specific sub-properties of classicality, one or more of which is used in each particular instance of a material '$\supset$' or Boolean '$\sim$'. A list of such sub-properties for '$\supset$' are: truth-preservation, contraposed truth-preservation, truth-preservation in an entailment context, entailment itself, enthymematic entailment, its contraposed form, not-or (with Boolean '$\sim$'), restricted quantification over a nonempty predicate (see (18)). Such sub-properties for '$\sim$' are: the LEM, the DS or MP for '$\supset$', the Explosion Rule. We then need to replace each of these sub-properties by incorporating them into an appropriate rule or formula, or by introducing a special assumption or rule. E.g. A rule arrow '$\Rightarrow$' would be used for truth-preservation, distinguishing classical and relevant deduction (as in (1)), and a meta-rule for contextual truthpreservation. (See [4] for meta-rules in a natural deduction context.)

The advantage of the rule '⇒' over the material implication is that we do not need to specify what happens when the antecedent is not true, as the rule only applies when the antecedent is fulfilled. This avoids the whole mess of having to say what happens when the antecedent is false, avoiding the current slippery slope into a material implication paradox — i.e. a false sentence implies any sentence — and then into the Explosion Rule.

There is a presumption, in proceeding with classical recapture, that almost all reasoning carried out by classical logicians, or as applied classical logic, would be able to operate within a weaker framework, such as that proposed in the above recapture. Classical logicians do not use the full force of classical logic. E.g. I cannot recall a single instance of the tautologies, $(A \supset B) \vee (B \supset A), (A\&B \supset C) \supset .(A \supset C) \vee (B \supset C)$ or $(A \supset B)\&(C \supset D) \supset .(A \supset D) \vee (C \supset B)$, being used, where '$\supset$' is used inferentially. Indeed, if some classical reasoning cannot be dealt with using the above recapture, there might be grounds for suspicion about the informal validity of the argument.

Certainly, sentences leading to paradox such as the Liar sentence, 'This very sentence is false', and 'Russell's class is a member of itself' are non-classical, being outside the suggested classical domain of (13) or any other reasoned classical domain. Further, the essential LEM for these sentences is not provable in the metacomplete way, as there is no way for either the Liar sentence or its negation to be provable in order to prove the LEM by disjunction introduction.

1.5   Quantifiers.

(15) Free and Bound Variables.

As in Gentzen systems, it is simpler to separate free and bound variables in Hilbert axiomatizations and in natural deduction. In Hilbert axiomatizations, only the Generalization Rule need have a condition on it, with no conditions on any of the standard quantificational axioms used in relevant logic. Allowing the same variables to be free and bound can cause unnecessary difficulties, e.g. in showing metacompleteness (see [17]). In natural deduction, it is simpler to eliminate and introduce quantifiers using distinct variables for the instantiations. (See 2.)

(16) Constants and Variables.

Arbitrary constants are needed in natural deduction arguments to indicate where universal generalization can be performed, and free variables generally suffice for the purpose. However, individual constants are not really needed, unless appropriate non-logical axioms are added to specify them. (The same applies for sentential constants, which could have the formal classicality properties as a non-logical axiom and a nonlogical rule, or they could just be axiomatized contingent sentences.) Otherwise, unspecified individual constants, i.e. without non-logical axiomatization, would behave in the same way as variables, one being substitutible for the other, as can be seen from the admissible rule, $A(c) \Rightarrow \forall x A(x)$, of Predicate Calculus. (See [2].) Further, to work the metacomplete way in (14), we need proper constants to be able to ensure the classicality of sentences, so that they can be formally represented by $\vdash A$ and $\dashv \sim A$, or by $\vdash \sim A$ and $\dashv A$.

(17) Duality between Universal and Existential Quantification.

As for conjunction and disjunction in (7), there should be a duality between the universal and existential quantifiers. The main problem for this occurs for Existential Instantiation in Hilbert-style systems. However, using rejection in a metacomplete logic, we can express it as: $\dashv A^a/_x \Rightarrow \dashv \exists x A$, for an unconstrained variable a, analogously with primeness for disjunction.

(18) Restricted Quantification.

For quantification over a non-empty sub-domain in classical Predicate Calculus, we require '$\supset$' and '$\&$' to express universal and existential domain restrictions in terms of their corresponding unrestricted quantifiers. For relevant logics, as argued in Ch. 13 of [11], the only way that seems to work is to use a primitive sub-domain restriction: $(\forall x Px)Ax$ and $(\exists x Px)Ax$, which is then axiomatized to ensure the logic still applies under the restriction. Here, we represent 'All S's are P's' as $\exists x Sx \Rightarrow (\forall x Sx)Px$, and 'Some S's are P's' as $\exists x Sx \Rightarrow (\exists x Sx)Px$, with the Aristotelian non-emptiness assumption applied to the restricting predicate Sx, being also appropriate for our restricted quantification to subdomains.

Importantly, from $\exists x Px \Rightarrow (\forall x Px)Ax$, the rule $Px \Rightarrow Ax$ follows, and conversely, linking universal restricted quantification to rules. A proof of this is as follows:

(I) Let $\exists x Px \Rightarrow (\forall x Px)Ax$.
$Px \Rightarrow \exists x Px \Rightarrow (\forall x Px)Ax \Rightarrow Ax$ (by universal instantiation, subject to Px holding).

(II) Let $Px \Rightarrow Ax$.
$\exists x Px \Rightarrow (\forall x Px)Px$ (by an identity axiom) $\Rightarrow (\forall x Px)Ax$ (by universally restricting the assumed rule to Px). $\square$

This easily enables one to establish the Barbara Syllogism of Aristotelian Logic. We use:

$\exists x M x \Rightarrow (\forall x M x) P x$ for MaP.
$\exists x S x \Rightarrow (\forall x S x) M x$ for SaM, and
$\exists x S x \Rightarrow (\forall x S x) P x$ for SaP.

This can easily be shown by converting it to:

$M x \Rightarrow P x$
$S x \Rightarrow M x$
$\therefore S x \Rightarrow P x.$

For Darii, we have:

$\exists x M x \Rightarrow (\forall x M x) P x$ for MaP.
$\exists x S x \Rightarrow (\forall x S x) M x$ for SiM, and
$\exists x S x \Rightarrow (\forall x S x) P x$ for SiP.

From the first premise, $M x \Rightarrow P x$, and then $(\exists x S x) M x \Rightarrow (\exists x S x) P x$ (by existentially restricting the above rule to $\exists x S x$).

For Conversion of SiP, we prove the following:

$\exists x S x \Rightarrow (\exists x S x) P x \leftrightarrow \exists x (S x \& P x).$

$L \to R$ By a special rule of the axiomatization.

$R \to L$ $S x \& P x \to P x \to (\exists x S x) P x$, by existential generalisation, given $S x$.

$\exists x (S x \& P x) \to (\exists x S x) P x$, by standard existential elimination.

Thus, $\exists x S x, \exists x P x \Rightarrow (\exists x S x) P x \leftrightarrow (\exists x P x) S x.$

We follow with two interesting notes:

Note 1. Note what happens when the usual classical translation, $\forall x (S x \supset P x)$, is used instead for SaP, etc. in the Barbara Syllogism:

$\forall x (M x \supset P x)$
$\forall x (S x \supset M x)$
$\therefore \forall x (S x \supset P x).$

Then one would expect to be able to show: $\sim M x \vee P x$, $\sim S x \vee M x \Rightarrow \sim S x \vee P x$. By the Deduction Theorem and by distribution of the two premises, we get:

$(\sim M x \& \sim S x) \vee (\sim M x \& M x) \vee (P x \& \sim S x) \vee (P x \& M x) \supset$
$. \sim S x \vee P x.$

What remains to be shown here is: $\sim Mx \& Mx \supset . \sim Sx \lor Px$. This requires LEM, and its rule form is an instance of Explosion.

Note 2. The null set being a subset of every set, i.e. $\emptyset \subseteq x$, is avoided with the definition of $x \subseteq y$ as $(\forall z \; z \in x)z \in y$, since, with the use of the restricted quantifier $(\forall z \; z \in x)$, we assume $\exists z \; z \in x$, which ensures the non-emptiness of x.

(19) Undecidability of Predicate Calculus.

Undecidability is a stumbling block for the Computer Age, is indicative of inner theoretical complexity and conflicts with the expected simplicity of concepts that are in general usage. Church's undecidability of Predicate Calculus in his [14] is a blight on a system of logic which is supposedly based on simple intuitive concepts. Thus, it would be nice if the quantified entailment logic is decidable. (See 3.)

1.6   Truth-Functional Semantics.

(20) Non-Recursive Domains.

The classical inductive evaluation process requires generalization over non-recursive domains when evaluating universally quantified statements. Such a generalization is physically impossible and evaluation can only be made in such a case on the basis of an unconstrained variable, representing an arbitrary element of the domain. This is typical of universal generalization in standard natural deduction systems of a pure logic.

For applied logics, we add a rule which enables the universal generalization to take place over specific domains. For a recursively infinite domain, such as that of the natural numbers, one can add a type of mathematical induction meta-rule, e.g. if $A(a) \Rightarrow A(a')$ then $A(0) \Rightarrow A(b)$, for unconstrained variables a, b over this domain. For finite domains, we can add a rule of form, $A(c_1), A(c_2), , A(c_n) \Rightarrow A(a)$, for an unconstrained variable a over the domain $\{c_1, c_2, \ldots, c_n\}$.

(21) Establishable Truth.

Truth in a semantics is generally used to *determine* validity and as such needs to be establishable, i.e. provable within a suitable logic or meta-logic of the logic. We have already seen problems in (12) and (20), suggesting that proof and truth should not be separated from each other. In (12), we saw that proof and truth were separated to allow for the re-introduction of a semantical LEM after this same principle was shown to fail in the proof system. In (20), we saw the impossibility of establishing a universal statement about a non-recursive domain from its elements, breaking down the classical inductive argument. Also, in (3), it was argued that, for logic, truth in a conceptual context is more important than truth in the world. Thus, if the concepts are captured in a logical system, such a logic would provide the conceptual context and truth should then be establishable within it.

That is, proof systems and their semantics are just different symbolic representations of a logic, yielding different modes of presentation of the laws of the logic. (See Meyer and Sylvan, "Extensional Reduction II", in [11], where this is strongly argued for.) Proof theories generally require rules of derivation from starting points to conclusions, whilst truth-functional semantics generally require some sort of inductive method of determining the value of composites in terms of their components.

(22) Worlds and Possible Worlds.

Do we need worlds? Meyer and Routley, Extensional Reduction II, in [11], have strongly argued against the use of worlds semantics. There it is seen as reducing what are really intensional concepts to extensional ones.

Another difficulty with worlds, other than the actual, is that they have to be built up so that they satisfy priming requirements, if they are relevant, and consistency requirements, if they are classical. Importantly, they also need to be maximal so as to be able to satisfy certain truth-conditions of a canonical model structure for the purpose of proving completeness. The problem is that worlds other than the actual usually have incomplete specifications, with the rest having to be made up in some arbitrary manner. Indeed, this is what happens in Henkin-style completeness arguments, where a denumerable set of formulae is considered in some arbitrary order.

Might we just use sets of hypotheses, as in natural deduction, with index sets in place of worlds? Accessibility relations would be replaced by relationships between index sets. Here, one would get what Fine calls a "theory" rather than a world, as whatever is derivable from hypotheses would satisfy provable-entailment preservation and the adjunction rule. (See his [15].) This leads us to ask the question of the next section.

(23) Whither Truth-functional Semantics?

Tense logic and **S5** seem to be the main contenders for the continued usage of possible worlds. In tense logic, each world represents a point in time and the accessibility relation, defined on these points in time, represents 'earlier than'. **S5** is the modal logic that captures necessity as universal quantification over a set of possibilities. Possible worlds may also be used for technical purposes, e.g. the proof of the Relevance Condition uses Belnap's or Meyer's matrix sets, both of which are equivalent to Routley-Meyer model structures defined over a finite set of worlds (see Ch.9 of [11]).

Algebraic-style semantics would be the main alternative. Of these, content semantics seems best[2], as it conceptualizes entailment, pins down the logic **MC**, and rejects what's not in **MC**. (See [8] and [13] for details of this.) This is preferable to representing algebraic properties in the semantics, for their own sake. Anyway, the algebraic properties, centering on fusion 'o', for **MC** are very minimal, to say the least.

Normalized natural deduction, given in 2., has an inductive character, given by the introduction and elimination rules for each of the connectives. (Note the absence of Anderson and Belnap's distribution rule in [1].) This should suffice to establish many of the technical results that one uses a truth-functional semantics for. One can also create a third style of semantics, which I call "free semantics", out of the rules of the natural deduction system. (See [12].) Since priming and existential elimination may fail under entailment hypothesis, the semantics is existentially free in not requiring a witness, i.e. a disjunct or existential instantiation. However, by meta-completeness, priming and $\exists E$ both hold for the main proof of a natural deduction system, where there are no hypotheses. (See [11].)

This ends the commentary, through which we have built up a rough picture of a unified approach in which we can deal with all 23 of these concerns. We next present in 2. the normalized natural deduction system for **MC** and the quantified logic **MCQ**, to enable us to further create this picture in 3.

2 Normalized Natural Deduction for MC and MCQ.

2.1 The Sentential Logic MC (a.k.a. DJ)

Primitives.
$\sim, \&, \vee, \rightarrow$ (connectives).
$p, q, r, \ldots$ (sentential variables).
Axioms.

1. $A \rightarrow A$.

2. $A\&B \rightarrow A$.

3. $A\&B \rightarrow B$.

4. $(A \rightarrow B)\&(A \rightarrow C) \rightarrow .A \rightarrow B\&C$.

5. $A \rightarrow A \vee B$.

---

[2]The content semantics referred to here is not the general content semantics of [5] and [6], which applies to all logics stronger than **BB** and **BBQ**, but the specific one which applies to **MC** and **MCQ**.

6. $B \to A \lor B$.

7. $(A \to C)\&(B \to C) \to .A \lor B \to C$.

8. $A\&(B \lor C) \to (A\&B) \lor (A\&C)$.

9. $\sim\sim A \to A$.

10. $A \to \sim B \to .B \to \sim A$.

11. $(A \to B)\&(B \to C) \to .A \to C$.

Rules.

1. $A, A \to B \Rightarrow B$.

2. $A, B \Rightarrow A\&B$.

3. $A \to B, C \to D \Rightarrow B \to C \to .A \to D$.

Note: To obtain classical logic, where '$\to$' would represent the material '$\supset$', we would add the following three axioms, and delete the redundant R3: $A \to B \to .B \to C \to .A \to C$; $A \to .A \to B \to B$; $A \to .B \to A$.

2.2   The Normalized Natural Deduction System NMC

We follow the Fitch-style natural deduction developed in [1], extended by Brady in [4], and normalized for some relevant logics by Brady in [10], where a complete detailed account of normalized natural deduction can be found. Index sets are sets of natural numbers, each representing a hypothesis. We use $\emptyset$ when no hypotheses are assumed. We introduce *signed formulae* and *structures* inductively as follows:

1. If A is a formula, TA and FA are signed formulae.

2. A signed formula is a structure.

3. If α and β are structures then so is (α, β).

Each signed formula within a structure is assigned the same index set. The index sets have either of the two forms: $\emptyset$ or a complete set, $\{j, \ldots, k\}$, for some natural numbers $k \geq 1$ and $1 \leq j \leq k$. [A complete set has no numerical gaps.]

We just present the rules of **NMC**, applying to what we call 'threads of proof' in [10], which are within a subproof or the main proof.

Hyp.

A signed formula of form TH may be introduced as the hypothesis of a new subproof, with a subscript $\{k\}$, where k is the depth of this new subproof in the main proof.

$T \rightarrow I.$

From a subproof with conclusion TB_a on a hypothesis $TA_{\{k\}}$, infer $TA \rightarrow B_{a-\{k\}}$ in its immediate superproof, where $a = \{j, \ldots, k\}$ and *either*:

(i) $a - \{k\} = \emptyset$ (with $j = k = 1$), *or*

(ii) $a - \{k\} = \{j, \ldots, k-1\}$ (with $k \geq 2, 1 \leq j \leq k-1$).

The conclusion and hypothesis need not be distinct in (i).

$T \rightarrow E.$

From TA_a and $TA \rightarrow B_b$, infer $TB_{a \cup b}$. (Direct version)

From FB_a and $TA \rightarrow B_b$, infer $FA_{a \cup b}$. (Contraposed version)

Whilst TA_a (or FB_a) and its conclusion $TB_{a \cup b}$ (or $FA_{a \cup b}$) are located in a proof P, either $TA \rightarrow B_b$ is located in the main proof or it is located in P's immediate superproof.

$T \rightarrow E$ carries the proviso that *either*:

(i) $b = \emptyset$, in which case $a \cup b = a$, *or*

(ii) $a = \{k\}, k \geq 2, b = \{j, \ldots, k-1\}, 1 \leq j \leq k-1$, *or*

(iii) $a = \{j, \ldots, k\}, k \geq 2, b = \{j, \ldots, k-1\}, 1 \leq j \leq k-1$.

$F \rightarrow I.$

From a derivation of $TB_{a \cup b}$ from TA_a or $FA_{a \cup b}$ from FB_a occurring within an immediate subproof, infer $FA \rightarrow B_b$.

All the signed formulae $FA \rightarrow B_b$, introduced by $F \rightarrow I$ from the subproof, form a structure, separated by commas.

$F \rightarrow I$ carries the proviso that *either*:

1. $b = \emptyset$ and $a = \{1\}$, *or*

2. $a = \{k\}, k \geq 2, b = \{j, \ldots, k-1\}, 1 \leq j \leq k-1$, *or*

3. $a = \{j, \ldots, k\}, k \geq 2, b = \{j, \ldots, k-1\}, 1 \leq j \leq k-1$.

Since $T \to E$(ii) and $F \to I$(ii) are the only rules that can be applied into a structure to change its index set, such applications of $T \to E$(ii) and $F \to I$ (ii) must be made *en bloc* to the signed formulae of the structure, so as to maintain its common index set. *En bloc* applications of $T \to E$(iii) and $F \to I$(iii) must also be made into any substructures that are new to the above structure.

$\underline{F \to E.}$

> From $FA \to B_{a-\{k\}}$, infer a subproof with conclusion TB_a on a hypothesis $TA_{\{k\}}$, where $a = \{j, \ldots, k\}$, $k \geq 2$, $1 \leq j \leq k-1$, and hence $a - \{k\} = \{j, \ldots, k-1\}$.
>
> The rules $F \to E$ and $F \to I$ work in tandem on the same subproof, which is called an F-subproof. Correspondingly, the rules $T \to E$ and $T \to I$ also work in tandem on a T-subproof.

$\underline{T \sim I.}$ From FA_a, infer $T \sim A_a$.
$\underline{T \sim E.}$ From $T \sim A_a$, infer FA_a.
$\underline{F \sim I.}$ From TA_a, infer $F \sim A_a$.
$\underline{F \sim E.}$ From $F \sim A_a$, infer TA_a.
$\underline{T\&I.}$ From TA_a and TB_a, infer $TA\&B_a$.
$\underline{T\&E.}$

> From $TA\&B_a$, infer TA_a.
>
> From $TA\&B_a$, infer TB_a.

$\underline{F\&I.}$

> From FA_a, infer $FA\&B_a$.
>
> From FB_a, infer $FA\&B_a$.

$\underline{F\&E.}$ From $FA\&B_a$, infer (FA_a, FB_a).
$\underline{T \vee I.}$

> From TA_a, infer $TA \vee B_a$.
>
> From TB_a, infer $TA \vee B_a$.

$\underline{T \vee E.}$ From $TA \vee B_a$, infer (TA_a, TB_a).
$\underline{F \vee I.}$ From FA_a and FB_a, infer $FA \vee B_a$.
$\underline{F \vee E.}$

> From $FA \vee B_a$, infer FA_a.
>
> From $FA \vee B_a$, infer FB_a.

$,E$. From SA_a, SA_a, infer SA_a.

A formula A is a *theorem of* **NMC** iff the structure $TA_\emptyset$ is provable in the main proof. Normal proofs of theorems have no elimination rules in the main proof, but these will be needed for deductions from assumptions. Further details about **NMC** can be found in [10].

We give a sample proof of $\sim(A \to C) \to .\sim(A \to B) \vee \sim(B \to C)$, which will involve using an F-subproof.

$$
\begin{array}{l}
\quad T\sim(A \to C)_{\{1\}} \\
\hline
\quad FA \to C_{\{1\}} \\
\qquad TA_{\{2\}} \\
\hline
\qquad TB_{\{1,2\}} \qquad [F \to I] \\
\qquad TC_{\{1,2\}} \qquad [F \to I] \\
\quad FA \to B_{\{1\}}, FB \to C_{\{1\}} \qquad [F \to E] \\
\quad T\sim(A \to B)_{\{1\}}, T\sim(B \to C)_{\{1\}} \\
\quad T\sim(A \to B)\vee\sim(B \to C)_{\{1\}}, T\sim(A \to B)\vee\sim(B \to C)_{\{1\}} \\
\quad T\sim(A \to B)\vee\sim(B \to C)_{\{1\}} \\
T\sim(A \to C) \to .\sim(A \to B)\vee\sim(B \to C)_\emptyset
\end{array}
$$

One can see here the use of the two $F\to$-rules acting in a contraposed fashion, in comparison with the corresponding use of the $T\to$-rules. The two basic negation properties of **MC** are double negation and contraposition; the use of $T\sim$- and $F\sim$-rules capture the former, whilst the $F\to$-rules, in conjunction with the other F-rules, taken in comparison with their corresponding T-rules, capture contraposition.

2.3 The Quantificational Logic MCQ

Additional Primitives.

$\forall, \exists$ (quantifiers).

$a, b, c, \ldots$ (free individual variables).

$x, y, z, \ldots$ (bound individual variables).

$f, g, h, \ldots$ (predicate variables).

Quantificational Axioms.

1. $\forall x A \to A^a/x$.

2. $\forall x(A \to B) \to .A \to \forall x B$.

3. $\forall x(A \vee B) \to .A \vee \forall x B$.

4. $A^a/_x \to \exists x A$.

5. $\forall x(A \to B) \to .\exists x A \to B$.

6. $A\&\exists x B \to \exists x(A\&B)$.

Quantificational Rule.

1. $A^a/_x \Rightarrow \forall x A$, where a does not occur in A.

2.4   The Normalized Natural Deduction System NMCQ

We distinguish *constrained and unconstrained free variable occurrences*, the constraints being applied by the hypotheses of subproofs, existential elimination and disjunction elimination, as indicated in the rules below.

We expand *structures* to include:

(4) If α is a structure and b is a variable then $(,_b \alpha)$ is a structure.

NMCQ.
Quantificational Rules.

<u>$T\forall I$.</u> From $TA^b/_{x_a}$, infer $T\forall x A_a$, provided b is unconstrained and does not occur in A.

<u>$T\forall E$.</u> From $T\forall x A_a$, infer $TA^b/_{x_a}$, where b can be constrained or unconstrained.

<u>$F\forall I$.</u> From $FA^b/_{x_a}$, infer $F\forall x A_a$, where b can be constrained or unconstrained.

<u>$F\forall E$.</u> From $F\forall x A_a$, infer $(,_b FA^b/_{x_a})$, where b is a constrained variable, does not occur in A, and has no other constraints.

<u>$T\exists I$.</u> From $TA^b/_{x_a}$, infer $T\exists x A_a$, where b can be constrained or unconstrained.

<u>$T\exists E$.</u> From $T\exists x A_a$, infer $(,_b TA^b/_{x_a})$, where b is a constrained variable, does not occur in A, and has no other constraints.

<u>$F\exists I$.</u> From $FA^b/_{x_a}$, infer $F\exists x A_a$, provided b is unconstrained and does not occur in A.

<u>$F\exists E$.</u> From $F\exists x A_a$, infer $FA^b/_{x_a}$, where b can be constrained or unconstrained.

$,_b E$. From $,_b SA_a$, infer SA_a, provided b does not occur in A.

All free variables which were constrained by the conclusion of the application of $F\forall E$ or $T\exists E$ that introduced the $,_b$ are no longer constrained by it in the conclusion of the $,_b E$ rule, but these variables may still be constrained by other means, appropriate to its new location. Further, hypothesis constraints are eliminated with the hypothesis, and constraints due to disjunction elimination are eliminated with the corresponding , E rule or when the variable concerned only occurs in one signed formula of the structure.

We generally start the conclusions of $T\forall E$ and $F\exists E$ with unconstrained free variables, new to the proof to this point. These may subsequently be replaced by another unconstrained or a constrained variable in order to make the two-premise rules, $T\&I$, $F\vee I$, $T\to E$ or $F\to I$, work as one might want. Again, a full detailed account is in [10].

3   Dealing with the 23 Concerns.

We briefly deal with each of the 23 concerns of 1, with reference to the normalized natural deduction systems **NMC** and **NMCQ**. That is, we add to the comments already made, points that refer to or require our systems **NMC** and **NMCQ**. Also, given the comments made in (21)-(23) of 1, we take the semantics of **MC** and **MCQ** to be the content semantics of [8] and [13], with the free semantics of [12] being an illuminating and useful adjunct.

(1&2) The Role of Meaning and Truth in Logic.

By abandoning truth-functional semantics in favour of content and free semantics, meaning and meaning containment are not thereby truth-functionally represented. Instead, meaning and meaning containment are captured using logical content in the content semantics of [8], [11] and [13]. Also, the meanings of the connectives and quantifiers can be captured in a deductive or operational sense in **NMC** (**Q**) and the corresponding free semantics, and both of these enable technical results to be proved, such as the validity/invalidity determinations and conservative extension results. This takes up work that might otherwise be done through use of a truth-functional semantics.

(3) Truth in the World.

The truth in the signed formula TA signifies A and the falsity in FA signifies $\sim A$, as in the completeness argument for **NMC** in [10]. Here, truth and falsity apply within the context of the logic and its application, through the natural deduction systems. This is in accordance with our earlier conceptual basis for logic, which does not require a relationship with the world.

(4) Material Implication.

MC is a weak relevant logic, with its main inference, entailment, interpreted as meaning containment. This can be seen from [8] and [13], where logical contents and content containment are introduced, which provide a

content semantics specifically for the logic **MC**, and also upon extension
for **MCQ**.

(5) The Fundamental Inferential Concept of Logic.

NMC, deductively at least, captures the fundamental concept of entailment. Truth-preserving rules of **MC**, represented by using '⇒', can also be
captured within the main proof of a natural deduction system **NMC**, by
deriving the conclusion from the assumptions.

(6) Perspicuous Characterization of Entailment.

As can be seen from 2, **NMC** has appropriate introduction and elimination rules for each connective, given the disjunctive interpretation of the
comma. In particular, $TA \rightarrow B$ is provable iff TB is derivable from TA,
subject to appropriate structuring of the proof.

(7) Duality between Conjunction and Disjunction.

This duality can be seen from the *I*- and *E*-rules of **NMC**, with the
comma representing disjunction and with separate signed formulae of the
same index set representing conjunction. This is formalised in [10] using
threads of proof, separated by a comma and understood disjunctively, and
strands of proof as signed formulae within a thread of proof, understood
conjunctively. Also, since **MC** is metacomplete (see [20]), duality can be
obtained for the Hilbert system with rejection.

(8) Assertion and Rejection Systems.

In [12], I import rejection into **NMC** by including rules for non-truth and
non-falsity, in subproofs as well as the main proof. This enables pieces of
proof to be contraposed and thus the reductio method to be employed using
these rejection rules. We could also add rejection to the Hilbert system **MC**.

(9) Rule of Explosion.

MC is a paraconsistent logic, as can be seen from [7], where Dialectical
Set Theory consisting of **MC** + LEM + naive set theory is shown to be
non-trivial.

(10) Consistency and Completeness Assumptions.

In contrast, there is no similar problem with meaning containment, which
is coded into the axiomatization of **MC** and continues to apply in any
extension of the logic. The key classicality properties of ∼- consistency and
∼-completeness and their formal analogues the DS and the LEM are not
generally assumed when working in **MC**, but the LEM and the DS can be
assumed or derived for particular sentences.

(11) Semantic and Set-Theoretic Paradoxes.

Since naive set theory, based on the logic **MCQ**, has been shown to be
simply consistent in [13][3], based on a consistency proof in [3] using the logic

---

[3]My book, Universal Logic, [13], was more-or-less complete in 1996 and accepted for
publication in 1999, and is hence referred to in the past tense.

TWQ (contained in the logic **CSQ** of the title), the use of **MCQ** provides a solution for the set-theoretic paradoxes. (See [9] for a general account of this). A similar simple consistency result has also been proved for the theory of truth and higherorder predicate theory in [13], and so **MCQ** can be said to solve the semantic paradoxes as well. It should be noted that both types of paradoxes involve a definition which is appropriately represented as a co-entailment.

(12) Incompleteness of Arithmetic.

Since The Gœdel sentence G has no free variables and is thus a constant, Gœdel's Theorem really shows, subject to simple consistency, that the LEM fails in an applied system of such complexity. Indeed, the system of arithmetic is not just $\sim$-incomplete, but incompletable, showing that the LEM cannot really be satisfied. The same results would still apply for weaker logics such as **MCQ**, and the advantage of such logics is that one can create models where the LEM is not satisfied. In such a case, we do not need to separate proof and truth in the Tarski manner and we can leave the LEM failure in place. Further, using the metacomplete way of classical recapture, one cannot derive $G \vee \sim G$ as neither G nor $\sim G$ is provable and so are unavailable for disjunction introduction.

(13) Meaning of Negation.

MC has mirror-image negation, since double negation and contraposition are both present, together with appropriate conjunction and disjunction laws. However, **MC** does not include the LEM nor the DS, otherwise naive set theory would be inconsistent (which it is not (see [9])) or inconsistent models would become trivial (which some do not (see [7])).

(14) The Classical Recapture in the Logic of Entailment.

We can proceed with the outlined classical recapture using **MC** as entailment logic, by setting up classical sentences using rejection, and proving the LEM and the DS for appropriate sentences by using disjunction introduction, metacompleteness and rejection. However, as stated previously, the LEM and the DS can often be bypassed when proving one's particular conclusion.

(15) Free and Bound Variables.

We use separate free and bound variables in the formalization of **MCQ**.

(16) Constants and Variables.

No individual constants are introduced in **MCQ**, free variables sufficing for the pure logic. They can be added for applied logics, suitably axiomatized.

(17) Duality between Universal and Existential Quantification.

As with (7), duality can be seen from the I- and E-rules of **NMCQ**. **MCQ** is metacomplete (see [11]); thus full duality is available for the Hilbert system **MCQ** with rejection.

(18) Restricted Quantification.

We can add restricted quantification to **MCQ** by axiomatizing it as in Ch.13 of [11].

(19) Undecidability of Predicate Calculus.

MCQ is decidable (by a recent result in [10]).

(20) Non-recursive Domains.

There is no such problem for **NMCQ** as we generalise on unconstrained free variables. For particular applications, structural rules can be added for particular recursive or finite domains.

(21) Establishable Truth.

Truth in the natural deduction system **NMCQ** and the accompanying free semantics (see [12]) is proved within the system itself through the proof of signed formulae of the form TA.

(22) Worlds and Possible Worlds.

If we replace truth-functional semantics by a suitable content semantics and by free semantics, which is a spin-off from **NMCQ**, then worlds, which are aligned to truth-functional semantics, are not needed for entailment. Sets of entailment hypotheses do a similar job in free semantics (see [12]).

(23) Whither Truth-functional Semantics?

NMCQ, and its associated free semantics, can largely do technical service for a truth-functional semantics, due to its inductive character. Content semantics suffices as a semantics and free semantics as a useful adjunct.[4]

BIBLIOGRAPHY

[1]   A.R. ANDERSON and Jr. N.D. BELNAP. *Entailment: The Logic of Relevance and Necessity, Vol.1.* Princeton U.P., 1975.

[2]   R.T. BRADY. Unspecified constants in predicate calculus and first-order theories. *Logique et Analyse*, 20:229–243, 1977.

[3]   R.T. BRADY. The simple consistency of a set theory based on the logic **CSQ**. *Notre Dame Journal of Formal Logic*, 24:431–449, 1983.

[4]   R.T. BRADY. Natural deduction systems for some quantified relevant logics. *Logique et Analyse*, 27:355–377, 1984.

[5]   R.T. BRADY. A content semantics for quantified relevant logics I. *Studia Logica*, 47:111–127, 1988.

[6]   R.T. BRADY. A content semantics for quantified relevant logics II. *Studia Logica*, 48:243–257, 1989.

[7]   R.T. BRADY. The non-triviality of dialectical set theory. In R. Routley G. Priest and J. Norman, editors, *Paraconsistent Logic*, pages 437–471. Philosophia Verlag, Munich, 1989.

[8]   R.T. BRADY. Relevant implication and the case for a weaker logic. *Journal of Philosophical Logic*, 25:151–183, 1996.

---

[4]I wish to thank the two referees whose valuable suggestions have made this paper much more readable. I also wish to thank my colleague, Manfred von Thun, for some incisive comments on this paper.

[9]   R.T. BRADY. Entailment, negation and paradox solution. In G. Priest J.-P. van Ben-
      degem D. Batens, C. Mortensen, editor, *Frontiers of Paraconsistent Logic*, pages 113–
      135. Research Studies Press, 2000.
[10] R.T. BRADY. Normalized natural deduction systems for some relevant logics -I, -
      II, -III. In *Australasian Association for Logic Conferences at Wellington in 2001,
      A.N.U. in 2002, at Adelaide in 2003, and at Dunedin in 2004, Part I will appear in
      the Journal of Symbolic Logic*, 2001-4.
[11] R.T. BRADY. *Relevant Logics and their Rivals, Vol.2: A Continuation of the Work
      of R. Sylvan, R. Meyer, V. Plumwood and R. Brady*. Ashgate, Aldershot, 2003.
[12] R.T. BRADY. Free semantics. In *presented to the Australasian Association for
      Philosophy Conference at South Molle Is.*, 2004.
[13] R.T. BRADY. *Universal Logic*. CSLI Publs, Stanford, forthcoming, 2005.
[14] A. CHURCH. A note on the entscheidungsproblem. *Journal of Symbolic Logic*, 1:40–
      41, 101–102, 1936.
[15] K. FINE. Sematics for quantified relevance logic. *Journal of Philosophical Logic*,
      17:27–59, 1988.
[16] R.K. MEYER. Entailment is not strict implication. *Australasian Journal of Philos-
      ophy*, 52:212–231, 1974.
[17] R.K. MEYER. Metacompleteness. *Notre Dame Journal of Formal Logic*, 17:501–516,
      1976.
[18] G. PRIEST. Motivations for paraconsistency: The slippery slope from classical logic
      to dialetheism. In G. Priest J.-P. van Bendegem D. Batens, C. Mortensen, editor,
      Frontiers of Paraconsistent Logic, pages 223–232. Research Studies Press, 2000.
[19] J. SLANEY. A metacompleteness theorem for contraction-free relevant logics. *Studia
      Logica*, 43:159–168, 1984.
[20] J. SLANEY. Reduced models for relevant logics without WI. *Notre Dame Journal of
      Formal Logic*, 28:395–407, 1987.

Ross Brady

La Trobe University, Melbourne, Victoria 3086, Australia.

Ross.Brady@latrobe.edu.au

Tautological Entailments and Their Rivals

FRANCESCO PAOLI

1 Introduction

Tautological entailments were first investigated in the late 1950s by Smiley and by Anderson and Belnap, in the context of their pioneering studies of relevance logics (see e.g. Anderson and Belnap, 1962 and, for an exhaustive survey of these early investigations, Anderson and Belnap, 1975). More precisely, tautological entailments were introduced and motivated as the provable *first-degree entailments* of the relevance logics $\mathbf{E}$ and $\mathbf{R}$, where a first-degree entailment is nothing but a formula of the form $A \to B$, in which the arrow stands for the relevant implication and both A and B contain no connective but negation ($\neg$) and the classical, extensional connectives $\wedge$ and $\vee$ (no nesting of arrows, therefore, is allowed). Alternatively, one can conceive of tautological entailments as expressing a relevant derivability relation between two classical formulae.

It soon turned out that the system of tautological entailments (henceforth $\mathbf{T}$) was an especially well-behaved fragment of the above-mentioned logics: nice normal form, variable sharing and interpolation results were readily proved and a characteristic matrix was provided. This model was subsequently given by Dunn (1976) and Belnap (1977) a perspicuous interpretation in terms of "epistemic truth values", a feature which made $\mathbf{T}$ particularly appropriate for handling problems of management of inconsistent and incomplete information (see also Routley and Routley, 1972). Since then, tautological entailments — and even more so their generalization known as *Belnap's four-valued logic*, where a more general derivability relation as well as new connectives are introduced - have become a favourite of computer scientists and artificial intelligence experts (see e.g. Levesque, 1984; Ginsberg, 1988; Patel-Schneider, 1989; Fitting, 1990, 1994; Weber, 1998).

In this paper, we shall survey some well-known results about Anderson's and Belnap's tautological entailments. Moreover, we shall try to highlight the connections between this system and other systems of first-degree entailments which share similar properties and are sometimes grounded on closely related intuitive motivations, although such similarities are not always stressed or made clear in the literature. As we shall see, these systems are directly drawn from, or anyway connected to, Lukasiewicz's three-valued logic, Kleene's strong and weak three-valued logics (Kleene, 1952), Priest's "logic of paradox" (Priest, 1979), Halldén's "logic of nonsense" (Halldén, 1949), as well as Epstein's logics of relatedness, equality of content, dependence and dual dependence (Epstein, 1990). Not all of these logics are paraconsistent, and many of them were introduced for quite different purposes than the logical investigation of inconsistency; some, moreover (e.g. Epstein's logics) have been given semantical and proof-theoretical presentations which render a comparison with Anderson's and Belnap's logic somewhat difficult. Therefore, we believe that encompassing them along with tautological entailments in a common framework might help to clarify some relationships or affinities between paraconsistent logics and other non-classical logics of many-valued or broadly relevant provenance.

This paper is structured as follows. After dispatching some preliminaries (Section 2), in Section 3 we shall consider four-valued characteristic matrices for tautological entailments and for the first-degree entailments of relatedness logic. From those tables we shall extract three-valued tables for a number of logics; inclusion relationships between such systems will be investigated. In Section 4 we shall present several uniform proof-theoretic formulations for some of the above systems. General algebraic semantics, relational semantics (in the style of Routley et al., 1982) or set-assignment semantics (in the style of Epstein, 1990) will not be discussed here.

Many results in this paper are not new, except perhaps in their formulations; the appropriate pointers to the literature will be duly provided. However, we shall also prove some original theorems, especially as regards Epstein's logics.

2   Preliminaries

The aim of this section is that of writing down some definitions which will turn out useful in the following, especially in Section 3.

DEFINITION 1 (First-degree entailment). Let $£$ be a propositional language containing a denumerable set $VAR(£)$ of variables and the connectives $\neg$ (unary), $\wedge, \vee$ (binary). The set $FOR(£)$ of well-formed formulae of $£$ and the absolutely free algebra $\mathcal{F}(£)$ of formulae of $£$ are constructed as usual. A *first-degree entailment (fde)* is an expression of the form $\phi \rightarrow \psi$, where $\phi, \psi \in FOR(£)$. If $\phi \in FOR(£)$, we set

$$Var(\phi) = \{p : p \in VAR(£) \& p \text{ occurs in } \phi\}.$$

DEFINITION 2 (FDE-matrix). A FDE-matrix for £ is an ordered pair $\mathcal{M} = \langle \mathcal{A}, R \rangle$, where $\mathcal{A} = \langle A, \neg^{\mathcal{A}}, \wedge^{\mathcal{A}}, \vee^{\mathcal{A}} \rangle$ is an algebra of type $\langle 1, 2, 2 \rangle$ ad $R \subseteq A^2$.

DEFINITION 3 (Interpretation). An *interpretation* for £ is an ordered pair $\mathcal{I} = \langle \mathcal{M}, v \rangle$, where $\mathcal{M} = \langle \mathcal{A}, R \rangle$ is a FDE-matrix for £ and $v : \mathrm{VAR}(£) \to A$ is a mapping which is extended in the customary way to a homomorphism from $\mathcal{F}(£)$ to $\mathcal{A}$ (which we shall denote by v as well, with a notational abuse).

DEFINITION 4 (Validity). We say that the fde $\phi \to \psi$ is *true* in the interpretation $\mathcal{I} = \langle M, v \rangle$, where $\mathcal{M} = \langle \mathcal{A}, R \rangle$ iff $\langle v(\phi), v(\psi) \rangle \in R$; we say that $\phi \to \psi$ is *valid* in the FDE-matrix $\mathcal{M}$ (in symbols: $\mathcal{M} \vDash \phi \to \psi$) iff it is true in every interpretation for £ whose first projection is $\mathcal{M}$.

We now introduce the important concept of *submatrix* of an FDE-matrix.

DEFINITION 5 (Submatrix). Let $\mathcal{M} = \langle \mathcal{A}, R \rangle$ be a FDE-matrix. A FDE-matrix $\mathcal{M}' = \langle \mathcal{A}', R' \rangle$ is said to be a *submatrix* of $\mathcal{M}$ iff $\mathcal{A}'$ is a subalgebra of $\mathcal{A}$ and $R' = R {\restriction} A'$.

It is expedient to recall the following facts:

LEMMA 6. *a) If $\mathcal{M}' = \langle \mathcal{A}', R' \rangle$ is a submatrix of $\mathcal{M} = \langle \mathcal{A}, R \rangle$, then for any $\phi, \psi \in \mathrm{FOR}(£)$, if $\mathcal{M} \vDash \phi \to \psi$ then $\mathcal{M}' \vDash \phi \to \psi$. b) If $\mathcal{M} = \langle \mathcal{A}, R \rangle, \mathcal{M}' = \langle \mathcal{A}, R' \rangle$ and $R \subseteq R'$, then for any $\phi, \psi \in \mathrm{FOR}(£)$, if $\mathcal{M} \vDash \phi \to \psi$ then $\mathcal{M}' \vDash \phi \to \psi$.*

3   Truth tables

We shall start by comparing the system **T** of tautological entailments and a system which we name **E** (after R.L. Epstein), corresponding to the first-degree entailment fragment of symmetric relatedness logic (Epstein, 1990). We introduce them semantically, viz. by means of appropriate FDE-matrices. The one for **T** was first suggested by T. Smiley (as reported in Anderson and Belnap, 1975), while the one for **E** stems from Paoli (1993). Such matrices involve four truth values — t, b, n, f. According to the Dunn-Belnap interpretation one can intuitively think of t as "true only", of b as "both true and false", of n as "neither true nor false" and of f as "false only". These labels must be understood epistemically, not ontologically — moreover, this reading is much more appropriate for the tables of **T** than for those of **E**, whose values still lack a convincing intuitive explanation.

DEFINITION 7 (The algebras $\mathcal{A}_T^4$ and $\mathcal{A}_E^4$). Let $V = \{t, b, n, f\}$. We set $\mathcal{A}_T^4 = \langle V, \neg_T^4, \wedge_T^4, \vee_T^4 \rangle$, where the operations are defined by the following tables:

$\neg_T^4$	
f	t
n	n
b	b
t	f

$\wedge_T^4$	f	n	b	t
f	f	f	f	f
n	f	n	f	n
b	f	f	b	b
t	f	n	b	t

$\vee_T^4$	f	n	b	t
f	f	n	b	t
n	n	n	t	t
b	b	t	b	t
t	t	t	t	t

We also set $\mathcal{A}_E^4 = \langle V, \neg_E^4, \wedge_E^4, \vee_E^4 \rangle$, where the operations are defined by the following tables:

$\neg_E^4$	
f	t
n	n
b	b
t	f

$\wedge_E^4$	f	n	b	t
f	f	b	b	f
n	b	n	b	b
b	b	b	b	b
t	f	b	b	t

$\vee_E^4$	f	n	b	t
f	f	b	b	t
n	b	n	b	b
b	b	b	b	b
t	t	b	b	t

DEFINITION 8 (Relations arising out of the tables). Let $\mathbf{S} \in \{\mathbf{T}, \mathbf{E}\}$. We define the following binary relations on V:

$$\leq_{\mathbf{S}}^{4l} = \{\langle x, y \rangle : x \wedge_{\mathbf{S}}^4 y = x\};$$
$$\leq_{\mathbf{S}}^{4u} = \{\langle x, y \rangle : x \vee_{\mathbf{S}}^4 y = y\};$$
$$\leq_{\mathbf{S}}^4 = \leq_{\mathbf{S}}^{4l} \cup \leq_5^{4u} .$$

LEMMA 9. *a)* $\leq_{\mathbf{T}}^{4l}, \leq_{\mathbf{T}}^{4u}, \leq_{\mathbf{E}}^{4l}, \leq_{\mathbf{E}}^{4u}$ *are semilattice orderings on v; b)* $\leq_{\mathbf{T}}^{4l} = \leq_{\mathbf{T}}^{4u} = \leq_{\mathbf{T}}^4$; *c)* $\leq_{\mathbf{T}}^4$ *is a lattice ordering on* V.

Proof. By inspection. Hasse diagrams for $\leq_{\mathbf{E}}^{4l}, \leq_{\mathbf{E}}^{4u}, \leq_{\mathbf{T}}^4$ are provided bleow.

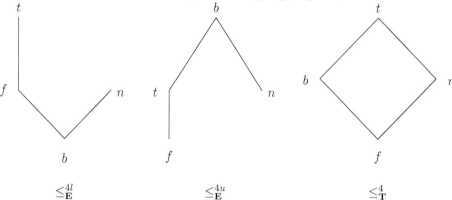

Remark that $\leq_{\mathbf{E}}^4$ is not even a preordering on V, since it is not transitive.

DEFINITION 10 (The FDE-matrices $\mathcal{M}_T, \mathcal{M}_E$). Let $\mathbf{S} \in \{\mathbf{T}, \mathbf{E}\}$. We set $\mathcal{M}_S = \langle \mathcal{A}_S^4, \leq_S^4 \rangle$.

Let us now consider more closely the four-valued tables for $\mathbf{T}$ and $\mathbf{E}$. It is interesting to remark that, if we restrict ourselves to the values t, b, and f, we respectively obtain the *strong* and the *weak* Kleene tables for $\neg, \wedge, \vee$ (Kleene, 1952; weak Kleene tables are also known as *Bochvar* tables). As far as $\mathbf{T}$ is concerned, restriction to $\{t, n, f\}$ would lead exactly to the same result; yet not for $\mathbf{E}$ — since the indicated subset would not even be a subuniverse of V. Three-valued logics based on the strong Kleene tables are especially important and are deeply investigated in the literature; they are called *natural* in Avron (1991). We are thus led to the following definitions:

DEFINITION 11 (The algebras $\mathcal{A}_T^3$ and $\mathcal{A}_E^3$). Let $\mathbf{S} \in \{\mathbf{T}, \mathbf{E}\}$, and let $V' = \{t, b, f\}$. We set $\mathcal{A}_S^3 = \langle V', \neg_S^3, \wedge_S^3, \vee_S^3 \rangle$, where each operation is the restriction to V' of the corresponding operation in $\mathcal{A}_S^4$.

It turns out that there are several natural ways to endow $\mathcal{A}_T^3$ and $\mathcal{A}_E^3$ with binary relations, so as to build up FDE-matrices. Let us examine some of them.

DEFINITION 12 (Relations arising out of the tables). Let $\mathbf{S} \in \{\mathbf{T}, \mathbf{E}\}$. We define the following binary relations on V':

$$\leq_S^{3l} = \{\langle x, y \rangle : x \wedge_S^3 y = x\};$$
$$\leq_S^{3u} = \{\langle x, y \rangle : x \vee_S^3 y = y\};$$
$$\leq_S^3 = \leq_S^{3l} \cap \leq_S^{3u}.$$

Remark that in Definition 12 $\leq_S^3$ is defined as the *intersection* of $\leq_S^{3l}$ and $\leq_S^{3u}$, while in Definition 8 we had used the union. However, this obviously makes no difference for $\leq_T^3$, whereas if we had chosen union in defining $\leq_E^3$ we would have obtained a useless FDE-matrix, satisfying exactly the classically valid fdes.

DEFINITION 13 (More FDE-matrices). We define:

$$\mathcal{M}_{\mathbf{M}} = \langle \mathcal{A}_T^3, \leq_T^3 \rangle;$$
$$\mathcal{M}_{\mathbf{Kl}} = \langle \mathcal{A}_T^3, \leq_T^3 \cup \leq_E^{3l} \rangle;$$
$$\mathcal{M}_{\mathbf{P}} = \langle \mathcal{A}_T^3, \leq_T^3 \cup \leq_E^{3u} \rangle;$$
$$\mathcal{M}_{\mathbf{EM}} = \langle \mathcal{A}_E^3, \leq_T^3 \rangle;$$
$$\mathcal{M}_{\mathbf{B}} = \langle \mathcal{A}_E^3, \leq_T^3 \cup \leq_E^{3l} \rangle;$$
$$\mathcal{M}_{\mathbf{H}} = \langle \mathcal{A}_E^3, \leq_T^3 \cup \leq_E^{3u} \rangle;$$
$$\mathcal{M}_{\mathbf{Eq}} = \langle \mathcal{A}_E^3, \leq_E^3 \rangle;$$
$$\mathcal{M}_{\mathbf{D}} = \langle \mathcal{A}_E^3, \leq_E^{3l} \rangle;$$
$$\mathcal{M}_{\mathbf{DD}} = \langle \mathcal{A}_E^3, \leq_E^{3u} \rangle.$$

DEFINITION 14 (Notions of validity). Let $\mathbf{S}, \mathbf{S}' \in \{\mathbf{E}, \mathbf{T}, \mathbf{M}, \mathbf{Kl}, \mathbf{P}, \mathbf{EM}, \mathbf{B}, \mathbf{H}, bfEq, D, DD\}$. We say that a fde $\phi \to \psi$ is $\mathbf{S}$-*valid* (in symbols, $\vDash_{\mathbf{S}} \phi \to \psi$)

iff $\mathcal{M_S} \vDash \phi \to \psi$. Occasionally, we shall refer to **S** as the logic determined by all **S**-valid fdes and write $\mathbf{S} \subseteq \mathbf{S}'$ just in case $\vDash_{\mathbf{S}} \phi \to \psi$ implies $\vDash_{\mathbf{S}'} \phi \to \psi$ for every ϕ, ψ in $\mathbf{FOR}(\mathcal{L})$.

It is noteworthy that $\vDash_{\mathbf{T}} \phi \to \psi$ iff $v(\psi) \in \{t, b\}$ whenever $v(\phi) \in \{t, b\}$. A proof of this fact can be found e.g. in Font (1997).

The logics introduced in Definition 14 are tightly related to several well-known logics:

1. **M** is related to Łukasiewicz's three-valued logic $\mathbf{L}_3$ in the following sense: for $\phi, \psi \in \mathbf{FOR}(\mathcal{L})$, $\phi \to \psi$ is **M**-valid just in case it is a valid fde of $\mathbf{L}_3$. Remark, however, that the standard semantic deduction theorem does not hold in $\mathbf{L}_3$ and thus it is not the case that $\phi \to \psi$ is **M**-valid iff ψ is a logical consequence of ϕ in $\mathbf{L}_3$ (i.e. $v(\psi) = T$ whenever $v(\phi) = t$). Also, $\phi \to \psi$ is **M**-valid just in case it is a valid fde of **RM**.[1] Avron (1991) considers two possible ways of extracting from $\mathcal{M_M}$ a consequence relation $\Gamma \vdash \Delta$ with multiple premises and conclusions, which he respectively dubs $\vdash_{Luk}$ and $\vdash_{Sob}$ (the latter stands for *Sobocinski's* logic). They collapse onto each other when Γ, Δ are singletons.

2. **Kl** is related to strong Kleene's three-valued logic (Kleene, 1952) in the following sense: for $\phi, \psi \in \mathbf{FOR}(\mathcal{L})$, $\phi \to \psi$ is **Kl**-valid just in case ψ is a logical consequence of ϕ in it (i.e. $v(\psi) = t$ whenever $v(\phi) = t$). Remark, however, that Kleene's logic has no tautologies and thus it is not the case that $\phi \to \psi$ is **Kl**-valid iff it is a valid fde of strong Kleene's logic. **Kl** is also known as Hao Wang's logic (Wang, 1961; Rose, 1963; Bolotov, 1986). Avron (1991) extracts from $\mathcal{M_{Kl}}$ a consequence relation called $\vdash_{Kl}$.

3. **P** is related to Priest's **LP** (Priest, 1979) and to D'Ottaviano's and Da Costa's $\mathbf{J}_3$ (D'Ottaviano and Da Costa, 1970) in the following sense: for $\phi, \psi \in \mathbf{FOR}(\mathcal{L})$, $\phi \to \psi$ is **P**–valid iff it is a valid fde of $\mathbf{J}_3$ iff ψ is a logical consequence of ϕ in **LP** (i.e. $v(\psi) \in \{t, b\}$ whenever $v(\phi) \in \{t, b\}$) Avron (1991) extracts from $\mathcal{M_P}$ a consequence relation called $\vdash_{Pac}$.[2]

4. **Eq, D, DD** respectively correspond to the fde fragments of Epstein's (1990) logics of *equality of contents, dependence* and *dual dependence*. Dependence logic is closely related, both in its motivation and in its

---

[1] Actually, we have so named **M** after R. K. Meyer, who investigated this logic as the first-degree entailment fragment of **RM**.

[2] Reasons to prefer **M** to **Kl** or **P** are provided in Seymour Michael (2002).

technical developments, to Parry's (1933) calculus of *analytic impli-cation* (for which see also Dunn, 1972). Its fdes are investigated e.g. in Deutsch (1981) or in Sinowjew and Wessel (1975). The fdes of dual dependence logic are motivated and studied in Paoli (1992). These approaches are extensively discussed and sharply criticized (under the name of "conceptivist" approaches to relevance) in Routley *et al.* (1982).

5. **B** bears the same relation to weak Kleene's three-valued logic, also known as Bochvar's logic, as Kl bears to strong Kleene's three-valued logic. On the other side, **H** bears the same relation to Halldén's (1949) *logic of nonsense* as **P** bears to **LP**.

LEMMA 15. *a)* $\leq_T^3, \leq_T^3 \cup \leq_E^{3l}, \leq_T^3 \cup \leq_E^{3u}, \leq_E^3, \leq_E^{3l}, \leq_E^{3u}$ *are preorderings on* v'; *b) they are all partial orderings, except* $\leq_T^3 \cup \leq_E^{3l}$ *and* $\leq_T^3 \cup \leq_E^{3u}$; *c) they are all linear, except* $\leq_E^3$.

Proof. By inspection. Hasse diagrams for these preorderings are provided below.

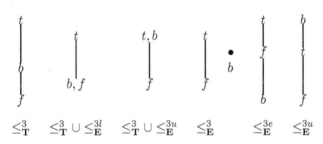

Epstein's logics of relatedness, dependence, dual dependence and equality of contents are called by Priest (2002) "filter logics", because they are the result of an attempt to single out a class of relevantly correct entailments (according to intuitive criteria of relevance which differ from one logic to another) by "filtering out" classically valid entailments through some kind of "sieve". On the level of fdes, such sieves are adequately characterized through appropriate variable sharing requirements, which also work for a wider range of logics. In fact, we have:

THEOREM 16 (Characterization of filter logics). *Let* $\mathbf{S} \in \{\mathbf{E}, \mathbf{EM}, \mathbf{B}, \mathbf{H}, \mathbf{EQ}, \mathbf{D}, \mathbf{DD}\}$. *Then* $\vDash_S \phi \to \psi$ *iff* $\phi \to \psi$ *is classically valid* ($\vDash_K \phi \to \psi$) *and:*

a) for **E**, $Var(\phi) \cap Var(\psi) \neq \varnothing$;

b) *for* **D**, $Var(\phi) \supseteq Var(\psi)$;

c) *for* **DD**, $Var(\phi) \subseteq Var(\psi)$;

d) *for* **Eq**, $Var(\phi) = Var(\psi)$;

e) *for* **EM**, *(either $\vDash_K \neg\phi$ or $Var(\phi) \supseteq Var(\psi)$) and (either $\vDash_K \psi$ or $Var(\phi) \subseteq Var(\psi)$)*;

f) *for* **B**, *either $\vDash_K \neg\phi$ or $Var(\phi) \supseteq Var(\psi)$*;

g) *for* **H**, *either $\vDash_K \psi$ or $Var(\phi) \subseteq Var(\psi)$*.

Proof. a) is proved in Paoli (1993); b) is proved e.g. in Deutsch (1981); c) is proved in Paoli (1992); a more general version of f) is proved in Urquhart (1986). We just prove e); the remaining items are taken care of similarly.

By Lemma 1, if $\vDash_{\mathbf{EM}} \phi \to \psi$, then $\phi \to \psi$ is classically valid. Now, suppose it is not the case that $\vDash_{\mathbf{K}} \neg\phi$ and it is not the case that $Var(\phi) \supseteq Var(\psi)$. Then there exists a classical valuation v such that $v(\phi) = t$. Now, let w coincide with v as to the variables in ϕ, while $w(p) = b$ if p does not occur in ϕ. Clearly, $\langle \mathcal{M}_{\mathbf{EM}}, w \rangle$ is an interpretation for £ such that $w(\phi) = t, w(\psi) = b$. Since $\langle t, b \rangle \not\leq^3_{\mathbf{T}}$, we conclude that it is not the case that $\vDash_{\mathbf{EM}} \phi \to \psi$. Likewise, if it is not the case that $\vDash_{\mathbf{K}} \psi$ and it is not the case that $Var(\phi) \subseteq Var(\psi)$, a similar argument shows that it is not the case that $\vDash_{\mathbf{EM}} \phi \to \psi$ either.

Conversely, suppose that it is not the case that $\vDash_{\mathbf{EM}} \phi \to \psi$. Then there is an interpretation $\langle \mathcal{M}_{\mathbf{EM}}, v \rangle$, such that either a) $v(\phi) = t, v(\psi) = f$ or b) $v(\phi) = t, v(\psi) = b$, or else c) $v(\phi) = b, v(\psi) = f$. If a) holds, then by inspection of the tables one sees that $v(p) \in \{t, f\}$ for any variable $p \in Var(\phi) \cup Var(\psi)$. This immediately yields a falsifying classical valuation for $\phi \to \psi$. If b) holds, once again an inspection of the tables shows that $v(p) \in \{t, f\}$ for any $p \in Var(\phi)$, while there must be q in $Var(\psi)$ such that $v(q) = b$. Hence it cannot be $Var(\psi) \subseteq Var(\phi)$. Moreover, let w coincide with v over the variables in ϕ, while $w(p) = f$ for any other variable. w is a classical valuation such that $w(\phi) = t$, which shows that $\neg\phi$ cannot be tautologous. Case c) is disposed of similarly. ∎

The next theorem provides an overview of the inclusion relationships among the logics defined above.

THEOREM 17 (Inclusion relationships among logics). *The inclusion relationships among the logics mentioned above are pictured below:*

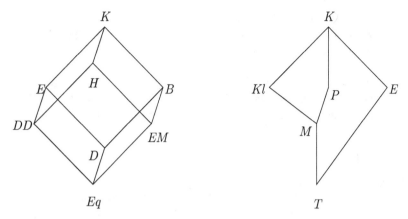

| *A cube of logics in the filter family …* | *… and how E relates to logics in the Belnap family.* |

Moreover, all such inclusions are proper.

Proof. For a start, remark that as a consequence of Lemma 6, $\mathbf{Kl}, \mathbf{P}, \mathbf{E}, \mathbf{B}, \mathbf{H} \subseteq \mathbf{K}$. For the same reason, we can also establish that $\mathbf{T} \subseteq \mathbf{M}, \mathbf{D} \subseteq \mathbf{B}, \mathbf{DD} \subseteq \mathbf{H}, \mathbf{Eq} \subseteq \mathbf{EM}, \mathbf{M} \subseteq \mathbf{Ke}, \mathbf{M} \subseteq \mathbf{P}, \mathbf{EM} \subseteq \mathbf{B}, \mathbf{EM} \subseteq \mathbf{H}$.

Theorem 16, given the fact that each formula in $FOR(\pounds)$ contains at least a propositional variable, implies that $\mathbf{Eq} \subseteq \mathbf{D}, \mathbf{Eq} \subseteq \mathbf{DD}, \mathbf{D} \subseteq \mathbf{E}, \mathbf{DD} \subseteq \mathbf{E}$.

It remains to prove that $\mathbf{T} \subseteq \mathbf{E}$. By Theorem 16, it suffices to show that if $\vDash_\mathbf{T} \phi \to \psi$, then $\phi \to \psi$ is classically valid and $\mathrm{Var}(\phi) \cap \mathrm{Var}(\psi) \neq \varnothing$. The former property follows from Lemma 6. As to the latter, suppose that $\mathrm{Var}(\phi) \cap \mathrm{Var}(\psi) = \varnothing$. Then let $v(p) = b$ for $p \in \mathrm{Var}(\phi)$ and $v(q) = n$ for $q \in \mathrm{Var}(\psi)$. Hence $v(\phi) = b, v(\psi) = n$ and, since $\langle b, n \rangle \not\leq^4_\mathbf{T}$, the interpretation $\langle \mathcal{M}_\mathbf{T}, v \rangle$ falsifies $\phi \to \psi$.

The next table shows that all the preceding inclusions are proper (Y means that the given formula is valid, N means that it is not):

	K	Ke	P	M	T	Eq	D	DD	E	H	EM	B
$\phi \wedge \neg\phi \to \psi$	Y	Y	N	N	N	N	N	N	N	N	N	Y
$\phi \to \psi \vee \neg\psi$	Y	N	Y	N	N	N	N	N	N	Y	N	N
$\phi \wedge (\neg\phi \vee \psi) \to \psi$	Y	Y	N	N	N	N	Y	N	Y	N	N	Y
$\phi \wedge \psi \to \phi$	Y	Y	Y	Y	Y	N	Y	N	Y	N	N	Y
$\phi \to \phi \vee \psi$	Y	Y	Y	Y	Y	N	N	Y	Y	Y	N	N
$\phi \wedge \neg\phi \to \psi \vee \neg\psi$	Y	Y	Y	Y	N	N	N	N	N	Y	Y	Y
$(\phi \wedge \neg\phi) \vee (\psi \wedge \neg\psi) \to \phi \wedge \psi$	Y	Y	N	N	N	Y	Y	Y	Y	Y	Y	Y
$\phi \vee \psi \to (\phi \vee \neg\phi) \wedge (\psi \vee \neg\psi)$	Y	N	Y	N	N	Y	Y	Y	Y	Y	Y	Y

∎

Such a table also yields additional nontrivial information: we can infer e.g. that **Eq** is not included in **T**, **H** is not included in **P**, **EM** is not included in **M**, and **B** is not included in **Kl**.

Of course, many more rivals to **T** than we mentioned could be examined within this framework. It could be worthwhile, for example, to find out what first-degree entailment systems could be extracted from the logics considered in Carnielli and Marcos (2002). We leave this as a task to the interested reader.

4   Proof theory

It is usually believed that logics in the filter family are difficult to compare with logics in the Belnap family, especially on the proof-theoretical side. In this section, we shall discuss several well-known proof-theoretic presentations of logics in the latter group, which however — interestingly enough — can be also adapted for at least one filter logic, **E**. We hope that such a perspective can facilitate further comparisons across both groups.

Remark, in any case, that the formalisms hereafter considered are by no means the sole employable tools: as regards **T** and its generalizations (such as Belnap's four-valued logic), for instance, several sequent calculi (e.g. Anderson and Belnap, 1975; Font, 1997; Arieli and Avron, 1998) as well as natural deduction calculi (Tamminga and Tanaka, 1999) have been suggested.

4.1   Hilbert-style systems

T admits of a simple Hilbert-style axiomatization (Anderson and Belnap, 1975), whence appropriate postulates can be extracted also for **Kl** (Rose, 1963), **P**, **M** (see e.g. Dunn, 1986) and **E** (Paoli, 1993).

DEFINITION 18 (Hilbert-type systems). The calculus **HT**, based on the language $\mathfrak{L}$, has the following axioms and rules:

A1.1	$\phi \wedge \psi \to \phi$	A1.2	$\phi \wedge \psi \to \psi$
A2.1	$\phi \to \phi \vee \psi$	A2.2	$\psi \to \phi \vee \psi$
A3.1	$\phi \to \neg\neg\phi$	A3.2	$\neg\neg\phi \to \phi$
A4	$\phi \wedge (\psi \vee \chi) \to (\phi \wedge \psi) \vee \chi$		
R1	$\phi \to \psi, \psi \to \chi \vdash \phi \to \chi$	R2	$\phi \to \psi, \phi \to \chi \vdash \phi \to \psi \wedge \chi$
R3	$\phi \to \chi, \psi \to \chi \vdash \phi \vee \psi \to \chi$	R4	$\phi \to \psi \vdash \neg\psi \to \neg\phi.$

Now, consider the following additional postulates:

A5	$\phi \wedge \neg\phi \to \psi \vee \neg\psi$	A6	$\phi \to \phi \wedge (\psi \vee \neg\psi)$
A7.1	$\neg(\phi \wedge \psi) \to \neg\phi \vee \neg\psi$	A7.2	$\neg\phi \vee \neg\psi \to \neg(\phi \wedge \psi)$
A7.3	$\neg(\phi \vee \psi) \to \neg\phi \wedge \neg\psi$	A7.4	$\neg\phi \wedge \neg\psi \to \neg(\phi \vee \psi)$
A8	$\phi \wedge \neg\phi \to \psi$	A9	$\phi \to \psi \vee \neg\psi$
R1′	$\phi \to \psi, \psi \to \chi \vdash \phi \to \chi$, if ϕ and χ share a variable.		

The calculus **HKl** can be obtained from **HT** by adding A7 and A8 and dropping R4. From **HKl** we get **HP** replacing A8 by A9. **HM** is obtained from **HT** by adjoining A5. Finally, **HE** differs from **HT** in that it has A6 as an extra axiom and the weaker rule R1' in place of R1.

The notions of proof and provability are defined as usual. We write $\vdash_{\mathbf{HS}}$ $\phi \to \psi$ to mean that the fde $\phi \to \psi$ is provable in the calculus **HS** (where $\mathbf{S} \in \{\mathbf{T}, \mathbf{Kl}, \mathbf{P}, \mathbf{M}, \mathbf{E}\}$). We shall also use the abbreviation $\vdash_{\mathbf{HS}} \phi \leftrightarrow \psi$ to mean that both $\vdash_{\mathbf{HS}} \phi \to \psi$ and $\vdash_{\mathbf{HS}} \psi \to \phi$.

LEMMA 19 (Theorems of **HT**, **HE**).

a) *The following theorems are provable in* **HT**: *(T1)* $\phi \to \phi$; *(T2)* $\phi \leftrightarrow \phi \wedge \phi$; *(T3)* $\phi \leftrightarrow \phi \vee \phi$; *(T4)* $\phi \wedge \psi \leftrightarrow \psi \wedge \phi$; *(T5)* $\phi \vee \psi \leftrightarrow \psi \vee \phi$; *(T6)* $\phi \wedge (\psi \wedge \chi) \leftrightarrow (\phi \wedge \psi) \wedge \chi$; *(T7)* $\phi \vee (\psi \vee \chi) \leftrightarrow (\phi \vee \psi) \vee \chi$; *(T8)* $\phi \vee (\psi \wedge \chi) \leftrightarrow (\phi \vee \psi) \wedge (\phi \vee \chi)$; *(T9)* $\phi \wedge (\psi \vee \chi) \leftrightarrow (\phi \wedge \psi) \vee (\phi \wedge \chi)$; *(T10)* $\neg(\phi \wedge \psi) \leftrightarrow \neg\phi \vee \neg\psi$; *(T11)* $\neg(\phi \vee \psi) \leftrightarrow \neg\phi \wedge \neg\psi$.

b) *Besides T1–T11, the following theorems are also provable in* **HE**: *(T12)* $\phi \wedge (\neg\phi \vee \psi) \to \psi$; *(T13)* $\phi \vee (\psi \wedge \neg\psi) \to \phi$; *(T14)* $\phi \wedge \neg\phi \to \psi$, *if ϕ and ψ share a variable; (T15)* $\phi \to \psi \vee \neg\psi$, *if ϕ and ψ share a variable.*

Proof. Proof sketches for some of the T1–T15 are provided e.g. in Anderson and Belnap (1975) or Paoli (1993). ∎

We also state without a proof the following results (Lemma 20 for **HT** can be found in Anderson and Belnap, 1975 and can be straightforwardly extended to **HE** using restricted transitivity, while Lemma 21 is proved in Paoli, 1993):

LEMMA 20 (Derivable rules). *Let* $\mathbf{S} \in \{\mathbf{T}, \mathbf{M}, \mathbf{E}\}$. *The following rules are derivable in* **HS**: *(R5)* $\phi \leftrightarrow \psi \vdash \neg\phi \leftrightarrow \neg\psi$; *(R6)* $\phi \leftrightarrow \psi, \chi \leftrightarrow \sigma \vdash \phi \wedge \chi \leftrightarrow \psi \wedge \sigma$; *(R7)* $\phi \leftrightarrow \psi, \chi \leftrightarrow \sigma \vdash \phi \vee \chi \leftrightarrow \psi \vee \sigma$.

LEMMA 21 (Derivable rules in HE).

a) *If* $(\mathrm{Var}(\phi_1 \cup \ldots \cup \mathrm{Var}(\phi_n)) \cap \mathrm{Var}(\psi) \neq \varnothing$ *and for some $i, j \leq n$ it is* $\phi_i = \neg\phi_j$, *then* $\vdash_{HE} \phi_1 \wedge \ldots \phi_n \to \psi$;

b) *If* $\mathrm{Var}(\phi) \cap (|\mathrm{Var}(\psi_1) \cup \ldots \cup \mathrm{Var}(\psi_n)) \neq \varnothing$ *and for some $i, j \leq n$ it is* $\psi_i = \neg\psi_j$, *then* $\vdash_{HE} \phi \to \psi_1 \vee \ldots \vee \psi_n$;

c) *If* $(\mathrm{Var}(\phi_1) \cup \ldots \cup \mathrm{Var}(\phi_n)) \cap \mathrm{Var}(\psi) \neq \varnothing$ *and each ϕ_i is such that either* $\vdash_{HE} \phi_i \to \psi$ *or it is a generalized conjunction* $\chi_1 \wedge \ldots \chi_m$ *where for some $i, j \leq m$ it is* $\chi_i = \neg\chi_j$, *then* $\vdash_{HE} \phi_1 \vee \ldots \vee \phi_n \to \psi$;

d) If $\mathrm{Var}(\phi) \cap (\mathrm{Var}(\psi_1) \cup \ldots \cup \mathrm{Var}(\psi_n)) \neq \varnothing$ *and each ψ_i is such that either $\vdash_{HE} \phi \to \psi_i$ or it is a generalized disjunction $\chi_1 \vee \ldots \chi_m$ where for some $i, j \leq m$ it is $\chi_i = \neg\chi_j$, then $\vdash_{HE} \phi \to \psi_1 \wedge \ldots \wedge \psi_n$.*

The following version of the replacement theorem is available in **HT, HM, HE**:

LEMMA 22 (Replacement). *Let $S \in \{\mathbf{T}, \mathbf{M}, \mathbf{E}\}$. $\vdash_{HS} \phi \leftrightarrow \psi$ implies $\vdash_{HS} \chi \leftrightarrow \chi[\phi/\psi]$, wher $\chi[\phi/\psi]$ is obtained by replacing zero or more occurrences of ϕ in χ by ψ.*

Proof. Induction on the complexity of χ, using Lemma 20. ∎

Remark, however, that stronger replacement results do not hold unrestrictedly for **HE**: for example, $\vdash_{HS} \phi \leftrightarrow \psi$ and $\vdash_{HS} \phi \to \chi$ do not necessarily imply $\vdash_{HS} \psi \to \chi$ (let e.g. ϕ be $p \wedge (q \vee \neg q)$, ψ be p, χ be $q \vee \neg q$). Anyway, this holds in case ϕ and ψ contain the same variables; this is just a special case of a more general result proved in Paoli (1996).

LEMMA 23 (Restricted strong replacement). *Let $S \in \{\mathbf{T}, \mathbf{M}, \mathbf{E}\}$. Let $\phi, \psi \in FOR(\pounds)$ and $\mathrm{Var}(\phi) = \mathrm{Var}(\psi)$. a) $\vdash_{HS} \phi \leftrightarrow \psi$ and $\vdash_{HS} \chi \to \sigma$ imply $\vdash_{HS} \chi[\phi/\psi] \to \sigma$; b) $\vdash_{HS} \phi \leftrightarrow \psi$ and $\vdash_{HS} \sigma \to \chi$ imply $\vdash_{HS} \sigma \to \chi[\phi/\psi]$.*

Appropriate versions of Lemmas 22 and 23 also hold for **HKl** and **HP**, although here there is a slight complication due to the fact that R5 fails in general. However, if $S \in \{\mathbf{Kl}, \mathbf{P}\}$, it is possible to prove that $\vdash_{HS} \phi \leftrightarrow \psi$ and $\vdash_{HS} \neg\phi \leftrightarrow \neg\psi$ together imply $\vdash_{HS} \chi \leftrightarrow \chi[\phi/\psi]$; also Lemma 23 can be modified accordingly (cf. Rose, 1963).

4.2   S-entailments

In the next definition we recall some basic syntactical notions. Apart from minor modifications, the terminology is drawn from Anderson and Belnap (1975).

DEFINITION 24 (Some syntactic notions). A *literal* is either a variable in $\mathrm{VAR}(\pounds)$ or the negation of such. $L, L' \ldots$ are used as metavariables for literals. By the *complementary L^c* of a literal L we mean $\neg p$ if L is the variable p, q if L is the negated variable $\neg q$. A *primitive conjunction (disjunction)* is a generalized conjunction (disjunction) of literals. A *conjunctive (disjunctive) normal form* is a generalized conjunction (disjunction) of primitive disjunctions (conjunctions). We shall often use the abbreviations "cnf" for "conjunctive normal form" and "dnf" for "disjunctive normal form".

In all of the systems considered above, given a formula ϕ, it is always possible to find a dnf ϕ^* and a cnf ϕ^{**} which are equivalent to ϕ and contain the same variables as ϕ. In fact, we have:

THEOREM 25 (Normal form). *Let $S \in \{\mathbf{T}, \mathbf{M}, \mathbf{Kl}, \mathbf{P}, \mathbf{E}\}$ and $\phi \in FOR(\mathcal{L})$.
Then there exist a dnf ϕ^* and a cnf ϕ^{**}, both containing the same variables
as ϕ, such that $\vdash_{HS} \phi \leftrightarrow \phi^*, \vdash_{HS} \phi \leftrightarrow \phi^{**}$.*

Proof. We proceed in the standard way using A3, A4, T4–T11 and the
appropriate versions of the replacement theorems. Remark that transitivity
is not impaired while performing normal form moves in **HE**, because in
all of the above-mentioned axioms and theorems both formulae in the (co-
)entailment contain exactly the same variables.   ∎

Applying canonical methods, we can select for each formula $\phi \in FOR(\mathcal{L})$,
among the dnfs and cnfs whose existence is guaranteed by the previous
theorem, a designated one. Thus, we shall henceforth speak of the dnf (cnf)
of the formula ϕ. Now we have all we need to introduce the important
notion of *S-entailment*.

DEFINITION 26 (Explicit S-entailment). Let ϕ be a primitive conjunction
and ψ a primitive disjunction, and consider the following conditions:

  a) ϕ and ψ share a literal;

  b) ϕ contains complementary literals;

  c) ψ contains complementary literals;

  d) ϕ and ψ share a variable.

A fde $\phi \to \psi$ is said to be a *primitive* **T**-*entailment* iff a) holds; a *primitive*
Kl-*entailment* iff either a) or b) holds; a *primitive* **P**-*entailment* iff either
a) or c) holds; a *primitive* **M**-*entailment* iff either a) holds or else both b)
and c) hold.

Now, let $\phi = \phi_1 \vee \ldots \vee \phi_n$ be a dnf, $\psi = \psi_1 \wedge \ldots \wedge \psi_m$ be a cnf, and
let $\mathbf{S} \in \{\mathbf{T}, \mathbf{M}, \mathbf{Kl}, \mathbf{P}\}$. The fde $\phi \to \psi$ is an *explicit S-entailment* iff for
every $i \leq n, j \leq m, \phi_i \to \psi_j$ is a primitive S-entailment. On the other hand,
$\phi \to \psi$ is called an *explicit* **E**-*entailment* iff: 1) for every $i \leq n, j \leq m$,
either a) or b) or c) hold of $\phi_i \to \psi_j$; 2) for some $k \leq n, l \leq m$, d) holds of
$\phi_k \to \psi_e$.

DEFINITION 27 (S-entailment). Let $\mathbf{S} = \{\mathbf{T}, \mathbf{M}, \mathbf{E}, \mathbf{Kl}, \mathbf{P}\}$. Let $\phi \to \psi$ be
a fde, and let $\phi^*(\psi^*)$ be the dnf(cnf) of ϕ (of ψ). We say that $\phi \to \psi$ is an
S-*entailment* iff $\phi^* \to \psi^*$ is an explicit **S**-entailment.

4.3   Tableaux

Several tableaux systems for **T** have been suggested in the literature, and
these are conveniently classified into two main approaches. Dunn (1976) and
Fitting (1994) employ a variant of Jeffrey's method of *coupled trees*; tableaux
of this kind will not be considered here. On the other hand, D'Agostino
(1990) and Priest (2001) investigate tableaux calculi of a more traditional
sort, which — as we shall show — can be extended also to **Kl**, **P**, **M**
and **E** (cf. also Bloesch, 1993, where something similar is done for **LP**).[3]
Throughout this subsection we shall mainly stick to Priest's notation, but
we shall occasionally resort to D'Agostino's terminology as well, providing
also a quick "translation method" from the former to the latter formalism.[4]

DEFINITION 28 (Signed formulae). By a *signed formula* we mean an or-
dered pair $\langle \phi, s \rangle$, where $\phi \in FOR(\pounds)$ and $s \in \{+, -\}$. A member of this
last set is called a *sign*. The *complementary* s^c of a sign s is $+$ if s is $-$
and $-$ otherwise. A signed formula is said to be *positive* iff it has the form
$\langle \phi, + \rangle$, *negative* otherwise. We also stipulate that:

- the *conjugate* of the signed literal $\langle L, s \rangle$ is $\langle L, s^c \rangle$;

- the *converse* of the signed literal $\langle L, s \rangle$ is $(\langle L^c, s^c \rangle$;

- the *weak conjugate* of the signed literal $\langle L, s \rangle$ is $\langle L^c, s \rangle$.

When referring to signed formulae, hereafter, angles are omitted whenever this is possible. As the reader will notice, the tableaux systems for **T**,
M, **Kl**, **P** and **E** contain exactly the same rules for the logical connectives;
they only differ in their closure rules.

DEFINITION 29 (fde-tableau). A *fde-tableau* is a labelled tree $\mathcal{T}$ whose
vertices are labelled by signed formulae. The signed formulae labelling the
root of $\mathcal{T}$ and its immediate successor(s) are called *initial formulae* of $\mathcal{T}$.
The other formulae occurring in $\mathcal{T}$ are called *non-initial*. Non-initial formulae in $\mathcal{T}$ are obtained from predecessor vertices by one of the rules below
(where the usual typographical conventions are adopted):

---

[3]Tableaux for full propositional symmetric relatedness logic are discussed in Carnielli
(1987).
  [4]Remark that these tableaux calculi are basically grounded on the same ideas as the
sequent calculi mentioned at the beginning or §4.

$$\frac{\neg\neg\phi, +}{\phi, +} \qquad \frac{\neg\neg\phi, -}{\phi, -} \qquad \frac{\phi \wedge \psi, +}{\begin{array}{c}\phi, + \\ \psi, +\end{array}} \qquad \frac{\phi \wedge \psi, -}{\phi, - \mid \psi, -} \qquad \frac{\phi \vee \psi, +}{\phi, + \mid \psi, +}$$

$$\frac{\phi \vee \psi, -}{\begin{array}{c}\phi, - \\ \psi, -\end{array}} \qquad \frac{\neg(\phi \wedge \psi), +}{\neg\phi, + \mid \neg\psi, +} \qquad \frac{\neg(\phi \wedge \psi), -}{\begin{array}{c}\neg\phi, - \\ \neg\psi, -\end{array}} \qquad \frac{\neg(\phi \vee \psi), +}{\begin{array}{c}\neg\phi, + \\ \neg\psi, +\end{array}} \qquad \frac{\neg(\phi \vee \psi), -}{\neg\phi, - \mid \neg\psi, -}$$

If $\phi \to \psi$ is a fde, we say that the fde-tableau $\mathcal{T}$ is *for* $\phi \to \psi$ iff its root is labelled by $\langle \phi, + \rangle$ and its sole immediate successor is labelled by $\langle \psi, - \rangle$. A branch of a fde-tableau $\mathcal{T}$ is called:

- *closed,* iff it contains conjugate vertices;[5]

- *+-semiclosed,* iff it contains positive weak conjugate vertices;

- *--semiclosed,* iff it contains negative weak conjugate vertices;

- *linked,* iff it contains either conjugate vertices or converse vertices.

DEFINITION 30 (Completed fde-tableau). A branch r of a fde-tableau $\mathcal{T}$ is called *complete* iff it contains both conclusions of a non-branching rule (α-rule, in Smullyan's standard jargon) and at least one conclusion of a branching rule (β-rule) provided it contains their premisses. A tableau is said to be *completed* iff all of its branches are complete.

DEFINITION 31 (**TS**-provability). We say that the fde $\phi \to \psi$ is:

- *provable in* **TT** (in symbols, $\vdash_{\mathbf{TT}} \phi \to \psi$) iff there is a completed fde-tableau $\mathcal{T}$ for $\phi \to \psi$ whose branches are all closed;

- *provable* in **TKl** (in symbols, $\vdash_{\mathbf{TKl}} \phi \to \psi$) iff there is a completed fde-tableau $\mathcal{T}$ for $\phi \to \psi$ whose branches are either closed or +-semiclosed;

- *provable* in **TP** (in symbols, $\vdash_{\mathbf{TP}} \phi \to \psi$) iff there is a completed fde-tableau $\mathcal{T}$ whose branches are either closed or --semiclosed;

- *provable* in **TM** (in symbols, $\vdash_{\mathbf{TM}} \phi \to \psi$) iff there is a completed fde-tableau $\mathcal{T}$ for $\phi \to \psi$ whose branches are either closed or else both +-semiclosed and --semiclosed;

---

[5]Throughout the rest of the paper we shall allow ourselves a notational abuse and identify vertices in fde-tableaux with the signed formulae which label them.

- *provable* in **TE** (in symbols, $\vdash_{\textbf{TE}} \phi \to \psi$) iff there is a completed fde-tableau $\mathcal{T}$ for $\phi \to \psi$ whose branches are either closed or +-semiclosed or −-semiclosed, and such that at least one branch is linked.

EXAMPLE 32. Here are some examples of fde-tableaux for fdes which are provable, respectively, in **TT, TM, TE**:

$$\phi \wedge (\psi \vee \chi), +$$
$$(\phi \wedge \psi) \vee (\phi \wedge \chi), -$$

$\phi, +$	$\phi \wedge \neg\phi, +$	$\phi \wedge (\neg\phi \vee \psi), +$
$\psi \vee \chi, +$	$\psi \vee \neg\psi, -$	$\psi, -$
$\phi \wedge \psi, -$	$\phi, +$	$\phi, +$
$\phi \wedge \chi, -$	$\neg\phi, +$	$\neg\phi \vee \psi, +$
$\diagup \quad \diagdown$	$\psi, -$	$\diagup \quad \diagdown$
$\phi, - \qquad \psi, -$	$\neg\psi, -$	$\neg\phi, + \quad \psi, +$
$\times \qquad \diagup \diagdown$	$\times$	$\times \quad \times$
$\phi, - \quad \chi, -$		
$\times \diagup \diagdown$		
$\psi, + \quad \chi, +$		
$\times \quad \times$		

Remark that, in order to obtain tableaux calculi in the style of D'Agostino (1990) for the logics at issue, it is sufficient to abide by the following translation rules:

- replace each label of the form $\langle \phi, + \rangle$ by a label of the form $t\phi$;

- replace each label of the form $\langle \phi, - \rangle$ by a label of the form $f\phi$;

- replace each label of the form $\langle \neg\phi, + \rangle$ by a label of the form $f * \phi$;

- replace each label of the form $\langle \neg\phi, - \rangle$ by a label of the form $t * \phi$.

The following definitions will be useful in what follows.

DEFINITION 33 (Faithful interpretation). Let $\mathcal{I} = \langle \mathcal{M}_{\textbf{T}}, v \rangle$ be an interpretation for £. We say that $\mathcal{I}$ is **T**-faithful to the signed formula $\langle \phi, s \rangle$ iff one of the following conditions holds:

- $\langle \phi, s \rangle$ has the form $\langle \psi, + \rangle$ and $v(\psi) \in \{t, b\}$;

- $\langle \phi, s \rangle$ has the form $\langle \psi, - \rangle$ and $v(\psi) \in \{f, n\}$;

- $\langle \phi, s \rangle$ has the form $\langle \neg\psi, + \rangle$ and $v(\psi) \in \{f, b\}$;

- $\langle \phi, s \rangle$ has the form $\langle \neg \psi, - \rangle$ and $v(\psi) \in \{t, n\}$;

An interpretation for £ $\mathcal{J} = \langle \mathcal{M_E}, v \rangle$ is called **Ke**-*faithful* (**P**-*faitful*) to the signed formula $\langle \phi, s \rangle$ iff one of the following conditions holds:

- $\langle \phi, s \rangle$ has the form $\langle \psi, + \rangle$ and $v(\psi) = t$ (and $v(\psi) \in \{t, b\}$);

- $\langle \phi, s \rangle$ has the form $\langle \psi, - \rangle$ and $v(\psi) \in \{f, b\}$ (and $v(\psi) = f$);

- $\langle \phi, s \rangle$ has the form $\langle \neg \psi, + \rangle$ and $v(\psi) = f$ (and $v(\psi) \in \{f, b\}$);

- $\langle \phi, s \rangle$ has the form $\langle \neg \psi, - \rangle$ and $v(\psi) = \{t, b\}$ (and $v(\psi) = t$);

A set S of signed formulae is called **T-** (**Kl-, P-**) *satisfiable* iff there is an interpretation $\mathcal{I}$ for £ which is **T-** (**Kl-, P-**) faithful to every signed formula in S; it is called **M**-*satisfiable* iff there is an interpretation $\mathcal{I}$ for £ which is either **Kl**-faithful to every signed formula in S or **P**-faithful to every signed formula in S.

DEFINITION 34 (Hintikka set). A set S of signed formulae is a **T**-*Hintikka set* iff the following conditions hold:

- for every literal L, it is not the case that both $\langle L, s \rangle$ and its conjugate $\langle L, s^c \rangle$ occur in S;

- if ϕ is a possible premiss for an α-rule of a fde-tableau, the conclusion(s) that would result from the application of the rule to ϕ occur in S;

- if ϕ is a possible premiss for a β-rule of a fde-tableau, at least one of the conclusions that would result from the application of the rule to ϕ occurs in S.

The definition of **Kl**-*Hintikka* set is identical, except for the fact that a further condition is added: for any literal L, it is not the case that both $\langle L, + \rangle$ and its weak conjugate $\langle L^c, + \rangle$ occur in S. **P**-Hintikka sets and **M**-Hintikka sets are defined with the obvious modifications.

4.4 Equivalence results

We are now in a position to prove the equivalence of the previous formulations. As regards T, the equivalence of (A), (B), (C) below can be found already in Anderson and Belnap (1975); the equivalence of (D) and the above was first shown by D'Agostino (1990). Some of the corresponding results for **Kl**, **P** and **M** can be found e.g. in Rose (1963), Muravitsky (1995), Priest (2001). As regards **E**, the equivalence of (A), (B), (C) has been established in Paoli (1993).

THEOREM 35 (Equivalence of the previous formulations). *Let $S \in \{\mathbf{T}, \mathbf{M}, \mathbf{Kl}, \mathbf{P}, \mathbf{E}\}$ and let $\phi \to \psi$ be a fde. The following are equivalent:*

(A) $\vdash_{HS} \phi \to \psi$;

(B) $\vDash_S \phi \to \psi$;

(C) $\phi \to \psi$ is an S-entailment;

(D) $\vdash_{TS} \phi \to \psi$.

Proof. (A) $\Rightarrow$ (B). The proof proceeds by a standard induction on the length of the proof of $\phi \to \psi$ in **HS**.

(B) $\Rightarrow$ (C) Let $\phi' = \phi_1 \vee \ldots \vee \phi_n$ ($\psi' = \psi_1 \wedge \ldots \wedge \psi_m$) be the dnf (cnf) of ϕ (of ψ). Now, suppose that $\phi \to \psi$ is not an S-entailment, whence $\phi' \to \psi'$ is not an explicit S-entailment. We shall show that $\phi' \to \psi'$ is not S-valid, whence $\phi \to \psi$ is not S-valid either, since conversion into normal form does not affect the value of a formula.

Let us consider the case of **T** first. If $\phi' \to \psi'$ is not an explicit **T**-entailment, there exist $i \leq n, j \leq m$ such that ϕ_i and ψ_j share no literal. Now, let $v : \mathcal{F}(\mathcal{L}) \to \mathcal{A}_{\mathbf{T}}^4$ be such that, for any variable p:

- if p occurs, $\neg p$ does not occur in ϕ_i, $v(p) = t$;

- if p occurs, $\neg p$ occurs in ϕ_i, $v(p) = b$;

- if p does not occur, $\neg p$ occurs in ϕ_i, $v(p) = f$;

- if p does not occur, $\neg p$ does occur in ϕ_i, $v(p) = n$.

Since ϕ_i and ϕ_j share no literal, one can see from the tables that $v(\phi_i) \in \{t, b\}$ and $v(\psi_j) \in \{f, n\}$. But then also $v(\phi') \in \{t, b\}$ and $v(\psi') \in \{f, n\}$. Since $\langle t, f \rangle, \langle t, n \rangle, \langle b, n \rangle, \langle b, f \rangle \not\leq_{\mathbf{T}}^4$, the interpretation $\langle \mathcal{M}_{\mathbf{T}}, v \rangle$ falsifies $\phi' \to \psi'$.

If $\phi' \to \psi'$ is not an explicit **Kl**-entailment, there exist $i \leq n, j \leq m$ such that ϕ_i and ψ_j share no literal, and moreover it is not the case that ϕ_i contains complementary literals. Now, let $v : \mathcal{F}(\mathcal{L}) \to \mathcal{A}_{\mathbf{T}}^3$ be such that, for any variable p:

- if p occurs, $\neg p$ does not occur in ϕ_i, $v(p) = t$;

- if p does not occur, $\neg p$ occurs in ϕ_i, $v(p) = f$;

- if p does not occur, $\neg p$ does not occur in ϕ_i, $v(p) = b$.

Since ϕ_i and ψ_j share no literal, one can see from the tables that $v(\phi_i) = t$ and $v(\psi_j) \in \{f, b\}$. But then also $v(\phi') = t$ and $v(\psi') \in \{f, b\}$. Since $\langle t, f \rangle, \langle t, b \rangle, \not\leq^3_{\mathbf{T}}$, the interpretation $\langle \mathcal{M}_{\mathbf{K1}}, v \rangle$ falsifies $\phi' \to \psi'$.

The cases of $\mathbf{P}$ and $\mathbf{M}$ are handled analogously, while the result for $\mathbf{E}$-entailments follows easily from Theorem 16 and well-known properties of classically valid fdes.

(C) $\Rightarrow$ (A) First remark that, by A1–A2, T1, R1, $\vdash_{\mathbf{HT}} \phi \to \psi$ whenever it is a primitive $\mathbf{T}$-entailment. Next, observe that by R2–R3 all explicit $\mathbf{T}$-entailments are provable in $\mathbf{HT}$. Finally, by R1 all $\mathbf{T}$-entailments are provable in $\mathbf{HT}$. The same proof also goes through for $\mathbf{M}$, $\mathbf{Kl}$, and $\mathbf{P}$, with minor differences. As for $\mathbf{E}$, let $\phi \to \psi$ be an explicit $\mathbf{E}$-entailment; then there are ϕ_i, ψ_j sharing a variable. For such formulae we also have that either ϕ_i and ψ_j share a literal, or at least one of ϕ_i, ψ_j contains complementary literals. Then, by Lemma 21, A1–A2, T1, R1', $\vdash_{\mathbf{HE}} \phi_i \to \psi_j$ and $\vdash_{\mathbf{HE}} \phi \to \psi$.

(B) $\Rightarrow$ (D). For $\mathbf{TT}$, we prove that every complete open branch of a fde-tableau is $\mathbf{T}$-satisfiable, whence the implication follows by the remark made after Definition 14. Since the set of the signed formulae occurring in a complete open branch of a fde-tableau is a $\mathbf{T}$-Hintikka set, it will be enough to show that any $\mathbf{T}$-Hintikka set S is $\mathbf{T}$-satisfiable. Now, let $\mathcal{I} = \langle \mathcal{M}_{\mathbf{T}}, v \rangle$ be constructed as follows:

- if $\langle p, + \rangle \in S$ and $\langle \neg p, + \rangle \notin S$, then $v(p) = t$;

- if $\langle p, + \rangle \in S$ and $\langle \neg p, + \rangle \in S$, then $v(p) = b$;

- if $\langle p, - \rangle \in S$ and $\langle \neg p, - \rangle \notin S$, then $v(p) = f$;

- if $\langle p, - \rangle \in S$ and $\langle \neg p, - \rangle \in S$, then $v(p) = n$;

- if $\langle p, - \rangle \in S$ and $\langle \neg p, - \rangle \notin S$, then $v(p) = t$;

- if $\langle p, + \rangle \in S$ and $\langle \neg p, + \rangle \notin S$, then $v(p) = f$;

Given Definition 34, $\mathcal{I}$ is well-defined because the cases envisaged above do not clash and exhaust all the alternatives. Moreover, $\mathcal{I}$ is $\mathbf{T}$-faithful to all the signed literals in S. Upon remarking that a valuation which is $\mathbf{T}$-faithful to the possible conclusion(s) of an α-rule is also $\mathbf{T}$-faithful to the respective premiss, while a valuation which is $\mathbf{T}$-faithful to at least one of the possible conclusions of a β-rule is also $\mathbf{T}$-faithful to the respective premiss, we conclude by induction that $\mathcal{I}$ is $\mathbf{T}$-faithful to all the signed formulae in S.

A similar argument works for $\mathbf{TKl}$. By hypothesis, a $\mathbf{Kl}$-Hintikka set S does not contain positive weak conjugate literals. The crucial interpretation is constructed as follows:

- if $\langle p, +\rangle \in S$ and $\langle \neg p, +\rangle \notin S$, then $v(p) = t$;

- if $\langle p, -\rangle \in S$ and $\langle \neg p, -\rangle \notin S$, then $v(p) = f$;

- if $\langle p, -\rangle \in S$ and $\langle \neg p, -\rangle \in S$, then $v(p) = b$;

- if $\langle \neg p, -\rangle \in S$ and $\langle p, -\rangle \notin S$, then $v(p) = t$;

- if $\langle \neg p, +\rangle \in S$ and $\langle p, +\rangle \notin S$, then $v(p) = f$;

Such an interpretation can be shown to be **Kl**-faithful to all the signed formulae in S. With **TP** and **TM**, we proceed similarly.

As to **TE**, suppose $\vDash_{\mathbf{E}} \phi \to \psi$. Then by Theorem 16 $\vDash_{\mathbf{K}} \phi \to \psi$, whence we can construct a closed completed classical tableau refuting $\neg(\phi \to \psi)$, which we can easily turn into a completed fde-tableau for $\phi \to \psi$ whose branches are either closed or +-semiclosed or −-semiclosed. Once more by Theorem 16, however, $\mathrm{Var}(\phi) \cap \mathrm{Var}(\psi) \neq \varnothing$. Hence at least one of such branches is linked; consequently, $\phi \to \psi$ is provable in **TE**.

(D) $\Rightarrow$ (C). For **TT**, just remark that any interpretation $\mathcal{I}$ which is **T**-faithful to the premiss of an α-rule is **T**-faithful to its conclusion(s) as well, while any interpretation $\mathcal{I}$ which is **T**-faithful to the premiss of a β-rule is **T**-faithful to at least one of its conclusions as well. Reasoning inductively, it follows that if $\mathcal{I}$ is **T**-faithful to $\langle \phi, +\rangle, \langle \psi, -\rangle$, then in any $\mathcal{T}$ for $\phi \to \psi$ there is at least a branch r such that $\mathcal{I}$ is **T**-faithful to all the signed formulae in r. However, no interpretation can be **T**-faithful to both $\langle \chi, +\rangle$ and $\langle \chi, -\rangle$, for any χ. Hence, if $\mathcal{T}$ is a fde-tableau for $\phi \to \psi$ all of whose branches are closed, then for any interpretation $\mathcal{I} = \langle \mathcal{M}_{\mathbf{T}}, v\rangle$, if $v(\phi) \in \{t, b\}$, then $v(\psi) \in \{t, b\}$. By the remark following Definition 14, then, $\vDash_{\mathbf{S}} \phi \to \psi$. Similar arguments apply to **TM**, **TKl**, **TP**. Finally, if $\phi \to \psi$ is provable in **TE**, it is easy to build a closed completed classical tableau refuting $\neg(\phi \to \psi)$. The fact that at least one branch must be linked yields variable sharing, whence by Theorem 16 $\vDash_{\mathbf{E}} \phi \to \psi$. ∎

4.5   Interpolation

Anderson and Belnap (1975) notice that **HT** admits of a "perfect" interpolation theorem: if $\phi \to \psi$ is provable in it, there is an interpolation formula χ such that both $\phi \to \chi$ and $\chi \to \psi$ are provable as well, and χ contains no variable which is not in both ϕ and ψ. The qualification "perfect" applies to it in contrast to the analogous result for classical logic — Craig's theorem — which establishes a weaker property: if either ϕ is contradictory or ψ is valid, we have no guarantee of the existence of the interpolation formula.

What about the other systems at issue? Clearly, **HKl** and **HP** cannot enjoy perfect interpolation, because they lack the variable sharing property.

Adapting an argument by R.K. Meyer, we shall see that **HM** does not allow even a weak Craig-style interpolation theorem. On the other hand, **HE** enjoys perfect interpolation; this shows that, among the subclassical logics, **E** is a "maximal perfect interpolation logic": in fact, if $E \subset S \subset K$, then **S** contains a fde $\phi \to \psi$ where ϕ and ψ share no variable, so that there can be no interpolation formula.

THEOREM 36 (Interpolation). *a) Let* $S \in \{T, E\}$. *If* $\vdash_{HS} \phi \to \psi$ *then there is* $\chi \in FOR(\mathcal{L})$ *such that* $\vdash_{HS} \phi \to \chi, \vdash_{HS} \chi \to \psi$ *and* $Var(\chi) \subseteq Var(\phi) \cap Var(\psi)$; *b) There are* $\phi, \psi \in FOR(\mathcal{L})$ *such that* $\vdash_{HM} \phi \to \psi, \phi$ *is not classically inconsistent,* ψ *is not classically valid and there is no* $\chi \in FOR(\mathcal{L})$ *such that* $\vdash_{HM} \phi \to \chi, \vdash_{HM} \chi \to \psi$ *and* $Var(\chi) \subseteq Var(\phi) \cap Var(\psi)$.

Proof. a) Suppose $\vdash_{HT} \phi \to \psi$. By Theorem 35 $\phi' \to \psi'$ (where $\phi' = \phi_1 \vee \ldots \phi_n$ is the dnf of ϕ and $\psi' = \psi_1 \wedge \ldots \wedge \psi_m$ is the cnf of ψ) is an explicit **T**-entailment. For $i \leq n$, let ϕ_i^* be a subconjunction of ϕ_i which contains just the literals shared by ϕ_i with each conjunct of ψ', and let χ be $\phi_1^* \vee \ldots \vee \phi_n^*$. By A1–A2, R1, R3, we obtain $\vdash_{HT} \phi' \to \chi$, whence by R1 $\vdash_{HT} \phi \to \chi$. Furthermore, by A1–A2, R1-R3, $\vdash_{HT} \chi \to \psi'$, whence by R1 $\vdash_{HT} \chi \to \psi$. Finally, given the fact that ϕ and ϕ' (respectively ψ and ψ') have the same variables, $Var(\chi) \subseteq Var(\phi) \cap Var(\psi)$.

A similar proof works also for **HE**. Thus, suppose $\vdash_{HE} \phi \to \psi$ and consider the fde $\phi' \to \psi'$ rewritten in normal form. In general, some ϕ_is and some ψ_js may contain complementary literals, other ones may not. For the sake of simplicity, suppose that for $p \leq i$ (for $q \leq j$) $\phi_p(\psi_q)$ contains no complementary literals, whereas for $p > i$ (for $q > j$) it does. In the following schema, primitive conjunctions and disjunctions which are free of complementary literals are printed in bold:

$$\boldsymbol{\phi_1} \vee \ldots \vee \boldsymbol{\phi_i} \vee \phi_{i+1} \vee \ldots \phi_n$$
$$\boldsymbol{\psi_1} \wedge \ldots \wedge \boldsymbol{\psi_j} \wedge \psi_{j+1} \wedge \ldots \wedge \psi_m$$

We distinguish four cases: a1) $i, j > 0$; a2) $i = 0, j > 0$; a3) $i > 0, j = 0$; a4) $i, j = 0$.

a1) Consider $\phi_1 \vee \ldots \vee \phi_i \to \psi_1 \wedge \ldots \wedge \psi_j$. By our definition of explicit **E**-entailment, each ϕ_p shares a literal with each ψ_q; hence the previous fde is an explicit **T**-entailment. Proceeding as above, thus, we get a formula χ interpolating between $\phi_1 \vee \ldots \vee \phi_i$ and $\psi_1 \wedge \ldots \wedge \psi_j$ (transitivity moves are countenanced by our variable sharing hypothesis). Should n be greater than i, by Lemma 21 we get $\vdash_{HE} \phi' \to \chi$, whence by R1' $\vdash_{HE} \phi \to \chi$; should m be greater than j, once again by Lemma 21 we get $\vdash_{HE} \chi \to \psi'$, whence by R1' $\vdash_{HE} \chi \to \psi$. Our construction also yields $Var(\chi) \subseteq Var(\phi) \cap Var(\psi)$.

a2) Consider ϕ_k, ψ_l with a variable in common, say p. The formula $\chi = p \wedge \neg p$ has no variables not in both ϕ' and ψ'; moreover, by Lemma 21 we have $\vdash_{\mathbf{HE}} \phi_k \to \chi$ and then $\vdash_{\mathbf{HE}} \phi' \to \chi$ (whence as usual $\vdash_{\mathbf{HE}} \phi \to \chi$). Applying Lemma 21 again, we get $\vdash_{\mathbf{HE}} \chi \to \psi'$ and thus $\vdash_{\mathbf{HE}} \chi \to \psi$.

a3) Dual of the above, with $\chi = p \vee \neg p$.

a4) We may proceed as in case a2) or as in case a3).

b) Let $\phi = s \vee (p \wedge q \wedge \neg q), \psi = (s \vee p) \wedge (s \vee r \vee \neg r)$. $\phi \to \psi$ is an explicit **M**-entailment, hence it is provable in **HM** by Theorem 35. It can be readily checked, too, that neither ϕ is classically contradictory nor is ψ classically valid. Now, suppose that there is $\chi \in FOR(\mathcal{L})$ such that $\mathrm{Var}((\chi) \subseteq \{p, s\}$ and $\vdash_{\mathbf{HM}} \phi \to \chi, \vdash_{\mathbf{HM}} \chi \to \psi$. By Theroem 35 it would be $\vDash_{\mathbf{M}} \phi \to \chi, \vDash_{\mathbf{M}} \chi \to \psi$. However, let v be such that $v(p) = t, v(q) = v(r) = b, v(s) = f$. Then $v(\phi) = v(\psi) = b, v(\chi) \in \{t, f\}$. Hence either $\langle v(\phi), v(\chi) \rangle \notin \leq_{\mathbf{T}}^3$ or $\langle v(\chi), v(\psi) \rangle \notin \leq_{\mathbf{T}}^3$, contradicting our hypothesis. ∎

BIBLIOGRAPHY

[1] Anderson A.R., Belnap N.D. (1962), "Tautological entailments", Philosophical Studies, 13, pp. 9-24.

[2] Anderson A.R., Belnap N.D. (1975), Entailment: The Logic of Relevance and Necessity, vol. 1, Princeton University Press, Princeton, NJ.

[3] Arieli O., Avron A. (1998), "The value of four values", Artificial Intelligence, 102, pp. 97-141.

[4] Avron A. (1991), "Natural 3-valued logics: characterization and proof theory", Journal of Symbolic Logic, 56, pp. 276-294.

[5] Belnap N.D. (1977), "How a computer should think", in G. Ryle (Ed.), Contemporary Aspects of Philosophy, Oriel Press, Boston, pp. 30-55.

[6] Bloesch A. (1993), "A tableau style proof system for two paraconsistent logics", Notre Dame Journal of Formal Logic, 34.

[7] Bolotov A. (1986), "Entailment in Hao Wang's logic: Semantical and syntactical analysis", in Logic and System Methods of the Analysis of Scientific Knowledge, Moscow.

[8] Carnielli W.A. (1987), "Methods of proof for relatedness and dependence logics", Reports on Mathematical Logic, 21, pp. 35-46.

[9] Carnielli W.A., Marcos J. (2002), "A taxonomy of C-systems", in W.A. Carnielli et al. (eds.), Paraconsistency: The Logical Way to the Inconsistent, Dekker, New York, pp. 1-94.

[10] Deutsch H.S. (1981), A Family of Conforming Relevant Logics, PhD Thesis, UCLA.

[11] D'Agostino M. (1990), "Investigations into the complexity of some propositional calculi", Oxford University Computing Laboratory Technical Monographs, Report PRG-88.

[12] D'Ottaviano I.M.L., Da Costa N.C.A. (1970), "Sur un problème de Jaskowski", C.R. Acad. Sc. Paris, 270 A, pp. 1349-1353.

[13] Dunn J.M. (1972), "A modification of Parry's analytic implication", Notre Dame Journal of Formal Logic, 13, pp. 195-205.

[14] Dunn J.M. (1976), "Intuitive semantics for first-degree entailments and coupled trees", Philosophical Studies, 29, pp. 149-168.

[15] Dunn J.M. (1986), "Relevance logic and entailment", in D. Gabbay, F. Guenthner (Eds.), Handbook of Philosophical Logic, Reidel, Dordrecht, vol. 3, pp. 117-224.

[16] Epstein R.L. (1990), The Semantic Foundations of Logic, Kluwer, Dordrecht.

[17] Fitting M. (1990), "Bilattices in logic programming", in G. Epstein (Ed.), XXth International Symposium of Multiple-Valued Logic, IEEE, pp. 238-246.

[18] Fitting M. (1994), "Kleene's three-valued logic and their children", Fundamenta Informaticae, 20, pp. 113-131.

[19] Font J.M. (1997), "Belnap's four-valued logic and De Morgan lattices", Logic Journal of the IGPL, 3, pp. 413-440.

[20] Ginsberg M.L. (1988), "Multivalued logics: A uniform approach to reasoning in artificial intelligence", Computational Intelligence, 4, pp. 265-316.

[21] Halldén S. (1949), The Logic of Nonsense, Uppsala Universitets Arsskrift, Uppsala.

[22] Kleene S. (1952), Introduction to Metamathematics, Van Nostrand, Amsterdam.

[23] Levesque H.J. (1984), "A logic of implicit and explicit belief", Proceedings of AAAI-84, Austin, TX, pp. 198-202.

[24] Muravitsky A.J. (1995), "On the first-degree entailment of two 3-valued logics", available at csdocs.cs.nyu.edu.

[25] Paoli F. (1992), "Regressive analytical entailments", Konstanzer Berichte zur Logik und Wissenschaftstheorie, Report no. 33.

[26] Paoli F. (1993), "Semantics for first-degree relatedness logic", Reports on Mathematical Logic, 27, pp. 81-94.

[27] Paoli F. (1996), "S is constructively complete", Reports on Mathematical Logic, 30, pp. 31-47.

[28] Parry W.T. (1933), "Ein Axiomensystem fuer eine neue Art von Implikation", Ergebnisse eines mathematischen Kolloquiums, 4, pp. 5-6.

[29] Patel-Schneider P.F. (1989), "A four-valued semantics for terminological logics", Artificial Intelligence, 38, 3, pp. 319-351.

[30] Priest G. (1979), "Logic of paradox", Journal of Philosophical Logic, 8, pp. 219-241.

[31] Priest G. (2001), An Introduction to Non-Classical Logic, Cambridge University Press, Cambridge.

[32] Priest G. (2002), "Paraconsistent logic", in D. Gabbay, F. Guenthner (Eds.), Handbook of Philosophical Logic, new edition, Kluwer, Dordrecht, vol. 6, pp. 287-393.

[33] Rose A. (1963), "A formalisation of the propositional calculus corresponding to Wang's calculus of partial predicates", Zeitschrift für Mathematische Logik und Grundlagen der Mathematik, 9, pp. 177-198.

[34] Routley R., Routley V. (1972), "The semantics of first degree entailment", Nous, 6, pp. 335-359.

[35] Routley R. et al. (1982), Relevant Logics and Their Rivals, vol. 1, Ridgeview, Atascadero.

[36] Seymour Michael F. (2002), "Entailment and bivalence", Journal of Philosophical Logic, 31, pp. 289-300.

[37] Sinowjew A., Wessel H. (1975), Logische Sprachregeln, Berlin, 1975.

[38] Tamminga A.M., Tanaka K. (1999), "A natural deduction system for first degree entailment", Notre Dame Journal of Formal Logic, 40, pp. 258-272.

[39] Urquhart A. (1986), "Many-valued logic", in D. Gabbay, F. Guenthner (Eds.), Handbook of Philosophical Logic, Reidel, Dordrecht, vol. 3, pp. 71-116.

[40] Wang H. (1961), "The calculus of partial predicates and its extension to set theory", Zeitschrift fuer Mathematische Logik und Grundlagen der Mathematik, 7, pp. 283-288.

[41] Weber S. (1998), Investigations in Belnap's Logic of Inconsistent and Unknown Information, PhD Thesis, University of Leipzig.

Francesco Paoli

Dipartimento di Scienze Pedagogiche e Filosofiche, Università di Cagliari - Via Is Mirrionis 1, 09123 Cagliari, Italy.

paoli@unica.it

A system of paraconsistent logic that has the notion of "behaving classically" in terms of the law of double negation and its relation to S5

Toshiharu Waragai and Tomoki Shidori

1 Introduction

If a theory has as its theorems, say A and $\neg A$ (i.e. the negation of A), it is called *contradictory*. If a theory derives all the formulas as its theorems, then it is called *trivial* or *overcomplete*. They are two different notions, even though they usually coincide with each other in the systems familiar to us. This distinction is one of the most important discoveries in the recent development of mathematical logic. A theory is *paraconsistent* if it is *contradictory but not trivial*.

The forerunners of paraconsistent logic are Jaśkowski and da Costa. Jaśkowski developed his *discussive logic* on the basis of S5. Da Costa developed his Cn-systems. Both had the same intention to construct systems in which formal contradictions do not in general lead to their triviality.

What interests us for the present study is the notion of "*behaving classically*" (for short BC) that plays a crucial role in da Costa's systems. In da Costa's systems, a proposition A behaves classically if it satisfies the law of non contradiction (and its nested form). Despite the importance of this notion of BC, it is not completely free from troubles: 1) it is complicated and difficult to capture its intention when n of Cn is greater than 2 and 2) the important property of BC is *required* as axioms in his systems. But it is clearly desirable to prove such axioms concerning BC from the axioms and rules that are easier to capture.

We will be concerned with this problem, taking the law of (elimination of) double negation (for short LDN) as the criterion of BC, and will show that some crucial axioms (A12 and A13) we find in da Costa's systems become derivable in the system we are going to propose.

Then we will show that this approach has a natural link to Jaśkowski's S5 based approach of paraconsistent logic.

One may ask for the reason why we took LDN as the criterion of BC. Needless to say, there are a couple of philosophical and technical reasons for our taking LDN as the criterion of BC. Here in this paper, however, instead of explaining them, we quote as an answer the following passage from da Costa, N. C. A., Béziau, J-Y, Bueno, O. [1995]:

> According to some writers, the first author [da Costa] has not spelled out the 'philosophical' rationale for including in his C-systems (presented, for instance, in da Costa [1974]), the constraint that this should not contain $\neg(\alpha \wedge \neg\alpha)$ as a logical truth. On this regard, we wish to point out that, from our viewpoint, when presenting a formal system, one does not need to be concerned with the formulation of philosophical rationales for mathematical constraints introduced. One should not confuse the mathematical development of logical systems with their philosophical interpretation: these are, in fact, quite distinct issues. (da Costa, N. C. A., Béziau, J-Y, Bueno, O. [1995], 115.)

We quote this passage in full accordance, for this standpoint is just ours which one of us expressed in an article[1] that itself has little to do with in our present context.

2   On da Costa's Cn $(1 \leq n \leq \omega)$

Our aim of this section is just to recapitulate da Costa's C-systems Cn for the sake of comparison. His system consists of the following 12 axioms and *modus ponens.* [2]

A1. $A \supset (B \supset A)$

A2. $(A \supset B) \supset ((A \supset (B \supset C)) \supset (A \supset C))$

A3. $A \supset (B \supset (A\&B))$

A4. $(A\&B) \supset A$

A5. $(A\&B) \supset B$

A6. $(A \supset C) \supset ((B \supset C) \supset (A \vee B \supset C))$

A7. $A \supset (A \vee B)$

---

[1] Waragai, T.[1999].
[2] Cf. da Costa[1974], Urbas, I. [1989]

A8. $B \supset (A \vee B)$

A9. $A, A \supset B/B$

A10. $A \vee \neg A$

A11. $\neg\neg A \supset A$

The system consisting of the above shown 10 axioms and *modus ponens* is called Cω. It should be noticed that the axioms *A1-8,A10,A11* together with *modus ponens* (*A9*) constitutes the system of Kleene [1967] with $(A \supset B) \supset ((A \supset \neg B) \supset \neg A)$ deleted.

DEFINITION 1.

$$A^o \iff \neg(A\&\neg A)$$

$$A^n \iff A^{\overbrace{oo\ldots o}^{n}}$$

$$A^{(n)} \iff A^1\&A^2\&A^3\&\ldots\&A^n$$

Da Costa gives the notion of "behaves classically" in terms of $A^{(n)}$ in the following way.

DEFINITION 2.

A proposition A behaves classically in Cn iff $A^{(n)}$ holds.

A12. $B^{(n)} \supset ((A \supset B) \supset ((A \supset \neg B) \supset \neg A))$

A13. $A^{(n)}\&B^{(n)} \supset (\neg A)^{(n)}\&(A \supset B)^{(n)}\&(A\&B)^{(n)}\&(A \vee B)^{(n)}$

The intention of A12 and A13 seems to be obvious, if $n=1$. If A and B behave classically, they are expected to be evaluated classically. $A12$ states, therefore, if B behaves classically and A deduces B and the negation of B ($\neg B$), then something is wrong with A, so that A is not the case, i.e. $\neg A$.

A13 assures that the propositions that behave classically make complex propositions that also behave classically. In case of $n=1$, it seems plausible, while if $1 < n$, the situation changes. It loses the intiuitive plausibility.

Just on the contrary, it will be shown that the formulas corresponding to A12 and A13 will be provable in our system to be proposed in this paper. They are proved as theorem 1 and theorem 3 in the section 5.

3   The Axioms and the Rule of Inference of PCL1

Now let us begin stating the axiom schemata and rules of inference of the system to be discussed in this paper. We name it PCL1.

AXIOM SCHEMATA 2. :

A1. $(A \supset B) \supset ((B \supset C) \supset (A \supset C))$

A2. $A \supset (A \vee B)$

A3. $B \supset (A \vee B)$

A4. $(A \supset C) \supset ((B \supset C) \supset ((A \vee B) \supset C))$

A5. $(A \wedge B) \supset A$

A6. $(A \wedge B) \supset B$

A7. $(C \supset A) \supset ((C \supset B) \supset (C \supset (A \wedge B)))$

A8. $(A \supset (B \supset C)) \supset ((A \wedge B) \supset C)$

A9. $((A \wedge B) \supset C) \supset (A \supset (B \supset C))$

A10. $NNA \supset A$

Axiom Schemata 1 constitute a subsystem of classical logic in Rasiowa and Sikorski [1970] with $(A \wedge \neg A) \supset B$, $(A \supset (A \wedge \neg A)) \supset \neg A$, $A \vee \neg A$ deleted and with *A10* added.

DEFINITION 1.

D1 $A \equiv B =_{\text{def}} (A \supset B) \wedge (B \supset A)$

D2 $A^I =_{\text{def}} A \equiv NNA$

AXIOM SCHEMATA 3. :

A11. $NA \supset NNNA$

A12. $A^I \supset ((NA \vee B) \equiv (A \supset B))$

A13. $A^I \wedge B^I \supset (N(A \wedge B) \supset (NA \vee NB))$

A14. $A^I \wedge B^I \supset (N(A \vee B) \supset (NA \wedge NB))$

A15. $A^I \supset ((A \wedge NB) \supset N(A \supset B))$

Now we accept the following rule of inference:

RULE OF INFERENCE 4. :

RA15. $A^I \vdash N(A \supset B) \supset (A \wedge NB)$

A11-A15 and RA15 will be used in order to prove that (N) and $(A*)$ behave classically if A and B behave classically, where $* \in \{\supset, \wedge, \vee\}$.[1]

Another rule of inference is *Modus Ponens*:

RULE OF INFERENCE 5. :
        Modus Ponens (MP): A, $A \supset B \vdash B$

At this point, an informal exposition of the formulas A11-A15 is desirable. If A and B behave classically, we may well expect that (NA) and $(A * B)$ where $* \in \{\supset, \wedge, \vee\}$ behave classically, so that the following formulas are expected to hold:

A11.1 $A^I \supset (NA)^I$

A12.1 $A^I \wedge B^I \supset ((NA \vee B) \equiv (A \supset B))$

A13.1 $A^I \wedge B^I \supset (N(A \wedge B) \equiv (NA \vee NB))$

A14.1 $A^I \wedge B^I \supset (N(A \vee B) \equiv (NA \wedge NB))$

A15.1 $A^I \wedge B^I \supset (N(A \supset B) \equiv (A \wedge NB))$

A11.1 is in fact provable in PCL1. The others are also provable in PCL1, but B^I appearing in A12.1 and A15.1 will prove not necessary, hence it is deleted. For the same reason, $\equiv$ appearing in A13.1 and A14.1 are replaced by $\supset$.

1   Some formulas provable in PCL1

It is easy to check that the following rule is admissible in PCL1.
        (syll.) $\vdash A \supset B, \vdash B \supset C \to \vdash A \supset C$ \hspace{2cm} (A1, MP, MP)
Now let us enumerate some formulas provable in PCL1.

T1 $A \supset (B \supset A)$ \hfill (A5, A9)

T2 $A \supset A$ \hfill (A9, A5, T1)

T3 $(B \wedge A) \supset (A \wedge B)$ (A5(A/B, B/A), A6(A/B, B/A), A7($C/B \wedge A$), MP, MP)

T4 $(A \supset (B \supset C)) \supset (B \supset (A \supset C))$ \hspace{0.5cm} (A1($A/ B \wedge A$, $B/A \wedge B$, C/C), T3, MP, A8, syll., A9, syll.)

---

[1]We are thankful to the anonymous referees for their careful reading of the original draft and the kind suggestions. In the original version, RA15 had the form:
$A^I \supset ((A \wedge NB) \equiv N(A \supset B))$,
which is not accurate, so that we split it into two, i.e. $A15$ and $RA15$.

T5 $NA \lor A$ (A10, A3 $(B/A, A/NA)$, A1, A2, A4, A11, D2, A12)

T6 $A^I \supset (NA \land A \supset B)$ (A12, A5$(A/(NA \lor B) \supset (A \supset B)$,
   $B/(A \supset B) \supset (NA \lor B))$, A2$(A/NA)$)

T7 $B^I \supset ((A \supset B) \supset (NB \supset NA))$ (T6$(A/B, B/NA)$, T1(A/NA),
   A4$(B/NA, C/NA)$, T5)

T8 $B^I \supset ((NA \supset NB) \supset (B \supset A))$ (T1$(B/NA, A/B)$, A7, T6$(A/B,$
   $B/A)$)

T9 $NN(A \supset B) \supset (NNA \supset NNB)$ (For a proof, cf. Appendix)

T9' $NN(A \supset B) \supset (A \supset B))$ ($A10(A/A \supset B)$)

T10 $A^I \supset (NA \supset ((A \lor B) \supset B))$ (A5, T7$(A/A \land B, B/A)$, A6,
    T7$(A/A \land B, B/B)$, A4)

T11 $A^I \land B^I \supset ((NA \lor NB) \supset N(A \land B))$ (A5, T7$(A/A \land B, B/A)$, A6,
    T7$(A/A \land B, B/B)$, A4)

T11' $A^I \land B^I \supset ((NA \lor NB) \equiv N(A \land B))$ (T11, A13)

T12 $A^I \land B^I \supset ((NA \land NB) \supset N(A \lor B))$ (T10, T7$(A/A \lor B, B/B)$)

T12' $A^I \land B^I \supset ((NA \land NB) \equiv N(A \lor B))$ (T12, A14)

T11' is A13.1 and T12' is A14.1.

2   PCL1 and Classical Logic

Now we will examine the notion of "behaving classically". It should be noticed that da Costa's C-systems need special axiom concerning this notion (Axioms 12 and 13). In PCL1, the logical situation is essentially different, for they are provable. The following formulas and a rule are provable in PCL1.

T13 $(NA)^I$ (A10, A11, D2)

T13' $A^I \supset (NA)^I$ (T13, T1$(A/NA, B/A^I)$)

T14 $(A^I \land B^I) \supset (A \land B)^I$ (A13, A12(B/NB), T7$(A/N(A \land B), B/NB)$)

T15 $(A^I \land B^I) \supset (A \lor B)^I$ (A14, A5$(A/NA, B/NB)$,
    T7$(A/N(A \lor B), B/NA)$, A6$(A/NA, B/NB)$, T7$(A/N(A \lor B), B/NB)$,
    A4$(C/NN(A \lor B)))$

T16 $(A^I \wedge B^I) \vdash (A \supset B)^I$ (D2(A/B), A12(B/NNB), T11(B/NNB), RA15, T14(B/NB), T7($A/N(A \supset B$), $B/A \wedge NB$), T9')

Notice that in T16, $\vdash$ is used instead of $\supset$. [1] T13' is A11.1. Thus we have the following theorem:

THEOREM 1. *Suppose* $\star \in \{\supset, \wedge, \vee\}$. *If A and B behave classically, then* (NA) *and* $(A \star B)$ *behave classically.*

Hence the claim corresponding to axiom 13 of Cn is provable in PCL1, though with a slight modification (i.e. the case of $\supset$).

The following formulas are provable in PCL1.

T17 $A^I \supset ((A \wedge NA) \supset B)$ (A2, T10)

T18 $(A \supset (A \wedge NA)) \supset NA$ (A6, A1($B/A \wedge NA, C/NA$), T4, A4($C/NA, B/NA$), T5)

T19 $B^I \supset ((A \supset B) \supset ((A \supset NB) \supset NA))$ (T7, A4($C/NA, B/NA$), T2, T5)

Now that we have T5, T17 and T18 as provable formulas, it is easy to see that we have established the following theorem:

THEOREM 2. *In the domain of propositions that behave classically, N,* $\supset$, $\wedge$, $\vee$ *are classical connectives. Thus PCL1 is rich enough to express the classical propositional logic in the domain of propositions that behave classically.*

Another theorem to be mentioned on the basis of T19 is:

THEOREM 3. *A12 of da Costa's Cn systems is provable in PCL1.*

3 A proof that PCL1 is paraconsistent

THEOREM 4. : *The following formulas are not provable in PCL1.*

- $(A \wedge NA) \supset B$

- $NA \supset (A \vee B \supset B)$

Proof
A proof of theorem 3 follows from the following truth tables with T and t as the designed values.

---

[1] Again we are very thankful for the kind comment of the anonymous referees. In the original draft, $\supset$ was used instead of $\vdash$, but the truth is that we must use $\vdash$, unless we change the notion of BC in another appropriate form.

N	
T	F
t	T
F	T

⊃	T	t	F
T	T	t	F
t	T	t	F
F	T	T	T

∧	T	t	F
T	T	t	F
t	t	t	F
F	F	F	F

∨	T	t	F
T	T	T	T
t	T	T	t
F	T	t	F

It is easy to see that modus ponens is truth preserving. An easy calculation shows that the values of the axioms of PCL1 are T or t, while if A takes the value t and if B takes the value F, then the value of $A \wedge NA \supset B$ takes the value F. Under the same valuation, the value of $NA \supset (A \vee B \supset B)$ takes the value F. Hence PCL1 is paraconsistent with respect to N. Thus the disjunctive syllogism does not hold generally in PCL1.

4   On the logical relation between PCL1 and S5

Let us consider the relationship that holds between PCL1 and S5. At first 1) let us add the classical negation to PCL1 that will be named PCL1C, and then 2) let us enrich PCL1C with a rule of inference (Rule of Necessitation).

4.1   PCL1 is included in S5

By means of the following transformation rules, we can translate the formulas of PCL1 to the formulas of S5.

TRANSFORMATION RULE 6. :
    $M\neg\alpha$ is the S5-transform of $N\alpha$
    $L\alpha$ is the S5-transform of $NN\alpha$

It is easy to see that the following theorems hold:

THEOREM 1. *The S5-transform of the axioms of PCL1 and the rule RA15 of PCL1 are provable and valid in S5, respectively.*

THEOREM 2. *The (S5-transforms of the) formulas provable in PCL1C are provable in S5,*

so that the following holds:

THEOREM 3. *(The S5-transform of) PCL1 is a subset of S5.*

0.2   PCL1C and S5

Let us call PCL1C the system that is obtained from PCL1 by adding classical negation and consider the translation of S5-formulas into PCL1C. We introduce the following two transformation rules between PCL1C and S5.

TRANSFORMATION RULE 7. :
    $NN\alpha$ is the PCL1C-transform of $L\alpha$
    $N\neg\alpha$ is the PCL1C-transform of $M\alpha$

Taking recourse to these transformation rules, let us consider the formulas K, T and E that are axioms of S5.

K	$L(A \supset B) \supset (LA \supset LB)$	(T9)
T	$LA \supset A$	(A10)
E	$MA \supset LMA$	(A11)

It is easy to show that each PCL1C-transform of K, T and E is provable in PCL1C. Hence we have:

THEOREM 1. *The PCL1C-transforms of the axioms of S5 are provable in PCL1C.*

0.3 PCL1CN and S5

Let us call the system PCL1CN that is obtained from PCL1C by adding the rule of necessitation.

RULE OF INFERENCE 8. Rule of Necessitation (RN): $\quad A \vdash NNA$

As (the S5-transform of) *RN* and (the S5-transform of) the rule *RA*15 are valid rules of inference in S5, PCL1CN has the same rules of inferences as S5 does. Therefore,

THEOREM 1. *PCL1C-transform of a formula provable in S5 is provable in PCL1CN.*

Thus we obtain from Theorem 8 and 9, we have:,

THEOREM 2. *S5 is inferentially equivalent to PCL1CN.*

Appendix

In this appendix, we give a proof of T9 in PCL1. That corresponds to the basic modal law K.

T9 $NN(A \supset B) \supset (NNA \supset NNB)$

1. $\vdash (NNA \supset A) \supset (NNA \supset (NNA \supset A))$		(T1($A/NNA \supset A$, B/NNA))
2. $\vdash NNA \supset (NNA \supset A)$		(A10, 1, MP)
3. $\vdash NNA \supset (A \supset NNA)$		(T1($A/NNA, B/A$))
4. $\vdash NNA \supset ((NNA \supset A) \wedge (A \supset NNA))$		(2, 3, A7, MP, MP)
5. $\vdash NNA \supset A^I$		(4, D2)
6. $\vdash NNA \supset A^I \wedge A$		(A10, 5, syll.)
7. $\vdash A^I \supset (A \supset (NB \supset N(A \supset B)))$		(A15, A9, syll.)

8. $\vdash A^I \wedge A \supset (NB \supset N(A \supset B))$ (7, A8, MP)

9. $\vdash NNA \supset (NB \supset N(A \supset B))$ (6, 8, syll.)

10. $\vdash (N(A \supset B))^I \supset ((NB \supset N(A \supset B)) \supset (NN(A \supset B) \supset NNB))$
    (T7(A/B, $B/N(A \supset B)$))

11. $\vdash (N(A \supset B))^I$ (A10($A/(A \supset B)$), A11($A/(A \supset B)$), D2)

12. $\vdash (NB \supset N(A \supset B)) \supset (NN(A \supset B) \supset NNB)$ (10, 11, MP)

13. $\vdash NNA \supset (NN(A \supset B) \supset NNB)$ (9, 12, syll.)

14. $\vdash NN(A \supset B) \supset (NNA \supset NNB)$ (13, T4, MP)

BIBLIOGRAPHY

[1] Arruda, A. I. et al. (eds.) [1978]: *Mathematical Logic: Proc. of the 1st Brazilian Conf.*, Marcel Dekker.
[2] Arruda, A. I., Chuaqui, R. and da Costa, N. C. A. (eds.) [1980]: *Mathematical Logic in Latin America*, North-Holland. (Abbreviated as *LA*).
[3] Arruda, A. I., da Costa, N. C. A. and Chuaqui, R. (eds.) [1977]: *Non-classical Logic, Model Theory and Computability*, North-Holland.
[4] Priest, G., Routley, R. and Norman, J. (eds.) [1989]: *Paraconsistent Logic*, Philosophia Verlag. (Abbreviated as *PL*).
[5] Arruda, A. I. [1989]: Aspect of the Historical Development of Paraconsistent Logic, in *PL*, 99—130.
[6] Béziau, J-Y.[2000]: What is Paraconsistent Logic?, *Frontiers of Paraconsistent Logic*, Batens, D., Mortensen. C., Priest, G., Van Bendegem, J-P., Research Studies Press LTD, 95-111.
[7] Chellas, B. F. [1980]: *Modal Logic: An Introduction*, Cambridge, Cambridge University Press.
[8] da Costa, N. C. A. [1974]: On the Theory of Inconsistent Formal Systems, *Notre Dame Journal of Formal Logic*, vol.15, No.4, 497—510.
[9] da Costa, N. C. A., Béziau, J-Y, Bueno, O. [1995]: Paraconsistent Logic in a Historical Perspective, *Logic et Analyse*, 111-125.
[10] Dubikajtis, L. [1975]: The Life and Works of Jaśkowski, *Studia Logica*, 38, vol.2.
[11] Dunn, J. M. [1986]: Relevance Logic and Entailment, in Gabby, D and Guenthner, F (eds.), *Handbook of Philosophical Logic*, vol.3, Reidel, 117–224.
[12] Jaśkowski, S. [1948]: Rachunek Zdań dla Systemów Dedukcyjnych Sprzecznych, *Studia Scientiarum Trunensis*, vol 1, Section A, 55—77.
[13] Jaśkowski, S. [1969]: Propositional Calculus for Contradictory Deductive Systems, *Studia Logica*, 24, 143—157.
[14] Kleene, S. C. [1967]: *Introduction to Metamathematics*, Amsterdam, North-Holland.
[15] Kotas, J. [1975]: Discussive Sentential Calculus of Jaśkowski, *Studia Logica*, 38, vol.2, 149 —168.
[16] Kotas, J. and da Costa, N. C. A. [1978]:On the Problem of Jaśkowski and the Logics of Łukasiewicz, in *BML*, 127—139.
[17] Kotas, J. and da Costa, N. C. A. [1989]: Problems of Modal and Discussive Logics, in *PL*, 227—224.
[18] Mortensen, C. [1983]: The Validity of Disjunctive Syllogism Is Not So Easily Proved, *Notre Dame Journal of Formal Logic*, 24, 35-40.

[19] Priest, G. and Routley, R. [1989a]: First Historical Introduction: A Preliminary History of Paraconsistent and Dialectic Approaches, in *PL*, 3—75.

[20] Priest, G. and Routley, R. [1989b]: Systems of Paraconsistent Logic , in *PL*, 151—186.

[21] Priest, G. and Routley, R. [1989c]: Applications of Paraconsistent Logic , in *PL*, 367—393.

[22] Prior, A. N. [1962]: *Formal Logic*, 2nd ed., Oxford.

[23] Rasiowa, H. and Sikorski, R. [1970]: *the Mathematics of Metamathematics*, Warszawa, Polish Scientific Publishers.

[24] Slater, B. H. [1995]: Paraconsistent Logics?, *Journal of Philosophical Logic*, 24, 451—454.

[25] Urbas, I.[1989]: Paraconsistency and the C-Systems of de Costa, *Notre Dame Journal of Formal Logic*, 30, No.4, 583-597.

[26] Waragai, T.[1999]: Aristotle's Master Argument about Primary Substance and Leśniewski's Logical Ontology: A Formal Character of Metaphysics, *La Docrine de la Science de lÂntiquité à L'Age Medieval*, Rashed, R. et Biard, J. (eds.), Peeters.

Toshiharu Waragai

Tokyo Institute of Technology, Nishi-9 bldg. Room 412, Tokyo, Japan.

twaragai@me.titech.ac.jp

Tomoki Shidori

Connectous Corp, Japan.

shidori@connectous.co.jp

On the Structure of Evidential Gluts and Gaps

DON FAUST[1]

ABSTRACT. If our knowledge is absolute and consistent, then we can use Classical Logic. If it is not absolute but remains consistent, then it is often the case that our knowledge is evidential in nature, indeed regularly involving evidential gluts and evidential gaps as well, and Evidence Logic (EL) provides an example of a suitable foundational framework. Of course, if the conflict we are dealing with rises to the level of contradiction, then EL is also inadequate and one must use a paraconsistent logic.

So EL provides a framework for this middle ground where evidential conflict commonly occurs yet no contradictions arise. EL also provides logical machinery helpful in studying further the concept of negation, which is certainly of foundational importance to paraconsistency. In EL, for a predication P, $P_c : e$ asserts that there is confirmatory evidence at the value e for P, while $P_r : e$ asserts refutatory evidence at value e. See "The Concept of Evidence", *Int. J. Intell. Systems* 15 (2000), 477-493 for precise definitions and theorems delineating EL and its Boolean Sentence Algebras, and "Conflict without Contradiction: paraconsistency and axiomatizable conflict toleration hierarchies in Evidence Logic", *Logic and Logical Philosophy* 9 (2001), 137-151 for initial examples of families of extensions of EL whose axioms reach out in a variety of ways toward domain-specific properties regarding evidential conflict.

In this paper we will look, mostly rather informally, at how evidential gluts and gaps are each structured in EL and also how they interact. In Section 6 we will formally examine three families of axiomatized extensions of EL: a previously introduced family of theories each allowing no conflict at some evidence value e, a new family each allowing no paucity at value e, and a new "conjunctive" family each *both* allowing no conflict at value e *and* allowing no paucity at value e'. Further, in Section 7, continuing formally and motivated

---

[1]I wish to thank Susan LaForais of Northern Michigan University for expert preparation of the typescript and Mike Gach of Range Telecommunications in Marquette, Michigan for numerous fruitful discussions about logic.

by the evidential contexts most likely in EL application domains, we
will generalize the notions of conflict and paucity and examine the
resulting theories. The Boolean Sentence Algebras are analyzed in
each case.

1   INTRODUCTION

In contexts where knowledge is absolute and no contradiction P AND (NOT
P) arises, Classical Logic is a reasonable tool. Further, in contexts where
some contradiction arises clearly one needs to turn to a paraconsistent logic.
But what about the vast "Middle Ground" of evidential conflict, certainly
a common occurrence, where we are dealing with the presence of both 'evidence in favor' and 'evidence against' some predication? It is with this
Middle Ground of evidential conflict that the logic Evidence Logic (EL)
deals.

Evidence Logic provides a logical framework for representing, and reasoning with, evidence confirmatory of a predication and evidence refutatory
of a predication. Hence it provides a framework for the vast intermediate
domains where absolute truth is lacking so Classical Logic is inadequate,
where evidence is often the best knowledge to be had and conflicting evidence abounds, and yet where this conflict does not rise to the level of
contradiction so the often complex machinery of paraconsistent logics is
not needed. By studying, in Evidence Logic, the interplay between evidential conflict and outright contradiction we can gain further insight into the
complex and foundational relationship between classical and paraconsistent
logics.

Since Evidence Logic provides machinery for expressing the presence of
both confirmatory and refutatory evidence, it can express conflict. Also,
since (pure) Evidence Logic contains no axioms asserting that any such
conflicting evidence is contradictory, it provides a logical environment which
allows such conflict while yet not being thereby explosive. Hence, Evidence
Logic allows for reasoning under, and indeed about, such conflict. Moreover, using, for example, a resolution-based implementation of Evidence
Logic provides an automated reasoning system with machinery for handling
evidential conflict.

In addition, since Evidence Logic is an extension of Classical Logic containing classical negation, Evidence Logic provides a logical framework helpful in further analysis of the concept of negation. Certainly the concept of
negation is perplexing, a concept which has been studied diligently for over
2000 years and which yet remains incompletely understood. With its inclusion of both classical negation and machinery for representing evidential
conflict, Evidence Logic provides an enriched environment where further

explication of the concept of negation is possible. This is particularly important for paraconsistency since the concept of negation is fundamental to our understanding of a number of the foundational issues of paraconsistency.

In three previous papers [4, 5, and 6] the fundamental structure of Evidence Logic has been defined and analyzed. In [4] theorems giving the Boolean Sentence Algebras of the various (pure) Evidence Logics, which vary according to the number and arities of the predicates included in the language, are stated and proven. In [5] theorems giving the Boolean Sentence Algebras of a variety of families of axiomatizable extensions of Evidence Logic, extensions which explore a variety of ways in which one can allow evidential conflict to attain, are stated and proven. In [6] initial observations are made concerning how the enriched framework of Evidence Logic may help further our understanding of negation especially as this relates to paraconsistency. In the present paper we will explore how evidential conflict and evidential paucity (evidential gluts and evidential gaps, respectively) interact. We will pay special attention to how these explorations might enrich our understanding of the concept of negation, and to how this enriched understanding of the concept of negation may help further our understanding of the foundational role negation plays in paraconsistency.

Throughout we keep in mind, and indeed interpret into EL, Aristotle's work on privatives which even 2000 years ago helped to further elucidation of the concept of negation. His opinion that non-P IMPLIES (NOT P) is closely related to the evidential gluts we define and study in EL, while the converse of this opinion connects with our evidential gaps in EL. By studying families of EL-theories containing a gradational version of the concept of privation, analysis of the interplay between privation and classical negation is achieved, and further insight into the general concept of negation is gained.

First, we turn to saying a few things about Evidence Logic, hopefully enough to make our heavy use of Evidence Logic throughout the paper easily understandable. For further explication of Evidence Logic, the reader is referred to the three papers cited above. For precise definitions, structure theorems, and proofs, the first two papers are particularly recommended.

2 EVIDENCE LOGIC (EL)

For any $n > 1$, let $E_n = \{i/(n-1) : i = 1, \ldots, n-1\} = \{e_1, \ldots, e_{n-1}\}$ be the Evidence Space of size $n - 1$, of *evidence values*, with smallest evidence value and evidence increment $\varepsilon = 1/(n-1) = e_1$. In general, the choice of n is bounded above by implementation considerations and bounded below by the granularity of the evidence itself. We will often find it more intuitive to refer to the *evidence levels* $d = 0, \ldots, n$ where: the level $d = 0$ corresponds to the informally always present evidence value 0, for each $d = 1, \ldots, n - 1$ the evidence level d corresponds to the evidence value $e_d = d\varepsilon$, and the level $d = n$ corresponds to the never attained evidence value $1 + \varepsilon$. (In deference to particularity and the decimal mind, the reader might like to sometimes think in terms of the case $n = 11$, having Evidence Space $E_{11} = \{.1, .2, \ldots, .9, 1\}$.)

The Evidence Logic EL_n, by definition, utilizes Evidence Space E_n and has, for each s-ary predicate symbol P and terms $t_1, \ldots, t_s$, atomic formulas

$$P_c t_1 \ldots t_s : e \qquad \text{and} \qquad P_r t_1 \ldots t_s : e \qquad \text{for} \qquad e \text{ in } E_n.$$

Here the former asserts that there is confirmatory evidence for $Pt_1 \ldots t_s$ at evidence value e, while the latter asserts refutatory evidence for $Pt_1 \ldots t_s$ at evidence value e. In addition to any usual axiomatization for Classical Logic, EL_n contains just a set of further axioms (finite in number if the language at hand has only a finite number of predicate symbols), easily supplied by the reader, which assure that "stronger evidence entails weaker evidence". Regarding semantics, any model for EL_n will contain an interpretation, for each t-ary predicate symbol, which is a pair of t-ary partial functions interpreting respectively the confirmatory predication P_c and the refutatory predication P_r (in each case, the function maps t-tuples over the universe of the model to elements of E_n). The reader is referred to [4] for a precise delineation of EL_n, along with detailed discussion and examples. We shall here use just EL, instead of EL_n, when no confusion arises thereby.

Clearly the structure of EL_n varies according to n. But just as clearly, it also varies according to the number and arities of the predicate symbols (and also the function symbols, which I intentionally suppress mentioning in the more informal context of the present paper) which are present. The exact results concerning all these structural variations are given in [4], and readers wishing to understand EL at a more detailed level are encouraged to go there. Suffice it here to say that the structure of EL varies interestingly across language types: the structure varies throughout the *monadic languages* having a stipulated finite number of propositions, constants, and unary predicates; there is a single structure for the *functional languages* having additionally one unary function; and there is a single structure for the *undecidable languages* having a finite number of predicates and functions including at least one predicate or function which is at least binary or at least two unary functions. Let us also note that as a result of the recursive isomorphism results in [4] we get immediately soundness and completeness theorems for EL.

Equipping EL with an array of predicates and functions is, as indicated above, not only possible but also absolutely routine in applications and often also very interesting in terms of the resulting structure of the logic. However, in the present paper (except in Sections 6 and 7), in order to focus more fluently on evidential gluts and gaps and the concept of negation, we choose to suppress this more general setting and consider just the simple case of a language which has just one 0-ary predicate symbol (proposition symbol) P. Hence, for example, $P_r : e$ asserts that there is refutatory evidence at evidence value e for P while NOT $P_c : e$ asserts the absence of confirmatory evidence at value e for P.

Actually, these examples illustrate a clarifying distinction which surfaces in EL and which we suggest the reader might want to pause and ponder a bit. In some sense, the former asserts an "evidence of absence" while the latter asserts an "absence of evidence". How, indeed, are "absence of evidence" and "evidence of absence" related? Will that relationship be in fact often domain dependent? More finely, won't that relationship depend upon some quantification of "absence of evidence"? For example, it would seem that an "absence of evidence" based on a more complete traversal of the search space of all (relevant) evidence should be knowledge worth more (and hence having a higher value) than an "absence of evidence" based on some less complete search through the evidence.

3   EVIDENTIAL GLUTS (E-gluts)

Let us say that the conjunction $P_c : e$ AND $P_r : e$ asserts an *evidential glut (E-glut) at evidence level d where e $= d\varepsilon$, i.e. where $d = e/\varepsilon$, $d = 1, \ldots, n$, and ε is the smallest evidence value $1/(n - 1)$.*

If we assert that there is no E-glut at level d, that there is no evidential conflict at level d, then we are, of course, asserting NOT ($P_c : e$ AND $P_r : e$), that $P_c : e$ and $P_r : e$ are contrary, that $P_c : e$ and $P_r : e$ are not both true. Now, Aristotle's opinion (cf. PRIOR ANALYTICS, chapter 46) about privation and its relation to negation is very interesting, and it is relevant to our considerations here. Let us interpret privation into our Evidence Logic by associating P with P_c and non-P with P_r. If we, further, generalize to the situation we have in EL of having evidence levels, we have the following situation. Aristotle's opinion that non-P IMPLIES (NOT P), a position he argued vigorously for in his PRIOR ANALYTICS, is in our setting, at level d, the assertion that $P_r : e$ IMPLIES (NOT $P_c : e$), which is (remembering that EL is built over Classical Logic) equivalent to NOT ($P_c : e$ AND $P_r : e$). That is, Aristotle's opinion, generalized to the evidence level machinery we have available in EL and particularized to evidence level d, is that "there is no E-glut" at level d.

In some domains it may be quite appropriate to axiomatize that evidential gluts will be allowed up to a certain level, but not allowed at that level or beyond that level. In other words, in such domains conflict at certain lower levels is tolerated while conflict at higher levels is not tolerated, a reasonable sort of *modus operandi* when working with evidence and wanting to be liberal enough to tolerate moderate amounts of evidential conflict while being careful enough not to allow 'too much' conflict. Indeed, the structure of EL entails a certain uniformity in regard to this matter. Namely, the assertion of the absence of E-gluts at a certain level, say level d, implies the absence of E-gluts at all levels $d' > d$ as well, as is shown by the following sequence of implications:

> $P_r : e'$ IMPLIES (by the EL-axiom that "the presence of stronger evidence implies the presence of weaker evidence") $P_r : e$, which IMPLIES (by our current assumption of the absence of any E-glut at evidence level d) (NOT $P_c : e$), which IMPLIES (since by contraposition of the just-cited EL-axiom, "the absence of weaker evidence implies the absence of stronger evidence") (NOT $P_c : e'$).

Thus: if there is no E-glut at level d, then there is also no E-glut at any of the levels d' for $d' > d$.

So, if we axiomatize in EL that evidential conflict in the form of E-gluts will be allowed to freely occur up to, say, level d but not at or above level d, then what we get is a theory, called $AL_n(d)$ in [5] (the "A" is for Aristotle), where E-gluts are uniformly disallowed at all evidence levels at or above level d. We will here call this theory an 'Evidential Conflict Logic', and denote it by $E\_C\_L_n(d)$.

Such axiomatizable extensions of EL, and others as well which explore different kinds of evidential conflict, are defined precisely in [5]. There one will find structure theorems, and their proofs, which delineate the Boolean Sentence Algebras of such axiomatized theories, sentence algebras which vary according to the size $n-1$ of the Evidence Space E_n and the number and arities of the predicates stipulated to be in the language. Suffice it to note, in the present context of having just a single proposition symbol P, that instead of the contribution of n^2 atoms to the ordered basis for the Boolean Sentence Algebra of this pure theory of EL, there is just a contribution of $d(2n - d)$ atoms to the ordered basis for the Boolean Sentence Algebra of the theory $E\_C\_L_n(d)$ (wherein there are "no E-gluts at any evidence level at or above level d").

4   EVIDENTIAL GAPS (E-gaps)

Let us say that the conjunction (NOT $P_c : e$) AND (NOT $P_r : e$) asserts an *evidential gap* (E-gap) *at level d where d is related to e via* $e = d\varepsilon$, $d = 0, \ldots, n - 1$.

The situation here, in the case of E-gaps, is dual to that for E-gluts in a sense analogous to the fact that the law of excluded middle is dual to the law of noncontradiction. If we assert that there is no E-gap at level d, that there is not an evidential paucity at level d, then we are, of course, asserting NOT (NOT $P_c : e$ AND NOT $P_r : e$), or equivalently that $P_c : e$ OR $P_r : e$, that $P_c : e$ and $P_r : e$ are subcontrary, that $P_c : e$ and $P_r : e$ are not both false. Note that we saw in the section above that Aristotle's opinion about the relation between privation and negation, interpreted into the present context of EL and particularized to evidence level d, is that $P_r : e$ IMPLIES (NOT $P_c : e$). So it is the converse of this opinion of Aristotle, namely that (NOT $P_c : e$) IMPLIES $P_r : e$, which is equivalent to the assertion that there is no E-gap at evidence level d. That is, it is the converse of Aristotle's opinion, generalized to the evidence level machinery we have in EL and particularized to evidence level d, which asserts that "there is no E-gap" at level d.

Analogous to the situation with respect to E-gluts, in some domains it may be appropriate to axiomatize that evidential gaps will be allowed down to a certain level, but not allowed at that level or below that level. This is certainly a reasonable sort of characteristic one may want a domain to possess, where an E-gap at high evidence levels may need to be tolerated simply because it is too much of a demand on the evidence to require no E-gaps at these high levels, while at some point lower down in the levels-of-evidence hierarchy it is reasonable to expect no E-gap. As with E-gluts and their upward uniformity, here with E-gaps there is a downward uniformity induced by the structure of EL. Namely, the assertion of the absence of E-gaps at a certain level, say level d, implies the absence of E-gaps at all levels $d' < d$ as well, as is proven by the following sequence of implications:

> (NOT $P_c : e'$) IMPLIES (by the contrapositive of the EL-axiom that "the presence of stronger evidence implies the presence of weaker evidence") (NOT $P_c : e$), which IMPLIES (by our current assumption of the absence of any E-gap at evidence level d) $P_r : e$, which IMPLIES (by the just-cited EL-axiom) $P_r : e'$.

That is, if there is no E-gap at level d, then there is also no E-gap at level d' for any $d' < d$.

So the only reasonable axiomatizations, in terms of E-gaps as we are working with them in the present context, are ones which allow E-gaps precisely in an evidential range from the highest evidence value (which is $(n-1)/(n-1) = 1$, corresponding to the evidence level $d = n - 1$ of, so to speak, 'evidential certainty') down through some level. Indeed, if we axiomatize in EL that evidential paucity in the form of E-gaps will be allowed to freely occur down to level d but not at level d (and hence also, as argued above, not at any level below level d either), then we get a theory where E-gaps are uniformly disallowed at all evidence levels at or below level d. Let us call this theory an 'Evidential Paucity Logic' and denote it by $E\_P\_L_n(d)$.

Such axiomatizable extensions of EL form families of Evidential Paucity Logics providing insight into the structure of evidential paucity. They are dual to those found in [5] which are now seen by way of the discussion in Section 3 above to be Evidential Conflict Logics providing insight into the structure of evidential conflict. For these Evidential Paucity Logics $E\_P\_L_n(d)$, there are (dual) structure theorems, and their proofs, which delineate the Boolean Sentence Algebras of such axiomatized theories, sentence algebras which vary according to the size $n-1$ of the Evidence Space E_n and the number and arities of the predicates stipulated to be in the language. We will look carefully at this phenomenon in Section 6 below. Suffice it here to note that, in the present context of having just a single proposition symbol P, instead of the usual contribution of n^2 atoms to the ordered basis for the Boolean Sentence Algebra of the pure theory of EL, there is in the theory $E\_P\_L_n(d)$ just a contribution of $n^2 - d^2$ atoms to the ordered basis for the Boolean Sentence Algebra of the theory (wherein there are "no E-gaps at or below evidence level d").

5 THE INTERPLAY OF EVIDENTIAL GLUTS AND EVIDENTIAL GAPS

First, regarding classical negation, note that two assertions are negations of each other just in case they are *both* contrary (not both true) *and* subcontrary (not both false). And, in his explication of privatives, Aristotle seems duly mindful of this. Relative to the present context into which we have interpreted Aristotle's privatives, if one were to assert that $P_c : e$ and $P_r : e$ are *both* contrary (as Aristotle did) *and* subcontrary (as Aristotle carefully did not), then they would be negations of each other. That is, in our context, we would then have (NOT $P_c : e$) logically equivalent to $P_r : e$. Back in Aristotle's context, this would yield non-P being logically equivalent to NOT P, something Aristotle vigorously argued to be not the case. (For related issues concerning how contemporary elementary logic handles/mishandles privatives, cf. [6, p. 506].)

Now consider the present context of E-gluts and E-gaps. Axiomatizing (NOT $P_c : e$) IFF $P_r : e$ is to assert (where as usual d is stipulated via $e = d\varepsilon$, $d = 1, \ldots, n-1$) *both* that there are no E-gluts at level d *and* that there are no E-gaps at level d. Well, what is such a theory like? First note that, on the basis of this axiom, $P_r : e$ is just exactly (equivalent to) the negation of $P_c : e$: sure enough, since here $P_c : e$ and $P_r : e$ form a pair which is both contrary and subcontrary. Second, such a theory has no E-gaps at or below level d *and also* no E-gluts at or above level d. That is, such a theory satisfies "evidential non-paucity" up through level d while allowing conflict up to level d, while it also satisfies "evidential non-conflict" from level d on up while allowing paucity above level d. Roughly speaking, this logic requires evidence to be *rich enough* to be gapless up through level d (while allowing gluts up to level d), *but not too rich* since the evidence is also required to be glutless from level d on up (while allowing gaps above level d).

For another example, the reader may wish to consider the details of the theory with an axiom asserting (NOT $P_c : e$) IFF $P_r : e$ for some 'interval' of levels. Note that it is easily seen, from our earlier arguments about the upward uniformity of "no E-gluts" and the downward uniformity of "no E-gaps", that if we have (NOT $P_c : f$) IFF $P_r : f$ and (NOT $P_c : g$) IFF $P_r : g$ with $f < g$ then we will also have (NOT $P_c : h$) IFF $P_r : h$ for all h between f and g, so our use of the term 'interval' here is well-motivated.

Consider also the theory not allowing any E-gluts or E-gaps at all. Clearly the two axioms $P_r : \varepsilon$ IMPLIES (NOT $P_c : \varepsilon$) and (NOT $P_c : 1$) IMPLIES $P_r : 1$ will do fine: from the first axiom we get E-glutlessness at all evidence levels all the way up, and from the second we will get E-gaplessness at all evidence levels all the way down. Note that here, at each evidence value e, (NOT $P_c : e$) is equivalent to $P_r : e$. Indeed, in this theory there are just the two atoms $P_c : 1 \wedge \neg P_r : \varepsilon \equiv P_c : 1 \equiv \neg P_r : \varepsilon$ and $P_r : 1 \wedge \neg P_c : \varepsilon \equiv P_r : 1 \equiv \neg P_c : \varepsilon$, and this theory is isomorphic to Classical Logic.

In the latter part of Section 6 below we will take a precise look at this interplay between E-gluts and E-gaps by defining, and proving a structure theorem for, a matrix of conjunctive theories which variously restrict *both* E-gluts *and* E-gaps.

6   THE STRUCTURE OF THE LOGICS

In this section we give precise definitions of, and structure theorems for, the "Evidential Conflict" logics $E\_C\_L_n(d)$ providing (as d ranges from 1 to n) a family of increasingly conflict-tolerant theories, and the "Evidential Paucity" logics $E\_P\_L_n(d)$ providing (as d ranges from $n - 1$ down to 0) a sequence of increasingly paucity-tolerant theories. Further, we conjunctively combine these logics to construct a family of n^2 theories $E\_C\&P\_L_n(d_1, d_2)$ which are, in varying degrees, both conflict-tolerant and paucity-tolerant. It is perhaps this family of theories which best allows us to view the two extremes: Evidence Logic, a "most tolerant logic", and Classical Logic, a "least tolerant logic".

In contrast to the above informal sections where we considered just the simple language with a single proposition symbol P, here in Section 6 we will consider arbitrary finitary languages, with equality. Let τ be a language similarity type. The Boolean Sentence Algebras (BSAs) of the theories we consider here vary through the following categorization of similarity types. First, let μ denote a *monadic* language consisting of p proposition symbols, k constant symbols, and u unary predicate symbols. Second, let μ' denote a *functional* language consisting of the symbols stipulated by μ plus one unary function symbol. Finally, let ν denote an *undecidable* language, stipulating a finite number of predicates/functions including at least one predicate or function which is at least binary or at least two unary functions. For a theory T, let $BSA(T)$ be the Boolean Sentence Algebra of T, let $BA(\alpha)$ be the Boolean Algebra with ordered basis of order type α, and let $\cong$ denote "recursive isomorphism". Finally, define the order type

$$\sigma(m) \stackrel{\text{def}}{=} \omega^{m^u} \cdot m^p \cdot \sum_{i=1}^{k} s_{ki} \cdot m^{ui},$$

where the s_{ki} are Stirling Numbers of the second kind.

In [4], it is proven that for monadic μ as stipulated above,

$$BSA(EL_{n,\mu}) \cong BA(\sigma(n^2)),$$

while the functional varieties of EL are all recursively isomorphic to the functional variety of CL and the undecidable varieties of EL are all recursively isomorphic to the universal Classical Logic $CL_{<2>}$ with one binary relation symbol.

Let us begin with an informal overview. In an analysis of the Boolean Sentence Algebra of any theory, one needs to focus on how exactly the variety of predicates available in the language of the theory variously contribute to the BSA of the theory. Consider, for example, the pure theory of a monadic language with similarity type μ as defined above, and consider first the case of Classical Logic. To simplify the discussion, let us refer to an atom of any proper reduct of a language L_τ as a preatom. For the monadic language L_μ, with respect to the reduct consisting of just the p proposition symbols, there will be 2^p preatoms, one for each of the p-ary conjunctions formable by choosing, for each proposition symbol Q, either Q or $\neg Q$. Conjoined with other preatoms resulting from, in our case here, the unaries and constants and their interplay, atoms of the theory are formed. Indeed, from the u unary relation symbols there will be an order type of ω^{2^u} preatoms deriving from the fact that u unaries splits the universe of any interpretation into 2^u subsets, so that adding consideration of the k constants (regarding which we refer the reader to [4]), one proves that $BSA(CL_\mu) \cong BA(\sigma(2))$.

Now, the case for Evidence Logic is more complicated than the case for Classical Logic above. For the p-ary conjunctions which yield the preatoms in the case of p proposition symbols, instead of the simple choice of Q or $\neg Q$, regarding the proposition symbol Q, which one has in Classical Logic, one has in EL the choice regarding Q of any one of the n^2 conjunctive pairings of one of the n confirmatory sentences

$$\alpha_{c,j} \stackrel{\text{def}}{=} \begin{cases} Q_c : 1 & \text{if } j = n - 1, \\ \neg Q_c : e_{j+1} \wedge Q_c : e_j & \text{if } j = n - 2, \ldots, 1, \\ \neg Q_c : e_1 & \text{if } j = 0 \end{cases}$$

with one of the n refutatory sentences

$$\alpha_{r,j} \stackrel{\text{def}}{=} \begin{cases} Q_r : 1 & \text{if } j = n - 1, \\ \neg Q_r : e_{j+1} \wedge Q_r : e_j & \text{if } j = n - 2, \ldots, 1, \\ \neg Q_r : e_1 & \text{if } j = 0, \end{cases}$$

resulting in $(n^2)^p = n^{2p}$ preatoms. Similarly, in EL each unary predicate symbol gives rise to a partitioning of the universe of any interpretation into n^2 subsets, one delineated by each of the appropriate n^2 conjunctive pairings. Hence, the u unaries split any universe into $(n^2)^u = n^{2u}$ subsets. While the details of this analysis, for the case of EL, are contained in [4], it is hoped that this informal overview has been helpful as we now turn our attention to the theories of "evidential conflict" and "evidential paucity" at hand.

First, then, we define the Evidential Conflict Logic $E\_C\_L_{n,\tau}(d)$ which has no gluts at level d or above.

DEFINITION 1. For $d \in \{1, \ldots, n\}$ define

$$E\_C\_L_{n,\tau}(d) \stackrel{\text{def}}{=} EL_{n,\tau}(\Phi_{1,d})$$

where $\Phi_{1,d}$ is the axiom given by the conjunction of the following sentences (where $e = d\varepsilon$):

$$\neg(Q_c : e \quad \text{AND} \quad Q_r : e)$$

for each proposition symbol Q stipulated by τ, and

$$\forall x_1 \ldots x_t \neg (R_c x_1 \ldots x_t : e \quad \text{AND} \quad R_r x_1 \ldots x_t : e)$$

for each t-ary predicate symbol R stipulated by τ.

THEOREM 2. *For each* $d \in \{1, \ldots, n\}$

$$BSA \ (E\_C\_L_{n,\mu}(d))$$

has ordered basis $\sigma(m)$ *with* $m = d(2n - d)$.

Since each theory $E\_C\_L_{n,\mu}(d)$ is the same as the theory $AL_{n,\mu}(d)$ of [5], the reader is referred there for the proof. Next we define the Evidential Paucity Logic $E\_P\_L_{n,\tau}(d)$ which has no gaps at level d or below.

DEFINITION 3. For $d \in \{0, \ldots, n - 1\}$ define

$$E\_P\_L_{n,\tau}(d) \overset{\text{def}}{=} EL_{n,\tau}(\Phi_{2,d})$$

where $\Phi_{2,d}$ is the axiom given by the conjunction of the following sentences (where $e = d\varepsilon$):

$$Q_c : e \quad \text{OR} \quad Q_r : e$$

for each proposition symbol Q stipulated by τ, and

$$\forall x_1 \ldots x_t (R_c x_1 \ldots x_t : e \quad \text{OR} \quad R_r x_1 \ldots x_t : e)$$

for each t-ary predicate symbol R stipulated by τ.

THEOREM 4. *For each* $d \in \{0, \ldots, n - 1\}$

$$BSA \ (E\_P\_L_{n,\mu}(d))$$

has ordered basis $\sigma(m)$ *with* $m = n^2 - d^2$.

Proof. It is sufficient, building upon earlier arguments in [4, 5], to consider just the situation with regard to a proposition symbol Q. Of the n^2 conjunctive pairings of the $\alpha_{c,j}$ and $\alpha_{r,k}$ defined above, exactly d^2 of them are inconsistent with axiom $\Phi_{2,d}$, namely the conjunctive pairings of any

$$\alpha_{c,j}, \ 0 \le j \le d - 1, \text{ with any } \alpha_{r,k}, \ 0 \le k \le d - 1$$

This is the case since it is just such conjunctive pairs which either directly assert or imply a gap at level d.

Hence, $m = n^2 - d^2$. //

Now we move on to conjunctive theories which axiomatize there being both a certain glutlessness from the top of the evidence hierarchy and a certain gaplessness from the bottom. That is, we wish to axiomatize a family of conjunctive "Conflict and Paucity" theories $E\_C\&P\_L_{n,\tau}(d_1, d_2)$ having no glut at or above level d_1 and no gap at or below level d_2, where $d_1 = 1, \ldots, n$ and $d_2 = 0, \ldots, n - 1$.

DEFINITION 5. For $(d_1, d_2) \in \{1, \ldots, n\} \times \{0, \ldots, n-1\}$ define

$$E\_C\&P\_L_{n,\tau}(d_1, d_2) \stackrel{\text{def}}{=} EL_{n,\tau}(\Phi_{3,d_1,d_2})$$

where

$$\Phi_{3,d_1,d_2} \quad \text{is the conjunction} \quad (\ \Phi_{1,d_1} \quad \text{AND} \quad \Phi_{2,d_2}\ ).$$

THEOREM 6. *For each* $(d_1, d_2) \in \{1, \ldots, n\} \times \{0, \ldots, n-1\}$

$$BSA\ (E\_C\&P\_L_{n,\mu}(d_1, d_2))$$

has ordered basis $\sigma(m)$ *with* $m = 2d_1(n - d_2) - (d_1 -_p d_2)^2$.

Proof. Here $d_1 -_p d_2$ is "proper subtraction": $a -_p b = a - b$ if $a \geq b$ while $a -_p b = 0$ if $a \leq b$. In the last proof we counted those of the n^2 preatoms which were *inconsistent* with the theory: there were d^2 of those, so the theory had $n^2 - d^2$ preatoms relative to the proposition symbol Q. In the present case we find it more convenient to count those of the n^2 preatoms which are *consistent* with the theory and hence persist as preatoms. We note, before starting, that our accounting will involve some double-counting, for which we will compensate at the end.

Well, for each $j = n - 1, \ldots, d_2$ and for each $k = 0, \ldots, d_1 - 1$, just the sentences ($\alpha_{c,j}$ AND $\alpha_{r,k}$) are consistent with this theory since it is just each such assertion which involves no glut at level d_1 or above and also no gap at level d_2 or below. So we get $d_1(n - d_2)$ preatoms. But also, interchanging the confirmatory and the refutatory here we get $d_1(n - d_2)$ further preatoms. So, altogether we get $2d_1(n - d_2)$ preatoms.

Finally, we need to consider the double-counting which may have occurred. Note that there are d_1 levels (namely levels $0, 1, \ldots, d_1 - 1$) below the first no-glut level at level d_1, and there are d_2 gapless levels (namely levels $1, \ldots, d_2$). So there is overlap (levels with no glut and no gap) if $d_1 \leq d_2$, or $d_1 - d_2 \leq 0$, and in these cases no double-counting has occurred. On the other hand, if $d_1 > d_2$ then no overlap occurred and $(d_1 - d_2)^2$ measures the amount of double-counting which has occurred. In any case, then, a subtractive term $(d_1 -_p d_2)^2$ is exactly what is needed, and we conclude that $m = 2d_1(n - d_2) - (d_1 -_p d_2)^2$. //

We note that it is also the case, analogous to the situation in [4], that for the theories we have defined here, the functional varieties are isomorphic to the functional variety of CL and the undecidable varieties are isomorphic to $CL_{<2>}$.

We have, then, a matrix of theories $E\_C\&P\_L_{n,\mu}(d_1, d_2)$ as given in Figure 1. (In this regard, the reader may also wish to refer to the particular case $n = 6$ with Evidence Space $E_6 = \{.2, .4, .6, .8, 1\}$ illustrated in Figure 2.) As one moves from the lower left corner of this matrix, whose theory is isomorphic to the Evidence Logic $EL_{n,\mu}$ with $m = n^2$, one passes through theories which are more and more restrictive regarding allowing some conflict and some paucity. Indeed, consider the main transverse diagonal in Figure 1, of "symmetric theories" allowing <u>both</u> conflict up to level d <u>and</u> paucity above level $n - d$ (see also Figure 3), which we denote as

$$E\_SymC\&P\_L_{n,\mu}(d) \overset{\text{def}}{=} E\_C\&P\_L_{n,\mu}(d, n-d),$$

for $d = n, \ldots, 1$. Note that $E\_SymC\&P\_L_{n,\mu}(d)$ has ordered basis

$$\sigma(m) \qquad \text{with} \qquad m = 2d^2 - (2d -_p n)^2.$$

Starting from $E\_C\&P\_L_{n,\mu}(n, 0) \cong EL_{n,\mu}$ where $m = n^2$, at the bottom of this diagonal in Figure 1, a march up this diagonal of theories takes us from

(1) the pure theory of $EL_{n,\mu}$ which has no restrictions at all on conflict or paucity, through

(2) theories which incrementally further restrict both conflict and paucity, to

(3) the most restrictive theory CL_μ which allows no conflict or paucity at all.

Seen from this perspective, when evidential knowledge is present EL offers a preferable logic. Indeed, CL is a logic allowing no conflict and no paucity, and yet often we have at least some of one or both of these evidential situations.

7   GENERALIZATION FOR APPLICATION

Consider how we might think about the behavior of evidence in a real-time environment. For each predication in the language constructed to deal with the domain at hand, there will be 'evidence packets' of confirmatory and refutatory evidence. Let's think about just one predicate and its evidence packet. Initially there is no evidence available at all: note that, therefore, initially there are E-gaps at every level and no E-gluts at all. As the system runs and evidence fluctuates, below a certain level there may sometimes be no E-gaps along with some E-gluts, while above a certain level there may be no E-gluts along with some E-gaps. For example, a superior evidence environment might provide a steadily increasing amount of refutatory evidence and no confirmatory evidence for a predication; in this case no E-gluts, and no E-gaps (except at the levels above the current level of the refutatory evidence), arise and the system heads toward greater and greater witnessing that the predication is refuted by the evidence (cf. [2] for more discussion of this).

With the above rough model of the fluctuation of evidence in a typical application domain, using an EL-based system for knowledge representation and knowledge processing would seem appropriate. Indeed, the evidence may fluctuate between being "dirtier" (having more conflict and/or paucity) and "cleaner" (having less conflict and/or paucity). That is, the conflict/paucity character of the evidence may vary (even very rapidly indeed) along the following *evidence continuum*:

dirtier. . . . . cleaner

The Evidence Continuum

Regularly monitoring this characteristic of the "current" state of the evidence, the EL-based system will, in order to optimize the inference engine doing the knowledge processing (e.g. decision generation, action recommendation), adjust the EL-theory so it is always a best-fit for the current state of the evidence.

With these considerations in mind, we set out the following generalization, noting that a conflict need not be "flat", need not involve the same confirmatory and refutatory levels (and similarly for a paucity), as has been the case up to now, in order to make more transparent the fundamental ideas about evidential conflict and paucity.

So let us say that the conjunction $Q_c : d\epsilon \land Q_r : g\epsilon$ asserts a (g, d) E-glut with respect to a proposition symbol Q, and similarly define a (g, d) E-glut for any t-ary predicate symbol R, $t \geq 1$. One now easily generalizes Definition 1 to

DEFINITION 7.

$$E\_C\_L_{n,\tau}(g, d) \stackrel{\text{def}}{=} EL_{n,\tau}(\Phi_{4,g,d})$$

where $\Phi_{4,g,d}$ asserts there are no (g, d) E-gluts. We then have the structure theorem

THEOREM 8. *For each* $(g, d) \in \{1, \dots n\}^2$

$$BSA(E\_C\_L_{n,\mu}(g,d))$$

has ordered basis $\sigma(m)$ *with* $m = ng + nd - gd$.

Further saying that the conjunction (NOT Q_c : $d\varepsilon$) $\wedge$ (NOT Q_r : $f\varepsilon$) asserts a (d, f) E-gap with respect to a proposition symbol Q, and similarly defining a (d, f) E-gap for any t-ary predicate symbol R, $t \geq 1$, one generalizes Definition 2 to

DEFINITION 9.

$$E\_P\_L_{n,\tau}(d, f) \stackrel{\text{def}}{=} EL_{n,\tau}(\Phi_{5,d,f})$$

where $\Phi_{5,d,f}$ asserts that there are no (d, f) E−gaps. Then we get

THEOREM 10. *For each* $(d, f) \in \{0, \ldots, n - 1\}^2$

$$BSA(E\_P\_L_{n,\mu}(d, f))$$

has ordered basis $\sigma(m)$ *with* $m = n^2 - df$.

Finally, moving on to the conjunctive theories which will have both no (g, d_1) E-gluts and no (d_2, f) E-gaps, Definition 3 generalizes to

DEFINITION 11.

$$E\_C\&P\_L_{n,\tau}(g, d_1, d_2, f) \stackrel{\text{def}}{=} EL_{n,\tau}(\Phi_{6,g,d_1,d_2,f})$$

where $\Phi_{6,g,d_1,d_2,f}$ asserts the conjunction of Φ_{4,g,d_1} and $\Phi_{5,d_2,f}$. And in this case we have

THEOREM 12. *For each* $(g, d_1, d_2, f) \in \{1, \ldots, n\}^2 \times \{0, \ldots, n - 1\}^2$,

$$BSA(E\_C\&P\_L_{n,\mu}(g, d_1, d_2, f))$$

has ordered basis $\sigma(m)$ *where* $m = g(n - d_2) + d_1(n - f) - (g -_p f)(d_1 -_p d_2)$.

Based on the proofs given earlier herein, and those given in [4] and [5], the reader can easily argue Theorems 4, 5, and 6. Note that, as with the earlier case of "flat" E-gluts and "flat" E-gaps in Theorems 1, 2, and 3, "no (g, d) E-gluts" projects upward in the evidence level hierarchy and "no (d, f) E-gaps" projects downward. With regard to the ordered basis for the conjunctive theory given in Theorem 6, certainly the most interesting, let us note the following (referring the reader to Figure 4): the term $g(n - d_2)$

counts preatoms consistent with refutatories below level g and confirmatories above or at level d_2; the term $d_1(n-f)$ counts preatoms consistent with confirmatories below level d_1 and refutatories above or at level f; and the subtracted term $(g -_p f)(d_1 -_p d_2)$ fixes any double-counting in the potential open middle region (above which there are no (g, d_1) E-gluts and below which there are no (d_2, f) E-gaps) which occurs when $g > f$ and $d_1 > d_2$, the subtracted term measuring this number of double-counted preatoms.

Also with regard to our generalization, Theorem 6, note that: if $f = d_2 = 0$ (i.e. no axiomatization regarding gaps) we get Theorem 4, and if further $g = d_1$ (so the "no gluts" condition is 'flat') then we get Theorem 1; if $g = d_1 = n$ (i.e. no axioms regarding gluts) we get Theorem 5, and if further $f = d_2$ (making the "no gaps" condition 'flat') then we get Theorem 2; and finally, if $g = d_1$ and $f = d_2$ (so both the "no gluts" and "no gaps" conditions are 'flat') then we indeed get Theorem 3.

We also note a relation between Standard Fuzzy Logic (SFL) and a certain axiomatizable extension of EL. By SFL is meant here any Fuzzy Logic which stipulates that

$$\text{Truth-value} (\neg \, \phi) \; = \; 1 \; - \; \text{Truth-value} (\phi).$$

Letting Φ_7 be the conjunction of the sentences $\neg \, Q_c : d\varepsilon \; \leftrightarrow \; Q_r : (n-d)\varepsilon$ for all $d = n - 1, \ldots, 1$ (plus, as usual, all similar sentences with respect to any t-ary predications, $t \geq 1$), we model SFL in EL as

DEFINITION 13.
$$E\_SFL\_L_{n,\tau} \; \overset{\text{def}}{=} \; EL_{n,\tau}(\Phi_7),$$

associated with which is the following structure theorem.

THEOREM 14. $BSA(E\_SFL\_L_{n,\mu})$ has ordered basis $\sigma(m)$ with $m = n$.

With this structure theorem we see illuminated the stipulative strength of this model of SFL. Namely, this model of SFL is stipulating a rather strong 'cleanliness' in regard to the nature of evidence: asserting that, for each $d = n - 1, \ldots, 1$, the evidence has no gluts at or above, and no gaps at or below, the $(d, n - d)$ level.

Indeed, this result casts SFL as a theory so strong that it reduces the BSA of EL, where $m = n^2$, to the much simpler BSA where $m = n$.

Certainly this result calls into question, or at least raises the need to study, the "naturalness" often argued for in regard to SFL.

Penultimately, let us remark about one of the motivating factors driving the character of Evidence Logic, that it distinguishes "absence of evidence" and "evidence of absence". Recall that in EL evidence annotations are provided only at the atomic level. What then can be said about an evidence valuation for an "absence of evidence"? For example, what about an evidence valuation for such as $\neg\, P_c : e$?

One possibility is the following: if, in the theory generated by the "current" evidence, there exists g such that $P_r : g\varepsilon$ is a weakest refutatory stronger than $\neg\, P_c : e$ and there exists f such that $P_r : f\varepsilon$ is a strongest refutatory weaker than $\neg\, P_c : e$ (see Figure 5), then we could assign the refutatory evidence interval $[f, g]$ as the evidence valuation interval for $\neg\, P_c : e$. However, what we have in $\neg\, P_c : e$ is the assertion of the absence of value e confirmatory evidence for P. To say that this situation is witnessed by the presence of refutatory evidence for P seems conflational of what we are in fact trying in EL to keep distinguished. In this sense this possiblity is not appealing to the writer.

Also note, regarding this possiblility of assigning the refutatory evidence interval $[f, g]$ as the evidence valuation for $\neg\, P_c : e$, that this assignment entails a considerable 'cleanliness' of the evidence. For, fix d such that $e = d\varepsilon$ and consider the Figure 5 diagram of the situation, along with Figure 4 with $d_1 = d_2 = d$. Then we easily see that such an assignment entails that there be no (g, d) E-gluts and no (d, f) E-gaps in the evidence, in fact the theory $E\_C\&P\_L_{n,\mu}(g, d, d, f)$ with ordered basis $\sigma(m)$ with $m = g(n-d)+d(n-f)$ (cf. Theorem 6). Analogous to our remark above concerning the treatment of negation in Standard Fuzzy Logic (SFL) as we modeled it in EL, this assignment of the refutatory evidence interval $[f, g]$ as the evidence valuation for $\neg\, P_c : e \stackrel{\text{def}}{=} \neg\, P_c : d\varepsilon$ involving as it does such strong 'cleanliness' of the nature of the evidence, raises questions about the "naturalness" of this assignment. Indeed, the strongest such interval assignment, $[f, g]$ with $f = g$, is the "point assignment" of level $f(= g)$ to the evidence for $\neg\, P_c : e$, which entails $\neg\, P_c : e \leftrightarrow P_r : f\varepsilon$, meaning that $P_c : e$ and $P_r : f\varepsilon$ are negations of each other, that they form a pair which is both contrary and subcontrary (cf. also the first two paragraphs in Section 5, involving the simpler context of "flat" conflict and paucity), which corresponds to the theory $E\_C\&P\_L_{n,\mu}(g, d, d, g)$.

A more appealing possibility would be to assign an evidence valuation to $\neg\, P_c : e$ which reflects the amount of the confirmatory evidence search space for P upon which this asserted absence is predicated. Possibly future work on EL, formalizing evidence space machinery and evidence space search measures, will provide a framework in which such valuations for $\neg\, P_c : e$ will be feasible.

Meanwhile, the writer finds the unvaluated assertion $\neg\, P_c : e$ quite satisfactory: it reflects well, thinking of negation as absence, an evidential-knowledge situation where one has plainly enough simply the *absence of* "evidence at value *e* confirming *P*".

Let us close with a general note regarding the structure of the extensions of EL we have studied both here and in [5]. In all cases any lack of conflict or paucity in evidence was stipulated to occur uniformly across all the proposition symbols and predicate symbols in L_τ. This construction can easily be generalized so that the axiomatized degree of lack of conflict or paucity can vary across the propositions and predicates. Consider the case of the monadic languages L_μ with p proposition symbols and u unary predicates (and k constants). Depending on the amount of lack of conflict or paucity, the resulting BSA had, in the uniform case, the ordered basis

$$\omega^{m^u} \cdot m^p \cdot \sum_{i=1}^{k} s_{ki} \cdot m^{ui}$$

for some m, $2 \le m \le n^2$.

However, in the generalization being suggested, where L_μ stipulates the p proposition symbols $Q_1, \ldots, Q_p$ and the u unary predicate symbols $V_1, \ldots, V_u$ (and k constants), we have the following. Regarding any stipulated axiomatization of this more general type, if q_i is the number of preatoms relative to Q_i (for each $i = 1, \ldots, p$) and v_i is the number of preatoms relative to V_i (for each $i = 1, \ldots, u$) then the associated BSA will have ordered basis

$$\omega^{\Pi v_i} \cdot \Pi q_i \cdot \sum_{j=1}^{k} s_{kj} \cdot (\Pi v_i)^j.$$

8   CONCLUSION

Evidence Logic (EL), then, is one example of a logical framework serving the commonly occurring circumstances where evidential conflict and/or evidential paucity abounds (so Classical Logic is inadequate) yet no contradiction has arisen (so a paraconsistent logic is not needed). Although not a paraconsistent logic, EL provides an environment for further analysis of the concept of negation, a concept certainly foundational to paraconsistency, as illustrated here with the analysis of evidential gluts and gaps and their interrelationships.

We particularly note the interplay of evidential gluts and gaps, and their relation to Aristotle's work on privation, discussed in Sections 3, 4, and 5. Also, the formal work of Section 6 led to evaluative remarks regarding Classical Logic and Evidence Logic. Namely, when evidential knowledge is present, Evidence Logic is preferable to Classical Logic. Indeed, Classical Logic is seen, by this analysis, to be deficient in that it allows no conflict and no paucity, while often we have at least some of one or both of these. Finally, the further formal work of Section 7 provided application-oriented generalizations, together with evaluative remarks concerning the concept of negation and "absence of evidence" versus "evidence of absence". Included was a modeling in Evidence Logic of the usual handling of negation in fuzzy logics, a model stipulating a rather strong "cleanliness" of evidence and thus indicating a considerable stipulative strength for the usual negation in fuzzy logics. This model calls into question, or at least raises the need to study further, the "naturalness" of this usual negation in fuzzy logics.

Evidence packeting is certainly an area where more work is needed. For example, some evidence would seem to be purely confirmatory, some purely refutatory, while the rest is both confirmatory and refutatory in one of the following two senses: the presence of the evidence is confirmatory while its absence is refutatory *or* the presence of the evidence is refutatory while its absence is confirmatory. An explication and quantification of these considerations, and others surely involved in developing any reasonable 'evidence preprocessor' for EL, may lead to efficacious frameworks for evidence search space construction and processing. These, in turn, may help with some of the problems surrounding a reasonable quantification of "absence of evidence" in terms of the 'amount' of the evidence space so far searched (cf. also the last paragraph in Section 2, and the penultimate remark in Section 7 concerning possible interval evidence valuations for negations of atomic EL-formulae). Other problem areas include enrichment of the evidence value spaces E_n used in EL, the use of interval-based evidence value spaces, and the exploration of implementations of EL in Knowledge-Based Systems.

BIBLIOGRAPHY

[1] Faust, Don, "The Boolean Algebra of Formulas of First-Order Logic", *Annals of Mathematical Logic* 23 (1982) (now *Annals of Pure and Applied Logic*), pp. 27-53.

[2] Faust, Don, "The Concept of Negation", *Logic and Logical Philosophy* 5 (1997), pp. 35-48.

[3] Faust, Don, "Conflict without Contradiction: Noncontradiction as a Scientific *Modus Operandi*", *Proceedings of the 20th World Congress of Philosophy*, at www.bu.edu/wcp/Papers/Logi/LogiFaus.htm, 1998.

[4] Faust, Don, "The Concept of Evidence", *Int'l J. of Intelligent Systems* 15 (2000), pp. 477-493.

[5] Faust, Don, "Conflict without Contradiction: Paraconsistency and Axiomatizable Conflict Toleration Hierarchies in Evidence Logic", *Logic and Logical Philosophy* 9 (2001), pp. 137-151.

[6] Faust, Don, "Between Consistency and Paraconsistency: Perspectives from Evidence Logic, in the proceedings of WCP2" (*Lecture Notes in Pure and Applied Mathematics* 228 (2002), pp. 501-510).

Figure 1

Matrix of the n^2 conjunctive theories $E\_C\&P\_L_{n,\mu}(d_1, d_2)$

$E\_C\&P\_L_{n,\mu}(d_1,d_2)$		$E\_P\_L_{n,\mu}(d_2)$							
		0	1	2	3	...(n-3)	(n-2)	(n-1)	n
	0								
	1	$\checkmark$	•	•	•	... •	•	•	$\boxed{CL_\mu}$
	2	$\checkmark$	•	•	•	... •	•	8	•
	3	$\checkmark$	•	•	•	... 18	•	•	•
$E\_C\_L_{n,\mu}(d_1)$	$\vdots$	$\vdots$	$\vdots$	$\vdots$	$\vdots$	$\nearrow$	$\vdots$	$\vdots$	$\vdots$
	n-3	$\checkmark$	•	•	$n^2 - 2\cdot 3^2$...	•	•	•	•
	n-2	$\checkmark$	•	$n^2 - 2\cdot 2^2$	•	... •	•	•	•
	n-1	$\checkmark$	$n^2 - 2$	•	•	... •	•	•	•
	n	$\boxed{EL_{n,\mu}}$	$\checkmark$	$\checkmark$	$\checkmark$	... $\checkmark$	$\checkmark$	$\checkmark$	

Note: $d_1 \in \{1, \ldots, n\}$ and $d_2 \in \{0, \ldots, n-1\}$.

$\checkmark$ with : $d_1 = n$: no gluts at level n: amounts to no axiom regarding gluts.

$\checkmark$ with : $d_2 = 0$: no gaps at level 0: amounts to no axiom regarding gaps.

Transverse diagonal values are values of m in the relevant ordered bases $\sigma(m)$.

For $EL_{n,\mu}$, $m = n^2$.

For CL_μ, $m = 2$.

Figure 2

Matrix of the 36 conjunctive theories $E\_C\&P\_L_{6,\mu}(d_1, d_2)$

$E\_C\&P\_L_{6,\mu}(d_1, d_2)$		$E\_P\_L_{6,\mu}(d_2)$						
		0	1	2	3	4	5	6
	0							
	1	11	10	8	6	4	$\boxed{CL_\mu}$	
	2	20	19	16	12	8	4	
$E\_C\_L_{6,\mu}(d_1)$	3	27	26	23	18	12	6	
	4	32	31	28	23	16	8	
	5	35	34	31	26	19	10	
	6	$\boxed{EL_{6,\mu}}$	35	32	27	20	11	

<u>Note</u>: Each entry is the value of m in the relevant ordered basis $\sigma(m)$.

For $EL_{6,\mu}$, $m = 36$.

For CL_μ, $m = 2$.

Figure 3

The n conjunctive theories
$$E\_SymC\&P\_L_{n,\mu}(d) \overset{\text{def}}{=} E\_C\&P\_L_{n,\mu}(d, n-d)$$

In each column d, the theory $E\_SymC\&P\_L_{n,\mu}(d)$ is indicated

	n	(n-1)	(n-2)	(n-3)	...	3	2	1
n-1		↑	↑	↑		↑	↑	↑↓
n-2			↑	↑		↑	↑↓	↑↓
n-3				↑		↑↓	↑↓	↑↓
Evidence levels within the theory　:	$EL_{n,\mu}$				...	::	::	::
3				↓		↑↓	↑↓	↑↓
2			↓	↓		↓	↑↓	↑↓
1		↓	↓	↓		↓	↓	↑↓

Note: ↑ indicates levels where no E-glut is allowed.
　　　 ↓ indicates levels where no E-gap is allowed.
　　　 Column $d = n$: $EL_{n,\mu}$, entailing no restrictions on E-gluts or E-gaps.
　　　 Column $d = 1$: CL_{μ}, allowing no E-gluts and no E-gaps.

Figure 4

The Theory $E\_C\&P\_L_{n,\mu}(g, d_1, d_2, f)$:
no (g, d_1) E-gluts and no (d_2, f) E-gaps

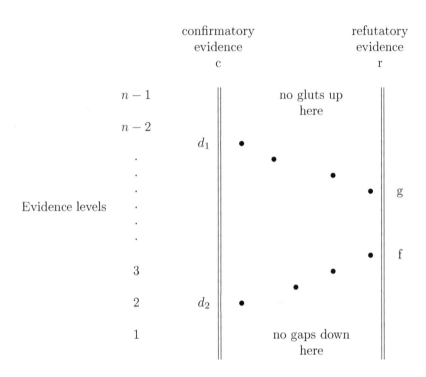

Note: (g, d_1, d_2, f) $\quad \varepsilon \quad \{1, \ldots, n\}^2 \times \{0, \ldots, n-1\}^2.$

Figure 5

The refutatory evidence interval [f,g] providing
an evidence valuation for $\neg P_c : e$

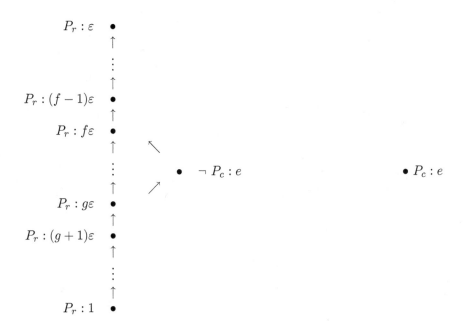

Note: $P_r : g\varepsilon$ is a weakest refutatory implying $\neg P_c : e$.
$P_r : f\varepsilon$ is a strongest refutatory implied by $\neg P_c : e$.

Don Faust
Department of Mathematics and Computer Science, Northern Michigan
University, Marquette, Michigan, USA.
dfaust@nmu.edu

Paraconsistent Provability Logic and Rational Epistemic Agents

CORRADO BENASSI AND PAOLO GENTILINI

ABSTRACT. A Paraconsistent Provability Logic is presented, developed inside a suitable sequent-formulated system of Paraconsistent Recursive Arithmetic **PCA**, based on paraconsistent **C**-systems. **PCA** allows a paraconsistent reasoning about standard numbers, so that a comparison with Classical and Intuitionistic Arithmetic is possible. **PCA** has a suitable expressive power and relevant paraconsistency properties, even if it rejects numerical absurdities such as 1=0. The formalization of metatheory in **PCA** is provided, and the provability predicate $\mathrm{Pr}_{\mathbf{PCA}}(.)$ is studied. The intensional character of the paraconsistent negation is investigated through provability logic tools. The non-triviality of a rec. ax. **PCA** -extension **T**, the negation-consistency of **T**, the classical consistency of **T**, are formalized inside **PCA**. It is proved that the provability predicate $\mathrm{Pr}_{\mathbf{PCA}}(.)$ preserves some of the standard properties of the classical provability predicates, but not all of them. A relevant novelty is that the well known Gödel Diagonal Lemma cannot hold for **PCA**; however, a Paraconsistent Diagonal Lemma for **PCA** is presented, where the local consistency assertions $°B$ of the paraconsistent **C**-systems play a central role. A possible weakened Hilbert's program for Paraconsistent Arithmetic is conjectured. In the second part of the paper a new application for the paraconsistent logical framework is proposed, through a logical formalization of rational agents: an agent $\mathbf{T}_i$ is a non trivial paraconsistent system of the form **PCA+AxT**. Then, the provability predicate $\mathrm{Pr}_{\mathbf{T}i}(.)$ is used as an epistemic predicate. In such a framework, a notion of relevant epistemic interaction between agents is expressed. Existence theorems of a denumerable infinity of societies made by complex and relevantly interacting agents are given, and a complexity hierarchy for the agent's different possible rationalities is proposed.

1 Introduction

In this paper a system of paraconsistent Arithmetic **PCA** is introduced, together with a Paraconsistent Provability Logic, that is a *provability pred-*

icate Pr$_\mathbf{T}$(.) for theories **T** based on **PCA**. In the last sections, an application to knowledge representation is also presented, which looks at the paraconsistent provability predicate as an epistemic predicate. We propose a Paraconsistent Arithmetic which is essentially new w.r.t. most of paraconsistent arithmetical theories existing in the relevant literature. Our aim is to give a Paraconsistent Arithmetic which can properly fit in the main stream Proof-Theory and Provability Logic of Classical Arithmetic and Intuitionistic Arithmetic (and, possibly, Classical and Intuitionistic Analysis): the idea is that paraconsistent reasoning, intuitionistic reasoning and classical rasonig can be compared as different ways to think about the *same* elementary mathematical objects. We wish to construct a paraconsitent reasoning about standard numbers, and, through the Paraconsistent Provability Logic, a formalized metatheory which can be formally and directly comparable with the formalized metatheories of Classical and Intuitionistic Arithmetic. Thus, our systems can support contradictions $B \wedge \neg B$ without trivializing, but reject identifications between different numbers, such as $0 = 1$ and so on. In our thinking, a logical contradiction and an identification between different objects are very different statements both from an ontological and an epistemological point of view. Therefore, we must underline a strong difference between the systems presented here and some other approaches to paraconsistent or inconsistent mathematics: e.g., the book by Mortensen [34] is fascinating but is conceived in a completely different perspective.[1] The paraconsistent logic which seemed best suited to build a paraconsistent Arithmetic with the features we have mentioned, is that provided by the sequent versions of the **C**-systems introduced and classified by Carnielli and Marcos in [16]. The **C**-systems seem to us the most natural environment to start from in order to formalize the (meta)reasoning about provability and consistency. Even at the propositional level, they allow a local expression of the consistency of a formula B by the *local consistency assertions* of the form $^\circ B$, having the intended meaning "B is consistent", i.e. a meaning close but not identical to that of $\neg(B \wedge \neg B)$. As is shown in Section 4, the local consistency assertions $^\circ B$ of the **C**-systems play a central role in establishing the paraconsistent version of some fundamental properties of classical Arithmetic and classical Provability Logic. Moreover, coherently with our objectives, the Paraconsistent Arithmetic **PCA** proposed here is endowed with an induction rule, and it maintains the explicit representation capabilities of both recursive functions and formalized

---

[1]Moreover, we can mention, among the others, some works of Meyer, Mortensen [32, 33,34,35] and Priest [38,39] . These are very interesting works, but the systems they introduce are drastically distinct from the system of the present paper. For example some of them are not trivialized by $S(x) = x$, others are not trivialized by $S(x) = 0$.

metatheory which are typical of classical systems for Arithmetic such as **PRA** or **PA**.

Paraconsistent Provability Logic is characterized both by subtler properties with respect to the classical Provability Logic of **PRA** and **PA**, and by some relevant substantial breaks with respect to the classical case. In the first class of results we may include the expression of the intensional character of the paraconsistent negation, as well as a different relationship between logical connectives and recursive relations: for example, implications of the form $\neg A \rightarrow \text{Pr}_{\mathbf{T}}(\#\neg A)$ do not hold in **PCA** for any quantifier-free non negated formula A. The second class of result is wider. First, the high representation power of **PCA** goes side by side with the fact that **PCA** is not transparent, i.e $u = v \rightarrow (A(u) \leftrightarrow A(v))$ does not hold in **PCA** for any arbitrary formula A and u, v arbitrary terms; second, the classical Gödel's Diagonal Lemma, stating that for each formula $A(u)$ with u free, a formula B exists such that **PRA** proves $A(\#B) \leftrightarrow B$, does not hold in **PCA**. However, the following counterparts of the mentioned classical properties do hold in the paraconsistent setting. A *weak trasparency property* for **PCA** is given, stating that $u = v \wedge \{(\forall x_j)^{\circ}C_k\}_{k=1,\ldots,r} \rightarrow (A(u) \leftrightarrow A(v))$ is **PCA**-provable, where the local consistency assertions $^{\circ}C_k$ are such that, essentially, $\neg C_k$ is an A-subformula; a *Paraconsistent Diagonal Lemma* is proved, stating that $\{(\forall x_j)^{\circ}D_s\}_{s=1,\ldots,d} \rightarrow (A(\#B) \leftrightarrow B)$ is **PCA**-provable,where the local consistency assertions $^{\circ}D_s$ are such that, essentially, $\neg D_s$ is an A-subformula. As a consequence, the proofs of both Gödel's incompleteness theorems cannot be the same as in the classical case, in a way that leads us to surmise that the notion of provability arising within the paraconsistent framework is more *constructive* than the classical one. Such consideration, together with the role played by the local consistency assertions $^{\circ}B$ in the previous results, suggests formulating the following conjecture, referred to the Full Paraconsitent Arithmetic **PCA**$_{\infty}$: *a minimal set* $\{^{\circ}F_j\}_{j=1,\ldots,m}$, $m \geq 2$, *of suitable local consistency assertions exists, such that the following sequent is* **PCA**$_{\infty}$*-provable*: $\{^{\circ}F_j\}_{j=1,\ldots,m} \vdash Non - Triv(\mathbf{PCA_{\infty}})$, being $Non - Triv(\mathbf{T})$ the formula expressing the non-triviality of any rec. ax. system **T**. If the $^{\circ}F'_j s$ should result as very weak w.r.t. $Non - Triv(\mathbf{PCA_{\infty}})$, we could speak of a weakened Hilbert's program for Paraconsistent Arithmetic.

We think that applications to knowledge representations, epistemic and cognitive agents, and interacting agents in games, are an important playground to test how fruitful the whole framework of paraconsistent logic can be. Thus, we show in the last part of the paper (sections 5,6, and 7) that Paraconsistent Provability Logic may have interesting applications to the fields of the logical formalization of rational agents in game theory and in social choice theory, and to the logical investigations about common knowledge and common rationality (for references on these fields see section 1.2). The central point of our proposal, is that *provability predicates can be used as epistemic predicates* in an intrinsic (and not merely descriptive) logical representation of a rational agent. Such "epistemic" Proof Theory is essentially based on the following identifications:

(rational agent i) $\equiv$ (formal theory $\mathbf{T}_i$, recursively axiomatized, undecidable, based on a suitably expressive predicate calculus);

(the rational agent i declaratively knows the idea or the state of the world described by the sentence A) $\equiv$ ($\mathbf{T}_i$ proves A);

(agent i believes that A) $\equiv$ ($\mathbf{T}_i$ proves $\mathrm{Pr}_{\mathbf{T}_i}(\#A)$;

(agent i believes that agent j knows B) $\equiv$ ($\mathbf{T}_i$ proves $\mathrm{Pr}_{\mathbf{T}_j}(\#B)$).

This epistemic characterization of Provability Logic is introduced in [4, 5, 6, 7] for a classical setting. However, game theory, social choice theory, etc. consider sets or societies of *interacting* agents. In [6] it is shown that if systems $\{\mathbf{T}_1, \ldots, \mathbf{T}_m\}$ are to represent one such, like a set of agents in a strategic game, the $\mathbf{T}_i$'s must be pairwise mutually inconsistent in a classical sense, that is $\mathbf{T}_i + \mathbf{T}_j$ must prove a contradiction $A \wedge \neg A$. Then, a very important limit affects the classical logic treatment: an agent's knowledge of the choices and opinions of other agents, and the interaction between agents, are relevant aspects that the theory must formalize; but if we extend agent $\mathbf{T}_i$ with the knowledge B of a different agent $\mathbf{T}_j$ (i.e., with a $\mathbf{T}_j$-theorem B), we have that in the classical setting $\mathbf{T}_i + B$ generally trivializes. Conversely, if we assume that the $\mathbf{T}_i$' s are paraconsistent extensions of **PCA**, we can use the non trivial extension $\mathbf{T}_i + B$ to express an epistemic interaction. *This is a strong reason for choosing paraconsistent systems in order to formalize epistemic agents.* Moreover, we can use a $\mathbf{T}_i$-theorem like $\mathrm{Pr}_{\mathbf{T}_j}(\#0 = 1) \wedge \neg \mathrm{Pr}_{\mathbf{T}_j}(\#0 = 1)$ in order to represent a conjecture of agent $\mathbf{T}_i$ about the rationality of agent $\mathbf{T}_j$.

1.1   Main results

In section 2 we introduce the sequent formulated system **PCA** of Paraconsistent Recursive Arithmetic, based on the paraconsistent sequent predicate calculus **BC,** which is the sequent version of the **C**-system **bC** presented in [16]. The cut-elimination property of **BC** is used in order to obtain relevant proof-theoretic properties of **PCA**. For example, in section 2.2. a free-cut elimination theorem for **PCA** is given, and the relationships between the classical **PRA**-provability and the **PCA**-provability are investigated. In section 3 it is shown that **PCA** has strong paraconsistency properties (e.g. contradictions of the form $m = m \wedge \neg m = m$ do not trivialize **PCA,** and so on) but it rejects numerical absurdities of the form $m = n$, m, n different numerals, and contradictions involving numerical absurdities. Moreover, *strong rationality conjectures* of the form $\mathrm{Pr_V}\,(\#0 = 1) \wedge \neg\,\mathrm{Pr_V}\,(\#0 = 1)$, **V** any rec.ax. **PCA**-extension, do not trivialize **PCA**. In section 4 the formalization of metatheory in **PCA** is provided, and the provability predicate $\mathrm{Pr_{PCA}}(.)$ is studied. The formulas *Non-Triv*(**T**), expressing the non triviality of a rec. ax. **PCA**-extension **T**, *Neg-Con*(**T**), expressing the negation consistency of **T**, and *Con*(**T**), expressing the classical consistency of **T**, are defined inside **PCA**. It is proved that the provability predicate $\mathrm{Pr_{PCA}}(.)$ preserves some of the standard properties of the classical provability predicates, but not all of them. The main *limiting results*, which are new w.r.t. the classical setting, and that are however very informative, are the following: *a*) **PCA** is not transparent, i.e. formulas of the form $r = t \wedge B(r) \rightarrow B(t)$ are in general not **PCA**-provable; *b*) the classical Gödel's Diagonal Lemma does not hold for **PCA**; *c*) formulas of the form $\neg A \rightarrow \mathrm{Pr_T}(\#\neg A)$, where A is any quantifier-free non negated formula, are in general not **PCA**-provable; *d*) formulas of the form $\neg x = y$ are not (**PCA+H**)-equivalent to any primitive recursive relation, for any added axiom set **H** formed by sequents of atomic formulas. The main peculiar *paraconsistent versions of some fundamental classical properties* are shown by the following results: *i*) the *Weak Transparency Lemma*, stating that the sequent $M : u = v, \{(\forall x_j)^\circ C_k\}_{k=1,\ldots,r},\, B(u) \vdash B(v)$ is **BC + EQ**-provable, **EQ** equality axioms, where each C_k can be obtained by a suitable term renaming from a proper subformula G_k of B such that $\neg G_k$ too is a B-subformula, and for each C_k, $(\forall x_j)^\circ C_k$ is the universal closure of a formula $^\circ F_k$ w.r.t each free variable different from the free variables occurring in $B(u) \vdash B(v)$; *ii*) the *Paraconsistent Diagonal Lemma* stating that if $A(u)$ is a **PCA**-formula in which the free variable u occurs, then there are a formula B and a finite set $\{^\circ D_s\}_{s=1,\ldots,d}$ of local consistency assertions such that the sequent $\{(\forall x_j)^\circ D_s\}_{s=1,\ldots,d} \vdash A(\#B) \leftrightarrow B$ is **PCA**-provable, $A(\#B)$ and B have exactly the same free-variable set V including the free variables occurring in $\{(\forall x_j)^\circ D_s\}_{s=1,\ldots,d}$ too, and each formula D_s can be obtained by term renaming from a proper subformula G_s of $A(u)$ such that $\neg G_s$ too is an $A(u)$-subformula. At he end of section 4 the conjecture on *the weak* **PCA**$_\infty$− *provability of the non triviality of the Full Paraconsistent Arithmetic* **PCA**$_\infty$ is formulated and discussed.

In section 5 the paraconsistent rational agent is defined, and different kinds of the agent's knowledge are formalized. In section 6 we introduce the notion of paraconsistent epistemic interaction, where the specificity of the paraconsistent framework, which allows for inconsistent but non trivial agents, plays a central role. The complexity level of the *inductive rationality* for the agent's reasoning is introduced, and existence theorems for a denumerable infinity of *intearctive societies* of complex agents, admitting relevant epistemic interactions, are given; such existence theorems fully employ the proof-theoretic properties of **PCA** stated in sections 2,3,4.

1.2 Related works

As to the fundamentals of paraconsistent logic, we refer to Batens et al. [3], Carnielli and Carnielli et al. [14, 15, 16, 17,18], Da Costa [19], Priest [37]. For the presentation of **C**-systems we refer to Carnielli-Marcos [16]. Sequent versions of paraconsistent systems are exposed in D'Ottaviano-Moura [20] and in Forcheri-Gentilini [22,23]; the system **BC** used here is also investigated in [22], where the formulation of the **C**-systems as sequent theories based on **BC** is given.

For the basic notion of Proof Theory and sequent calculus, we refer to Buss [12], Troelstra-Schwichtenberg [43], Takeuti [42]. For the proof-theory of classical systems of Arithmetic we refer to Buss[13],[11], Girard[30]; for selfreference in Arithmetic and for Provability Logic we refer to Boolos [10], Bernays-Hilbert [8], Japaridze-De Jongh [31] and Smorynski [40,41]; for the proof-theoretic approach to Provability Logic we refer to Gentilini [24,25,26,27,28].

For the comparison with the notion of conjecture arising from the Informational Logic setting see Forcheri-Gentilini [22,23] and Gentilini-Forcheri-Molfino [29].

As to the formalization of rational agents, we refer to Aumann [1], Fagin *et al.* [21], for the analysis of the agent's knowledge based on extensions of standard epistemic logic, which is a descriptive external approach to agents; conversely, the intrinsic approach to the logical formalization of agents proposed in this paper is introduced for the classical logic setting in Benassi-Gentilini [4, 5,6,7].We must emphasize that our intrinsic approach is very different from the standard descriptive one. For the discussions on the epistemic conditions of game equilibria and for the various interactions between logic and game theory see Aumann-Brandenburger [2], Binmore [9] and Van Benthem [45,46]. A remarkable peculiar contribution to the application of logic to game theory is given in Tsuj-Da Costa-Doria [44], though it considers the whole theory and not the formalization of agents.

2   A sequent system for Paraconsistent Arithmetic

2.1   The sequent calculus BC for paraconsistent C-systems

In order to develop an expressive Provability Logic in the paraconsistent setting, we require that the paraconsistent arithmetical system formalizing the metatheory admits of a handy sequent version, based on a sequent predicate calculus for which cut elimination holds.

Then, as predicate sequent calculus we choose the calculus **BC**, that is the sequent version of the **C**-system **bC** of [16]. Our references for the general properties of the sequent calculi will be [12, 43, 42, 30]. As is well known (see [25,12]), a sequent S is an expression of the form $X \vdash Y$ where X and Y are finite (possibly empty) sets of formulas. Each element of $X \cup Y$ is called an *isolated formula* of the sequent. X is called the *antecedent* of S, Y the *succedent* of S. We will use the symbols $X, Y, \Lambda, \Gamma, \ldots$ as meta-expressions for sets of formulas, $A, B, C, D, \ldots$ for formulas. The intended meaning of a sequent $A_1, A_2, \ldots, A_n \vdash B_1, B_2, \ldots, B_m$ is $\wedge_i A_i \longrightarrow \vee_j B_j$ and such equivalence holds both in a classical and in a paraconsistent setting. Given a rule $\dfrac{S_1 \ldots S_n}{S}$, the sequents $S_1, \ldots, S_n$ are the *premises* of the rule, the sequent S is the *conclusion* of the rule. The proofs are trees, whose leaves are *axioms*, , and whose branches are formed by sequent rules. Axioms are also called *initial sequents*. In a proof-tree P a *branch* is a maximal linearly ordered set of sequents in it, having an axiom as the first element, where each sequent is a premise of the successive sequent. A rule occurrence in a proof P is also called an *inference in P*.

We also use the writings $\wedge X$, ($\vee Y$) to indicate the conjunction (resp. disjunction) of the elements of X, (resp. Y). We call the formula $\wedge X \longrightarrow \vee Y$ the *positive translation* of the sequent $X \vdash Y$. The writing Λ, Γ stands for $\Lambda \cup \Gamma$. If $X \subset \Lambda$ and $Y \subset \Gamma$ we say that $X \vdash Y$ is a *sub-sequent* of $\Lambda \vdash \Gamma$.

We say that a sequent formulated system or theory *trivializes* if and only if it proves each sequent of the form $\vdash A$. A **BC**-based system trivializes if and only if it proves the empty sequent $\vdash$. We recall that a formula A is in *prenex form* if A is $Q_1 \ldots Q_n B$ where each $Q_j \in \{\forall x_j, \exists x_j\}_{j=1,\ldots,n}$ and B is quantifier-free; in general, any formula F is not **BC**-equivalent to a formula D in prenex form, since the interdefinability of quantifiers does not hold in a paraconsistent setting. In order to reduce the number of parentheses in the writings of formulas, we adopt the convention that $\vee, \wedge, \neg$ link more than $\rightarrow$, and that $\rightarrow$ links more than $\leftrightarrow$.

The sequent system **BC** is given by:

BC−*Axioms*: $A \vdash A$

BC−*Positive propositional logical rules:*

$$\dfrac{B, \Gamma \vdash \Delta}{A \wedge B, \Gamma \vdash \Delta} \wedge{-}L \qquad \dfrac{B, \Gamma \vdash \Delta}{B \wedge A, \Gamma \vdash \Delta} \wedge{-}L \qquad \dfrac{\Gamma \vdash \Delta, A \qquad \Lambda \vdash X, B}{\Gamma, \Lambda \vdash \Delta, X, A \wedge B} \wedge{-}R$$

$$\dfrac{\Gamma \vdash \Delta, A}{\Gamma \vdash \Delta, A \vee B} \vee{-}R \qquad \dfrac{\Gamma \vdash \Delta, A}{\Gamma \vdash \Delta, B \vee A} \vee{-}R \qquad \dfrac{A, \Gamma \vdash \Delta \qquad B, \Lambda \vdash X}{A \vee B, \Gamma, \Lambda \vdash \Delta, X} \vee{-}L$$

$$\frac{A,\Gamma \vdash \Delta, B}{\Gamma \vdash \Delta, A \to B} \longrightarrow -R \qquad \frac{\Gamma \vdash \Delta, A \qquad B, \Lambda \vdash X}{A \to B, \Gamma, \Lambda \vdash \Delta, X} \longrightarrow -L$$

BC−*Negation rules:*

$$\frac{A, \Gamma \vdash \Delta}{\neg\neg A, \Gamma \vdash \Delta} \neg - L1 \qquad \frac{{}^{\circ}A, \Gamma \vdash \Delta, A}{{}^{\circ}A, \neg A, \Gamma \vdash \Delta} \neg - L3$$

$$\frac{A, \Gamma \vdash \Delta}{\Gamma \vdash \Delta, \neg A} \neg - R$$

BC−*Quantifier rules:*

$$\frac{[t/x]\, A, \Gamma \vdash \Delta}{\forall x A, \Gamma \vdash \Delta} \forall - L \qquad \frac{\Gamma \vdash \Delta, [b/x]\, A}{\Gamma \vdash \Delta, \forall x A} \forall - R$$

$$\frac{[b/x]\, A, \Gamma \vdash \Delta}{\exists x A, \Gamma \vdash \Delta} \exists - L \qquad \frac{\Gamma \vdash \Delta, [t/x]\, A}{\Gamma \vdash \Delta, \exists x A} \exists - R$$

where t is an arbitrary term and b is a free variable which does not occur in Γ, Δ. Moreover, t may be not fully quantified while b must be uniformly replaced by x (see [42]).

BC−*Structural rules:*

Weakening rules:

$$\frac{\Gamma \vdash \Delta}{\Gamma \vdash \Delta, A} W - R \qquad\qquad \frac{\Gamma \vdash \Delta}{A, \Gamma \vdash \Delta} W - L$$

Cut rule:

$$\frac{\Gamma \vdash \Delta, A \qquad A, \Lambda \vdash X}{\Gamma, \Lambda \vdash \Delta, X} Cut$$

We call the formula ${}^{\circ}A$ in the rule $\neg - L3$ *constraint formula* of the rule. It is well known (see [16]) that in **bC** and in the **C**−systems the connective ${}^{\circ}(.)$, having the intended meaning "A is consistent", is not definable starting from the other logical connectives. We also call formulas of the form ${}^{\circ}A$ *local consistency assertions*. We remark that in the calculus **BC**, the axioms $A \vdash A$ cannot be restricted to the atomic case only. For example, due to the constraints in $\neg - L3$, we cannot prove $\neg A \vdash \neg A$ from $A \vdash A$. We observe that the classical predicate calculus **LK** [12,42] can be obtained from **BC** by replacing the negation rule set with the following rules:

$$\frac{\Gamma \vdash \Delta, A}{\neg A, \Gamma \vdash \Delta} \neg - L2, \qquad \frac{A, \Gamma \vdash \Delta}{\Gamma \vdash \Delta, \neg A} \neg - R.$$

A formula without any occurrence of the connective $\neg$ is called a *positive formula*. A proof tree segment is *positive* if no negation rule occurs in it. A theory **T** based on a sequent calculus **W** is given by the deduction apparatus of **W** plus a (possibly empty) proper axiom set **AxT** and a (possibly empty) proper rule set **RuT**, expressed by sequents. We also write **W** + **AxT** + **RuT** for **T**. The proofs of **T** are trees of sequents; the theorems of **T** are the roots of the trees. We also say that a formula A is a theorem of **T** if the sequent $\vdash A$ is a theorem of **T**. A **BC**-based theory **T** is *trivial* if it proves the empty sequent. A sequent S *trivializes* the theory **T** if **T** plus S is a trivial theory. **T** is *paraconsistent* if any formula B exists such that **T** plus $\vdash B \wedge \neg B$ is a non trivial theory. **T** is *negation inconsistent* if it proves any theorem of the form $\vdash B \wedge \neg B$, otherwise *negation consistent*. We say that the sequents S and L are **T**-equivalent if their positive translations are **T**-equivalent.

We need to recall some notions of proof-theory; our definitions are similar but not identical to that of Buss [12], Troelstra-Schwichtenberg [43] and, subordinately, to that of Takeuti [42]:

DEFINITION 1. i) The *depth* or *height* $h(P)$ of a tree P in **BC** is the highest number of proof-lines in a branch. The *grade* $g(A)$ of a formula A is the number of occurrences of logical symbols in it.

ii) In a rule occurrence R in a proof-tree P in **BC** we call: *auxiliary formulas* the formula occurrences in the premises on which the rule acts; *principal formula*, or *formula introduced by the rule*, the formula occurrence produced by the rule in the conclusion. Each formula in the conclusion of R is called the *successor* of the formulas in the premises corresponding to it, that are called its *predecessors*.

iii) In a branch of a proof-tree P in **BC** we say that the formula occurrence B is an *ancestor* of the formula occurrrence C occurring below B in the branch, called a *descendant* of B, if they are connected by a sequence of predecessor-successor relations alongside the branch. C is called an *integral descendant* of B if B and C are the same formula; if C is an integral descendant of B then B is called a *direct ancestor* of C. If B has a direct ancestor which is the principal formula of an inference R or occurring in an axiom S, then we say that B *is intoduced by R (resp. by S) in* P .

We have the following result:

THEOREM 2. *Cut elimination holds for* **BC** .

Proof. We consider an uppermost cut C in a proof-tree P in **BC** and show that it can be replaced by **BC**−deductions without cuts. The proof is by induction on the cutrank with a subinduction on the level of C, that are so defined: the *cutrank* of C is the grade of the cut formula in C, the *level* of C is the sum of the depths of the deductions of the premises.

Let C be the following cut in P:

$$\frac{\begin{matrix} Q1 \\ \Gamma \vdash \Delta, A \end{matrix} \qquad \begin{matrix} Q2 \\ A, \Lambda \vdash X \end{matrix}}{\Gamma, \Lambda \vdash \Delta, X}$$

where Q1 and Q2 are the P−sub proofs of the premises. The only cases that we have to examine are the ones in which the cut formula A has either the form $\neg B$ or the form ${}^\circ B$. As to the other cases, we refer to the cut elimination proofs for classical predicate calculus. We previously establish the following *preliminary reductions* on the tree P: if one of the cut-formulas of C has the form $\neg B$ and is the descendant of the principal formula F of a weakening rule in P, we delete such weakening and each rule in P acting on a descendant of F, by getting a cut-free proof of a sub-sequent of $\Gamma, \Lambda \vdash \Delta, X$; analogously, if the left cut formula of C has the form ${}^\circ B$ and is the descendant of the principal formula of a weakening rule, we delete such weakening, getting a cut-free proof of a sub-sequent of $\Gamma, \Lambda \vdash \Delta, X$.

After such preliminary reductions, we consider these different cases:

1. At least one of Q1, Q2 is an axiom. Let the left premise be an axiom of the form $A \vdash A$. Then, the conclusion of C has the form: $A, \Lambda \vdash X$. Then, we replace C by the right premise. If the right premise is an axiom, we replace the cut by the left premise.

2. Neither Q1 nor Q2 is an axiom and the cut formula is not principal in at least one of the premises. If the cut formula is not on the both sides principal, let us consider for example the premise $Z \vdash W, A$ of the one-premise rule R in Q1 having the left cut premise $\Gamma \vdash \Delta, A$ as conclusion. Then we produce the following proof:

$$\frac{\dfrac{Z \vdash W, A \qquad A, \Lambda \vdash X}{Z, \Lambda \vdash W, X} \; R}{\Gamma, \Lambda \vdash \Delta, X}$$

where the level of the introduced cut is lower than that of C, and the cutrank is the same. If R is a two-premise rule the reduction is similar, and so it is for the sub-cases in which the right cut formula is not principal, *with the exception of the case in which A has the form ${}^\circ B$ and the right cut premise is the conclusion of a $\neg - L3$ rule K having ${}^\circ B$ as constraint formula*. In this case let us consider the sub-proof Q1 of the left C−premise $\Gamma \vdash \Delta, {}^\circ B$ in P. We observe that, by hypotheses, the occurrence of ${}^\circ B$ in such premise cannot be the principal formula of any rule in Q1 and then it must be introduced by a set of axiom occurrences of the form ${}^\circ B \vdash {}^\circ B$;

moreover, it must be the integral descendant of the right formula of each axiom. If we replace each axiom occurrence by the proof of the C-premise $\circ B, \Lambda \vdash X$, we have a cut free proof of $\Gamma, \Lambda \vdash \Delta, X$, after possible suitable renaming of free variables in the branches. We observe that no $(\neg - L3)$-rule constraints are broken in the reduction, since a $\circ B$ in the succedent cannot be a constraint formula in any $\neg - L3$.

3. The cut formula is principal in both the premises of the cut C. If the cut formula has the form $\circ B$, then both the premises of the cut are the conclusions of weakening rules, and this case has already been solved by the preliminary reductions. Let us consider the cases in which the cut formula has the form $\neg B$. By hypotheses, both occurrences of $\neg B$ are the principal formulas of negation rules.

3.1 The left premise is the conclusion of a $\neg - R$ rule and the right premise is the conclusion of a $\neg - L1$ rule:

$$\frac{\dfrac{\neg B, \Gamma \vdash \Delta}{\Gamma \vdash \Delta, \neg\neg B} \qquad \dfrac{B, \Lambda \vdash X}{\neg\neg B, \Lambda \vdash X}}{\Gamma, \Lambda \vdash \Delta, X}$$

then, we replace the cut C with the following proof:

$$\frac{\dfrac{B, \Lambda \vdash X}{\Lambda \vdash X, \ \neg B} \qquad \neg B, \Gamma \vdash \Delta}{\Gamma, \Lambda \vdash \Delta, X}$$

where the cutrank is lower.

3.2 The left premise of C is the conclusion of a $\neg - R$ rule and the right premise is the conclusion of a $\neg - L3$ rule:

$$\frac{\dfrac{B, \Gamma \vdash \Delta}{\Gamma \vdash \Delta, \ \neg B} \qquad \dfrac{\circ B, W \vdash X, B}{\circ B, \neg B, W \vdash X}}{\circ B, \Gamma, W \vdash \Delta, X}$$

then we replace the cut C with the following proof:

$$\frac{\circ B, W \vdash X, B \qquad B, \Gamma \vdash \Delta}{\circ B, \Gamma, W \vdash \Delta, X}$$

where the cutrank is lower.

We will explicitly refer to the proof of the Theorem 2 in the next sections.

2.2 Theories based on the Paraconsistent Recursive Arithmetic PCA

We want to select a paraconsistent arithmetical system with one fundamental property: that of expressing a paraconsistent reasoning about numbers as they are considered by standard mathematics, that is, about standard numbers. In this way, the proofs of such system can be directly compared with those of classic and intuitionistic Arithmetic, as proofs of statements about the same elementary objects. Morevoer, the system we look for should also have the following features:

- it must explicitly represent all primitive recursive functions;
- it must be endowed by an induction rule or axiom;
- it must be recursively axiomatized and undecidable;
- it must support without trivializing a wide class of contradictions having different complexities, so that relevant paraconsistent, and possibly negation inconsistent theories **T**, proving relevant contradictions, can be based on it;
- it must include also the propositional local consistency assertions $°B$ allowed by the **C**-systems proposed and classified in [16].

The system of *Paraconsistent Recursive Arithmetic* **PCA** is so defined. The language of **PCA** is the same as that of Primitive Recursive Arithmetic **PRA**, plus the monadic propositional connective $°(.)$ of the **C**-systems. For the proof theory of classical Arithmetic we refer to [12, 40, 41, 42,30].

We choose a version of **PRA** with the only predicate $= (.,.)$, the individual constant 0 and a function letter for each primitive recursive function. We assume as identified the writings $= (t, s)$ and $t = s$. All the primitive recursive predicates R different from $= (.,.)$ are expressed by their characteristic function X_R, and $X_R(t_1, ..., t_n) = 1$ means that $R(t_1, ..., t_n)$ holds, $X_R(t_1, ..., t_n) = 0$ means that $R(t_1, ..., t_n)$ does not hold. Moreover, we establish that each proper axiom set we shall present in this work is closed under term substitution.

PCA is given by the system **BC** plus the set **AxPCA** of proper axioms and the rule **Ind**.

AxPCA is the following axiom set:

1) Arithmetical axioms defining primitive recursive function (in the following sequents all the explicitly indicated variables x_i, y_j, are free):

1j) Definitions of the basic recursive functions (*zero function, successor function, projection function*):

$\vdash Z_k (x_1, ..., x_k) = 0$ for each $k \geq 1$ (Z_k *zero* function);

$S(x) = 0 \vdash$ (S *successor* function);

$S(x) = S(y) \vdash x = y$

$\vdash P_i^k (x_1, ..., x_k) = x_i$ for each $k \geq 1$, $i \leq k$ (P_i^k *projection* function);

1jj) Composition schema:

$\vdash f (x_1, ..., x_m) = h (g_1 (x_1, ..., x_m) , ..., g_m (x_1, ..., x_m))$

where $g_1, ..., g_m$ are n-ary function letters and h is a m-ary function letter.

1jjj) Recursion schema :

$\vdash f (x_1, ..., x_n, 0) = g (x_1, ..., x_n)$

$\vdash f (x_1, ..., x_n, S (y)) = h (x_1, ..., x_n, y, f (x_1, ..., x_n, y))$

where g is a n-ary function letter and h is a $n + 1$-ary function letter.

2) Equality axioms:

$\vdash x = x$

$$x_1 = y_1, ..., x_n = y_n \vdash f(x_1, ..., x_n) = f(y_1, ..., y_n)$$
$$x_1 = y_1, x_2 = y_2, x_1 = x_2 \vdash y_1 = y_2$$

Convention:

for each natural number m we write $\overline{m}$ for the term $S...S(0)$, with m occurences of S on the left. The terms of the form $\overline{m}$ are called *numerals*. We will briefly write $1, 2, ...$,for numerals, each time no confusion arises.

The induction rule **Ind** of **PCA** has the following form:

$$\frac{F(x), X \vdash Y, F(S(x))}{F(0), X \vdash Y, F(t)} \quad \textbf{Ind}$$

where $F(x)$ is an atomic formula; the free variable x, called the *eigenvariable of the rule*, does not occur in X, Y, t; t is an arbitrary term which we say *introduced by* **Ind**; $F(0), F(t)$ are the *principal formulas* of **Ind**; $F(x)$, $F(S(x))$ are the *auxiliary formulas* of **Ind**.

Note that the usual definitions by recursion of sum and product are included in the schemata of **AxPCA**. Moreover, the axiom $S(x) = 0 \vdash$ gives rise to a set of bottom particles (see [16]) for **PCA**, so that any formula of the form $S(t) = 0$ trivializes **PCA**.

We note that the deduction apparatus of classical Primitive Recursive Arithmetic **PRA** is given by the classical predicate calculus **LK** plus **AxPCA** and the rule **Ind**. We call *full Classical Arithmetic* **PA** the extension of **PRA** by induction rules admitting arbitrary induction formulas. We can suppose that we have preserved in **PCA** a relevant arithmetical representation power, compatibly with a paraconsistent deduction. However, due to the proof theoretical strength of the system, we have to prove in detail that it is really paraconsistent in a significant manner. Moreover, note that the **PRA**-language and the **PCA**-language are not identical, due to the occurrence of the connective $°(.)$ in the **PCA**-language. If E is an expression in the **PCA**-language we write $\#E$ for its gödel-number; if V is a recursively axiomatized theory we write $\mathrm{Pr}_V(.)$ for the canonical provability predicate of V (see [31],[40],[41]) and if L is a formula or a sequent in the V-language $\mathrm{Pr}_V(\#L)$ is a formula in the **PCA**-language having the intended meaning "L is V-provable". We recall that a term t of the **PCA**-language is called a *ground term* if no variables occur in it. We write Δ_0 for the class of primitive recursive relations, Σ_1 for the class of formulas $\exists x A(x)$ with $A \in \Delta_0$, Π_1 for the class of formulas $\forall x B(x)$ with $B \in \Delta_0$ (see also [13],[40]). We will write **N** for the standard classical model of **PRA** and $\mathbb{N}$ for the natural number set. We call *full Paraconsistent Arithmetic* **PCA**$_\infty$ the extension of **PCA** with an induction rule admitting arbitrary auxiliary formulas.

As usual, we assume the consistency of classical arithmetic **PA**, with the following qualfications:

- this assumption is actually too strong, i.e. it is not really necessary for many of our results, both at the theory level, and at the level of meta-theoretical reasoning;[2]

- as should be obvious, the consistency of **PCA** and **PCA**$_\infty$, as well as their Provability Logic, do not depend on the consistency and Provability Logic of **PRA** and **PA**. Thus, it must be remarked that it is indeed a relevant objective, that of looking for direct proofs of the consistency of **PCA** or **PCA**$_\infty$, without assuming the consistency of **PRA** or **PA**.

We are interested in studying arithmetical theories **T** of the form **W**+ **AxT** + **Ind**, **AxT** including **AxPCA**, where **W** $\in$ {**LK**, **BC**}. It is important to establish whether the proofs in **T** admits of the elimination of some class of cut inferences. According to the canonical exposition of Buss [12], p 43, we employ the notion of *free cut*:

DEFINITION 3. Let P be a proof in **W** + **AxT** + **Ind**, **W** $\in$ {**LK**, **BC**}. We say that a formula occurrence B in P is *anchored* if B is the direct descendant either of a formula occurring in an initial sequent belonging to **AxT** or of a principal formula of an induction rule. A cut inference in P is called a *free cut* if either : i)the cut formula is not atomic and both the cut formula occurrences in the premises are not anchored, or ii) at least one cut formula occurrence in the premises is not anchored and is introduced in P by weakenings or logical axioms only. A cut inference which is not free is said to be *anchored*.

In Buss [12], pp 43-47, the following result is proven:

THEOREM 4. *(Classical free-cut elimination theorem) Let* **T** $\equiv$ **LK** + **AxT** + **Ind**, *be a theory of arithmetic with* **AxT** *closed under term substitution. Then* **T** *admits of free-cut elimination.*

DEFINITION 5. We call *reduced translation* of a **PCA**-formula B into **PRA** the formula B^* obtained from B by replacing each occurrence of a B-subformula of the form $°F$ with $\neg(F \wedge \neg F)$. We call *reduced* **PRA**-*translation* of a **PCA**-tree Q the tree Q^* in the conservative extension **PRA** + $\neg - L1$ obtained from Q by replacing each formula with its reduced translation.

PROPOSITION 6. *i)* **PCA** *is recursively axiomatized; ii)* **PCA** *is non trivial and negation consistent.*

Proof. Point *i*) is straightforward. As to point *ii*), we note that from a **PCA**-proof tree Q of the empty sequent we would obtain, by the reduced

---
[2]We note that it is usual in Proof Theory to get results on the basis of minimal assumptions on the consistency of the systems involved: for example, in order to develop the Provability Logic of **PA** it is not necessary to assume the consistency of **PA**, but that of Robinson's arithmetic **Q** suffices (see e.g. Buss [13]). We remark that the epistemological difference between these two assumptions is huge.

PRA-translation, a proof tree Q^* of the empty sequent in $\mathbf{PRA} + \neg - L1$, and this is absurd under our assumptions.   ■

DEFINITION 7. We call *standard reduced interpretation* of a **PCA** formula B the interpretation of the reduced translation B^* in the standard classical model **N** of **PRA**.

The reduced translation and the reduced interpretation of a **PCA**-formula B do not reflect the peculiar meaning of the formulas of the form $^{\circ}C$ defined inside the **C**-systems setting. Thus, they do not give a real interpretation of **PCA**-formulas into **PRA**, even if they can be useful technical tools. Indeed, it is easy to see that the reduced **PRA**-interpretations of the **PCA**-rules are sound w.r.t. **N**, and then we say that the reduced **PRA**-interpretation of **PCA** preserves the standard classical model **N**.

The following is a canonical result which follows directly from the classical free-cut elimination theorem mentioned above:

PROPOSITION 8. *Cut-elimination of non atomic cuts holds in* **PRA**.

Proof. **PRA** is a theory of the form $\mathbf{LK} + \mathbf{AxPRA} + \mathbf{Ind}$, which satisfies the hypotheses of the classical free-cut elimination theorem. The thesis follows from the fact that each non atomic cut is free, since in **AxT** only atomic formulas occur, and only atomic induction formulas are admitted in the **PRA**-induction rule.   ■

We have now to investigate the extension of the free-cut elimination theorem to the paraconsistent case. The relevant part of the problem is solved by the following Lemma:

LEMMA 9. *Let* $\mathbf{V} \equiv \mathbf{BC} + \mathbf{AxV}$ *with* $\mathbf{AxV}$ *closed under term substitution. Then* **V** *admits of free-cut elimination.*

Proof. Without loss of generality we consider only trees in **V** where the preliminary reductions mentioned in the proof of Theorem 2 have been performed. Then, we consider an uppermost free cut D in a proof P and show that it can be replaced by **V**-deductions without free cuts. The new cases that we must examine w.r.t. the classical free-cut elimination proof for systems of the form $\mathbf{LK} + \mathbf{AxV}$ are those where the cut formula A has either the form $^{\circ}B$ or the form $\neg F$. We can employ the same transformations on subproofs of P already produced on the cut C in the proof of the cut-elimination theorem for **BC** (Theorem 2), with the following further proviso: it must be proven that the P-subproof transformations do not introduce new free cuts. It is easy to check that the subproof transformations presented at points 1., 3., and at point 2. for cut-formulas different from $°B$, do not introduce free cuts. One has to examine the transformations performed at point 2. in the case in which the cut formula of D has the form $°B$ and the right cut-premise is the conclusion of a $\neg - L3$ rule K having $°B$ as constraint formula. Let us assume that D has the same form of C in Theorem 2 with the left premise $\Gamma \vdash \Delta, °B$ (which is the root of the P-subproof Q1) and the right premise $°B, \Lambda \vdash X$ (which is the root of the P-subproof Q2). Replace in Q1 the logical axioms $°B \vdash° B$ containing a direct ancestor of the left cut formula of D in the succedent with the proofs Q2 of $°B, \Lambda \vdash X$, thus getting a proof Q* of the sequent $\Gamma, \Lambda \vdash \Delta, X$. We note that no free cuts are introduced in Q*, since no free cuts occur in Q1 and Q2 and the Q1-segment in Q* only acts on descendants of the $°B$ in the antecedent of the sequents $°B, \Lambda \vdash X$, while all the descendants of the formulas of Λ and X remain integral alongside Q*: then, if a free cut occurred in Q*, then it should also occur in Q1 too, which is absurd. ∎

THEOREM 10. *(Free-cut elimination theorem) Let* $\mathbf{T} \equiv \mathbf{BC} + \mathbf{AxT} +$ **Ind** *be a theory of arithmetic with* $\mathbf{AxT}$ *closed under term substitution. Then* $\mathbf{T}$ *admits of the elimination of free cuts.*

Proof. As in the classical case, the first part of the thesis (case i) of Definition 3) follows from Lemma 9 and from the remark that possible $\mathbf{Ind}$−rule occurrences in the considered tree P do not change the transformations for the reduction of the free cuts. The second part of the thesis (case ii) of Definition 3) is straightforward. ∎

PROPOSITION 11. *Cut-elimination of non atomic cuts holds in* $\mathbf{PCA}$.

Proof. The proof is the same as in the classical case (Proposition 8). ∎

DEFINITION 12. Let $\mathbf{T} \equiv \mathbf{U} + \mathbf{AxT}$, $\mathbf{U} \in \{\mathbf{PRA}, \mathbf{PCA}\}$, $\mathbf{AxT}$ possibly empty axiom set. Then a proof P in $\mathbf{T}$ is called *normal* if free cuts do not occur in P.

LEMMA 13. *Let* $X \vdash Y$ *be a sequent such that* X *and* Y *are not both empty and include at most atomic formulas. Then* $X \vdash Y$ *is* $\mathbf{PCA}$-*provable if and only if it is* $\mathbf{PRA}$-*provable.*

Proof. If $X \vdash Y$ is $\mathbf{PCA}$-provable, then, by the cut elimination of non atomic cuts, it admits of a $\mathbf{PCA}$-proof Q in which only atomic formulas occur; therefore, Q is a $\mathbf{PRA}$-proof too. On the other hand, if $X \vdash Y$ is $\mathbf{PRA}$-provable it admits of a $\mathbf{PRA}$-proof P in which only atomic formulas occur, and P is also a $\mathbf{PCA}$-proof. ∎

Lemma 13 confirms that **PCA** remains a powerful mathematical system, since, through the full primitive recursive function apparatus, sequents of open atomic formulas can express relevant and essentially complex mathematical statements. For example, Fermat's theorem is the following open atomic sequent:

$$(x+1)^{u+3} + (y+1)^{u+3} = (z+1)^{u+3} \vdash$$

where x, y, z, u are free variables. Obviously, we do not know if it is **PCA**-provable.

We introduce the notion of odered proof that will be useful in section 6:

DEFINITION 14. Let **T** $\equiv$ **U** $+$ **AxT**, **U** $\in$ {**PRA**, **PCA**}, **AxT** non empty axiom set. Then a proof P in **T** is called *ordered* if it can be divided into two segments: an upper segment P1 where axiom of **AxT** do not occur; a lower segment P2, where the only inferences are cuts such that: the left premise has the form $\vdash (\wedge\Delta \rightarrow \vee\Pi)^*$, where $\Delta \vdash \Pi$ is an axiom of **AxT**, F^* universal closure of the formula F.

PROPOSITION 15. *Let* **T** $\equiv$ **U** $+$ **AxT**, *such that* **U** $\in$ {**PRA**, **PCA**} *and* **AxT** *is a non empty axiom set whose element has the form* $\vdash \Pi$, Π *set of closed formulas. Then each* **T**-*provable* $X \vdash Y$ *admits of an ordered* **T**-*proof.*

Proof. Let Q be any **T**-proof of $X \vdash Y$. We replace in Q each occurrence of an axiom $\vdash \Pi_j$ with $\vee\Pi_j \vdash \Pi_j$, without changing any inference. Then we get a **U**-proof P1 of $\{\vee\Pi_j\}_{j=1,\dots m}, X \vdash Y$, where $\{\vdash \vee\Pi_j\}_{j=1,\dots m}$ are the **AxT**$-$axiom occurrences of Q. By applying m cuts having the left premises of the form $\vdash \vee\Pi_j$, by starting from the P1-root, we produce the P2-segment of the **T**-ordered proof of $X \vdash Y$.   ■

3   Paraconsistency properties of PCA

It is well known that the system **BC** is trivialized by an infinity of non classical contradictions $A \wedge \neg A$, where A includes the connective $°(.)$, and, on the other hand, it is also well known that **BC** proves an infinity of non classical formulas of the form $\neg(B \wedge \neg B)$. However, no contradiction in the classical language is either proved or rejected by **BC**. As to Arithmetic **PCA** things are less clear-cut, as **PCA** should provide non standard (i.e. paraconsistent) reasoning about standard numbers, in such a way as to be comparable with both classical and intuitionistic Arithmetic. For this reason **PCA** rejects identification among different numbers. As we have seen, this option is expressed by the axiom $S(x) = 0 \vdash$, which imposes the rejection of *numerical absurdities*- the latter to be defined formally as follows:

DEFINITION 16. A *numerical absurdity* is an atomic formula $s = t$ where s and t are ground terms which are respectively **PCA**-provably equal to m and n, with m, n different numerals.

First, we explore the kind of paraconsistent reasoning of **PCA**:

THEOREM 17. *Sequents of the form* $\vdash (m = m) \land \neg(m = m)$, m *any numeral, do not trivialize* **PCA**.

Proof. It must be proven that a proof of the empty sequent in the system $\mathbf{V} \equiv \mathbf{PCA} + \vdash \neg(m = m) \equiv \mathbf{BC} + \mathbf{AxPCA} + \vdash \neg(m = m) + \mathbf{Ind}$ cannot exist. Let us suppose ad absurdum that the empty sequent is **V**-provable; then, it is the root of a normal **V**-proof Q in which at least one **V**-proper axiom of the form $\vdash \neg(m = m)$ occurs, since Q cannot be a **PCA**-proof. Therefore, in Q at least one anchored cut C having $\neg(m = m)$ as cut-formula must occur. C has the form:

$$\frac{X \vdash Y, \neg(m = m) \qquad \neg(m = m), \Gamma \vdash \Delta}{\Gamma, X \vdash Y, \Delta}$$

where the right cut formula cannot have all its direct ancestors introduced by weakenings or by logical axioms; then it must be introduced also by $\neg - L3$ inferences having a constraint formula $°(m = m)$ in the antecedents. But each possible descendant $F(°(m = m))$ of $°(m = m)$ must be cut in Q by a cut D that cannot be anchored, since induction rules in **PCA** cannot have any formula of the form $F(°(m = m))$ as a principal formula, and formulas such as $F(°(m = m))$ do not occur in any proper axiom of **V**. Then D should be a free cut, and this is absurd due to the normality of Q. ∎

PROPOSITION 18. *Sequents of the form* $\vdash \neg(t = t \rightarrow t = t)$, t *any term, do not trivialize* **PCA**.

Proof. It must be proven that the system $\mathbf{U} \equiv \mathbf{PCA} + \vdash \neg(t = t \rightarrow t = t)$ is not trivial; the considerations are similar to that in the proof of Theorem 17 above. ∎

As we shall see in section 5, sentences of the form $\mathrm{Pr}_{\mathbf{V}}(\#A \land \neg A) \land \neg \mathrm{Pr}_{\mathbf{V}}(\#A \land \neg A)$,...etc, speaking about the various kind of consistency of a system **V**(see section 4 below) may have an interesting conjectural character in the cases in which **V** represents an epistemic agent. We call them *strong rationality conjectures*. It is then relevant the following property:

THEOREM 19. *Strong rationality conjectures of the form* $\vdash \mathrm{Pr}_{\mathbf{V}}(\#A \land \neg A) \land \neg \mathrm{Pr}_{\mathbf{V}}(\#A \land \neg A)$, *where* **V** *is any non trivial recursively axiomatized system extending* **PCA** , *do not trivialize* **PCA**.

Proof. It is known that $\mathbf{U} \equiv \mathbf{PCA} + \vdash \mathrm{Pr_V}(\#A \wedge \neg A)$ is a classically consistent system. Then it must be proven that $\mathbf{U} + \vdash \neg \mathrm{Pr_V}(\#A \wedge \neg A)$ is not trivial, under the hypothesis that $\mathbf{U}$ cannot prove the empty sequent. The thesis is given by the free-cut elimination theorem as in the proof of Theorem 17. ∎

PROPOSITION 20. *There is a denumerable infinity of formulas A such that* $\mathbf{PCA}$ *does not prove any sequent of the form* $\vdash \neg(A \wedge \neg A)$.

Proof. Let $\vdash B \wedge \neg B$ any sequent such that B has one of the following forms: $m = m$, $t = t \rightarrow t = t$, $\mathrm{Pr_V}(\#A \wedge \neg A)$ as mentioned in Proposition 18 and Theorems 17 and 19. By construction, $\vdash B \wedge \neg B$ does not trivialize $\mathbf{PCA}$. Let us assume ad absurdum that $\vdash \neg(B \wedge \neg B)$ is the root of a normal $\mathbf{PCA}$-proof Q. The root formula $\neg(B \wedge \neg B)$ cannot be introduced in Q only by weakenings since $\mathbf{PCA}$ is not trivial and we delete each weakening introducing a direct ancestor of the root formula, getting a normal proof H with the same end sequent. Then, the root formula of H must be the integral descendant either of principal formulas of $\neg - R$ inferences with premises of the form $B \wedge \neg B, X \vdash Y$; or of succedents of logical axioms of the form $\neg(B \wedge \neg B) \vdash \neg(B \wedge \neg B)$. But in such last case, the descendants of the antecedents $\neg(B \wedge \neg B)$ of the axioms should be cut by free cuts, against the normality of H; then, the root formula is introduced in H only by $\neg - R$ inferences. But this is absurd, since sequents $\vdash B \wedge \neg B$ would result as $\mathbf{PCA}$-trivializing sequents, against the hypotheses: indeed, if we cut each mentioned $\neg - R$ premise $B \wedge \neg B, X \vdash Y$ by $\vdash B \wedge \neg B$ we obtain from H a proof H§ of the empty sequent in the system $\mathbf{PCA} + \vdash B \wedge \neg B$. ∎

We explore now the implication of of the rejection of numerical absurdities as possible $\mathbf{T}$-theorems of a $\mathbf{PCA}$ based theory $\mathbf{T}$. *We have that classical contradictions of a peculiar class can trivialize* $\mathbf{PCA}$.

LEMMA 21. *Each sequent of the form* $\vdash m = n$, *m, n different numerals, trivializes* $\mathbf{PCA}$.

Proof. ¿From the successor axioms and equality axioms we easily prove $m = n \vdash$ in $\mathbf{PCA}$. Note that the axiom $S(x) = 0 \vdash$ is always strictly necessary in the proof. ∎

COROLLARY 22. *Numerical absurdities are bottom particles for* $\mathbf{PCA}$.

LEMMA 23. $\mathbf{PCA}$ *proves sequents of the form* $\vdash \neg(A \wedge \neg A)$, *where only classical connectives at most occur in A.*

Proof. We have:
$$\frac{\dfrac{S(x) = 0 \vdash}{S(x) = 0 \wedge \neg(S(x) = 0) \vdash}}{\vdash \neg(S(x) = 0 \wedge \neg(S(x) = 0))}$$ ∎

THEOREM 24. *Let* $\vdash \neg(A \land \neg A)$ *be a* **PCA**-*provable sequent such that in A classical connectives at most occur, through a normal proof Q. Then* $\neg(A \land \neg A)$ *must be a descendant in Q of a sequent* $s = t \vdash$ *such that* $s = t$ *has at least one instance which is a numerical absurdity, and atomic formulas having numerical absurdities as instances necessarily occur in A.*

Proof. First, we uniformly replace in Q each free variable occurring in A with a numeral, by imposing that different variables are replaced by different numerals, getting a proof Q* of a sequent $\vdash (B \land \neg B)$ as the root. Then, we proceed by induction on the grade of B.

Basis: let B be an atomic formula. The Q*-root has the form $S \equiv \; \vdash \neg(s = t \land \neg(s = t))$. Through considerations analogous to that employed in the proofs of the theorems above, we have that, by the normality of Q*, $s = t \land \neg(s = t) \vdash$ must be the premise of S in Q*. By the normality of Q*, $s = t \land \neg(s = t)$ can be introduced only by weakenings and at least one $\land - L$ rule occurrence. Recalling that no descendant of the introduced $s = t \land \neg(s = t)$'s can be an auxiliary or constraint formula of any **Ind** or $\neg - L3$ rule, we delete the introducing rules and obtain from Q* a normal proof Q** having the same atomic formulas occurring as isolated in the sequents, and $\neg(s = t) \vdash$ or $s = t \vdash$ as possible roots. But the first case would be absurd, since neither a $\neg - L3$ rule nor a logical axiom can introduce $\neg(s = t)$ in Q**. Then $s = t \vdash$ is the Q**-root. But the ground terms s and t cannot be **PCA**-provably equal to the same numeral m: if so it is, by the **PCA**-theorem $\vdash m = m$ we have a proof of the empty sequent, against the non triviality of **PCA**. Then, $s = t$ is a numerical absurdity, and by construction the sequent $s = t \vdash$ occurs in Q* in a way such that $s = t$ is an ancestor of the root formula $\neg(B \land \neg B)$.

Induction step: analogously to the basis case, we have that Q* has a $\neg - R$ as end-rule and that $B \land \neg B \vdash$ is the premise of the root; we delete in Q* the introducing rules of $B \land \neg B$ and conclude that a proof Q** of $B \vdash$ is obtained from Q*. According with the outermost connective of B, we have that B can be introduced in Q* only by weakenings and by at least one logical rule of the class $(.) - L$ and different from $\neg - L3$. By deleting the introducing rules from Q**, we get a normal proof P of a sequent $F \vdash$ where F is a proper subformula of B having a lower grade. Then, we can obviously extend P to a proof P' of a sequent $\vdash \neg(F' \land \neg F')$, also having suitably replaced in P possible free-variables occurring in F, so that F' is a closed instance of F. Then, by induction hypothesis, the thesis holds for P' and $\neg(F' \land \neg F')$, so that, by construction, it holds for Q* and $B \land \neg B$. ∎

Thus, paraconsistent arithmetical reasoning of **PCA** rejects classical contradictions involving numerical absurdities, while can support classical contradictions involving true numerical relations. A **PCA**-based theory **T** can represent an epistemic agent who considers contradictions about true numerical relations as a piece of useful information (i.e. allowing interesting conjectures), but sees as as meaningless contradictions about numerical absurdities. We moreover remark that the elimination of $S(x) = 0 \vdash$ from the axioms **AxPCA** would give an arithmetic **W** which is not trivialized by numerical absurdities, that could support theories having, e.g., equations of the kind $1 = 3$ as theorems, but this is not compatible with the kind of paraconsistent mathematics we are interested in building.

4   Incompleteness, Selfreference and Provability Logic for PCA

In this section we introduce the provability predicate for **PCA**, and prove that the minimal standard properties of the provability predicates are preserved in the paraconsistent setting. However, introducing the provability predicate in a paraconsistent framework based on **C** -systems, turns out to yield interesting general results, such as:

a) It is possible to show the intensional charactyer of the paraconsistent negation in a very expressive way;

b) The Gödel's Diagonal Lemma in its classical form does not hold in **PCA**, but a Paraconsistent Diagonal Lemma is provable, with the local consistency assertions playing a key role – all of which makes it inevitable that Gödel's incompleteness theorems for **PCA** cannot be proved in the same way as in the classical case. As a consequence, a sort of weakened Hilbert's program could be investigated for full Paraconsistent Arithmetic **PCA**$_\infty$ based on **C**-systems: is it possibile for **PCA**$_\infty$ to prove its own non triviality, under the assumption of sufficiently weak local consistency assertions? (see Conjecture 42).

As canonical references for classical Provability Logic we indicate [8, 10, 31, 40, 41]. For the proof-theoretic approach see [24, 25, 26,27,28].

As usual, we proceed to the formalization of metatheory inside **PCA** through the recursive function apparatus. Being **PCA** a recursively axiomatized theory, the basic steps of this process (i.e. those which do not employ the diagonal lemma) are identical to those for **PRA**. The writing $\#E$ stands for the gödel-number of any expression E of the **PCA**-language. Therefore, we can define a binary primitive recursive predicate $Prov_{\mathbf{PCA}}(.,.)$ such that $Prov_{\mathbf{PCA}}(m,n)$ holds iff m is the gödel-number of a **PCA**- proof of the sentence with gödel-number n. We recall that in our language each primitive recursive predicate R is expressed by the characteristic function X_R, so that we employ the fuction $X_{\mathrm{Prov-PCA}}$, and $Prov_{\mathbf{PCA}}(m,n)$ corresponds to $X_{\mathrm{Prov-PCA}}(m,n) = 1$ in the **PCA**-language. However, we will briefly write $Prov_{\mathbf{PCA}}(m,n)$ for $X_{\mathrm{Prov}-\mathbf{PCA}}(m,n)=1$, and so on. The formula $\exists y X_{\mathrm{Prov}-\mathbf{PCA}}(y,\#B)=1$ means "the sentence B is $\mathbf{PCA}$-provable" and we also write it as $\mathrm{Pr}_{\mathbf{PCA}}(\#B)$. In general if K is a recursive relation between terms $t_1,...,t_n$ we formally express it by $X_K(t_1,...,t_n)=1$, X_K characteristic function, and its recursive complementary relation by $X_K(r_1,...,r_n)=0$: *note that we do not establish in* $\mathbf{PCA}$ *any link between* $X_K(r_1,...,r_n)=0$ *and the formula* $\neg(X_K(r_1,...,r_n)=1)$, since we expect that the $\mathbf{PCA}$-negation is not the boolean negation of classical $\mathbf{PRA}$. $\mathrm{Pr}_{\mathbf{PCA}}(\cdot)$ is the non recursive Σ_1-provability predicate of $\mathbf{PCA}$. We recall that a slightly different canonical predicate $\mathrm{Pr}_{\mathbf{PCA}}[.]$ is definable, such that if B is any open formula with free variables $x,y,...$, then $\mathrm{Pr}_{\mathbf{PCA}}[\#B]$ has the same $x,y,...$as free variables and means "the formula B is $\mathbf{PCA}$-provable"; $\mathrm{Pr}_{\mathbf{PCA}}[\#B]$ coincides with $\mathrm{Pr}_{\mathbf{PCA}}(\#B)$ if B is closed. We briefly write $\mathrm{Pr}_{\mathbf{PCA}}(\#B)$ even if B is open, with the convention that $\mathrm{Pr}_{\mathbf{PCA}}(\#B)$ has the same free variables as B. We extend in a straightforward way the godelization to sequents, and so $\mathrm{Pr}_{\mathbf{PCA}}(\#X \vdash Y)$ means "sequent $X \vdash Y$ is $\mathbf{PCA}$-provable" and it is evident, since we are working with sequent formulated systems, that $\mathrm{Pr}_{\mathbf{PCA}}(\#B)$ has the same meaning as $\mathrm{Pr}_{\mathbf{PCA}}(\# \vdash B)$. For each recursively axiomatized theory $\mathbf{T}$ of the form $\mathbf{PCA}+\mathbf{AxT}$, $\mathbf{AxT}$ proper axiom set, we can analogously define a provability predicate $\mathrm{Pr}_{\mathbf{T}}(.)$ for $\mathbf{T}$.

The formalization of metatheory allows a system to speak about its own consistency or inconsistency. However, in the paraconsistent framework, if $\mathbf{T}$ is any paraconsistent extension of $\mathbf{PCA}$, we have to express formally in $\mathbf{T}$ the non triviality of $\mathbf{T}$ and the negation consistency of $\mathbf{T}$ as separate notions. The formula $\exists x \neg \mathrm{Pr}_{\mathbf{T}}(x)$ means "$\mathbf{T}$ is not trivial" and we often shorten it by $Non-Triv(\mathbf{T})$. If the theory $\mathbf{T}$ is $\mathbf{BC}$-based, we also employ $\neg \mathrm{Pr}_{\mathbf{T}}(\# \vdash)$ as $Non-Triv(\mathbf{T})$, and $\mathrm{Pr}_{\mathbf{T}}(\# \vdash)$ as the formal expression of "$\mathbf{T}$ is trivial", that we often shorten by $Triv(\mathbf{T})$. Then, for each $\mathbf{PCA}$-based $\mathbf{T}$, $Non-Triv(\mathbf{T})$ can be taken as $\neg \mathrm{Pr}_{\mathbf{T}}(\# \vdash)$, stating the unprovability of the empty sequent. Since in a $\mathbf{BC}$-based theory $\mathbf{T}$ we have a $\mathbf{T}$-proof of $\vdash A \wedge \neg A$ if and only if we have a $\mathbf{T}$-proof of $\vdash A$ and a $\mathbf{T}$- proof of $\vdash \neg A$, the formula $\forall x \neg (\mathrm{Pr}_{\mathbf{T}}(x) \wedge \mathrm{Pr}_{\mathbf{T}}(h(x)))$ means "$\mathbf{T}$ is negation consistent", where we establish that $h(x)$ is a recursive term such that if x is the code of any formula then $h(x)$ is the code of its negation. We often shorten $\forall x \neg (\mathrm{Pr}_{\mathbf{T}}(x) \wedge \mathrm{Pr}_{\mathbf{T}}(h(x)))$ by $Neg-Con(\mathbf{T})$; the formula $\exists x (\mathrm{Pr}_{\mathbf{T}}(x) \wedge \mathrm{Pr}_{\mathbf{T}}(h(x)))$ means "$\mathbf{T}$ is negation inconsistent" and we shorten it by $Neg-Inc(\mathbf{T})$. We observe that if $\mathbf{T}$ cannot prove $\neg(\mathrm{Pr}_{\mathbf{T}}(\#1=1) \wedge \mathrm{Pr}_{\mathbf{T}}(\#\neg 1=1))$ then it cannot prove $Neg-Con(\mathbf{T})$. If $\mathbf{V}$ is any recursively axiomatized classical system extending $\mathbf{PRA}$, then $\neg \mathrm{Pr}_{\mathbf{V}}(\#0=1)$ means "$\mathbf{V}$ is consistent" and we often shorten it by $Con(\mathbf{V})$.

Even in the investigation of the basic properties of $\mathrm{Pr}_{\mathbf{PCA}}(.)$, we must be careful. In fact, it is true that **PCA** can be seen as isomorphic to a proper sub-system of **PRA**, but this holds in a very radical sense: indeed, the underlying logic **BC** is a proper sub-logic of the logic **LK** of **PRA**. An elementary but relevant break w.r.t. the classical case is the following:

PROPOSITION 25. *Let $B(z)$ any formula in which the free variable z occurs; then the sequent $r = t$, $B(r) \vdash B(t)$ is in general not **PCA**-provable, and, following [34], we also say that **PCA** is not transparent.*

Proof. Let $f(z) = k$ any atomic formula, let $f(\cdot)$, s, t , k be such that $s = t \vdash \neg f(t) = k$ is not **PRA**-provable, and s, t are different terms; then the sequent J: $r = t, \neg(f(r) = k) \vdash \neg(f(t) = k)$, is not **PCA**-provable. Indeed, if a normal **PCA**-proof Q of J exists, then $\neg f(r) = k$ in the antecedent should be introduced by $\neg - L3$ rules, or by logical axioms, but this is impossible since descendants of formulas of the forms $^\circ(f(r) = k)$, $\neg(f(r) = k)$, cannot be cut in Q. As a particular example for J take the sequent $x = y, \neg S(x) = 5 \vdash \neg S(y) = 5$, where x, y are free variables.   ∎

Conversely, $x = y, \neg S(x) = 5 \vdash \neg S(y) = 5$ is obviously **PRA**-provable. Note however that, following the terminology of [34], **PCA** is *functional*, since $x_1 = y_1, ..., x_n = y_n \vdash f(x_1, ..., x_n) = f(y_1, ..., y_n)$ is a **PCA**-axiom, by which, for example, we can prove $x = y, S(x) = 5 \vdash S(y) = 5$ in **PCA**.

However, a very interesting fact is that we can have a weak transparency property through the local consistency assertions $^\circ F$ of the **C**-systems, that is:

LEMMA 26. *Let $B(z)$ be any formula in which the free variable z occurs and let $S \equiv u = v$, $B(u) \vdash B(v)$ be a sequent in the **PCA**-language with u, v free variables; then, a proof Q of S exists in the classical predicate calculus **LK** plus equality axioms **EQ** such that in Q non-atomic cuts do not occur.*

Proof. The thesis follows from the standard properties of classical predicate calculus with equality axioms, see e.g. Takeuti [42] p 38.   ∎

LEMMA 27. *(Weak Transparency Lemma) Let $B(z)$ be any formula in which the free variable z occurs, and let $S \equiv u = v$, $B(u) \vdash B(v)$ be a sequent in the **PCA**-language with u, v free variables, such that a proof Q of S without non-atomic cuts exists in **LK** + **EQ**. Then the sequent M:*
$$u = v, \{(\forall x_j)^\circ C_k\}_{k=1,...,r} , B(u) \vdash B(v)$$
*is provable in **BC** + **EQ** by a proof P, where:*

i)each C_k can be obtained by a suitable term renaming from a proper subformula G_k of B such that $\neg G_k$ too is a B-subformula;

ii)for each C_k, $(\forall x_j)^\circ C_k$ is the universal closure of a formula $^\circ F_k$ w.r.t each free variable different from the free variables occurring in $B(u) \vdash B(v)$;
iii)the set $\{(\forall x_j)^\circ C_k\}_{k=1,\ldots,r}$ is univocally fixed by the proof Q.

Proof. Starting from the uppermost $\neg - L2$ rules in Q downwards, we replace each $\neg - L2$ rule occurrence $\dfrac{\Gamma \vdash \Delta, A}{\neg A, \Gamma \vdash \Delta}$ with the following proof-segment H:

$$\cfrac{\cfrac{\cfrac{\Gamma \vdash \Delta, A}{^\circ A, \Gamma \vdash \Delta, A} \; weak}{^\circ A, \neg A, \Gamma \vdash \Delta} \; \neg - L3}{\cdots\cdots\cdots\cdots\cdots} \forall - L$$

$$(\forall x_j)^\circ A, \; \neg A, \; \Gamma \vdash \Delta$$

where the $\forall - L$ rules quantify in $^\circ A$ the free variables different from those freely occurring in the Q-root. After the replacement, the same quantifier rule occurrences as in Q can be applied below H, since no new free variable occurrences have been added. The obtained proof P has the sequent M as the root, and it is a **BC + EQ**-proof. By construction, in P non-atomic cuts cannot occur. Then, a descendant D of the principal formula $\neg A$ of each $\neg - L3$ rule must occur in the sub-sequent $B(u) \vdash B(v)$ of the P-root, such that D includes a subformula E which differs from $\neg A$ only by term renaming.Thus, for each $\neg - L3$ constraint formula $^\circ A$ in P, the formula B in the root includes a proper subformula which differs from A only by term renaming. Moreover, the set $\{(\forall x_j)^\circ C_k\}_{k=1,\ldots,r}$ is univocally determined by the $\neg - L2$ occurrences in Q. ∎

COROLLARY 28. *Let the sequent $S \equiv u = v$, $B(u) \vdash B(v)$ be given with the same hypotheses of the lemma above. Then $u = v$, $\{(\forall x_j)^\circ D_s\}_{s=1,\ldots,q} \vdash B(u) \leftrightarrow B(v)$ is provable in **BC + EQ**, and each D_s can be obtained by a suitable term renaming from a proper subformula G_k of B such that $\neg G_k$ too is a B-subformula.*

Proof. If we uniformly replace u with v and v with u in the proof P mentioned in the previous lemma we have a proof P' of the sequent $u = v$, $\{(\forall x_j)^\circ E_k\}_{k=1,\ldots,r}$, $B(v) \vdash B(u)$; then, considering the roots of P and P', by obvious $\rightarrow -R$ and $\wedge - R$ instances, and by the elementary equality theorem $u = v \leftrightarrow v = u$, we have the thesis. ∎

We remark that in the thesis of the Weak Transparency Lemma the universal closures of the introduced local consistency assertions w.r.t the free variables different from that freely occurring in the Q-root, are not eliminable. Consider for example S of the form: $u = v, \exists x \neg A(x, u) \vdash$

$\exists x \neg A(x, v)$ with A atomic formula. The corresponding sequent M must necessarily be:
$$u = v, \forall x(^{\circ}A(x, u)), \exists x \neg A(x, u) \vdash \exists x \neg A(x, v)$$

COROLLARY 29. *Let* $S \equiv u = v, B(u) \vdash B(v)$ *be a sequent in the* **PCA**-*language as in the lemmas above, such that* S *is* **LK**$+$**EQ**-*provable through a proof* Q *where* $\neg - L2$ *rules do not occur. Then* S *is* **BC** $+$ **EQ**-*provable and* **PCA**-*provable too.*

In particular, Corollary 29 above holds for each positive sequent of **PCA**.

However, the lack of full tranparency does not prevents **PCA** from proving some relevant Provability Logic statements:

PROPOSITION 30. *If* **PCA** *proves* $\vdash B$, *then* **PCA** *proves* $\vdash \mathrm{Pr}_{\textbf{PCA}}(\#B)$.

Proof. **PCA** proves all true closed recursive relations, since it has exactly the same defining axioms for recursive functions as **PRA**. Then if **PCA** proves $\vdash B$ by a proof Q, it must prove $\vdash Prov_{\textbf{PCA}}(\#Q, \#B)$ from which, by the $\exists - R$ rule, $\vdash \mathrm{Pr}_{\textbf{PCA}}(\#B)$ is obtained. ∎

PROPOSITION 31. *The following sequents are* **PCA** -*provable:*
  (D1) $\mathrm{Pr}_{\textbf{PCA}}(\#A \to B) \vdash \mathrm{Pr}_{\textbf{PCA}}(\#A) \to \mathrm{Pr}_{\textbf{PCA}}(\#B)$
  (D2) $\mathrm{Pr}_{\textbf{PCA}}(\#A \to B) \wedge \mathrm{Pr}_{\textbf{PCA}}(\#A) \vdash \mathrm{Pr}_{\textbf{PCA}}(\#B)$
  (D3) $\vdash \mathrm{Pr}_{\textbf{PCA}}(\#A \wedge B) \leftrightarrow \mathrm{Pr}_{\textbf{PCA}}(\#A) \wedge \mathrm{Pr}_{\textbf{PCA}}(\#B)$

Proof. Classical **PRA** proves the mentioned properties for the provability predicate $\mathrm{Pr}_{\textbf{T}}(.)$ of each recursively axiomatized theory **T** having the same *positive* inference rules as the classical predicate calculus **LK**, and this holds for **PCA**. Moreover, we note that, due to the elimination of non atomic cuts in **PRA**, negation rules cannot occur in normal **PRA**-proofs of the considered sequents. Then, such proofs are **PCA**-proofs too. ∎

PROPOSITION 32. *The following sequent is* **PCA**-*provable:*
  (D4) $\mathrm{Pr}_{\textbf{PCA}}(\#B) \vdash \mathrm{Pr}_{\textbf{PCA}}(\# \mathrm{Pr}_{\textbf{PCA}}(\#B))$.

Proof. We recall the classical proof in **PRA** of
  $\mathrm{Pr}_{\textbf{PRA}}(\#B) \to \mathrm{Pr}_{\textbf{PRA}}(\# \mathrm{Pr}_{\textbf{PRA}}(\#B))$
as presented by sequent formalism in [24], by rephrasing the original proof of Bernays-Hilbert [8]; the relevant part is the production of a proof-tree having as the root the following sequent:
  $X_{\mathrm{Prov-}\textbf{PRA}}(a, \#B) = 1 \vdash \mathrm{Pr}_{\textbf{PRA}}(\#X_{\mathrm{Prov-}\textbf{PRA}}(a, \#B) = 1)$
  a free variable; the proof involves only positive inferences and induction rules, and the only hypothesis on $X_{\mathrm{Prov-}\textbf{PRA}}$ is its primitive recursiveness. Then, we can have $X_{\mathrm{Prov-}\textbf{PCA}}(a, \#B) = 1 \vdash \mathrm{Pr}_{\textbf{PCA}}(\#X_{\mathrm{Prov-}\textbf{PCA}}(a, \#B) = 1)$

in **PCA** as the root of a similar tree, and by applying a $\exists - R$ rule we have: $\mathrm{Pr_{PCA}}\,(\#B) \vdash \exists y\,\mathrm{Pr_{PCA}}(\#X_{\mathrm{Prov-PCA}}(y, \#B) = 1)$. Since by **BC**-inferences and the properties D1, D2 mentioned in Proposition 31 we have a **PCA**-proof of:

$$\exists y\,\mathrm{Pr_{PCA}}(\#X_{\mathrm{Prov-PCA}}(y, \#B) = 1) \vdash \mathrm{Pr_{PCA}}(\#\exists y X_{\mathrm{Prov-PCA}}(y, \#B) = 1),$$

by a cut inference we obtain the thesis.   ∎

The thesis of Proposition 32 is remarkable. In fact, in the classical setting it is possible to obtain the following property:

PROPOSITION 33. $B \vdash \mathrm{Pr_{PRA}}\,(\#B)$ *is* **PRA**-*provable for any formula* B *which is a* Σ_1 *or a quantifier-free formula.*

Proof. The thesis follows from two classical results: the first one (see e.g. [8, 24, 41]) states that the sequent $f(x_1, ..., x_n) = q \vdash \mathrm{Pr_{PRA}}(\#f(x_1, ..., x_n) = q\,)$ is **PRA**-provable for each function letter f in the **PRA**-language, i.e. for each primitive recursive function represented by f, and $x_1, ..., x_n$ free variables; the second one (see e.g. [30], p 67) establishes that for each quantifier free formula A in the **PRA**-language, with free variables among $y_1, ..., y_n$, a primitive recursive function g exists such that **PRA** proves $\vdash g(y_1, ..., y_n) = 1 \leftrightarrow A$, possibly after a very simple extension of **AxPRA** by sequents of atomic formulas.   ∎

The last proposition in general *does not hold* for **PCA**:

PROPOSITION 34. *The sequent* $B \vdash \mathrm{Pr_{PCA}}\,(\#B)$ *is in general not* **PCA**-*provable for any arbitrary quantifier free formula* B.

Proof. Let A be any quantifier free formula such that $\vdash \neg A$ is not a **PCA**-theorem and A is not a negated formula. Consider the sequent S: $\neg A \vdash \mathrm{Pr_{PCA}}\,(\#\neg A)$. We state that it is not **PCA**-provable. First, we prove that $\neg A$ in the antecedent cannot be introduced by weakenings only, since $\vdash \mathrm{Pr_{PCA}}\,(\#\neg A)$ is not **PCA**-provable. We can prove directly this last fact. Let us assume ad absurdum that $\vdash \mathrm{Pr_{PCA}}\,(\#\neg A)$ is **PCA**-provable. Then, since **PCA** is not trivial, a normal proof P of $\vdash \mathrm{Pr_{PCA}}\,(\#\neg A)$ exists, such that the root succedent is introduced in P also by a $\exists$-R inference having the premise of the form:

$$\vdash X_{\mathrm{Prov-PCA}}(t_1, \#\neg A) = 1, ..., X_{\mathrm{Prov-PCA}}(t_1, \#\neg A) = 1,$$

with $t_1, ..., t_m$, closed terms, $m \geq 1$. But each formula in such premise is a false ground recursive relation, and the normal proof of such premise is a **PRA**-proofs too. This is absurd, since **PRA** does not prove any disjunction of false ground recursive relations. Therefore, if a normal **PCA**-proof Q of S exists, the $\neg A$ in the root antecedent must be introduced also by $\neg - L3$

inferences or by logical axioms; but this in not possible, since descendants of formulas of the forms $°A$ or $\neg A$ cannot be cut in Q.   ∎

Note the relevance of the hypothesis assuming that A is not a negated formula in the Proposition 34 above: otherwise, from the axiom $S(t) = 0 \vdash$ by $\neg - L1$ rule we have $\neg\neg(S(t) = 0) \vdash$ from which by a weakening $\neg\neg(S(t) = 0) \vdash \mathrm{Pr}_{\mathbf{PCA}}(\#\neg\neg(S(t) = 0))$.

As a particular case of Proposition 34, take for example the sequent $\neg 1 = 1 \vdash \mathrm{Pr}_{\mathbf{PCA}}(\#\neg 1 = 1)$. We think that the **PCA**-unprovability of sequents of the form $\neg A \vdash \mathrm{Pr}_{\mathbf{PCA}}(\#\neg A)$ with A quantifier free is an intersting expression *of the intensional character of the paraconsistent negation of* **BC** *and of the* **C***-systems in general.* *The remarkable point is that the paraconsistent negation of a recursive predicate in the form* $X_R(t_1, ..., t_m) = r$ *is in general not* **T***-equivalent to a recursive predicate, for each* **T** *extending* **PCA** *only through sequents of atomic formulas.* Indeed, the sequent $1 \neq 1 \vdash \mathrm{Pr}_{\mathbf{PCA}}(\#1 \neq 1)$, being $\neq (.,.)$ the complement of the relation $= (.,.)$, that is, in the chosen formal language, $X_=(1,1) = 0 \vdash \mathrm{Pr}_{\mathbf{PCA}}(\#X_=(1,1) = 0)$, is **PCA**-provable, where $X_=$ is the characteristic function of the recursive relation $= (.,.)$. Let us examine why $\neg 1 = 1 \vdash \mathrm{Pr}_{\mathbf{PCA}}(\#\neg 1 = 1)$ and $X_=(1,1) = 0 \vdash \mathrm{Pr}_{\mathbf{PCA}}(\#X_=(1,1) = 0)$ results as **PRA**-equivalent but not **PCA**-equivalent sequents. If we add to the **PRA** or **PCA**-axioms (recall that the arithmetical parts of the two axiom sets are equal) the following sequents of atomic formulas: $\vdash x = y, X_=(x,y) = 0$, $x = y, X_=(x,y) = 0 \vdash$ we have that the so extended **PRA** can derive, by the **LK**-negation rules, both $\neg x = y \vdash X_=(x,y) = 0$ and $X_=(x,y) = 0 \vdash \neg x = y$, that is $\vdash \neg x = y \leftrightarrow X_=(x,y) = 0$, while the so extended **PCA**, by the **BC**-negation rules, cannot derive $\neg x = y \vdash X_=(x,y) = 0$. Therefore, even if the axiomatic information is the same as in the classical case, the implication $\neg x = y \rightarrow X_=(x,y) = 0$ is not admitted by the paraconsistent setting.

Even in the paraconsistent framework we have the unprovability of non-triviality:

THEOREM 35. **PCA** *does not prove its own non triviality* $Non-Triv(\mathbf{PCA})$, *nor it proves* $Triv(\mathbf{PCA})$.

Proof. We employ $Non - Triv(\mathbf{PCA})$ in the form $\neg \mathrm{Pr}_{\mathbf{PCA}}(\# \vdash)$ and $Triv(\mathbf{PCA})$ in the form $\mathrm{Pr}_{\mathbf{PCA}}(\# \vdash)$. By the previous results, we have that **PCA** proves the empty sequent if and only if classical **PRA** proves the empty sequent, and the possible normal proofs of the empty sequent may be assumed identical in the two systems. Then, by the **PRA**-representation of Provability Logic, we have that **PRA** proves the sequent $\vdash \mathrm{Pr}_{\mathbf{PRA}}(\# \vdash) \leftrightarrow \mathrm{Pr}_{\mathbf{PCA}}(\# \vdash)$ and then $\vdash \neg \mathrm{Pr}_{\mathbf{PCA}}(\# \vdash) \rightarrow \neg \mathrm{Pr}_{\mathbf{PRA}}(\# \vdash)$. If we suppose ad absurdum that **PCA** proves $\vdash \neg\mathrm{Pr}_{\mathbf{PCA}}(\#\vdash)$, then also **PRA** proves $\vdash \neg\mathrm{Pr}_{\mathbf{PCA}}(\#\vdash)$ and then $\vdash \neg\mathrm{Pr}_{\mathbf{PRA}}(\#\vdash)$ should be **PRA**-provable. This is absurd, by Gödel theorems for **PRA**, since $\vdash \neg\mathrm{Pr}_{\mathbf{PRA}}(\#\vdash)$ also expresses the classical consistency of **PRA**. As to **PCA**-unprovability of $\mathrm{Pr}_{\mathbf{PCA}}(\#\vdash)$, we can conclude in a similar way, starting from the **PRA**-provability of $\vdash \mathrm{Pr}_{\mathbf{PRA}}(\#\vdash) \leftrightarrow \mathrm{Pr}_{\mathbf{PCA}}(\#\vdash)$. The **PCA**-unprovability of $\vdash \mathrm{Pr}_{\mathbf{PRA}}(\#\vdash)$ can be proven also in a direct manner, as it is shown in the proof of Proposition 34.   ∎

Theorem 35 is also an incompleteness theorem for **PCA**, with the remark that in the paraconsistent setting to state the unprovability of B is not the same as to state the unprovability of $\neg\neg B$, and it is not sound to consider $\mathrm{Pr}_{\mathbf{PCA}}(\#\vdash)$ as the negation of $\neg\mathrm{Pr}_{\mathbf{PCA}}(\#\vdash)$. So we have shown that a formula B exists, which is *false* in the standard model **N**, and such that **PCA** does not prove neither B nor its negation $\neg B$. However, as to incompleteness, it is possible to conclude also in a different way, which provides a denumerable infinity of formulas B such that **PCA** neither proves B nor $\neg B$:

PROPOSITION 36. *Let* **T** *be any non trivial recursively axiomatized extension of classical* **PRA**, *preserving the standard model* **N**. *Then neither* $Non - Triv(\mathbf{T})$, *nor* $Triv(\mathbf{T})$ *are* **PCA**-*provable.*

Proof. Assuming $Non - Triv(\mathbf{T})$ and $Triv(\mathbf{T})$ respectively as $\neg\mathrm{Pr}_{\mathbf{T}}(\#\vdash)$ and $\mathrm{Pr}_{\mathbf{T}}(\#\vdash)$, we have that they cannot be **PRA**-provable, since they affirm the consistency or inconsistency of a consistent **PRA**-extension. Then, they cannot be **PCA**-provable.   ∎

We now meet a further important break between classical and paraconsistent Provability Logic, i.e. the fact that Gödel's Diagonal Lemma, in its standard formulation, cannot hold for **PCA**. Thus, the main tool for the formalized selfreference seems to fail in the paraconsistent setting. Gödel's Diagonal Lemma for **PRA** (see e.g.[40],[41]) states that:

LEMMA 37. *(Classical Diagonal Lemma) Let* $A(u)$ *be a* **PRA**-*formula in which the free variable* u *occurs. Then there is a formula* B *such that* $A(\#B) \leftrightarrow B$ *is* **PRA**-*provable. Moreover,* $A(\#B)$ *and* B *have exactly the same free variables.*

Proof. It is possible to define the following primitive recursive functions: a function $subst(.,.,.)$ such that for each **PRA** formula $D(u)$ with the free variable u, $subst(\#D(u), \#u, \#t) = \#D(t)$ is **PRA**-provable for each term t; a function $num(.)$ that links a term t of the **PRA**-language with its gödel-number $\#t$, such that $num(0) = \#0$, $num(x + 1) = g(num(x))$, where g is a primitive recursive function depending on the godelization. Then, we introduce the primitive recursive function $sub(.,.)$ such that, if x, y, u, z are free variables, $sub(x, y) = subst(x, \#u, num(y))$. Thus, given the formula $A(u)$, $sub(\#A(u), z)$ is a term preserving z as free variable, and such that $sub(\#A(u), n)$ is $\#A(n)$ for each numeral n. We define in **PRA**: $E(u) \leftrightarrow A(sub(u, u))$, $m = \#E(u)$. Therefore, we have in **PRA**: $sub(\#E(u), \#E(u)) = \#E(m)$, and then, by classical properties of the $= (.,.)$ predicate, $A(sub(\#E(u), \#E(u))) \leftrightarrow A(\#E(m))$ from which, by construction of E, $E(\#E(u)) \leftrightarrow A(\#E(m))$ which gives, by definition of m, $E(m) \leftrightarrow A(\#E(m))$. If B is taken as $E(m)$ the thesis is obtained. Note that if variables u_1, u_2,..., different from u, occur in A, then, by construction, they occur in B also. ∎

Conversely, we have that:

THEOREM 38. *Classical Diagonal Lemma does not hold for* **PCA***.*

Proof. First, we prove that if $A(u)$ is an arbitrary **PCA**-formula in which the free variable u occurs, then a sequent $\vdash A(\#B) \leftrightarrow B$ is in general not **PCA**-provable. Let us consider $A(u)$ as $\neg \mathrm{Pr_{PCA}}(u)$. Then $\vdash \neg \mathrm{Pr_{PCA}}(\#B) \leftrightarrow B$ is not **PCA**-provable for any B which is not a **PCA**-theorem. Let us assume ad absurdum that a normal **PCA**-proof Q of $\neg \mathrm{Pr_{PCA}}(\#B) \vdash B$ exists. By hypothesis, the antecedent cannot be introduced by weakenings only; then it must be introduced also by at least one logical axiom or by at least one $\neg - L3$ inference. But this would imply that descendants of formulas of the forms $\neg \mathrm{Pr_{PCA}}(\#B)$, $^\circ(\mathrm{Pr_{PCA}}(\#B))$ must be cut in Q, and this is absurd by the normality of Q, that allows only atomic cuts.

Secondly, we examine in what points the classical proof of the diagonal lemma fails for **PCA**. The answer is simple: in the classical proof we employ inferences such as $<<$ if $s = t$ then $A(s) \leftrightarrow A(t) >>$, soundly assuming they hold in **PRA**: they are not admitted in **PCA**. Indeed, as we have shown in Proposition 25, **PCA** is in general not transparent. ∎

We observe that $\neg \mathrm{Pr_{PCA}}(\#B) \leftrightarrow B$ is that instance of the Classical Diagonal lemma which is employed in the classical proofs of Gödel theorems.

However, a kind of paraconsistent version of the Diagonal Lemma is provable by employing the local consistency assertions $^\circ F$ provided by the C-systems

THEOREM 39. *(Paraconsistent Diagonal Lemma) Let $A(u)$ be a* **PCA**-*formula in which the free variable u occurs. Then there are a formula B and a finite set $\{^\circ D_s\}_{s=1,...,d}$ of local consistency assertions such that:*
   i)the sequent $\{(\forall x_j)^\circ D_s\}_{s=1,...,d} \vdash A(\#B) \leftrightarrow B$ *is* **PCA**-*provable;*

ii) $A(\#B)$ and B have exactly the same free-variable set V and the free variables occurring in $\{(\forall x_j)^\circ D_s\}_{s=1,\ldots,d}$ belong to V;

iii) each formula D_s can be obtained by term renaming from a proper subformula G_s of $A(u)$ such that $\neg G_s$ too is an $A(u)$-subformula.

Proof. Consider the proof of Lemma 37. We note that the definition of the recursive functions *subst, num, sub* is the same both in **PRA** and in **PCA**; moreover, without any loss of generality, we can consider the defining axioms $\vdash E(u) \leftrightarrow A(sub(u,u))$, $\vdash m = \#E(u)$ for E and m. By the definition of the recursive function *sub*, we have in **PCA**: (*) $\vdash sub(\#E(u), \#E(u)) = \#E(m)$. Indeed, we must recall that the pure equational parts of **PRA** and **PCA**, where no logical connectives occur, have the same theorems. Then, by the Weak Transparency Lemma (Lemma 27) we have in **PCA**:

($\cdot$) $\quad sub(\#E(u), \#E(u)) = \#E(m), \{(\forall x_j)^\circ F_h\}_{h=1,\ldots,p} \vdash$
$A(sub(\#E(u), \#E(u))) \leftrightarrow A(\#E(m))$

where each F_h can be obtained by term renaming from a proper subformula G_h of A such that $\neg G_h$ too is an A-subformula, and the free variables occurring in $\{(\forall x_j)^\circ F_h\}_{h=1,\ldots,p}$ belong to the free variable set of
$A(sub(\#E(u), \#E(u))) \leftrightarrow A(\#E(m))$. Then , by cut between ($\cdot$) and (*) we have:

$\{(\forall x_j)^\circ F_h\}_{h=1,\ldots,p} \vdash A(sub(\#E(u), \#E(u))) \leftrightarrow A(\#E(m))$;

but $\vdash E(\#E(u)) \leftrightarrow A(sub(\#E(u), \#E(u)))$ is an axiom instance, and so, by **BC**-rules, we have:

($\S$) $\quad \{(\forall x_j)^\circ F_h\}_{h=1,\ldots,p} \vdash E(\#E(u)) \leftrightarrow A(\#E(m))$.

On the other hand, through the Weak Transparency Lemma we have:

($\cdot\cdot$) $\qquad m = \#E(u), \{(\forall x_j)^\circ C_r\}_{r=1,\ldots,q} \vdash E(\#E(u)) \leftrightarrow E(m)$

where each C_r can be obtained by term renaming from a proper subformula H_r of E such that $\neg H_r$ too is an E-subformula, and the free variables occurring in $\{(\forall x_j)^\circ C_r\}_{r=1,\ldots,q}$ belong to the free variable set of $E(\#E(u)) \leftrightarrow E(m)$. We note that, by definition of E, if $E(u)$ is taken as identical to $A(sub(u,u))$, the C_r 's can result as obtained by term renamings from proper subformulas of A. By cut between ($\cdot\cdot$) and the axiom $\vdash m = \#E(u)$ we have:

($\S\S$) $\qquad\qquad \{(\forall x_j)^\circ C_r\}_{r=1,\ldots,q} \vdash E(\#E(u)) \leftrightarrow E(m)$

Then, by **BC**-rules , from ($\S$) and ($\S\S$) we have:

$\qquad \{(\forall x_j)^\circ C_r\}_{r=1,\ldots,q}, \{(\forall x_j)^\circ F_h\}_{h=1,\ldots,p} \vdash E(m) \leftrightarrow A(\#E(m))$

which is the thesis if we take B as $E(m)$. ∎

Example: Consider $A(u)$ as $\neg \Pr_{\mathbf{PCA}}(u)$. Then , the Paraconsistent Diagonal Lemma ensures us that a closed **PCA**-formula B exists such that the sequent

$\{(\forall x_j)^\circ \operatorname{Pr}_{\mathbf{PCA}}(t_s)\}_{s=1,\dots,d} \vdash \neg \operatorname{Pr}_{\mathbf{PCA}}(\#B) \leftrightarrow B$ is **PCA**-provable,
with the t_s's suitable closed terms.

We remark that a direct proof of the **PCA**-incompleteness could not be identical to the classical proof of the first Gödel theorem.

Thus, local consistency assertions play a crucial role in Paraconsistent Provability Logic. Let us comment the obtained result: it essentially means that the diagonalization property for the formula A holds in the paraconsistent setting under the assumption of the local consistency of a suitable set of proper A-subformulas, which moreover must occur in negated A-subformulas. Such set strictly depends on the structure of A, and in particular on the set of $\neg - L2$ rule instances that must be employed in order to prove sequents of the form: $u = v,\ A(u) \vdash A(v)$ in the classical logic **LK** + **EQ**. Moreover, in a sense, the complexity of the set $\{(\forall x_j)^\circ D_s\}_{s=1,\dots,d}$ in the Paraconsistent Diagonal Lemma thesis for A, could be a measure of the distance between A and the classical meaning of negation. In particular, we have the following straightforward Corollary:

COROLLARY 40. *Let $A(u)$ be a **PCA**-formula in which the free variable u occurs, such that any sequent of the form $z = v,\ A(z) \vdash A(v)$ is* **LK** + **EQ**-*provable through a proof Q where $\neg - L2$ rules do not occur. Then the thesis of the Diagonal Lemma holds for A in the classical way.*

We wish to point out the specific features of the paraconsistent framework which prevent the classical proofs of Gödel's theorems for **PRA** and **PA** from working (see e.g. Smorynsky [40] pp 827, 828). The basic constraint is the different form of the Paraconsistent Diagonal Lemma. Two special facts which stand in the way of the classical proofs are:

PROPOSITION 41. *i)* $\vdash \neg \operatorname{Pr}_{\mathbf{PCA}}(\#B) \to B$ *is not* **PCA**-*provable for any B which is not a **PCA**-theorem; ii) Assuming that $\neg \operatorname{Pr}_{\mathbf{PCA}}(\#B)$ is not **PCA**-provable, $\vdash \neg \operatorname{Pr}_{\mathbf{PCA}}(\#0 = 1) \to \neg \operatorname{Pr}_{\mathbf{PCA}}(\#B)$ is not **PCA**-provable for any B different from $0 = 1$ which is not a **PCA**-theorem.*

Proof. Point i) is already proved in the Theorem 38 above. Point ii) can be proved through analogous considerations. ∎

Since $0 = 1$ is a bottom particle for **PCA**, $\neg \operatorname{Pr}_{\mathbf{PCA}}(\#0 = 1)$ is the same as $\neg \operatorname{Pr}_{\mathbf{PCA}}(\# \vdash)$ i.e. $Non - Triv(\mathbf{PCA})$. Then, in the paraconsistent setting there is a formal distinction between $Non - Triv(\mathbf{PCA})$ and the unprovability statements on arbitrary unprovable sentences. Conversely, we know that $\vdash \neg \operatorname{Pr}_{\mathbf{PRA}}(\#0 = 1) \leftrightarrow \neg \operatorname{Pr}_{\mathbf{PRA}}(\#C)$ is **PRA**-provable for C varying in an infinite set of non **PRA**-provable sentences. The suggestion is that the proof of $Non - Triv(\mathbf{PCA})$ requires weaker instruments than that of the **PRA**-consistency $Con(\mathbf{PRA})$. We surmise that this difference becomes relevant if one compares the full arithmetics **PA** e **PCA**$_\infty$. This remark, together with the form of the Paraconsistent Diagonal Lemma, suggests the following conjecture:

CONJECTURE 42. *Let us consider the full Paraconsistent Arithmetic* **PCA**$_\infty$. *Then a minimal set* $\{^\circ F_j\}_{j=1,\dots,m}$, $m \geq 2$, *of suitable local consistency assertions exists, such that the following sequent is* **PCA**$_\infty$-*provable:*
$$\{^\circ F_j\}_{j=1,\dots,m} \vdash Non - Triv(\mathbf{PCA}_\infty)$$

Note that the minimality condition would impose that no sequent of the form $^\circ F_k \vdash Non - Triv(\mathbf{PCA}_\infty)$ would result as **PCA**$_\infty$-provable for any $k \in \{1, \dots, m\}$. Moreover, we suppose that non atomic cuts and non atomic inductions necessarily occur in the proof.

In essence, one might summarize this as follows: one could conjecture that the consistency of **PCA**$_\infty$ be provable within **PCA**$_\infty$, by assuming local consistency assertions, each of which is weaker than the assumption of non-triviality of **PCA**$_\infty$. If these $^\circ F'_j$ s turned out to be very weak with respect to $Non - Triv(\mathbf{PCA}_\infty)$, then one could speak of the possibility of a weakened Hilbert's program for paraconsistent Arithmetic.

5   Rational epistemic agents as paraconsistent systems

As already proposed in [4,5,6,7] we introduce an intrinsic formalization of a rational agent i by a theory $\mathbf{T}_i$ based on a suitable predicate calculus and including a fragment of any formal number theory that endows the agent with the basic computation capabilities. Here we propose agents of the form $\mathbf{W} + \mathbf{AxT}_i$, $\mathbf{W}$ paraconsistent arithmetical system, $\mathbf{AxT}_i$ agent proper axiom set. Our approach allows an a priori definition of rationality:

DEFINITION 43. A *paraconsistent rational agent* is a paraconsistent system $\mathbf{T}_i$ of the form $\mathbf{W} + \mathbf{AxT}_i$ such that:

i) $\mathbf{W}$ is a paraconsistent arithmetical system at least including $\mathbf{AxPCA}$, $\mathbf{AxT}_i$ is a recursive set of axioms, so that a Σ_1-provability predicate can be defined;

ii) $\mathbf{T}_i$ may be negation inconsistent, i.e. contradictions $\vdash A \wedge \neg A$ may be proved by $\mathbf{T}_i$, under the following conditions: iia) $\mathbf{T}_i$ does not prove any sequent of the form $\vdash B \wedge \neg B$ where B is a Δ_0-formula; iib) There exists a denumerable set of undecidable and recursively axiomatized systems $\mathbf{V}_j$ such that the *strong rationality conjectures* of the form $\vdash \mathrm{Pr}_{\mathbf{V}_j}(\# \vdash) \wedge \neg \mathrm{Pr}_{\mathbf{V}_j}(\# \vdash)$ do not trivialize $\mathbf{T}_i$.

The condition *iia*) imposes that ground computations of the agent cannot be contradictory, i.e. a sentence $A \wedge \neg A$ such that A is $t = s$, t and s ground terms representing the same number, cannot be stated by a rational agent.

Conversely, $\mathbf{T}_i$-theorems of the form $A \wedge \neg A$ where A is not Δ_0 express the complexity of the $\mathbf{T}_i$-rationality: in particular, they allows conjectural statements about the rationality of an agent, as indicated in the definition below. The knowledge of agents is so defined:

DEFINITION 44. i) We say that *agent* $\mathbf{T}_i$ *declaratively knows* the opinion or the state of the world described by the sentence B of the language if $\mathbf{T}_i$ proves $\vdash B$;

ii) We say that *agent* $\mathbf{T}_i$ *thinks that* B *holds*, or that he *believes that* B *holds*, if $\mathbf{T}_i$ proves $\vdash \mathrm{Pr}_{\mathbf{T}_i}(\#B)$;

iii) We say that *agent* $\mathbf{T}_i$ *believes that agent* $\mathbf{T}_j$ *knows* B if $\mathbf{T}_i$ proves $\vdash \mathrm{Pr}_{\mathbf{T}_j}(\#B)$;

iv) We say that $\mathbf{T}_i$ *declaratively knows that agent* $\mathbf{T}_j$ *is rational*, if $\mathbf{T}_i$ proves $Non - Triv(\mathbf{T}_j)$ and $\mathbf{T}_i$ does not prove $Triv(\mathbf{T}_j)$;

v) We say that *agent* $\mathbf{T}_i$ *believes that agent* $\mathbf{T}_j$ *is rational but he cannot declaratively know whether agent* $\mathbf{T}_j$ *is rational*, if $\mathbf{T}_i$ proves the strong rationality conjecture $\vdash \mathrm{Pr}_{\mathbf{V}_j}(\# \vdash) \wedge \neg \mathrm{Pr}_{\mathbf{V}_j}(\# \vdash)$.

Epistemically, this definition endows the agent with the possibility of inferring all knowledge through a well defined set of assumptions and inference rules: we can say that an agent is able to distinguish between premises and conclusions of his reasoning, and between what he postulates (or takes as given from his environment) and what he draws from that through inference. In fact, we interpret the *proofs* of $\mathbf{T}_i$ as the *reasoning* of the agent. This is particularly evident if we assume $\mathbf{T}_i$ to be formulated in Gentzen sequent version: the agent's inferential knowledge is then given by sequent proof trees whose leaves are logical axioms and the axioms in $\mathbf{AxT}_i$: the latter collects the knowledge data base of $\mathbf{T}_i$ on the states of the world and his specific *assumptions* (such as preferences) – in short, the postulates of his subjective thinking. Through the provability predicates $\mathrm{Pr}_{\mathbf{T}_i}(.)$ any agent $\mathbf{T}_i$ can reason about his proofs, that is, about his reasoning and conclusions: therefore, he is naturally endowed by a *self-reference capability*.

As follows from the Provability Logic results obtained in section 4, a paraconsistent rational agent cannot declaratively know its own rationality.

In many relevant applications, notably game theory, we must consider sets of epistemically interacting rational agents. In order to have significant societies of agents we must impose that agents may disagree on at least one statement, and so we formally define a democratic society:

DEFINITION 45. The set $\{\mathbf{T}_1, ..., \mathbf{T}_m\}$ of paraconsistent rational agents is an *interactive society* if for each pair $\mathbf{T}_i$, $\mathbf{T}_j$, $i \neq j$, at least one non Δ_0 -sentence A exists, such that $\mathbf{T}_i$ proves $\vdash A$ and $\mathbf{T}_j$ proves $\vdash \neg A$.

We note that if $\{\mathbf{T}_1, ..., \mathbf{T}_m\}$ is an interactive society then each system

$\mathbf{T}_i + \mathbf{T}_j, i \neq j$, is negation inconsistent.

As we shall see in section 6, the paraconsistent setting allows to formalize epistemic interactions in interactive societies in a way that could be not possible in a classical setting.

In several applicative contexts, agents must choose an action in a set of possible action alternatives or strategies. Such rational agent that knows and chooses, can be so defined:

DEFINITION 46. We say that *a paraconsistent rational agent* $\mathbf{T}_i \in \{\mathbf{T}_1, ..., \mathbf{T}_m\}$ *has a choice capability*, if the axioms $\mathbf{AxT}_i$ defines a *choice function* $f : \{1, ..., m\} \times \mathbb{N} \to \mathbb{N}$ such that the intended meaning of $f(i, x) = q$ is: << agent i chooses at time x, or relatively to its memory cell x, the strategy coded by q >>. We call a $\mathbf{T}_i$- *possible choice* each $\mathbf{T}_i$ -theorem of the form $\vdash f(i, x) = q$.

6   The level of inductive rationality of agents and the paraconsistent epistemic interaction

As we have already established in [4,6,7], rational agents $\mathbf{T}_i$'s which are able to represent the players of a strategic game, must have a suitable complexity level of rationality that, essentially, can be measured by the proof-theoretic strength of the system $\mathbf{T}_i$. The required level, is the level of the *inductive rationality*. In order to specify the upper complexity bound for the inductive rationality level we must recall some basic notions on transfinite induction. The consistency of full Arithmetic $\mathbf{PA}$ is provable by transfinite induction up to the countable ordinal ε_0, which is so defined: *if* $\omega(0) \equiv 1$ and $\omega(n+1) \equiv \omega^{\omega(n)}$, *then* $\varepsilon_0 \equiv \sup_n \{\omega(n)\}$. It is possible to formalize ordinals below ε_0 inside the $\mathbf{PA}$−language, by means of a canonical bijection ϕ between the set $\{\alpha : \alpha \prec \varepsilon_0\}$ and the non-negative integer set $\mathbb{N}$ (see [43], p 262); ϕ allows to define in $\mathbb{N}$ *a well ordering of order type* ε_0, which reproduces in $\mathbb{N}$ the well ordering of ε_0. We denote by $\angle$ such ε_0-well ordering in $\mathbb{N}$, and we introduce $\angle$ in the PA language. Moreover, following [43], p.264, we add to $\mathbf{PA}$ a set of axioms that we denote $\mathbf{Ord}$ describing the properties of $\angle$, and hence of the $\mathbf{PA}$-translation of ordinals below ε_0. For simplicity, we also use the Greek letters η, ξ, ...to indicate the elements of $\mathbb{N}$ representing infinite ordinals according to the bijection ϕ, but we should recall that they are in fact the numerals $\phi(\eta)$, $\phi(\xi)$,..., of the $\mathbf{PA}$ language. Given all this, the axiom of transfinite induction up to ε_0 in the $\mathbf{PA}$ language has the form:

$$\forall x (\forall y \angle x F(y) \longrightarrow F(x)) \longrightarrow \forall x F(x)$$

the axiom of transfinite induction up to α, $\omega\angle\alpha\angle\varepsilon_0$, in the PA language has the form:

$$\forall x \left(\forall y \angle x F\left(y\right) \longrightarrow F\left(x\right)\right) \longrightarrow \forall x \angle \alpha F\left(x\right)$$

We also indicate with $I(\varepsilon_0)$, $I(\alpha)$ the axioms above.

DEFINITION 47. We say that a rational paraconsistent agent $\mathbf{T}_i$ *is at the level of the inductive rationality* if the following conditions hold:

i) $\mathbf{T}_i$ includes $\mathbf{PCA}$ and then is endowed with an induction rule or axiom, such that $\mathbf{T}_i$ has the form $\mathbf{U} + \mathbf{AxT}_i$ where $\mathbf{U}$ is an over-system of $\mathbf{PCA}$;

ii) $\mathbf{T}_i$ has a choice capability such that $\mathbf{AxT}_i$ axiomatizes a non recursive choice function f, and, moreover, the choice theorems of $\mathbf{T}_i$ are not $\mathbf{U}$-theorems;

iii) $\mathbf{T}_i$ may be endowed with a transfinite induction axiom at most over denumerable ordinals up to ε_0:

$\forall x \left(\forall y \angle x F\left(y\right) \longrightarrow F\left(x\right)\right) \longrightarrow \forall x F\left(x\right)$, where $\angle$ is a well ordering of order type ε_0. The proof-theoretic strength of $\mathbf{T}_i$ respects the following upper bound: $\mathbf{T}_i$ at most proves the consistency of full Paraconsistent Arithmetic $\mathbf{PCA}_\infty$.

LEMMA 48. *It is possible to construct a denumerable infinity of systems* $\mathbf{T}_i$*'s, such that each* $\mathbf{T}_i$ *is a paraconsistent rational agent at the level of the inductive rationality.*

Proof. Let $i \geq 2$ be a fixed agent index. Let $n = 10^i$ and $h = 2^n$ be fixed. Let $\mathbf{V}_k \equiv \mathbf{PCA} + \mathbf{Ind}_k + J_k$ where $\mathbf{Ind}_k$ is the induction rule $\dfrac{F\left(x\right), X \vdash Y, F\left(S\left(x\right)\right)}{F\left(0\right), X \vdash Y, F\left(t\right)}$ such that F is a first order formula formula in prenex form (section 2.1) with at most $k \geq 1$ quantifiers. J_k is defined as follows. The language of the $\mathbf{V}_k$'s is extended by adding second order variables $X, Y, \ldots$ for arbitrary first order predicates definable starting from the $\mathbf{PCA}$-language, without introducing any second order quantifier; for each $r > 0$ let us consider the quantifier sequences so defined: $(\exists\forall)_1 \equiv \exists x_1 \forall x_2$, $(\exists\forall)_2 \equiv \exists x_1 \forall x_2 \exists x_3 \forall x_4, \ldots, (\exists\forall)_r \equiv (\exists\forall)_{r-1} \exists x_{2r-1} \forall x_{2r}$;then J_k is the following induction axiom in the $\mathbf{V}_k$'s language :

$\vdash [(\exists\forall)_k X(0, x_1, x_2, \ldots, x_{2k}) \wedge \forall y((\exists\forall)_k X(y, x_1, x_2, \ldots, x_{2k}) \rightarrow$
$(\exists\forall)_k X(S(y), x_1, x_2, \ldots, x_{2k})] \rightarrow \forall y[(\exists\forall)_k X(y, x_1, x_2, \ldots, x_{2k})]$

The systems $\mathbf{V}_k$ are proper extensions of $\mathbf{PCA}$, and the graph of the characteristic function of their theorems not only is non-recursive, but also not $\mathbf{PCA}$-provable; moreover, $\mathbf{V}_k$ cannot prove J_{k+1} and then $\mathbf{V}_{k+1}$ properly extends $\mathbf{V}_k$ for each k (see also [42], p. 116 and [41], p. 58). Let the following function ψ be defined over $\mathbb{N}$, having fixed i :

if z is the gödel-number of a formula and $\mathrm{Pr}_{\mathbf{V}_i}(z)$ holds, then $\psi(z) = i$;

if z is the gödel-number of a formula and $\mathrm{Pr}_{\mathbf{V}_i}(z)$ does not hold, then $\psi(z) = i + n\psi(\#1 = 1)$;

otherwise, $\psi(z) = i + h\psi(\#1 = 1)$.

Then, the choice function for agent i is defined as follows:

$f(i,x) = 2^{2x}5^i$ iff $\psi(x) = i$

$f(i,x) = 3^{2x}7^i$ iff $\psi(x) = i + n\psi(\#1 = 1)$

$f(i,x) = 0$ iff $\psi(x) = i + h\psi(\#1 = 1)$

The term $\psi(\#1 = 1)$ is introduced for technical reasons and does not have any substantial meaning; by construction, $\psi(\#1 = 1)$ results as equal to i. By the properties of prime numbers, the values of $f(i,x)$, yield two disjoint sets different from $\emptyset$ and from $\{0\}$, and hence $f(i,x)$ as defined is a version of the characteristic function of the $\mathbf{V}_i$-theorems. In order to formalize the definition above into a formal system $\mathbf{W}_i$, we add to the **PCA**-language the new funtion letters ψ and f, that do not correspond to any recursive function, and write the following axiom set $\boldsymbol{\Omega}_i$:

(1) $\vdash \forall z[fmla(z) \wedge \mathrm{Pr}_{\mathbf{V}_i}(z) \rightarrow [(z \neq \gamma) \rightarrow \psi(z) = i]]$

(2) $\vdash \forall z[fmla(z) \wedge \neg\mathrm{Pr}_{\mathbf{V}_i}(z) \rightarrow [(z \neq \gamma) \rightarrow \psi(z) = i + n\psi(\gamma)]]$

(3) $\vdash \forall z[\neg fmla(z) \rightarrow [(z \neq \gamma) \rightarrow \psi(z) = i + h\psi(\gamma)]]$

(4) $\vdash \forall x[f(i,x) = 2^{2x}5^i\psi(\gamma) \leftrightarrow x \neq \gamma \rightarrow \psi(x) = i\,]$

(5) $\vdash \forall x[f(i,x) = 3^{2x}7^i\psi(\gamma) \leftrightarrow x \neq \gamma \rightarrow \psi(x) = i + n\psi(\gamma)]$

(6) $\vdash \forall x[f(i,x) = 0 \leftrightarrow x \neq \gamma \rightarrow \psi(x) = i + h\psi(\gamma)]$

where: $fmla(.)$ is the recursive predicate such that $fmla(b)$ means "b is the code of a **PCA**-formula"; γ is an abbreviation for the writing $\#1 = 1$; the implications $(z \neq \gamma) \rightarrow ...$, $x \neq \gamma \rightarrow ....$ and the further factors $\psi(\gamma)$ are technical devices that does not takes any generality from the axiomatization of the choice function. The writings of the forms $z \neq \gamma$, $a^b = c$, $e + f = r$, $st = u$ are obvious abbreviations for the **PCA**-formulas $X_=(z,\gamma) = 0$, $exp(a,b) = c$, $+(e,f) = r$, $\bullet(s,t) = u$.

We introduce the system $\mathbf{W}_i \equiv \mathbf{PCA} + \boldsymbol{\Omega}_i$, and we state that:

I) $\mathbf{W}_i$ *is non-trivial, and this can be proven by a constructive syntactic proof* : let us assume as absurdum that a proof P of the empty sequent exists in $\mathbf{W}_i$. By Proposition 15 we can assume that P is an ordered proof having a P1-segment which is a **PCA**-proof with a root that is a subsequent of a **PCA**-provable sequent S of the form:

$\forall z[fmla(z) \wedge \mathrm{Pr}_{\mathbf{V}_i}(z) \rightarrow [(z \neq \gamma) \rightarrow \psi(z) = i]], \forall z[fmla(z) \wedge \neg\mathrm{Pr}_{\mathbf{V}_i}(z) \rightarrow [(z \neq \gamma) \rightarrow \psi(z) = i + n\psi(\gamma)]],$

$\forall z[\neg fmla(z) \rightarrow [(z \neq \gamma) \rightarrow \psi(z) = i + h\psi(\gamma)]], \forall x[f(i,x) = 2^{2x}5^i\psi(\gamma) \leftrightarrow x \neq \gamma \rightarrow \psi(x) = i],$

$\forall x[f(i,x) = 3^{2x}7^i\psi(\gamma) \leftrightarrow x \neq \gamma \rightarrow \psi(x) = i + n\psi(\gamma)], \forall x[f(i,x) = 0 \leftrightarrow x \neq \gamma \rightarrow \psi(x) = i + h\psi(\gamma)] \vdash$

in the most general case. By the **BC**-theorems $\forall y F(y) \vdash \forall y(A(y) \to F(y))$ and $\forall y(B(y) \to C(y)), \forall y(C(y) \to B(y)) \vdash \forall y(B(y) \leftrightarrow C(y))$ we have from S that the following sequent:

$\forall z((z \neq \gamma) \to \psi(z) = i), \forall z((z \neq \gamma) \to \psi(z) = i + n\psi(\gamma)), \forall z((z \neq \gamma) \to \psi(z) = i + h\psi(\gamma)), \forall x(f(i, x) = 2^{2x}5^i\psi(\gamma)),$

$\forall x(x \neq \gamma \to \psi(x) = i), \forall x(f(i, x) = 3^{2x}7^i\psi(\gamma)), \forall x(x \neq \gamma \to \psi(x) = i + n\psi(\gamma)), \forall x(f(i, x) = 0),$

$\forall x(x \neq \gamma \to \psi(x) = i + h\psi(\gamma)) \vdash$

must be **PCA**-provable, and it must admits of a normal **PCA**- proof Q. The most difficult case is that where each $\forall x F(x)$ in the antecedent has been introduced by $\forall - L$ rules. If so it is, a sequent of the form:

$(a \neq \gamma) \to \psi(a) = i, (b \neq \gamma) \to \psi(b) = i + n\psi(\gamma), (c \neq \gamma) \to \psi(c) = i + h\psi(\gamma), f(i, d) = 2^{2d}5^i\psi(\gamma),$

$s \neq \gamma \to \psi(s) = i, f(i, e) = 3^{2e}7^i\psi(\gamma), p \neq \gamma \to \psi(p) = i + n\psi(\gamma),$

$f(i, q) = 0, r \neq \gamma \to \psi(r) = i + h\psi(\gamma) \vdash$

must be **PCA**− provable by a normal proof H, where $a, b, c, d, e, p, q, r, s$ are suitable terms that we can assume as closed terms.

Since no **PCA**-proper axioms exist that define the function letters f and ψ, we uniformly replace in H each occurrence of f with the zero function Z and each occurrence of ψ with the primitive recursive function g so defined : $g(m) = i$ for each m different from γ, $g(\gamma) = 0$. Then we obtain that the sequent M :

$(a \neq \gamma) \to g(a) = i, (b \neq \gamma) \to g(b) = i + ng(\gamma), (c \neq \gamma) \to g(c) = i + hg(\gamma), Z(i, d) = 2^{2d}5^i g(\gamma),$

$s \neq \gamma \to g(s) = i, Z(i, e) = 3^{2e}7^i g(\gamma), p \neq \gamma \to g(p) = i + ng(\gamma),$

$Z(i, q) = 0, r \neq \gamma \to g(r) = i + hg(\gamma) \vdash$

is **PCA**-provable. But all formulas in M are elementary **PCA**-theorems, since the defining axiom of g and Z belong to **AxPCA**. From M, by obvious cuts, the empty sequent should result as **PCA**-provable, against the non triviality of **PCA**.

II) *The systems* $\mathbf{U}_i \equiv \mathbf{W}_i + \{\vdash \neg \mathrm{Pr}_{\mathbf{V}_k}(\#J_{k+1})\}_{k \in \mathbf{N}}$ *preserves as classical model an expansion of the standard model* $\mathbf{N}$ *and then is not trivial:* indeed the axiom set Ω_i in $\mathbf{W}_i \equiv \mathbf{PCA} + \Omega_i$ describe a version of the characteristic function of the theorems of a **PA**-subsystem, under the assumption of the consistency of full classical Arithmetic **PA** ; then, they must be true in a suitable expansion of **N**. As to the **N**-soundness property of $\mathbf{W}_i$, it is preserved since no rule has been added to **PCA**-rules, and the **PCA**-rules are **N**-sound. Furthermore, we note that each formula $\neg \mathrm{Pr}_{\mathbf{V}_k}(\#J_{k+1})$ is obviously true in **N**.

III) *The system* $\mathbf{T}_i \equiv \mathbf{U}_i + \vdash \psi(\gamma) = 1$ *is non trivial, recursively axiomatized, and really proves a denumerable infinity of different formulas of both the forms* $f(i,x) = 2^{2x}5^i$, $f(i,x) = 3^{2x}7^i$. We note that the axioms (1),(2),(3) of Ω_i do not define the value of the non recursive function ψ on the term γ; therefore, $\vdash \psi(\gamma) = 1$ cannot trivialize the system $\mathbf{U}_i$. The $\mathbf{U}_i$-axioms are a recursive set, since sequents J_{k+1} are effectively and univocally determined. On the other hand, by varying t in the set $\{\#A : A \text{ is a } \mathbf{V}_i\text{-theorem}\}$ or in the set $\{\#J_{k+1}\}_{k \in \mathbf{N}}$ we have by standard Provability Logic and **PCA**-inferences applied to $\Omega_i + \vdash \psi(\gamma) = 1$ $+ \{\vdash \neg \Pr_{V_k}(\#J_{k+1})\}_{k \in \mathbf{N}}$ the searched for infinity of $\mathbf{T}_i$-proven relations $f(i,t) = 2^{2t}5^i$, $f(i,t) = 3^{2t}7^i$. Thus, an infinity of non recursive $\mathbf{T}_i$-possible choices is provided.

IV) *The system* $\mathbf{T}_i$ *defined in* III) *can be paraconsistently extended by the induction axiom schema* $I(\varepsilon_0)$ *on denumerable ordinals up to* ε_0: indeed $\mathbf{T}_i + \mathbf{Ord} + I(\varepsilon_0)$ admits as classical model a suitable expansion of the standard model $\mathbf{N}$. Moreover, $\mathbf{T}_i$ remains paraconsistent, as the considerations at point V) below show.

V) *There is a denumerable set of strong rationality conjectures that do not trivialize* $\mathbf{T}_i$: let L_j be the sequent $\vdash \Pr_{\mathbf{G}_j}(\# \vdash) \wedge \neg \Pr_{\mathbf{G}_j}(\# \vdash)$ where the denumerable class of systems $\{G_j\}$ is so defined : $G_0 = \mathbf{ZF}$; $G_{i+1} = G_i + Con(\mathbf{ZF})$, $\mathbf{ZF}$ Zermelo-Fraenkel set theory. Then $\mathbf{T}_i + L_j$ is not trivial even if $\mathbf{T}_i$ is endowed by transfinite induction $I(\varepsilon_0)$. Indeed, we observe that $\mathbf{T}_i$ cannot prove $\neg \Pr_{\mathbf{G}_j}(\# \vdash)$ and that $\mathbf{T}_i + \vdash \Pr_{\mathbf{G}_j}(\# \vdash)$ is not trivial. Then, the thesis follows through the considerations already employed in section 3. ∎

THEOREM 49. *(Existence of interactive societies with agents at the level of the inductive rationality) Given* $m \geq 2$ *it is possible to construct a denumerable infinity of sets* $\{\mathbf{T}_1, ..., \mathbf{T}_m\}$ *such that each* $\mathbf{T}_i$ *is a paraconsistent rational agent at the level of the inductive rationality, and* $\mathbf{T}_i + \mathbf{T}_j$ *is negation inconsistent for each* $i \neq j$.

Proof. Let $i \in \{1, ..., m\}$, $m \geq 2$; consider the systems $\mathbf{V}_k$, defined for each natural number k as in the proof of Lemma 48, and the sequents J_k defined as in the proof of the same Lemma. Let $h = 2^m$ be fixed. Let $\ell(i) = 2^{hi}$.

We define the $\mathbf{T}_i$'s almost as in the proof of Lemma 48, in a way such that each particular $\mathbf{T}_i, i \in \{1, ..., m\}$, gives a different axiomatization of the function letter ψ belonging to the common language of the society:

For $i = 1$, if $fmla(z)$ and $\Pr_{\mathbf{V}_1}(z)$ hold, then $\psi(z) = 1$; if $fmla(z)$ holds and $\Pr_{\mathbf{V}_{\ell(1)}}(z)$ does not hold then $\psi(z) = 1 + m$; otherwise, $\psi(z) = 1 + h$.

For $i = 2$, if $fmla(z)$ holds and $\Pr_{\mathbf{V}_2}(z)$ does not hold, and $\Pr_{\mathbf{V}_{\ell(2)}}(z)$ holds, then $\psi(z) = 2$; if if $fmla(z)$ holds and $\Pr_{\mathbf{V}_{\ell(2)}}(z)$ does not hold, $\psi(z) = 2 + m$; otherwise, $\psi(z) = 2 + h$.

....

For $i = m$, if $fmla(z)$ holds and $\mathrm{Pr}_{\mathbf{V}_m}(z)$ does not hold, and $\mathrm{Pr}_{\mathbf{V}_{\ell(m)}}(z)$ holds, then $\psi(z) = m$; if $fmla(z)$ holds and $\mathrm{Pr}_{\mathbf{V}_{\ell(m)}}(z)$ does not hold, then $\psi(z) = m + m$; otherwise, $\psi(z) = m + h$.

Moreover, we consistently add the axiom:

$\forall x \forall y (X_=(x, y) = 0 \to \neg x = y)$.

For each i, we axiomatize the choice function f, given the axiomatization of ψ, with proper axioms Ω_i as in the proof of Lemma 48. We claim that the agents $\mathbf{T}_i$ are now pairwise inconsistent. Indeed, recalling that the theorems of system $\mathbf{V}_k$ are strictly included in those of $\mathbf{V}_{k+1}$, there certainly exists a gödel-number s such that $\mathbf{T}_1$ proves $\vdash \psi(s) = m + 1$, while $\mathbf{T}_2$ proves $\psi(s) = 2$; e.g. we have this if s is the gödel-number of $J_{\ell(2)}$; then $\mathbf{T}_2$ proves $\vdash \neg(\psi(s) = m + 1)$. We moreover observe that ψ is in each $\mathbf{T}_i$ defined as composed by characteristic functions of the theorems of **PCA**-extensions, and then formulas of the form $\psi(t) = u$ are in general not equivalent to any recursive relation.

It is straightforward to obtain, with further slight changes in the definitions of ψ and f, a denumerable infinity of different sets $\{\mathbf{T}_1, ..., \mathbf{T}_m\}$ with the required properties. ∎

Paraconsistency allows to express the knowledge changes of epistemic interacting agents in an interactive society. That is, we can define a kind of $\mathbf{T}_i$- declarative knowledge of the choices or opinions of a different agent $\mathbf{T}_j$, without trivializing $\mathbf{T}_i$:

DEFINITION 50. (Paraconsistent epistemic interaction) Let $\{\mathbf{T}_1, ..., \mathbf{T}_m\}$ be an interactive society of paraconsistent rational agents at the level of the inductive rationality. Let A_i be a $\mathbf{T}_i$ -theorem and let A_j be a $\mathbf{T}_j$ -theorem, $i \neq j$. Let $\mathbf{T}_i^* = \mathbf{T}_i + A_j$ and $\mathbf{T}_j^* = \mathbf{T}_j + A_i$. Then the pair $< \mathbf{T}_i^*, \mathbf{T}_j^* >$ is called a *paraconsitent epistemic interaction* if $\mathbf{T}_i^*$ and $\mathbf{T}_j^*$ are not trivial and A_i, A_j are not Δ_0-formulas. We say that the epistemic interaction is *relevant* if $\mathbf{T}_j^*$ proves a sequent S and $\mathbf{T}_j^*$ proves a sequent L such that $L, S \in \{\vdash A_i \wedge \neg A_i , \vdash A_j \wedge \neg A_j \}$.

The relevant epistemic interaction expresses the actual updating in the agent's knowledge when it establishes, through interacting, new opinions and conjectures on the state of the world. We remark that the classical setting does not allow such a direct and powerful notion of knowledge increasing through other's opinions, ideas or actions. In an interactive society of classical agents, the systems $\mathbf{T}_i^*, \mathbf{T}_j^*$ are generally trivial.

THEOREM 51. *It is possible to construct a denumerable infinity of interactive societies* $\{\mathbf{T}_1, ..., \mathbf{T}_m\}$ *of rational paraconsistent agents at the level of the inductive rationality such that each pair* $\mathbf{T}_i, \mathbf{T}_j$, $i \neq j$, *admits of a relevant epistemic interaction* .

Proof. Let $\{\boldsymbol{\Theta}_1, \boldsymbol{\Theta}_2\}$ be an interactive society obtained from $\{\mathbf{T}_1, \mathbf{T}_2\}$ as constructed in the proof of Theorem 49 ,with the following further specifications: $\boldsymbol{\Theta}_1 = \mathbf{T}_1 + \vdash Con(\mathbf{ZF})$, $\boldsymbol{\Theta}_2 = \mathbf{T}_2 + \vdash \neg Con(\mathbf{ZF})$. Then $\boldsymbol{\Theta}_1^* = \mathbf{T}_1 + \vdash Con(\mathbf{ZF}) + \vdash \neg Con(\mathbf{ZF})$, $\boldsymbol{\Theta}_2^* = \mathbf{T}_2 + \vdash Con(\mathbf{ZF}) + \vdash \neg Con(\mathbf{ZF})$ are not trivial, through the same considerations already mentioned in the proof of Lemma 48, point V), and satisfy the thesis.   ∎

7   Work in progress:paraconsistent agents and epistemic conditions of Nash equilibria in games

The Epistemic Provability Logic presented in the previous sections, allows to contribute to the formal investigation on the epistemic conditions of Nash equilibria (see [36]) in strategic games. The discussion about such topic has obtained a remarkable emphasis in the game theory community in the past few years (see [2,4, 36, 45]). First we note that, as it also shown in [7], our setting allows an explicit and complete formalization of the epistemic features of the game: players are a set $\{\mathbf{T}_1, ..., \mathbf{T}_m\}$ of rational agents in the sense established in the sections 5,6. The game rules are expressed as a first order theory $\mathbf{G}$ based on the *classical* predicate calculus $\mathbf{LK}$.

We assume that: the game $\mathbf{G}$ is classically consistent and the system $\mathbf{T}_i + \mathbf{G}$ is paraconsistent for each i. Moreover, each $\mathbf{T}_i$ *knows the consistency of the game*, that is $\mathbf{T}_i$ proves $Con(\mathbf{G})$ for each i. This also implies that the logical complexity of the players is higher than that of the game, and this seems to us a very realistic assumption. The strategic choices of the player $\mathbf{T}_i$ in the game will be expressed by the choice theorems of $\mathbf{T}_i$. An interesting possibility of the provability logic setting is that of formalizing the Nash equilibrium conditions for $\mathbf{G}$ through formulas of the $\mathbf{T}_i$'s language. One can define a m-place predicate $Nash_{i,\mathbf{G}}$ such that $Nash_{i,\mathbf{G}} (\#B_i, < \#B_r >_{r \neq i})$ states that "B_i is a Nash-choice for $\mathbf{T}_i$, fixed the choices B_r , $r \neq i$, $r \in \{1, ..., m\}$ of the other players" and a m-place predicate $Nash_{\mathbf{G}}(\#B_1, ..., \#B_m) \leftrightarrow \wedge Nash_{i,\mathbf{G}}(\#B_i, < \#B_r >_{r \neq i})$ that states that "the choice theorems $B_1, ..., B_m$ are a Nash Equilibrium of $\mathbf{G}$". In the paraconsistent framework the following problems can be explored:

1) Already in the classical setting it can be discussed whether the player $\mathbf{T}_i$ really knows that its choice is a Nash-choice. In fact, it is not necessarily given that $\mathbf{T}_i$ proves $\vdash Nash_{i,\mathbf{G}} (\#B_i, < \#B_r >_{r \neq i})$ since $Nash_{i,\mathbf{G}}$ results in general as a complex predicate. Moreover, note that, as it is usual in the provability logic setting, even if $\mathbf{T}_i$ proves $\vdash Nash_{i,\mathbf{G}} (\#B_i, < \#B_r >_{r \neq i})$ this does not imply that $\mathbf{T}_i$ proves B_k, i.e. the sentences in the predicate

argument. Anyway, what are the consequences on the effective equlibrium choices of an epistemic conjectural situation such as:

$\mathbf{T}_i$ proves $\vdash Nash_{i,\mathbf{G}}(\#B_i, < \#B_r >_{r \neq i}) \wedge (B_r \wedge \neg B_r)$ for any index $r \in \{1, ..., m\}$?

2) The question 1) acquires a stronger meaning in the cases where we consider a possible epistemic interaction between players, so that each player knows the other's choices. In this cases we may have a priori of any interaction, that:

$\mathbf{T}_i$ proves $\vdash Nash_{i,\mathbf{G}}(\#B_i, < \#B_r >_{r \neq i}) \wedge B_i$ and that $\mathbf{T}_i$ does not prove $\vdash \neg B_i$, but possibly that :

$\mathbf{T}_i^*$ proves $\vdash Nash_{i,\mathbf{G}}(\#B_i, < \#B_r >_{r \neq i}) \wedge (B_i \wedge \neg B_i)$

after a relevant paraconsistent epistemic interaction (Definition 50) through the knowledge of the other's choices. Such question may be relevant for the analysis of the repeated games and for the exploration of the epistemic conditions that may really drive the player to deviate from the expected Nash equilibrium choice.

The questions above suggest the investigation of a possible formulation of the actual $\mathbf{G}$-Nash equilibrium having the form:

$$°B_1 \wedge ... \wedge° B_m \wedge Nash_\mathbf{G}(\#B_1, ..., \#B_m)$$

that is, where the local consistency of the choices B_k is explicitly imposed.

BIBLIOGRAPHY

[1] R.J. Aumann, 'Interactive epistemology I: Knowledge', International Journal of Game Theory, 28, 1999, 263-300.

[2] R.J. Aumann, A. Brandenburger, 'Epistemic Conditions for Nash Equilibrium', Econometrica, 63,1995, 1161-1180.

[3] D. Batens, C. Mortensen, G. Priest, J.P. Van Bendegem (eds.), *Frontiers of Paraconsistent Logic*, Proceedings of WCP I, Research Studies Press Limited, Baldock, UK, 2000.

[4] C. Benassi, P. Gentilini, 'Can a Nash Equilibrium be Known ?' in *LGS3 Extended Abstracts*, S. Vannucci ed., University of Siena, Siena, 2003, 193-204.

[5] C. Benassi, P. Gentilini, 'Rational Economic Agents as Formal Logical Systems', Collana di Teoria Economica del Dipartimento di Scienze Economiche dell' Università di Bologna, n.53, Bologna, 1997.

[6] C. Benassi, P. Gentilini, 'Common Knowledge and Common Rationality through Provability Logic', in H. De Swart ed. *Tutorials and Position Papers-LGS 99 - International Conference on Logic, Game Theory and Social Choice*, University of Tilburg, Tilburg, 1999, 1-21.

[7] C. Benassi, P. Gentilini, 'Agents and Games as Logical Systems: Rationality and Complexity', to appear.

[8] P. Bernays, D. Hilbert, *Grundlagen der Mathematik*, Springer Verlag, Berlin, 1939.

[9] K. Binmore 'Modeling Rational Players: Part I', Economics and Philosophy, 3, 1987, 179-214.

[10] G. Boolos, *The Logic of Provability*, Cambridge University Press, 1993.

[11] S.R. Buss (ed.), *Handbook of Proof Theory*, Elsevier, Amsterdam, 1998.

[12] S.R. Buss, 'An Introduction to Proof Theory', in S.R. Buss ed., *Handbook of Proof Theory*, Elsevier, Amsterdam, 1998, 1-78.

[13] S.R. Buss, 'First Order Proof Theory of Arithmetic', in S.R. Buss ed., *Handbook of Proof Theory*, Elsevier, Amsterdam, 1998, 79-147.

[14] W.A. Carnielli, M.E. Coniglio, I.M.L. D'Ottaviano (eds.), *Paraconsistency: the logical way to the Inconsistent*, Proceedings of WCP 2000, Dekker, New York, 2002.

[15] W.A. Carnielli, 'On sequents and tableaux for Many-Valued Logics', The Journal of non Classical Logic, vol. 8, 1, 1991, 59-78.

[16] W.A. Carnielli, J. Marcos, 'A taxonomy of **C**-systems' in *Paraconsistency:the logical way to the Inconsistent*, Proceedings of WCP 2000, W.A.Carnielli, M.E.Coniglio, I.M.L. D'Ottaviano eds., Dekker, New York, 2002, 1-94.

[17] W.A. Carnielli, J. Marcos, 'Limits for paraconsistent calculi', Notre Dame Journal of Formal Logic, 40, 3, 1999.

[18] W.A. Carnielli, J. Marcos, 'Tableau systems for logics of formal inconsistency', in *Proceedings of the 2001 International Conference on Artificial Intelligence*, H.R.Arabnia ed., vol.II, CSREA Press, USA, 2001, 848-852

[19] N.C.A. Da Costa, 'On the theory of inconsistent formal systems', Notre Dame Journal of Formal Logic, vol. XV, n.4, 1974, 497-510.

[20] I.M.L. D'Ottaviano, J.E. Moura, 'On NCGω: a paraconsistent sequent calculus' in *Paraconsistency: the logical way to the Inconsistent*, Proceedings of WCP 2000, W.A.Carnielli, M.E.Coniglio, I.M.L. D'Ottaviano eds., Dekker, New York, 2002, 227-240.

[21] R. Fagin, J.Y. Halpern, Y. Moses, M.Y. Vardi, *Reasoning about knowledge*, MIT Press, Cambridge Mas.,1995.

[22] P. Forcheri, P. Gentilini, 'Paraconsistent Informational Logic', Journal of Applied Logic, 3, 2005, 97-118.

[23] P. Forcheri, P. Gentilini, 'Paraconsistent Conjectural Deduction based on Logical Entropy Measures I: the C-Systems as Non Standard Inference Framework', Journal of Applied Non Classical Logics, vol. 15, 3, 2005, 285-319.

[24] P.Gentilini, *Teoremi di completezza aritmetica della logica modale: una trattazione sintattica*, PHD Thesis in Mathematics, Genova, 1990.

[25] P. Gentilini, 'Proof-Theoretic Modal PA-Completeness I: a System-Sequent Metric', Studia Logica,vol. 63, 1999, 27-48

[26] P. Gentilini, 'Proof-theoretic Modal PA- Completeness II: the Syntactic Countermodel', Studia Logica, vol. 63, 1999, 245-268.

[27] P. Gentilini, 'Proof-theoretic Modal PA- Completeness III: the Syntactic Proof', Studia Logica, vol.63, 1999, 301-310.

[28] P. Gentilini, 'Provability Logic in the Gentzen Formulation of Arithmetic', Zeitschrift für Mathematische Logik und Grundlagen der Mathematik, vol. 38, 1992, 535-550.

[29] P. Gentilini, P. Forcheri, M.T. Molfino, 'Conjectural Provability Logic Based on Logical Information Measures' Bullettin of Symbolic Logic, vol. 3, 2, 1997, 259-260.

[30] J.Y. Girard, *Proof-Theory and logical complexity*, Bibliopolis, Napoli,1987.

[31] G. Japaridze, D. de Jongh, 'The Logic of Provability', in S.R. Buss ed., *Handbook of Proof Theory*, Elsevier, Amsterdam, 1998, 476-546.

[32] R. K. Meyer, C. Mortensen, 'Inconsistent Models for Relevant Arithmetics', The Journal of Symbolic Logic, 49, 1984, 917-929.

[33] C. Mortensen, 'Inconsistent Nonstandard Arithmetic', The Journal of Symbolic Logic, 52, 1987, 512-518.

[34] C. Mortensen, *Inconsistent Mathematics*, Kluwer Academic Publishers, Dordrecht, 1995.

[35] C. Mortensen, 'Prospect for Inconsistency' in *Frontiers of Paraconsistent Logic*, D.Batens, C.Mortensen, G.Priest, Van Bendegem eds., Research Studies Press Limited, Baldock, UK, 2000, 203-208.

[36] M.J. Osborne, A. Rubinstein , *A Course in Game Theory*, MIT Press, Cambridge Mas.,1996.

[37] G. Priest, *In Contradiction*, Kluwer, Dordrecht, 1987.
[38] G. Priest, 'Inconsistent Models of Arithmetic', Journal of Philosophical Logic, vol. 26, 2, 1997, 223-235.
[39] G. Priest, 'On a Paradox of Hilbert and Bernays', Journal of Philosophical Logic, vol. 26, 1, 1997, 45-56.
[40] C. Smorynski, 'The Incompleteness Theorems', in *Handbook of Mathematical Logic*, J.Barwise ed., North Holland, Amsterdam, 6th printing 1991, 821-895.
[41] C. Smorynski, *Selfreference and Modal Logic*, Springer Verlag, New York, 1985.
[42] G. Takeuti, *Proof Theory*, North-Holland, Amsterdam, 1987.
[43] A.S. Troelstra, H. Schwichtenberg, *Basic Proof Theory*, Cambridge University Press, 1996.
[44] M. Tsuji, N. C. A. Da Costa, F.A. Doria, 'The Incompleteness of Theories of Games', Journal of Philosophical Logic, vol. 27, 6, 1998, 553-568.
[45] J. Van Benthem, (ed.), *Theoretical Aspects of Rationality and Knowledge*, Proceedings of the VIII Conference TARK, Morgan Kaufmann Publishers, S. Francisco (CA), 2001.
[46] J. Van Benthem, *Logic in Games*, ILLC, Amsterdam, 2000.

Corrado Benassi
Dipartimento di Scienze Economiche, Alma Mater Studiorum - Università di Bologna, 40126 Bologna Italy.
benassi@economia.unibo.it

Paolo Gentilini
Istituto Regionale di Ricerca Educativa - IRRE Liguria, Via Assarotti 15, 16122 Genova, Italy, and
Istituto di Matematica Applicata e Tecnologie Informatiche, Consiglio Nazionale delle Ricerche
Via De Marini 6, 16149 Genova, Italy.
gentilini@ima.ge.cnr.it

Paraconsistency, Entailment and Truth

FRED SEYMOUR MICHAEL

ABSTRACT. In this paper, I give an account of a paraconsistent logic that is an enhanced version of the relevant system **E** and is therefore called '**EE**'. Although **EE** is a weak paraconsistent logic, the principal dialetheic theses can be embedded in it. **EE** has three values, T, F and O, where O is a value between T and F. Its most distinctive feature is the conception of entailment, called 'full entailment', defined as follows: $p \vdash q$ iff for all valuations v, $v(p) \leq v(q)$. In a three valued logic, full entailment takes the following form: $p \vdash q$ iff for all valuations v, if $v(p) =$T, $v(q) =$T; and if $v(q) =$F, $v(p) =$F. With this notion of entailment, *ex falso quodlibet*, $p \& \neg p \vdash q$, does not hold. **EE** has an intensional conditional, defined in terms of full entailment and a truth operator, 'T', such that $v(Tp) =$T if $v(p) =$T; otherwise $v(Tp)=$F. All other logical operators and connectives are normal three valued extensions of their classical counterparts. Contraposition holds both for full entailment and the conditional in **EE**. The properties of negation, conjunction and disjunction are much the same as in classical logic. But the widely accepted truth schema, $Tp \leftrightarrow p$ does not hold either in **EE**, nor in any logic with full entailment. We may now introduce into **EE**, by definition, a truth operator dual to the one **EE** already contains, one with most of the properties of the truth operator in dialetheic logic. Using this operator, which I call the weak truth operator, the principal features of dialetheic logic hold in **EE** and **EE** allows for true contradictions in Priest's sense. But also every dialetheic thesis in **EE** is logically equivalent to a thesis that is not dialetheic. The same results are shown to hold in Priest's *In Contradiction* as well.

1 Standard Logic and Relevance

As originally conceived, a relevant logic is a standard logic, not markedly different from classical logic, except that it remedies some of classical logic's inadequacies, in particular, it does not admit the paradoxes of C. I. Lewis:

$$p \& \neg p \vdash p^1$$

---

[1] In this paper p, q, r, ... are allowed to stand for any formulae, atomic or not. I use the same letters schematically, relying on context to make intended use clear.

$$p \vdash q \vee \neg q$$

These paradoxes have been known since the medieval period. The first is known as *ex falso quodlibet* (EFQ); the second does not have a standard designation. Both are theses of classical logic and calling them paradoxes implies only that they are counterintuitive. Still, developing a logic in which these classical theses does not hold seems a worthwhile enterprise.

The initial goal of relevant logic, which was to do basically the same thing as classical logic does, only better, has not been realized. There are a number of reasons for this. First, there were initially three principal systems of relevant logic, **T** , **E** and **R** with **E** intended as the system to replace classical logic, but these relevant systems were soon joined by others, so that now there is an alphabet soup of relevant logics, and there is no consensus about which of them, if any, should take the place of classical logic. Second, in the early development of the principal relevant logics, semantics was given little attention. But the semantics of classical logic had been similarly undeveloped in the forty or so years after the pioneering work of Frege and Peirce. The axiomatic proof theory of classical logic was however later to be provided with an intuitively satisfying semantics that would justify it. When the same thing was attempted with relevant logics however, the forms of semantics proposed were highly counterintuitive and emerged as a barrier to their widespread acceptance.

It is not hard to see why this happened. The problem that had to be faced was how to falsify the Lewis paradoxes. Two largely equivalent strategies emerged to achieve this objective. One, known as the Australian plan semantics, allows situations in which p and $\neg p$ are both true and others in which both are false, so that in those situations there are literally true contradictions and false tautologies. The other, the American plan semantics, achieves the same objective by introducing two values between T and F; B for 'both true and false' and N for 'neither true nor false'([RLR], 192ff.). Third, with neither of these strategies likely to gain popular support, a new justification of the relevant logic project was offered. We are to think of the propositions of relevant logic as information and since we cannot reasonably expect that information obtained will always be free of inconsistencies, a logic that tolerates contradictions is of value, not only for its own sake, but for the sake of its applications to information technology and computer science ([ABD], 142ff.). What justifies the counterintuitive semantics of relevant logics are these applications. This is certainly a valuable use for relevant logics, but to justify relevant logic in this way is to give up the conception of it as a standard logic, improving on classical logic, and to conceive of it instead as a non-standard logic useful for special purposes. When relevant logic is thought of as substructural, it is being seen in this way. That relevant logic is non-standard is the prevailing view at present. Finally, the principal relevant logics have been shown to be undecidable ([ABD], 348-75.), lending further support to the view that they are non-standard and not suitable as substitutes for classical logic.

The logic that is the principle focus of this paper I call **EE**, or enhanced **E**, because all of the theorems of the Anderson-Belnap system **E** hold in it. The reasons to consider it an enhancement of **E** will emerge later. A key feature of **EE** is a generalized notion of entailment that restricts logical consequence. Like any relevant logic it is paraconsistent. Paraconsistent logics are said to come in two strengths. Weak paraconsistent logics enable us to falsify or otherwise reject *ex falso quodlibet*, the thesis that from a contradiction anything follows. They allow for inconsistent theories that are non-trivial. Strong paraconsistent logics are supposed to go further, allowing inconsistent theories that are true. **EE** is a weak paraconsistent logic. Nonetheless, there is no difficulty in introducing a truth operator into **EE** with most of the properties of the truth operator in Graham Priest's In Contradiction [IC], although the logic in IC is said to be a strong paraconsistent logic.

2 Ontological Neutrality

EE is a three valued logic. In addition to the values T and F, there is an intermediate value O. 'O' stands for the English word 'out' and is to be understood in the sense of extra mundane. In the actual world, or any other possible world, w, with respect to an object that exists at w or a state of affairs that obtains at w, every proposition is either true or false. But if we do not assume that worlds are maximal, as is commonly thought, then we allow objects and states of affairs not at w. Certainly we can talk about objects or states of affairs not in the actual world. If p is not at w, the value of p at w is O. So for example in the actual world, statements about Kant's wife or the third world war in the twentieth century have the value O. In the actual world, there are countless states s, such that neither s nor any alternative has been realized, at least up to now, so that if 'p' asserts that s, both p and $\neg p$ have the value O. We think of material things that appear to be entirely solid as mainly void space. In the same way, we can think of the actual world, despite appearances, as in the main logically indeterminate, in which only a fragmentary part of what is logically possible ever is realized. Possibilities here are logical possibilities, not real or metaphysical possibilities. It is not the business of logic to tell us what is real or what is really possible. Logic ought as far as possible be ontologically neutral. It is a defect or limitation of classical logic that it is not. It is worth stressing that the value 'O' is not ontologically suspect. Having the value

'O' in our logic adds nothing to our ontology; on the contrary, having this value makes it possible to free logic from the ontological entanglements in which classical logic is ensnared.

I do not make much use of possible worlds in this paper. My main purpose here has been to give some intuitive explanation of what it means for a proposition to have the value O. At the same time, **EE** formally requires only that there be a value between T and F, not that it have the intuitive content given above. Also while **EE** is designed to accommodate empty terms such as 'the golden mountain', it is limited by logical possibility. In its semantics, there is no place for 'the round square' or 'the circle squared, using only ruler and compass.' The distinctive features of **EE** are: 1. Its conception of entailment. 2. Its conception of tautology. 3. Its use of the truth operator. 2. Its conditional.

3 Generalizing Entailment

If q is derived from p in the sense of being a logical consequence of p, then following the usage of Anderson and Belnap, I will say that p entails q, and will represent this situation by '$p \rightarrow q$'. Where $\rightarrow$ is the necessary relevant conditional of **E**, then by the deduction theorem, in the form that it holds in **E**, $p \vdash q$ is logically equivalent to $\vdash p \rightarrow q$ ([AB], 278), and formulae of the form '$p \rightarrow q$' are also called entailments. The axioms of **E** are all entailments and all are logical truths. This however is not the case for extensional tautologies in **E** tautologies that are not entailments. That creates the following problem. Consider the entailment, $p \vdash p$. Since this is intuitively stronger than the material conditional, $p \supset p$, logically equivalent classically to $\neg p \lor p$, the inference from $p \rightarrow p$ to $\neg p \lor p$ ought to be valid. But as is clear from Smiley's characteristic matrices for first degree entailments, when p has a value other than T or F, $p \rightarrow p$ is true but $\neg p \lor p$ is not ([AB],161-62). The inference does not preserve truth. This is not entirely unprecedented. It is true of the conditional of Heyting's intuitionistic calculus as well([6],312). But it does raise questions about how classical two valued logic can be a subsystem of **E**. In **E**, T is the only designated value. This means that the conception of semantic consequence in **E** is the same as it is in classical logic. In Priest's "The Logic of Paradox", both T and P (paradoxical) are designated values, F is not, with the result that non-falsity is preserved, but truth is not ([LP], 227). Designating more than one value preserves only their disjunction and so weakens semantic consequence. **EE** uses a notion of semantic consequence I call 'full entailment' that preserves both truth and non-falsity. Full entailment preserves not just values but the order of values as well. If '$p \Vdash q$' represents 'p fully entails q,' then in a three valued logic, for instance, **EE**, such that T $>$ O $>$ F, full entailment takes the following form:

$$p \Vdash q \text{ iff for all valuations } v, \quad \text{if} \quad v(p) =T, v(q) =T; \text{ and}$$
$$\text{if} \quad v(q) = F, v(p) =F$$

If 'p classically entails q' is represented by '$p \Vdash_c q$', then full entailment can be defined in terms of classical entailment as follows:

$$p \Vdash q \text{ iff } p \Vdash_c q \text{ and } \neg q \Vdash_c \neg p$$

In a sense, full entailment is just truth preservation, but it is full truth preservation, truth preservation both for the positive and the contrapositive. There is also a general form of full entailment, applicable to logics with any number of values and any value ordering:

$$p \Vdash q \text{ iff for all valuations, } v, v(p) \leq v(q).$$

This means that for the inference from p to q to hold, q may not take a value lower than p.[2] So if F is the lowest value, then q can take the value F only if p takes the value F. Full entailment is readily seen to have two properties. First, contraposition plainly holds. Second, full entailment is paraconsistent; to falsify the sequent $p, \neg p \Vdash q$, let $v(p) =O$ and $v(q) =F$.

4   Normal Logic

A normal logic is a logic that satisfies certain conditions, whether bivalence holds or not. Classical logic is usually thought of as normal, non-classical logics are generally considered deviant or non-standard logics, logics that deviate from the classical norm in order to achieve any of a variety of special purposes. But classical logic is not the norm of logic; it is just one realization of normal logic, a realization in which bivalence holds. There are also normal three valued logics, four valued logics, normal logics with any number of values. Normal structure is exemplified by Kleene's strong three valued tables for negation, conjunction and disjunction ([IM], 334). Kleene's tables are:

---

[2]Belnap uses this notion of entailment in his "useful four valued logic". See [ABD], 519.

	$p\&q$		$p \vee q$

p	$\neg p$		$p\backslash q$	TOF		$p\backslash q$	TOF
T	F		T	TOF		T	TTT
O	O		O	OOF		O	TOO
F	T		F	FFF		F	TOF

That is,

Negation: $v(\neg p) =$ T, if $v(p) = $ F
F, if $v(p) = $ T
O, otherwise

Conjunction: $v(p\&q) =$ T, if $v(p) = $ T, $v(q) = $ T
F, if $v(p) = $ F or $v(q) = $ F
O, otherwise

Disjunction: $v(p \vee q) =$ T, if $v(p) = $ T or $v(q) = $ T
F, if $v(p) = $ F, $v(q) = $ F
O, otherwise

Normal conjunction and disjunction satisfy the following conditions:

1. the value of a conjunction is always the minimum of the value of its conjuncts,

2. the value of a disjunction is always the maximum of the value of its disjuncts.

Under these conditions, the following familiar Gentzen natural deduction rules for conjunction and disjunction hold:

& **Introduction.**	$p, q \vdash p\&q$
& **Elimination.**	$p\&q \vdash p$; $p\&q \vdash q$
$\vee$ **Introduction.**	$p \vdash p \vee q$; $q \vdash p \vee q$
$\vee$ **Elimination.**	If $p \vdash r, q \vdash r$, then $p \vee q \vdash r$

Normal negation satisfies the following conditions:

Double Negation.	$v(\neg\neg p) = v(p)$
Contraposition.	$v(p \vdash q) = v(\neg q \vdash \neg p)$
TF polarity.	$v(\neg p) = $ T iff $v(p) = $ F

If however entailment is classical, contraposition fails. For instance, if $v(p) = O$ and $v(q) = F$, then although $p \vdash q$ holds, $\neg q \Vdash \neg p$ does not. If contraposition is to hold in a non-bivalent logic then, entailment cannot be the same as it is in classical logic. These properties of normal negation are very like properties of negative integers in mathematics. If m and n are integers, then $(--m) = m$ and $m \leq n$ iff $-n \leq -m$. This implies that $p \vdash q$ holds in a normal logic just in case there is a relation of partial order between the values of p and q, and so in a non-bivalent normal logic just in case full entailment, not classical entailment, is the consequence relation. In a two valued logic, double negation and contraposition suffice for complementation; but in a logic with more than two values, they do not. The normal negation of p is the complement of p only if p is bivalent. Negation, satisfying the above conditions, has come to be known as De Morgan negation.

A normal logic is then a logic with full entailment, normal conjunction, disjunction, and negation. These may seem like relatively modest requirements but most of the well known non-classical logics are not normal logics; they fail to satisfy the above conditions in one respect or other. There are few non-bivalent logics with full entailment. In Heyting's intuitionistic logic, negation is not normal; one cannot infer a formula from its double negation ([6],312) . Negation is not normal in Da Costa's paraconsistent logic either; from a formula one cannot infer its double negation (see [4], 499). Nor is negation normal in logics in which truth alone or truth and non-falsity both are designated values, since then contraposition does not hold for entailment. In Bochvar's logic disjunction is not normal, conjunction is not normal in the discursive logic of Jaskowski and neither is normal in Belnap's logic 'four'.[3] In partial logics with supervaluations, assuming that supervaluations are valuations, neither conjunction nor disjunction is normal; $p \vee \neg p$ has the value T, even if neither p nor $\neg p$ has the value T; and even when neither has the value F, that is the value $p \& \neg p$ has ([1], 40-47). In non-classical logics, normal logic has not been the norm. There have however been normal non-classical logics. Stephen Blamey's partial logic is one example (see [2]), and there is a paper of Dana Scott in which a

---

[3] See [3], 91, [7], 154 and [ABD], 515. The clearest indication that Bochvar's system is not normal is that the value of a disjunction is not the maximum of the value of its disjuncts. When $v(p) = T$ and $v(q) = N$ (his third value), $v(p \vee q) = N$. In Jaskowski's logic, to infer $p \& q$ from its conjuncts is assimilated to inferring that $p \& q$ is possible given that p is possible and q is possible, an inference which it is easy to falsify. In Belnap's four valued logic, if $v(p) = N$ (for neither) and $v(q) = B$ (for both), then $v(p \& q) = F$. But F is lower than either N or B, so the value of a conjunction here is not the minimum of the value of its conjuncts. One could satisfy this condition by recognizing a generic third value, equivalent to a disjunction of all third values, including N and B.

similar logic is proposed [11]. It may come as something of a surprise that the logic in Graham's Priest's *In Contradiction*, turns out to be normal, on the conditions for normality proposed here.[4]

A normal logic has two significant properties. The first is that Kleene's strong tables are generalizations of classical semantics and normal logic semantics is a generalization of Kleene's strong tables, so a normal logic should be seen not as a deviant or alternative logic; it is rather a generalization of classical logic. Logics with more than two values are simply more finely grained than classical logic, enabling us to discriminate between states that are not classically distinguished. The failure of paraconsistency, as well as other well known anomalies and limitations of classical logic, suggest that the semantics of classical logic is not fine grained enough. Secondly, any many valued logic with a normal subsystem is a conservative extension of classical logic. Any logic with full entailment, with normal conjunction, disjunction and negation replacing their classical counterparts, is a conservative extension of classical logic. Proof: Classical logic is normal, so if p holds in classical logic, it holds in one normal logic. Now consider an n-valued normal logic. Then given a true conjunction in that logic, all of its components must be true making its classical counterpart true as well. If its classical counterpart is false, then at least one of its components must be false in which case the n-valued conjunction is also false. By similar arguments, we can show that disjunctions and negations in an n-valued normal logic entail their classical counterparts. In addition, if $p \vdash q$ in any normal logic, then $p \vdash_c q$. An entailment in a normal logic must hold classically. Full entailment restricts classical entailment; it does not extend it. Classical logic is a subsystem of normal logic. So whatever holds in normal logic must hold classically as well.

5 Negations

In classical logic, the normal negation of p is the complement of p. In non-bivalent normal logics however, the two notions do not coincide. But we do not have to choose between them. We can extend our normal logic by adding to it the complement of p, true if p is not true and false otherwise. If we represent the complement of p by '$*p$' then the conditions determining what semantic value should be assigned to $*p$, its semantic value conditions,

---

[4]Although entailment in the logic of IC appears to be classical, the use of {1}, {0,1} and O as values, with {1} representing T and {0} representing F makes a critical difference ([IC], 99). From a proposition that is true only, one cannot infer a proposition that is both true and false, since that involves an inference from true to false, and from one that is both true and false, for the same reason, one cannot infer one that is false only. This will not work however where there are four values with the fourth value represented by { }, for neither true nor false.

or, more simply, value conditions, can be expressed as follows:

$$v(*p) = \text{T, if } v(p) \neq \text{T}; \ v(*p) = \text{F, otherwise.}$$

Whereas normal negation is De Morgan negation, this is commonly known as Boolean negation. It is not a form of normal negation; we do not have $**p$ iff p; we have instead $**p$ iff $\neg *p$. Moreover, given these two kinds of negation, we can define 'Tp' where 'T' is the truth operator, as follows: Tp iff $\neg *p$; Tp is also equivalent to $**p$, the double Boolean negation of p. Given the value conditions for $*p$, the value conditions for Tp are

$$v(Tp) = \text{T, iff } v(p) = \text{T}; \ \ v(Tp) = \text{F, otherwise.}$$

Robert K. Meyer has shown that Boolean negation can be added conservatively to many relevant logics. In fact, it can be conservatively added to any logic with De Morgan negation, because $\neg p \vdash *p$, so that for any Γ such that $\Gamma \vdash q$, if r is the result of replacing any occurrence of $*p$ by $\neg p$ in q, then $\Gamma \vdash r$. But for some in the relevant logic community, Boolean negation should be avoided at any cost, because to accept it makes valid some instances of EFQ (see [ABD], 488-98). Indeed, $*p, \neg *p \vdash q$ cannot be falsified in our extended normal logic. But then EFQ cannot be falsified if there is even a single bivalent instance of p and we do not want to go from the extreme of claiming that *all* propositions are bivalent to the at least equally unacceptable extreme of denying that *any* proposition is bivalent. If EFQ is to be avoided, it must be avoided in some other way than by banishing all bivalent propositions from a logic. I will have more to say about this later. Assuming that we allow that there are bivalent propositions, then there is no reason, so far as I can tell, for rejecting Boolean negation. Given De Morgan and Boolean negation, the intuitionistic negation of Heyting's system ([6], 326) can be defined; the intuitionistic negation of p, AKA its *pseudo-complement*, which we may represent by '$\dashv p$', is just the dual of Boolean negation, That is, $\dashv p$ iff $\neg * \neg p$.

From now on, I will generally avoid the somewhat confusing multiplicity of different non-normal negation notions, and will use the truth operator instead, representing the Boolean negation of p, when needed, by $\neg Tp$ and intuitionistic negation by $T \neg p$. The truth operator and the various kinds of negation are represented in the tables below.

p	$\neg p$	$*p$	$\dashv p$	Tp
T	F	F	F	T
O	O	T	F	F
F	T	T	T	F

EE is a normal logic with the truth operator (or equivalently, with both De Morgan and Boolean negation).

6   Tautologies and Logical Truth

If a logical truth is understood as a formula that is true for all values of its propositional variables, then it is already implicit in **T**, **E** and **R** that there should be a distinction between tautologies and logical truths. $p \vee \neg p$ is a tautology of the TV subsystem of **T**, **E** and **R**([AB], 283-85), but it is not a logical truth in any of these systems. The usual way to show that a formula is a tautology is to show that it cannot be falsified. If p cannot be false in a bivalent logic, then it must be true. In classical logic, there is no distinction between a formula never being false and always being true. In a non-bivalent logic, however, there is such difference. That faces us with three choices: one is to hold that none of the tautologies of classical logic are tautologies of our non-bivalent logic. The second is to use supervaluations to preserve the classical tautologies, even though they are not true for all values of their propositional variables. We take a third option, which is to accept non-falsifiability as the test of being a tautology, whether in a bivalent or a non-bivalent logic, allowing for a class of formulae that are tautologies because they are not falsifiable but are still not logical truths.

We will say that p is *satisfied* by some valuation v, iff the value of p on v is not F. We will say that p is a *tautology* iff p is not falsifiable, that is, there is no valuation v on which $v(p)$ is F. A proposition that is satisfied then need not be true; when it is true we will say that it is *verified* and a valuation on which p has the value T will be said to verify p. In case every valuation verifies p, p will be said to be a *logical truth*. In classical logic, satisfaction and verification coincide; they do not coincide however in **EE** or other logics with more than two values.

With this we are able to show that in **EE**, unlike the principal relevant logics, **T**, **E** and **R**, not only does the rule of disjunctive syllogism fail, rule (γ) fails as well. According to rule (γ) , if $\neg p$ and $p \vee q$ are provable, so is q ([AB], 300), that is,

from $\vdash \neg p$ and $\vdash p \vee q$, it follows that $\vdash q$.

To falsify (γ) in **EE** let $v(p) = \mathrm{O}$ and $v(q) = \mathrm{F}$. Rule (γ) and disjunctive syllogism both hold however provided that p is bivalent.

7   The Truth Operator

The truth operator that has been introduced into **EE** is a logical constant in the object language and should not be confused with Tarski's metalinguistic truth predicate. If we represent the Tarski truth predicate by 'Tr', then

Tarski's truth schema, which we may represent as $Tr'p' \leftrightarrow p$, asserts a correspondence in value between the metalinguistic $Tr'p'$ and its object language counterpart, p. If they do correspond, then any function of $Tr'p'$ will correspond to the same function of p. So we will have $\neg Tr'p' \leftrightarrow \neg p$. But we do not have $\neg Tp \leftrightarrow \neg p$; to falsify this, just let $v(p) = $O. Tp is a function of the value of p, but its value need not correspond to the value of p. It is bivalent, partitioning propositions into two classes, those whose value is T, and those that have some other value. $Tr'p'$ is not necessarily bivalent. In order for the truth schema to be satisfied, $Tr'p'$ and p must have the same value; so if the value of p is O, the 'value of $Tr'p'$ must be O as well. I am not arguing against Tarski's theory of truth or the Tarskian truth predicate; I am arguing that accepting the Tarskian predicate does not mean rejecting a truth operator such as the one that has been added to **EE**. They are not the same, but they are not incompatible either, so we do not have to choose between them.

The truth operator of **EE** is simply an extensional logical constant the meaning of which is exhausted by the valuation conditions given above. D. A. Bochvar [3] in 1937 used ' $\vdash$ ' as an external assertion operator with these value conditions, in which with $\vdash a$ is to be read as 'a is true'; but even earlier in 1935 Bruno de Finetti [5] had used the term '$thèse$' for an operator with the same value conditions; the best known account of the truth operator is probably the one by Timothy Smiley [12]. However many values p may take, Tp is always bivalent. It is, or ought to be, a basic resource of non-classical logic.

The truth operator 'T' greatly expands the expressive power of **EE**. Representing 'p is false' by 'Fp', defined as '$T\neg p$', the principle of bivalence, can be expressed in **EE** by $Tp \vee Fp$, for any p, or equivalently, by $Tp \vee T\neg p$, for all p. Using the principle of bivalence, we are also able to express formally in **EE** the conditions under which $\neg p, p \vee q \vdash q$, disjunctive syllogism, holds: it holds just in case bivalence holds for p, that is, $Tp \vee T\neg p$. We may also introduce a satisfaction operator 'S' such that Sp iff $\neg T\neg p$, that is, p is satisfied iff p is not false. Tp and Sp are duals. Since a tautology cannot be false, if p is a tautology, Sp (that is, $\neg T\neg p$) is always logically true.

We have spoken of the Boolean negation of p as the complement of p. In fact, p may be said to have two complements in **EE**, a complement with respect to truth, its T-complement, and the S-complement of p, its complement with respect to satisfaction. Boolean negation, that is, $*p$ (or $\neg Tp$) is its T-complement; its S-complement is $\dashv p$ (or $T\neg p$). $*p$ is the T-complement of p because 1) $p \vee *p$ is a logical truth and 2) one and only one of its disjuncts is true. $\dashv p$ is the S-complement of p because 1) every valuation satisfies $p \vee \dashv p$, but 2) on no valuation are both disjuncts satisfied;

in every case, one is satisfied, the other is false. As already mentioned, $\dashv p$ and $*p$ are duals. We should not forget however that accepting the truth operator comes at the price of admitting paradoxical forms of inference such as:

$$Tp, \neg Tp \vdash q$$
$$p \vdash Tq \vee \neg Tq.$$

With **EE** as we have it now, we do not have either complete paracon-sistency or complete relevance. There is however a way to attain complete paraconsistency and relevance even in a system that admits bivalent propo-sitions. It would take us too much out of our way to take up this issue in detail here, but to put it briefly, what is needed is an intensional constraint on entailment. If p is to entail q intensionally, we must require the value of q to depend upon (or be a function of) the value of p ; more specifically, if p is to entail q, it must be a necessary condition, that every assignment of values to the propositional variables of p sufficient to fix the value of p also fixes the value of q. By adding this to the value conditions for entailment, complete paraconsistency and complete relevance can be realized even in classical logic.

The following forms of inference hold for the truth operator.

T1 $Tp \vdash p$

T2 If $p \vdash_c$ q, then $Tp \vdash Tq$

T3 $Tp \vdash TTp$

T4 $\neg T \neg Tp \vdash Tp$

T5 $T(p \vee q) \vdash Tp \vee Tq$

While, $p \vdash Tp$ holds classically, it does not hold in **EE**; to falsify it, let $v(p) = \mathrm{O}$. T2 enables us to import into **EE** counterparts of every classical theorem and entailment. But if we are to have complete paraconsistency and relevance in **EE**, the intensional constraint on entailment must apply not only to $p \vdash q$, but also to $p \vdash_c q$, so that, for instance, $p \& \neg p \vdash_c q$, would not hold. T3 really is just an instance of T2; $p \vdash_c Tp$ is classically valid, so by T2, T3 follows. T4 is double Boolean negation. T4 could also be expressed, $STp \vdash Tp$, that is, if Tp is satisfied (is not false) then Tp and this plainly implies that the truth operator is bivalent. Because the truth operator is bivalent, if every propositional variable in the premises of an argument occurs in the scope of an occurrence of T (or any of the operators defined using it), the logic of the argument is entirely classical. The converses of T3 and T4 hold and are easily proven. The structure of the axioms for the truth operator should look familiar. The axioms, T1-T4, are isomorphic to a complete set of axioms for the modal system, S5. The modal counterpart to T5 does not hold classically, but does hold in a logic in which one of p and $\neg p$ must always be true; that is, it holds in a logic in which $p \vee \neg p$ is not logically true, such as Heyting's intuitionistic logic or **EE**. The fact that T5 does not hold classically, whereas the counterparts of all the other T-axioms do hold, shows that T5 is independent of these other axioms. The converse of T5 holds and has a straightforward proof.

8 Conditionals

As we have already noted, the conditional, $p \supset q$, on the strong Kleene tables, has exactly the same value conditions as $\neg p \vee q$. It is a three valued version of the material conditional. If we were to adopt the value conditions of this conditional in **EE**, then given full entailment, from $\vdash p \supset q$, we could not infer $p \vdash q$. By letting $v(p) = $ O and $v(q) =$ F we falsify $p \vdash q$, but not $\vdash p \supset q$ and non-falsifiability suffices to preserve theoremhood. If we use the Kleene strong tables for the conditional, then, the deduction theorem fails. Moreover, while EFQ, in the form, $p\&\neg p \vdash q$, can be falsified, it cannot be falsified when expressed in this form: $(p\&\neg p) \supset q$. The Kleene conditional plainly will not do. Although the deduction theorem in its classical form, $\Gamma, p \vdash q$ iff $\Gamma \vdash p \rightarrow q$, does not hold in **EE** either, we can adapt the Kleene conditional so as to reflect full entailment and, where p $\rightarrow$ q represents the adjusted conditional, the deduction theorem holds in a restricted form: $p \vdash q$ iff $\vdash p \rightarrow q$. This corresponds to the entailment theorem for system **E** ([AB] (277ff). If the deduction theorem is restricted in this manner, then EFQ is falsifiable, not only when it has the form of an inference, but also when it has the form of a proposition. The modified version of the Kleene conditional has the following value conditions:

$$v(p \rightarrow q) \quad = \quad \text{T, if } v(p) = \text{F or } v(q) = \text{T}$$
$$\text{O, if } v(p) = v(q) = \text{O}$$
$$\text{F, otherwise (i.e., if } v(p) > v(q))$$

This differs from the Kleene conditional in that $v(p \rightarrow q) = $ F when $v(p) =$ T and $v(q) =$ O, and also when $v(p) =$ O and $v(q) =$ F, cases in which the Kleene conditional has the value O.

$$p \supset q \qquad\qquad\qquad p \to q$$

$p\backslash q$	TOF
T	TOF
O	OOF
F	FFF

$p\backslash q$	TOF
T	TTT
O	TOO
F	TOF

The Kleene strong conditional and the conditional of $\mathbf{RM_3}$.

To reduce the need for parentheses, we adopt the convention that assigns the conditional greater scope than conjunction or disjunction. The revised conditional corresponds exactly to the conditional yielded by the Sugihara matrices in the three valued case, which have the same matrices for negation, conjunction and disjunction as the Kleene strong tables. This is also the conditional of $\mathbf{RM_3}$ in which the axioms of $\mathbf{R}$ plus mingle,

$$p \to (p \to p)$$

hold. In fact it is easily confirmed that given 1) the $\mathbf{RM_3}$ conditional, 2) negation, conjunction and disjunction, with the value conditions of the Kleene strong tables, and 3) full entailment: in the resulting system the axioms of $\mathbf{RM_3}$ can all be verified in the sense that none are falsifiable. But this system is not $\mathbf{RM_3}$. It is a restricted version of $\mathbf{RM_3}$ that we may call $\mathbf{RRM_3}$. $\mathbf{RRM_3}$ differs from $\mathbf{RM_3}$ in at least three respects: 1) the consequence relation of $\mathbf{RRM_3}$ is full entailment not classical entailment; 2) $\mathbf{RM_3}$ uses the deduction theorem in its full classical strength, but the deduction theorem in $\mathbf{RRM_3}$ is restricted; and 3) there is no truth operator in $\mathbf{RM_3}$. Meyer's proofs of completeness, consistency and decidability for $\mathbf{RM_3}$ carry over with little change into $\mathbf{RRM_3}$ ([AB],400-15). The restriction on the deduction theorem in $\mathbf{RRM_3}$ has no effect, since Meyer's proofs do not use the deduction theorem. The classical deduction theorem, $\Gamma, p \vdash q$ iff $\Gamma \vdash p \to q$ is falsified in $\mathbf{RRM_3}$ given that all of the propositions in Γ are true and $v(p) = v(q) = O$. This allows the unrestricted deduction theorem to be falsified even if the propositions in Γ are logical truths. The classical deduction theorem can be falsified so long as the value of the propositions in Γ are independent of the value of p and therefore so long as these propositions do not share some propositional variable with p. That is, weak relevance is a necessary condition for the deduction theorem to hold in its classical form. Otherwise, the deduction theorem holds only in the form: $p \vdash q$ iff $\vdash p \to q$.

Although $\mathbf{RM}$ certainly has formally desirable features, it also has problems serious enough to make it unacceptable to many in the relevant logic community ([AB], 429). Notably, the following two propositions are provable in **RM**:

$$\neg(p \rightarrow p) \rightarrow (q \rightarrow q)$$
$$(p \rightarrow q) \vee (q \rightarrow p).$$

The first of these formulae is hopelessly irrelevant and the second is one of the paradoxes of material implication that C. I. Lewis was concerned to avoid. The same two formulae unfortunately are provable in **RRM₃**. But the first formula holds in **RRM₃** only because it has too few values. Add an additional value between T and F to **RRM₃**, then this formula is falsifiable, if p has one of these values and q the other. Consider, for instance, a system that is like **RRM₃** except that it has two values, B and N, between T and F. Assume that B and N are not ordered amongst themselves. The order of values is then T > B, N > F. Suppose also that propositions with value B and N are not comparable in that they do not entail one another.

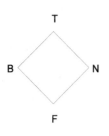

This is the value structure of the familiar diamond lattice used, for instance, in Belnap's "useful four valued logic." [ABD], 516) In this scenario, the conditional would have the following value conditions:

$$v(p \rightarrow q) = \quad \begin{array}{l} \text{T, if } v(p) = F \text{ or } v(q) = T \\ \text{B, if } v(p) = v(q) = B \\ \text{N, if } v(p) = v(q) = N \\ \text{F, otherwise.} \end{array}$$

In this case, $\neg(p \rightarrow p) \rightarrow (q \rightarrow q)$ is falsified by letting $v(p) = B$ and $v(q) = N$. Meyer's proof of this formula ([AB], 429) uses the unrestricted deduction theorem that does not hold in **RRM**.

That takes care of one of the two offending formulae. But the other, $(p \rightarrow q) \vee (q \rightarrow p)$, is not so easily dealt with. It cannot be falsified in **RRM** however many values we may have. The problem is with the **RRM** conditional. Although different from the Kleene conditional, it is still a material conditional and with such a conditional, we cannot prevent formulae of the form $(p \rightarrow q) \vee (q \rightarrow p)$ from being validated. C. I. Lewis's solution to this kind of problem was to replace the material conditional by strict implication. The corresponding manoeuvre in the present case is to introduce a strict version of the **RRM** conditional. There is no need to introduce a necessity operator as a new primitive. Instead, let $p \Rightarrow q$ be a conditional such that $p \Rightarrow q = _{df} \vdash (p \rightarrow q)$. Now in the spirit of C. I. Lewis, $\Box p$ can be identified with $\vdash p$ (see [8], 244f). Value conditions for this conditional are:

$$v(p \Rightarrow q) \quad = \quad \text{T, if for all } v, v(p) = \text{F or } v(q) = \text{T}$$
$$\text{O, if for all } v, v(p) \leq v(q) \text{ and for some } v, v(p) = v(q) = \text{O}$$
$$\text{F, otherwise (that is, if for some } v, v(p) > v(q)).$$

Then unlike $(p \rightarrow q) \lor (q \rightarrow p)$, $(p \Rightarrow q) \lor (q \Rightarrow p)$ can be falsified. $(p \Rightarrow q)$ is a generalization of $p \rightarrow q$ and is false provided it has a falsifying instance. Since there are propositions of both the forms $p \Rightarrow q$ and $q \Rightarrow p$ that have falsifying instances, the disjunction of them is false. This is an entailment conditional just as the conditional in **E**. It is this conditional rather than the **RRM** conditional that will be used in **EE**. With this conditional, it is easy to see that all the axioms of **E** hold in **EE**. Mingle, however, does not hold and the axioms of **R** do not all hold either.[5] Mingle,

$$p \Rightarrow (p \Rightarrow p)$$

fails because $p \Rightarrow p$ has instances in which it has the value O, so that $p \Rightarrow p$ can have no value higher than O, but even though $p \Rightarrow p$ has the value O, p can have the value T, falsifying mingle in **EE**. In a similar way, it can be seen that the **R** rule R3,

$$p \Rightarrow ((p \Rightarrow q) \Rightarrow q)$$

is false in **EE**. **EE** is obviously closely related to **E**, but while **E** has been demonstrated not to be $\mathbf{R}^{\Box}$ ([AB], 351-52), that is, **R**, but with the relevant conditional of **R** replaced by a necessary relevant conditional, **EE** is, by construction, $\mathbf{RRM}^{\Box}$.

Just as full entailment in a three valued logic has been analysed as the conjunction of two classical entailments, a classical entailment and its contrapositive, the conditional in $\mathbf{RRM_3}$ can be given a similar analysis. Just as we say

$$p \vdash q \text{ iff } p \vdash_c q \text{ and } \neg q \vdash_c \neg p,$$

we can also say

$$p \rightarrow q \text{ iff } (p \rightarrow_c q) \& (\neg q \rightarrow_c \neg p)$$

---

[5]The most convenient listing of the axioms of **E**, **R** and **RM** is at the beginning of [ABD], xxivff.

Examination of the conditional of $\mathbf{RRM}_3$ shows that $p \rightarrow_c q$ is equivalent to $\neg(p\&\neg Tq)$, in which case the biconditional above is equivalent to

$$p \rightarrow q \text{ iff } \neg(p\&\neg Tq)\&\neg(\neg q\&\neg T\neg p)$$

Since $\neg(p\&\neg Tq)$ is equivalent to $\neg p \vee Tq$, we also have

$$p \rightarrow q \text{ iff } (\neg p \vee Tq)\&(T\neg p \vee q)$$

The result is that the $\mathbf{RRM}_3$ conditional proves to be nothing more than a variant of the classical material conditional. This is not surprising since it is a modification of the Kleene conditional and that is the classical material conditional extended to logics with more than two values. Of course, $p \rightarrow q$ can be defined *directly* in terms of negation, conjunction and the truth operator; there is no need for intermediary formulae of the form, $p \rightarrow_c q$, at all.

In addition, $\neg p \vee Tq$ and $T\neg p \vee q$ are true under exactly the same conditions; that is, $\neg p \vee Tq \vdash_c T\neg p \vee q$ and conversely. The proof uses the fact that $p \vdash_c Tp$, although of course $p \vdash Tp$ does not hold. By T1, $Tp \vdash p$, we have $\neg p \vee Tq \vdash_c \neg p \vee q$ from which we then infer, by $p \vdash_c Tp$, $T\neg p \vee q$. The same reasoning shows that $T\neg p \vee q \vdash_c \neg p \vee Tq$. But while $p \rightarrow_c q$, $T\neg p \vee q$ and $\neg p \vee Tq$ are true under the same conditions, one can easily satisfy onself that they are not false under the same conditions; their negations are not equivalent. $\neg(p \rightarrow_c q)$ is equivalent to $\neg(T\neg p \vee q)$, but neither is equivelent to $\neg p \vee Tq$, which has the value O when $v(p) = \mathrm{O}$ and $v(q) = \mathrm{F}$, whereas the other two have the value T. Thus we have

$$p \rightarrow q \vdash_c \neg p \vee Tq \vdash_c T\neg p \vee q, \text{ and conversely.}$$

$$\neg(p \rightarrow q) \vdash_c \neg(T\neg p \vee q), \text{ and conversely.}$$

This being the case, one may represent $p \rightarrow q$ both in positive and negative classical sequents by $T\neg p \vee q$ or its De Morgan equivalent, $\neg(\neg T\neg p\&\neg q)$ and that is what we will do. It is now possible to give proofs of sequents that otherwise might have to be taken as primitive, sequents such as:

$$
\begin{array}{ll}
p \rightarrow q, q \rightarrow r \vdash p \rightarrow r & \text{(transitivity)} \\
p \rightarrow (q \rightarrow r) \vdash q \rightarrow (p \rightarrow r) & \text{(permutation)} \\
p \rightarrow (q \rightarrow r), p \rightarrow q \vdash p \rightarrow r & \text{(monotonicity)} \\
p \rightarrow (p \rightarrow q) \vdash p \rightarrow q & \text{(absorption or contraction)}
\end{array}
$$

Theses corresponding to three of these are taken as primitive in the axiomatization of the pure calculus of entailment given by Anderson and Belnap ([AB], 20). The proofs, lengthy but not difficult, are left to the interested reader. The reduction of the conditional given above means that

RRM and **EE** need only one logical constant in addition to those available in classical logic, the truth operator. **RRM** is at bottom just a normal logic (or, more accurately, a class of normal logics) with the truth operator; **EE** is its strict extension.

9 Rules of Inference for EE

When **EE** is set up as a set of Gentzen style natural deduction introduction and elimination rules, what is striking is how little it differs from the set of rules Gentzen gives for intuitionistic and classical logic. In retrospect, this similarity between the rules for **EE** and those for classical logic should not be surprising, since classical logic, like **EE**, is a normal logic. The rules for conjunction, disjunction and the conditional in **EE** are very much like their classical counterparts. There is no negation elimination rule, since that is just EFQ, acceptable in no paraconsistent logic. Except for the introduction rules for negation and the conditional, all introduction and elimination rules may have parametric constituents. Structural rules are given principally to make it plain that **EE** is not substructural. All structural rules hold, but those for weakening and cut are formulated so as to ensure that they do not have sequents with empty antecedents as premises. A double negation rule not in Gentzen's original set is added. The rules for the truth operator are of course new. For completeness, T5 is needed, though it does not introduce or eliminate a logical constant but is rather a rule to distribute the truth operator over disjunction. For convenience, we should have a collection of derived equivalences, as we commonly have in classical logic. In **EE**, these would include the following equivalent pairs: Tp and TTp, Tp and $\neg T \neg Tp$ (double Boolean negation), $\neg Tp$ and $T \neg Tp$, $T(p \vee q)$ and $Tp \vee Tq$. Using the first three of these pairs, it is easy to show that there are only four distinct truth modalities, Tp (corresponding to : p is necessary), $T \neg p$ (p is impossible), $\neg T \neg p$ (p is possible) and $\neg Tp$ (p is not necessary, or possibly not p). This corresponds to the situation in Lewis' S5. We also need, of course, the equivalences between $p \Rightarrow q$ and $\vdash p \rightarrow q$, and between $p \rightarrow q$ and $\neg(p \& \neg Tq) \& \neg(\neg T \neg p \& \neg q)$.

Natural Deduction in EE.

Structural Rules.

$$\frac{p \vdash q}{\Gamma, p \vdash q} \qquad \text{Weakening}$$

$$\frac{\Gamma, p, q, \triangle \vdash r}{\Gamma, q, p, \triangle \vdash r} \qquad \text{Permutation}$$

$$\frac{\Gamma, p, p, \triangle \vdash q}{\Gamma, p, \triangle \vdash q} \qquad \text{Contraction}$$

$$\frac{\begin{array}{c}\Gamma, p \vdash q \\ \triangle, q \vdash r\end{array}}{\Gamma, \triangle, p \vdash r} \qquad \text{Cut}$$

Introduction and Elimination Rules.

$$\frac{p,\ q}{p\&q}\quad \&\ \mathrm{I} \qquad\qquad \frac{p\&q}{p}\qquad \frac{p\&q}{q}\quad \&\ \mathrm{E}$$

$$\frac{p}{p\vee q}\qquad \frac{q}{p\vee q}\quad \vee\ \mathrm{I} \qquad\qquad \begin{array}{c} p\vdash r \\ q\vdash r \\ \hline p\vee q\vdash r \end{array}\quad \vee\ \mathrm{E}$$

$$\frac{p\vdash q}{\vdash p\Rightarrow q}\quad \Rightarrow\!\mathrm{I} \qquad\qquad \frac{p,\ p\Rightarrow q}{q}\quad \Rightarrow\ \mathrm{E}$$

$$\frac{p\vdash_c q}{Tp\vdash Tq}\quad \mathrm{T\ I} \qquad\qquad \frac{Tp}{p}\quad \mathrm{T\ E}$$

$$\frac{\neg\neg p}{p}\quad \neg\neg\ \mathrm{E}$$

$$\begin{array}{c} p\vdash r \\ q\vdash \neg r \\ \hline p\vdash \neg q \end{array}\quad \neg\ \mathrm{I} \qquad\qquad \frac{\neg T\neg Tp}{Tp}\quad \mathrm{DBE}$$

Apart from the difference in their conditionals, **RRM** and **EE** are the same. Both at bottom are just normal logics with the truth operator added. They are not so much individual systems as classes of systems. With respect to the propositional calculus of the three valued case, which is all that I have investigated in this paper, both **RRM** and **EE** are complete. The proofs of completeness are lengthy and are not given here. The tables for the logical constants of $\mathbf{RRM}_3$ enable us to determine for a given formula, whether or not it is a tautology and so, given the completeness of $\mathbf{RRM}_3$, whether or not it is provable. That makes $\mathbf{RRM}_3$ decidable (see [ABD], 365). Given $\vdash p \to q$ in $\mathbf{RRM}_3$, we can infer $p \vdash q$ by the deduction theorem, so the solution to the decision problem for $\mathbf{RRM}_3$ is also a solution to the deducibility problem. These results are easily extended to **EE**.

10   Dialetheic Properties

Because **EE** is paraconsistent, it has no problem with inconsistent theories, inconsistent descriptions and the like. But that does not mean that it allows for inconsistent objects or states of affairs. For that, one might think, we need a real dialetheic logic. The surprising thing is that for the most part we can get the same results in **EE** that we get from the dialetheic logic **IC** (the logic of *In Contradiction*). Central to Priest's conception of truth are two principles, the principle of exhaustion, that holds if there are truth value gluts, and the principle of exclusion, that holds if there are truth value gaps.([**IC**], 88-89) Priest argues for the first and against the second. By comparison with the truth operator in **EE**, we may call truth in **IC** 'weak truth' and represent it by 'T^-'. Then the exhaustion principle and the exclusion principle are,

$$\neg T^- p \Rightarrow T^- \neg p \text{ (exhaustion principle)}$$

$$T^- \neg p \Rightarrow \neg T^- p \text{ (exclusion principle)}$$

Since $F^- p \Leftrightarrow T^- \neg p$ also holds, the exhaustion principle is equivalent to $\neg T^- p \Rightarrow F^- p$, or $\neg F^- p \Rightarrow T^- p$. Using the semantics of **EE**, if the exhaustion principle is to be satisfied, but the exclusion principle is to fail, it would appear that weak truth would have to have the following value conditions: $v(T^- p) =$F, if $v(p) =$F; otherwise, $v(T^- p) =$T. It then would have the same value conditions as $\neg T \neg p$, but that is just Sp. So $T^- p$ can be introduced by definition into **EE**, defined as $\neg T \neg p$ (or Sp). With these conditions, the exhaustion principle holds in **EE**, and the exclusion principle is falsified by letting $v(p) =$O, just as Priest maintains. Priest's claim that there are two notions antithetical to truth, falsity and untruth, may be represented in **EE**: the falsity of p is represented by $T^- \neg p$, equivalent to $\neg T p$, the weak, or Boolean negation of p; the untruth of p, $\neg T^- p$, equivalent to $T \neg p$, the strong, or intuitionistic negation of p. To verify Priest's most striking claim, the claim that there are true contradictions, that is, true instances of $T^- p \& F^- p$, we need only let $v(p) =$O.

There is one claim in IC however that definitely is not verified in EE, the claim ([IC], 90-91) that there are some propositions for which the exclusion principle both holds and does not hold. Priest gets this result by using the truth schema, 'p' is true iff p, where p is $\neg T^- p$, the untruth of p. Given $T^- p \Leftrightarrow p$ and $p \Leftrightarrow \neg T^- p$, we get

$$T^- p \Leftrightarrow \neg T^- p.$$

So, as Priest expresses it, p can be both true and untrue. That is,

$$T^- p \ \& \neg T^- p, \text{ for some p}$$

The exclusion principle then does not hold for propositions that are true iff untrue. But given the exhaustion principle, the exclusion principle can be proven to hold for these propositions.

Proof. Let p be a propositions such that $T^- p \Leftrightarrow \neg T^- p$. Then

1. $\neg T^- p \Rightarrow T^- \neg p$ exhaustion principle
2. $\neg T^- \neg p \Rightarrow T^- p$ 1, contraposition
3. $\neg\neg T^- \neg p \Rightarrow \neg T^- p$ 2, applying $T^- p \Leftrightarrow \neg T^- p$
4. $T^- \neg p \Rightarrow \neg T^- p$ exclusion principle

There are therefore propositions for which the exclusion principle both holds and does not hold. But if the truth schema should rather be represented by $Tr'p'$ iff p, where $Tr'p'$ and p always correspond in value, as I have earlier argued and as Priest himself holds in his early paper, "The Logic of Paradox" ([LP], 238), then applying this to untruth, that is, $\neg T^- p$, yields

$$Tr'(p)' \Leftrightarrow \neg T^- p.$$

Using this formula, the exclusion principle cannot be proven. The inconsistency in **IC** arises because Priest does not distinguish between weak truth and Tarski's semantic concept. For Tarski's concept, the exclusion principle holds; for weak truth, it does not. Priest combines the two concepts of truth and is prepared to accept the consequent formal inconsistency. That far I am not prepared to go.

It is worth exploring further the relation between weak and strong truth. If we replace ' T^-' by its definition, we get

$$T^- p \& F^- p \Leftrightarrow \neg Tp \& \neg Fp.$$

That is, there are true contradictions and so truth value gluts relative to ' T^-' iff there are truth value gaps relative to ' T.' Proof: $T^- p \Leftrightarrow \neg T^- \neg p \Leftrightarrow \neg Fp, F^- p \Leftrightarrow T^- \neg p \Leftrightarrow \neg T^- \neg\neg p \Leftrightarrow \neg Tp$, so $T^- p \& F^- p \Leftrightarrow \neg Tp \& \neg Fp$. Similarly, since $\neg T^- p \Leftrightarrow T^- \neg p$ and $T^- \neg p \Leftrightarrow \neg Tp$, then $\neg T^- p \Rightarrow T^- \neg p \Leftrightarrow T^- \neg p \Rightarrow \neg Tp$, that is, the T^--exhaustion principle is logically equivalent to the T-exclusion principle and likewise the T^--exclusion principle is logically equivalent to the T-exhaustion principle. Dialetheic principles can be embedded in **EE** but these are logically equivalent to non-dialetheic principles. These correspondences however do not suffice for embedding the whole dialetheic content of **IC** into **EE**. Dialetheia based on the semantic paradoxes are not covered.[6]

There is a great deal of similarity between **EE** and **IC**, but there are also significant differences. In **IC** the truth schema holds, the conditional is bivalent, tautologies are not distinguished from logical truths and there are notational differences. Nonetheless, just as we can embed weak truth in **EE**, we can embed strong truth in **IC**, since $Tp \Leftrightarrow \neg T^- \neg p$ holds. Tp and $T^- p$

---

[6]All that appears to be lacking for this purpose is a truth predicate that can take more than two values, which as we have seen can be added to **EE**, giving it the resources to represent the semantic paradoxes and Priest's treatment of them.

are duals and, just as in **EE**, the exhaustion principle for T^- is the dual of the exclusion principle for T, the exclusion principle for T^- is the dual of the exhaustion principle for T and $T^-p\&F^-p$ is the dual of $\neg Tp\&\neg Fp$, the dialetheic principles holding exactly when their non-dialetheic duals hold, the same is true in **IC**.[7] This leaves one to wonder if there is enough difference between the two systems to explain why **IC** should be a strong paraconsistent logic, given that **EE** is definitely weak.

11   Conclusion

RRM and **EE** are normal logics with a truth operator. Such logics have at their base, a generalized notion of entailment and a generalization of the Kleene strong tables that are themselves three valued extensions of the truth conditions for the classical logical connectives. A normal logic with four or more values but without the truth operator is not subject to EFQ; it is paraconsistent and relevant. Problems arise when we add the truth operator, as we do in **RRM** and **EE**, or admit bivalent propositions in any guise at all. With the truth operator, the whole of classical logic can be embedded as a subsystem of **EE** and within this bivalent subsystem, the paradoxes of C. I. Lewis reappear. The solution I suggest is not to reject the truth operator but to add an intensional condition to entailment. The truth operator has many uses in **EE**. With the truth operator or some equivalent, the **RRM** and **EE** conditionals do not have to be taken as primitive; we can represent the principle of bivalence, several useful notion of negation, can embed in **EE** many of the major principles of Priest's dialetheic logic. The replacement of classical entailment by full entailment, and the use of the truth operator are the key enhancements of both **RRM** and **EE** with respect to **RM** and **E**. When **RRM** and **EE** are compared to **E**, **R**, **RM** and Belnap's 'four', none of these latter systems have a truth operator, none is comparable in expressive power and the conditional in each has to be taken as primitive. Only 'four' has full entailment, but in this system neither conjunction nor disjunction is normal. Of the paradoxes of **RM**, one is falsifiable in any version of **RRM** with more than three values, both are falsifiable in **EE** assuming that it has more than three values. If the tables on p. 161-62 of [AB] are correct, then the conditional in first degree formulae of **E** and therefore in **E** as a whole does not entail the material conditional. Given that **E** entails **R**, the same is true of **R**. For those who adhere to the original conception of relevant logic, as a logic to replace classical logic, a logic like **EE** should be seen as providing a promising new

---

[7]When **EE** is extended to four values, T^-p is no longer logically equivalent to $\neg T\neg p$, Tp and T^-p are not duals and the parallels between weak and strong truth break down. The reasons we had these parallels in the first place is that **EE** had too few values.

direction to explore.

BIBLIOGRAPHY

[AB] A. R. Anderson and N. D. Belnap, *Entailment*, volume I, Princeton University Press, Princeton, 1975.

[ABD] A. R. Anderson, N. D. Belnap and J. M. Dunn, *Entailment*, volume II,Princeton University Press, Princeton, 1992.

[IM] Stephen C. Kleene, *Introduction to Metamathematics*, Van Nostrand, Princeton, N.J., 1952.

[IC] Graham Priest, *In Contradiction: Essays on the Trans-consistent*, Martinus Nijhoff, Dordrecht, 1987

[LP] Graham Priest, "The Logic of Paradox", *Journal of Philosophical Logic*, T8, (1979), pp. 219-241

[RLR] Richard Routley, R. K. Meyer, V. Plumwood and R. T. Brady, *Relevant Logics and their Rivals*, I, Ridgeview, Atascadero, Ca., 1982.

[1] E. Bencivenga, K. Lambert and B. C. Van Fraassen, *Logic, Bivalence and Denotation*, Ridgeview, Atascadero, Ca., 1991

[2] Stephen Blamey, "Partial Logic," in D. Gabbay and F. Guenthner, eds., *Handbook of Philosophical Logic*, vol. III, Reidel, Dordrecht, Holland, 1986, pp. 1-70

[3] D. A. Bochvar, "On a Three Valued Logical Calculus," Merrie Bergmann, trans., *History and Philosophy of Logic*, 2, 1981, pp. 87-112

[4] Newton Da Costa, "On the Theory of Inconsistent Formal Systems," *Notre Dame Journal of Formal Logic*, XV, 1974, pp. 497-510

[5] Bruno De Finetti, "The Logic of Probability," R. B. Angell, trans., *Philosophical Studies*, 77, pp. 181-190

[6] Arend Heyting, "The Formal Rules of Intuitionistic Logic," in Paolo Mancosu, ed., *From Brouwer to Hilbert*, Oxford University Press, Oxford, 1998, pp. 311-327

[7] Stanislaw Jaskowski, "Propositional Calculus For Contradictory Deductive Systems," *Studia Logica*, XXIV, 1969, pp. 143-157

[8] C. I. Lewis and C. H. Langford, *Symbolic Logic*, 2nd ed., Dover, New York, 1959

[9] Fred S. Michael, "Entailment and Bivalence," *Journal of Philosophical Logic*, 31 (2002), 289-300

[10] *Paraconsistent Logic: Essays on the Inconsistent* (Graham Priest, Richard Routley, Jean Norman, editors), Philosophia Verlag, Munich, 1989

[11] Dana Scott, "Combinators and Classes," in C. Bohm, ed., *λ-Calculus and Computer Science*, Springer Verlag, Heidelberg, 1975, pp. 1-26

[12] Timothy Smiley, "Sense Without Denotation," *Analysis*, (1960), 20 pp. 125-135

Fred Seymour Michael

Department of Philosophy, Brooklyn College, CUNY, USA.

fmichael@brooklyn.cuny.edu

Part III

Concepts and Tools for Paraconsistent Logics

Non-deterministic Semantics for Families of Paraconsistent Logics

ARNON AVRON

ABSTRACT. We investigate four large families of paraconsistent logics. Two are obtained from the positive fragment of classical logic (with or without a bottom element) by adding to it the classical $(\Rightarrow \neg)$ rule, as well as various standard Gentzen-type rules for combinations of negation with other connectives. Two others are similarly obtained from the positive fragment of intuitionistic logic (with or without a bottom element). We provide for all the systems simple semantics which is based on non-deterministic three-valued structures, and prove soundness and completeness for all of them. We show how the semantics can be used for proving interesting proof-theoretical properties of some of these systems. We also determine what version of the cut-elimination theorem obtains in each case. Among the 6144 logics included in these families there are famous paraconsistent logics like $CluN, CLuNs, C_{min}, C_\omega, PI^*$ and J_3.

1 Introduction

In paraconsistent logics we want to reject inferences of the form:

$$\neg p, p \vdash q$$

where p and q are distinct atomic formulas. Intuitively this means that we would like to allow situations in which somehow both p and $\neg p$ are considered to be true, while q is false. This naturally leads to two different ways in which a proposition φ may be true. φ may be *consistently true* (or absolutely true), and φ may be *inconsistently true* (or contradictory, or paradoxical). In the first case φ is true and its negation is false, while in the second case both φ and its negation are true. Since on the other hand we would like to retain at least the validity of the law of excluded middle *LEM* (in order for our negation to at least partially deserve this name), this intuition may formally be reflected by the use of *three* truth values: t (for "consistently true"), $\top$ (for "inconsistently true"), and f (for "not true"). Given some meta-notion of truth and falsity, we accordingly expect a valuation v in $\{t, f, \top\}$ to satisfy:

- $v(\varphi) = t$ if φ is true and $\neg\varphi$ is false
- $v(\varphi) = \top$ if φ is true and $\neg\varphi$ is true
- $v(\varphi) = f$ if φ is false (and so $\neg\varphi$ is true, by LEM)

Given the truth value of φ, what do these principles tell us about the truth-value of its negation? Well, it is easy to see that they dictate the following derived principles (and nothing stronger):

- If $v(\varphi) = t$ then $v(\neg\varphi) = f$
- If $v(\varphi) = \top$ then $v(\neg\varphi) \in \{t, \top\}$
- If $v(\varphi) = f$ then $v(\neg\varphi) \in \{t, \top\}$

It follows that the truth-value of φ does not fully determine the truth-value of $\neg\varphi$. Hence *nondeterministic* semantics seems to be appropriate here.

In this paper we explore the application of these ideas for large families of paraconsistent logics. We concentrate on logics which are easily and naturally defined by using Gentzen-type systems with various standard, very common, rules for negation. The differences between the 6144 different systems we investigate are with respect to:

The underlying logic : We consider four possibilities: positive classical logic, positive classical logic with a bottom element (the falsehood constant **ff**), positive intuitionistic logic, and positive intuitionistic logic with a bottom element (again, the falsehood constant **ff**).

The rules for negation : We include the $(\Rightarrow \neg)$ rule (which corresponds to LEM) in *all* the systems we consider (including the intuitionistic ones). In addition, our systems may have various rules for combining negation with other connectives (taken from a list given below).

We provide simple non-deterministic semantics for all the systems we consider, and prove for all of them soundness and completeness with respect to this semantics. We use this semantics for showing various properties of the systems, including their being conservative over their underlying logics, and an appropriate version of the cut-elimination theorem in each case.

2   Preliminaries

In what follows p, q, r denote atomic formulas, $A, B, C, \psi, \varphi, \phi$ denote arbitrary formulas, and Γ, Δ denote finite sets of formulas. A sequent has the form $\Gamma \Rightarrow \Delta$. Following tradition, we write Γ, φ and Γ, Δ for $\Gamma \cup \{\varphi\}$ and $\Gamma \cup \Delta$ (respectively). By a *logic* we shall mean a pair $\langle \mathcal{L}, \vdash \rangle$, in which $\mathcal{L}$ is a propositional language, and $\vdash$ is a consequence relation on $\mathcal{L}$.

2.1   The Standard Positive Logics

THE SYSTEM LK^+

Axioms: $A \Rightarrow A$

Structural Rules: Cut, Weakening

Logical Rules:

$$(\supset\Rightarrow) \quad \frac{\Gamma \Rightarrow \Delta, A \quad B, \Gamma \Rightarrow \Delta}{A \supset B, \Gamma \Rightarrow \Delta} \qquad\qquad \frac{\Gamma, A \Rightarrow B, \Delta}{\Gamma \Rightarrow A \supset B, \Delta} \quad (\Rightarrow\supset)$$

$$(\wedge\Rightarrow) \quad \frac{\Gamma, A, B \Rightarrow \Delta}{\Gamma, A \wedge B \Rightarrow \Delta} \qquad\qquad \frac{\Gamma \Rightarrow \Delta, A \quad \Gamma \Rightarrow \Delta, B}{\Gamma \Rightarrow \Delta, A \wedge B} \quad (\Rightarrow\wedge)$$

$$(\vee\Rightarrow) \quad \frac{\Gamma, A \Rightarrow \Delta \quad \Gamma, B \Rightarrow \Delta}{\Gamma, A \vee B \Rightarrow \Delta} \qquad\qquad \frac{\Gamma \Rightarrow \Delta, A, B}{\Gamma \Rightarrow \Delta, A \vee B} \quad (\Rightarrow\vee)$$

THE SYSTEM LK: This is the system obtained from LK^+ by adding to it the following axiom:

$$\mathbf{ff} \Rightarrow$$

THE SYSTEMS LJ^+ and LJ: These are the systems which are obtained from LK^+ and LK (respectively) by weakening the $(\Rightarrow\supset)$ rule to:

$$\frac{\Gamma, A \Rightarrow B}{\Gamma \Rightarrow A \supset B}$$

Notes:

1. LK is a standard Gentzen-type calculus for propositional classical logic, while LK^+ is its purely positive fragment. The system LJ is the propositional fragment of a well-known (see [Takeuti, 1975]) multiple-conclusion version of Gentzen's original sequent calculus for intuitionistic logic, while LJ^+ is its purely positive fragment. The systems are sound and complete for these logics and admit cut-elimination.

2. In both LK and LJ it is possible to define the usual negation connective of the corresponding logics by letting $\sim\varphi =_{Df} \varphi \supset \mathbf{ff}$ (for intuitionistic logic this is in fact the common procedure). We shall nevertheless take all four systems as "positive" logics, since our principal goal is to investigate the systems which are obtained from them by adding to their language an independent, paraconsistent negation connective $\neg$. In the case of LK and LJ this would mean that we are extending the propositional classical and intuitionistic logics with an extra connective, not definable in their language.

2.2   Standard Rules for Negation and Corresponding Systems

The two standard Gentzen-type rules for classical negation are:

$$(\neg \Rightarrow) \quad \frac{\Gamma \Rightarrow \Delta, A}{\neg A, \Gamma \Rightarrow \Delta} \qquad\qquad \frac{A, \Gamma \Rightarrow \Delta}{\Gamma \Rightarrow \Delta, \neg A} \quad (\Rightarrow \neg)$$

Now the rule $(\neg \Rightarrow)$ forces any contradiction to entail every formula. This rule should therefore be rejected in the framework of paraconsistent logics. In order for a connective $\neg$ to still be called "negation", paraconsistent logics usually retain the other rule (corresponding to LEM, the law of excluded middle). This choice leads to the following four basic paraconsistent systems:

DEFINITION 1. The systems PLK, $PLK^{\mathbf{ff}}$, PLJ, and $PLJ^{\mathbf{ff}}$ are obtained from LK^+, LK, LJ^+, and LJ (respectively) by enriching their language with the unary connective $\neg$, and adding $(\Rightarrow \neg)$ to their sets of rules.

Instead of $(\neg \Rightarrow)$ (and sometimes instead of both classical rules of negation) many paraconsistent logics and relevance logics employ rules for introducing combinations of negation with other connectives. The most common rules used for this task are the following:

$$(\neg\neg \Rightarrow) \quad \frac{A, \Gamma \Rightarrow \Delta}{\neg\neg A, \Gamma \Rightarrow \Delta} \qquad\qquad \frac{\Gamma \Rightarrow \Delta, A}{\Gamma \Rightarrow \Delta, \neg\neg A} \quad (\Rightarrow \neg\neg)$$

$$(\neg \supset \Rightarrow) \quad \frac{A, \neg B, \Gamma \Rightarrow \Delta}{\neg(A \supset B), \Gamma \Rightarrow \Delta} \qquad \frac{\Gamma \Rightarrow \Delta, A \quad \Gamma \Rightarrow \Delta, \neg B}{\Gamma \Rightarrow \Delta, \neg(A \supset B)} \quad (\Rightarrow \neg \supset)$$

$$(\neg\vee \Rightarrow) \quad \frac{\Gamma, \neg A, \neg B \Rightarrow \Delta}{\Gamma, \neg(A \vee B) \Rightarrow \Delta} \qquad \frac{\Gamma \Rightarrow \Delta, \neg A \quad \Gamma \Rightarrow \Delta, \neg B}{\Gamma \Rightarrow \Delta, \neg(A \vee B)} \quad (\Rightarrow \neg\vee)$$

$$(\neg\wedge \Rightarrow) \quad \frac{\Gamma, \neg A \Rightarrow \Delta \quad \Gamma, \neg B \Rightarrow \Delta}{\Gamma, \neg(A \wedge B) \Rightarrow \Delta} \qquad \frac{\Gamma \Rightarrow \Delta, \neg A, \neg B}{\Gamma \Rightarrow \Delta, \neg(A \wedge B)} \quad (\Rightarrow \neg\wedge)$$

Now in the original formulation of Gentzen ([Gentzen, 1969]) the rules ($\land \Rightarrow$) and ($\Rightarrow \lor$) were split into two rules, each with only one side formula. To make our investigation finer we do the same here to ($\neg\lor \Rightarrow$), ($\Rightarrow \neg\land$) and ($\neg \supset \Rightarrow$). So instead of these three rules we will consider the following six:

$$(\neg \supset\Rightarrow)_1 \quad \frac{A, \Gamma \Rightarrow \Delta}{\neg(A \supset B), \Gamma \Rightarrow \Delta} \qquad\qquad \frac{\neg B, \Gamma \Rightarrow \Delta}{\neg(A \supset B), \Gamma \Rightarrow \Delta} \quad (\neg \supset\Rightarrow)_2$$

$$(\neg\lor \Rightarrow)_1 \quad \frac{\Gamma, \neg A \Rightarrow \Delta}{\Gamma, \neg(A \lor B) \Rightarrow \Delta} \qquad\qquad \frac{\Gamma, \neg B \Rightarrow \Delta}{\Gamma, \neg(A \lor B) \Rightarrow \Delta} \quad (\neg\lor \Rightarrow)_2$$

$$(\Rightarrow \neg\land)_1 \quad \frac{\Gamma \Rightarrow \Delta, \neg A}{\Gamma \Rightarrow \Delta, \neg(A \land B)} \qquad\qquad \frac{\Gamma \Rightarrow \Delta, \neg B}{\Gamma \Rightarrow \Delta, \neg(A \land B)} \quad (\Rightarrow \neg\land)_2$$

DEFINITION 2. Let NR be the union of the following sets of rules:

$$\{(\neg\neg \Rightarrow), (\Rightarrow \neg\neg), (\Rightarrow \neg \supset), (\Rightarrow \neg\lor), (\neg\land \Rightarrow)\}$$

$$\{(\neg \supset\Rightarrow)_1, (\neg \supset\Rightarrow)_2, (\neg\lor \Rightarrow)_1, (\neg\lor \Rightarrow)_2, (\Rightarrow \neg\land)_1, (\Rightarrow \neg\land)_2\}$$

For $\mathbf{L} \in \{PLK, PLK^{\mathbf{ff}}, PLJ, PLJ^{\mathbf{ff}}\}$ and $S \subseteq NR$ we denote by $\mathbf{L}(S)$ the system which is obtained from $\mathbf{L}$ by adding to it the rules in S.

Historical Notes: Some of the logics introduced in Definition 2 have already been studied in the literature. PLK itself and $PLK(NR)$ were introduced in [Batens, 1980], where they were called PI and PI^s (respectively). Their names have later been changed by Batens to $CLuN$ and $CLuNs$, respectively (see e.g. [Batens, 2000]). $PLK(NR)$ was independently introduced (together with the 3-valued deterministic semantics described below) in [Avron, 1986; Avron, 1991; Rozoner, 1989]. In [Avron, 1991] it was called PAC (this name is adopted in [Carnielli and Marcos, 2002]). $PLK^{\mathbf{ff}}(NR)$ was originally introduced in [Schütte, 1960]. Later it was reintroduced (together with its 3-valued deterministic semantics) in [D'Ottaviano and da Costa, 1970; D'Ottaviano, 1985], and was called there J_3 (see also [Epstein, 1990]). In [Carnielli and Marcos, 2002] it is called $LFI1$. The system $PLK(\{(\neg\neg \Rightarrow)\})$ was studied under the name C_{min} in [Carnielli and Marcos, 1999]. $PLK(\{(\Rightarrow \neg\neg), (\Rightarrow \neg\lor)\})$ was again introduced in [Batens, 1980] (under the name PI^*). $PLJ(\{(\neg\neg \Rightarrow)\})$ is Raggio's formulation (in [Raggio, 1968]) of da Costa famous logic C_ω (see [da Costa., 1974]).

2.3 Corresponding Hilbert-type Systems

Some of the logics mentioned above (like C_ω and C_{min}) have originally been introduced using Hilbert-type systems. Such systems can easily be given for every logic $\mathbf{L}(S)$. We start with some standard Hilbert-type system for $\mathbf{L}$ (having MP as the only rule of inference), and add to it the axiom $\varphi \vee \neg\varphi$, together with the axioms which correspond to the negation rules in S. Here is the list of axioms that correspond to our 11 rules (details are left for the reader):

$(\neg\neg \Rightarrow):$ $\quad \neg\neg\varphi \supset \varphi$

$(\Rightarrow \neg\neg):$ $\quad \varphi \supset \neg\neg\varphi$

$(\neg \supset \Rightarrow)_1:$ $\quad \neg(\varphi \supset \psi) \supset \varphi$

$(\neg \supset \Rightarrow)_2:$ $\quad \neg(\varphi \supset \psi) \supset \neg\psi$

$(\Rightarrow \neg \supset):$ $\quad (\varphi \wedge \neg\psi) \supset \neg(\varphi \supset \psi)$

$(\neg\vee \Rightarrow)_1:$ $\quad \neg(\varphi \vee \psi) \supset \neg\varphi$

$(\neg\vee \Rightarrow)_2:$ $\quad \neg(\varphi \vee \psi) \supset \neg\psi$

$(\Rightarrow \neg\vee):$ $\quad (\neg\varphi \wedge \neg\psi) \supset \neg(\varphi \vee \psi)$

$(\neg\wedge \Rightarrow):$ $\quad \neg(\varphi \wedge \psi) \supset (\neg\varphi \vee \neg\psi)$

$(\Rightarrow \neg\wedge)_1:$ $\quad \neg\varphi \supset \neg(\varphi \wedge \psi)$

$(\Rightarrow \neg\wedge)_2:$ $\quad \neg\psi \supset \neg(\varphi \wedge \psi)$

2.4 Nondeterministic Matrices

Our main semantical tool in what follows will be the following generalization from [Avron and Lev, 2001; Avron and Lev, 2005] of the concept of a matrix:[1]

DEFINITION 3.

1. A *non-deterministic matrix* (*Nmatrix* for short) for a propositional language $\mathcal{L}$ is a tuple $\mathcal{M} = \langle \mathcal{T}, \mathcal{D}, \mathcal{O} \rangle$, where:

   (a) $\mathcal{T}$ is a non-empty set of *truth values*.

---

[1] A special two-valued case of this definition has essentialy been introduced in [Batens *et al.*, 1999]. Another particular case of the same idea, using a similar name, has been used in [Crawford and Etherington, 1998]. It should also be noted that Carnielli's "possible-translations semantics" (see [Carnielli, 2000]) was originally called "non-deterministic semantics", but later the name has been changed to the present one. It is known that the the semantics of non-deterministic matrices used here and possible-translations semantics are not identical, but it seems obvious that there are strong connections between them. The exact relations between the two types of semantics has not been clarified yet.

(b) $\mathcal{D}$ is a non-empty proper subset of $\mathcal{T}$.

(c) For every n-ary connective $\diamond$ of $\mathcal{L}$, $\mathcal{O}$ includes a corresponding n-ary function $\tilde{\diamond}$ from $\mathcal{T}^n$ to $2^{\mathcal{T}} - \{\emptyset\}$.

We say that $\mathcal{M}$ is *(in)finite* if so is $\mathcal{T}$.

2. Let $\mathcal{F}$ be the set of formulas of $\mathcal{L}$. A *(legal) valuation* in an Nmatrix $\mathcal{M}$ is a function $v : \mathcal{F} \to \mathcal{T}$ that satisfies the following condition for every n-ary connective $\diamond$ of $\mathcal{L}$ and $\psi_1, \ldots, \psi_n \in \mathcal{F}$:

$$v(\diamond(\psi_1, \ldots, \psi_n)) \in \tilde{\diamond}(v(\psi_1), \ldots, v(\psi_n))$$

3. A valuation v in an Nmatrix $\mathcal{M}$ is a *model* of (or *satisfies*) a formula ψ in $\mathcal{M}$ (notation: $v \models^{\mathcal{M}} \psi$) if $v(\psi) \in \mathcal{D}$. v is a *model* in $\mathcal{M}$ of a set Γ of formulas (notation: $v \models^{\mathcal{M}} \Gamma$) if it satisfies every formula in Γ.

4. $\vdash_{\mathcal{M}}$, the consequence relation induced by the Nmatrix $\mathcal{M}$, is defined as follows:

$\Gamma \vdash_{\mathcal{M}} \Delta$ if for every v such that $v \models^{\mathcal{M}} \Gamma$ there exists $\varphi \in \Delta$ such that $v \models^{\mathcal{M}} \varphi$

5. A logic $\mathbf{L} = \langle \mathcal{L}, \vdash \rangle$ is *sound* for an Nmatrix $\mathcal{M}$ (where $\mathcal{L}$ is the language of $\mathcal{M}$) if $\vdash \subseteq \vdash_{\mathcal{M}}$. $\mathbf{L}$ is *complete* for $\mathcal{M}$ if $\vdash \supseteq \vdash_{\mathcal{M}}$. $\mathcal{M}$ is *characteristic* for $\mathbf{L}$ if $\mathbf{L}$ is both sound and complete for it (i.e.: if $\vdash = \vdash_{\mathcal{M}}$).

Note: We shall identify an ordinary (deterministic) matrix with an Nmatrix the functions in $\mathcal{O}$ of which always return singletons.

DEFINITION 4. Let $\mathcal{M}_1 = \langle \mathcal{T}_1, \mathcal{D}_1, \mathcal{O}_1 \rangle$ and $\mathcal{M}_2 = \langle \mathcal{T}_2, \mathcal{D}_2, \mathcal{O}_2 \rangle$ be Nmatrices for a language $\mathcal{L}$. $\mathcal{M}_2$ is called a *refinement* of $\mathcal{M}_1$ if $\mathcal{T}_1 = \mathcal{T}_2$, $\mathcal{D}_1 = \mathcal{D}_2$, and $\tilde{\diamond}_{\mathcal{M}_2}(\vec{x}) \subseteq \tilde{\diamond}_{\mathcal{M}_1}(\vec{x})$ for every n-ary connective $\diamond$ of $\mathcal{L}$ and every $\vec{x} \in \mathcal{T}_2^n$.

The following proposition can easily be proved:

PROPOSITION 5. *If $\mathcal{M}_2$ is a refinement of $\mathcal{M}_1$ then $\vdash_{\mathcal{M}_1} \subseteq \vdash_{\mathcal{M}_2}$. Hence if $\mathbf{L}$ is sound for $\mathcal{M}_1$ then $\mathbf{L}$ is also sound for $\mathcal{M}_2$.*

3   The Classical Case

3.1   The General Semantics

Classical Logic has of course the semantics of the usual two-valued deterministic matrix. This semantics can however easily be generalized as follows.

DEFINITION 6.

1. Let $\mathcal{M} = \langle \mathcal{T}, \mathcal{D}, \mathcal{O} \rangle$ be an Nmatrix for a language which includes that of LK^+. We say that $\mathcal{M}$ is *suitable* for LK^+ if the following conditions are satisfied:

   - If $a \in \mathcal{D}$ and $b \in \mathcal{D}$ then $a\tilde{\wedge}b \subseteq \mathcal{D}$
   - If $a \notin \mathcal{D}$ then $a\tilde{\wedge}b \subseteq \mathcal{T} - \mathcal{D}$
   - If $b \notin \mathcal{D}$ then $a\tilde{\wedge}b \subseteq \mathcal{T} - \mathcal{D}$

   - If $a \in \mathcal{D}$ then $a\tilde{\vee}b \subseteq \mathcal{D}$
   - If $b \in \mathcal{D}$ then $a\tilde{\vee}b \subseteq \mathcal{D}$
   - If $a \notin \mathcal{D}$ and $b \notin \mathcal{D}$ then $a\tilde{\vee}b \subseteq \mathcal{T} - \mathcal{D}$

   - If $a \notin \mathcal{D}$ then $a\tilde{\supset}b \subseteq \mathcal{D}$
   - If $b \in \mathcal{D}$ then $a\tilde{\supset}b \subseteq \mathcal{D}$
   - If $a \in \mathcal{D}$ and $b \notin \mathcal{D}$ then $a\tilde{\supset}b \subseteq \mathcal{T} - \mathcal{D}$

2. Let $\mathcal{M} = \langle \mathcal{T}, \mathcal{D}, \mathcal{O} \rangle$ be an Nmatrix for a language which includes the language of LK. We say that $\mathcal{M}$ is *suitable* for LK if it is suitable for LK^+, and the following condition is satisfied:

   - $\tilde{\mathbf{ff}} \subseteq \mathcal{T} - \mathcal{D}$

THEOREM 7. *LK (LK^+) is sound for any Nmatrix $\mathcal{M}$ which is suitable for it. Moreover: it is complete for the relevant fragment of $\mathcal{M}$.*

Proof. We leave the easy proof for the reader. ∎

Note: A more general formulation of the last theorem is that an adequate model for LK is a triple $\langle \mathcal{T}, \mathcal{D}, v \rangle$, where $\emptyset \subset \mathcal{D} \subset \mathcal{T}$, and v is a valuation in $\mathcal{T}$ satisfying:

- $v(\varphi \wedge \psi) \in \mathcal{D}$ iff $v(\varphi) \in \mathcal{D}$ and $v(\psi) \in \mathcal{D}$
- $v(\varphi \vee \psi) \in \mathcal{D}$ iff $v(\varphi) \in \mathcal{D}$ or $v(\psi) \in \mathcal{D}$
- $v(\varphi \supset \psi) \in \mathcal{D}$ iff $v(\varphi) \notin \mathcal{D}$ or $v(\psi) \in \mathcal{D}$
- $v(\mathbf{ff}) \notin \mathcal{D}$

Convention: For convenience, we use henceforth the same symbol for a connective and for a corresponding nondeterministic operation in a given Nmatrix. We shall also denote by the same symbol (usually $\mathcal{O}$) the set of connectives of a language $\mathcal{L}$ and the corresponding set of operations of an Nmatrix for $\mathcal{L}$.

We turn now to Nmatrices for paraconsistent negation which are based on the basic three truth values described in the introduction.

DEFINITION 8. Let $\mathcal{M}_P$ ($\mathcal{M}_P^{\mathbf{ff}}$) be the unique Nmatrix $\langle \mathcal{T}, \mathcal{D}, \mathcal{O} \rangle$ for the language $\{\neg, \wedge, \vee, \supset\}$ ($\{\neg, \wedge, \vee, \supset, \mathbf{ff}\}$) which satisfies the following conditions:

- $\mathcal{T} = \{t, \top, f\}$

- $\mathcal{D} = \{t, \top\}$

- $\mathcal{M}_P$ ($\mathcal{M}_P^{\mathbf{ff}}$) is suitable for LK^+ (LK)

- $\widetilde{\diamond}(x, y) \in \{\mathcal{D}, \{f\}\}$ for $x, y \in \mathcal{T}$ and $\diamond \in \{\wedge, \vee, \supset\}$

- $\widetilde{\neg} t = \{f\}$ $\quad \widetilde{\neg} f = \neg\top = \mathcal{D}$

For the reader's convenience, here are the nondeterministic truth tables of $\mathcal{M}_P$ (for $\mathcal{M}_P^{\mathbf{ff}}$ one just has to add the condition $\widetilde{\mathbf{ff}} = \{f\}$, i.e: that $v(\mathbf{ff}) = f$ for any legal valuation v).

$\vee$	**f**	$\top$	**t**
f	f	$t, \top$	$t, \top$
$\top$	$t, \top$	$t, \top$	$t, \top$
t	$t, \top$	$t, \top$	$t, \top$

$\wedge$	**f**	$\top$	**t**
f	f	f	f
$\top$	f	$t, \top$	$t, \top$
t	f	$t, \top$	$t, \top$

$\supset$	**f**	$\top$	**t**
f	$t, \top$	$t, \top$	$t, \top$
$\top$	f	$t, \top$	$t, \top$
t	f	$t, \top$	$t, \top$

$\neg$	**f**	$\top$	**t**
	$t, \top$	$t, \top$	f

Note: It is straightforward to see that a valuation v in $\mathcal{T} = \{t, \top, f\}$ is $\mathcal{M}_P$-legal iff it satisfies the following four classical conditions:

$\wedge$: $v(\varphi \wedge \psi) = f$ iff $v(\varphi) = f$ or $v(\psi) = f$

$\vee$: $v(\varphi \vee \psi) = f$ iff $v(\varphi) = f$ and $v(\psi) = f$

$\supset\ :\ v(\varphi \supset \psi) = f$ iff $v(\varphi) \neq f$ and $v(\psi) = f$

$\neg\ :\ v(\neg\psi) = f$ iff $v(\psi) = t$

v is a legal $\mathcal{M}_P^{\mathbf{ff}}$-valuation iff in addition $v(\mathbf{ff}) = f$.

PROPOSITION 9. *PLK and PLK$^{\mathbf{ff}}$ are sound for $\mathcal{M}_P$ and $\mathcal{M}_P^{\mathbf{ff}}$ (respectively).*

Proof. This follows from Theorem 7 and the fact that $\neg f \subseteq \mathcal{D}$.   ■

Notes:

1. By proposition 5, PLK and $PLK^{\mathbf{ff}}$ are sound also for any refinement of $\mathcal{M}_P$ and $\mathcal{M}_P^{\mathbf{ff}}$ (respectively). On the other hand these matrices are *not* the most nondeterministic matrices based on $\mathcal{T} = \{t, \top, f\}$ for which PLK and $PLK^{\mathbf{ff}}$ are sound. From the proof of Proposition 9 it is clear that we could have taken $\neg t = \neg\top = \mathcal{T}$. This would have left us with no difference between t and $\top$, and with semantics which is equivalent to the two-valued nondeterministic semantics for these logics presented in [Avron and Lev, 2001; Avron and Lev, 2005] (and implicit already in [Batens, 2000] and elsewhere). However, the difference between t and $\top$ will be crucial below for the various extensions of PLK and $PLK^{\mathbf{ff}}$ with other negation rules.

2. We shall later see that PLK and $PLK^{\mathbf{ff}}$ are also complete for $\mathcal{M}_P$ and $\mathcal{M}_P^{\mathbf{ff}}$ (respectively).

3.2   The Effects of the Negation Rules

We turn now to the effects of the various negation rules. We shall show that to each of them corresponds a condition which leads to a certain refinement of $\mathcal{M}_P$ (or $\mathcal{M}_P^{\mathbf{ff}}$). These conditions are independent of each other, but never contradict each other. To see how these conditions are obtained, take $(\neg \supset\Rightarrow)_2$ as an example. This rule is equivalent to the validity of $\neg(\varphi \supset \psi) \vdash \neg\psi$. It means therefore that we should have $v(\neg(\varphi \supset \psi)) = f$ in case $v(\neg\psi) = f$. By the truth tables of $\mathcal{M}_P$, this means that $v(\varphi \supset \psi)$ should be t in case $v(\psi) = t$. This is therefore the condition that corresponds to this rule, and it turns 3 possible nondeterministic choices in $\mathcal{M}_P$ (or $\mathcal{M}_P^{\mathbf{ff}}$) to deterministic ones. Similar analysis can be done for the other ten rules. The resulting list of conditions is listed in the next Definition.

DEFINITION 10. The refining conditions induced by the negation rules:

$C(\neg\neg\Rightarrow):\qquad \tilde{\neg}f = \{t\}$

$C(\Rightarrow\neg\neg):\qquad \tilde{\neg}\top = \{\top\}$

$C(\neg\supset\Rightarrow)_1:\qquad f\tilde{\supset}x = \{t\}$

$C(\neg\supset\Rightarrow)_2:\qquad x\tilde{\supset}t = \{t\}$

$C(\Rightarrow\neg\supset):\qquad t\tilde{\supset}\top = \top\tilde{\supset}\top = \{\top\}$

$C(\neg\vee\Rightarrow)_1:\qquad t\tilde{\vee}x = \{t\}$

$C(\neg\vee\Rightarrow)_2:\qquad x\tilde{\vee}t = \{t\}$

$C(\Rightarrow\neg\vee):\qquad f\tilde{\vee}\top = \top\tilde{\vee}f = \top\tilde{\vee}\top = \{\top\}$

$C(\neg\wedge\Rightarrow):\qquad t\tilde{\wedge}t = \{t\}$

$C(\Rightarrow\neg\wedge)_1:\qquad \top\tilde{\wedge}t = \top\tilde{\wedge}\top = \{\top\}$

$C(\Rightarrow\neg\wedge)_2:\qquad t\tilde{\wedge}\top = \top\tilde{\wedge}\top = \{\top\}$

DEFINITION 11.

1. For $S \subseteq NR$, let $C(S) = \{C(r) \mid r \in S\}$

2. For $S \subseteq NR$, let $\mathcal{M}_P[S]$ and $\mathcal{M}_P^{\mathrm{ff}}[S]$ be the weakest refinements of $\mathcal{M}_P$ and $\mathcal{M}_P^{\mathrm{ff}}$ (respectively) in which the conditions in $C(S)$ are satisfied.[2]

The following proposition can now easily be proved:

PROPOSITION 12. *If $S \subseteq NR$ then $PLK(S)$ $(PLK^{\mathrm{ff}}(S))$ is sound for $\mathcal{M}_P[S]$ $(\mathcal{M}_P^{\mathrm{ff}}[S])$.*

We simultaneously prove now the completeness of the 2^{12} systems considered in Proposition 12 and the cut-elimination theorem for them.

THEOREM 13. *Let $S \subseteq NR$, and assume that $\Gamma \Rightarrow \Delta$ does not have a cut free proof in $PLK(S)$ $(PLK^{\mathrm{ff}}(S))$. Then $\Gamma \nvdash_{\mathcal{M}_P[S]} \Delta$ $(\Gamma \nvdash_{\mathcal{M}_P^{\mathrm{ff}}[S]} \Delta)$. In other words: there is a valuation v in $\mathcal{M}_P[S]$ $(\mathcal{M}_P^{\mathrm{ff}}[S])$ such that $v(\varphi) \in \{t, \top\}$ if $\varphi \in \Gamma$, while $v(\psi) = f$ if $\psi \in \Delta$.*

---

[2]It is easy to see that the conditions in NR cannot cause any conflict, so this is well-defined.

Proof. We prove the case of $PLK(S)$ (the proof in the case of $PLK^{\text{ff}}(S)$ is almost identical).

Call a sequent $\Gamma \Rightarrow \Delta$ *saturated* if it closed under the inverses of the rules of $PLK(S)$. More precisely: $\Gamma \Rightarrow \Delta$ is saturated if it satisfies the following conditions:

1. If $\varphi \wedge \psi \in \Gamma$ then $\varphi, \psi \in \Gamma$

2. If $\varphi \wedge \psi \in \Delta$ then $\varphi \in \Delta$ or $\psi \in \Delta$

3. If $\varphi \vee \psi \in \Gamma$ then $\varphi \in \Gamma$ or $\psi \in \Gamma$

4. If $\varphi \vee \psi \in \Delta$ then $\varphi, \psi \in \Delta$

5. If $\varphi \supset \psi \in \Gamma$ then $\varphi \in \Delta$ or $\psi \in \Gamma$

6. If $\varphi \supset \psi \in \Delta$ then $\varphi \in \Gamma$ and $\psi \in \Delta$

7. If $\neg\varphi \in \Delta$ then $\varphi \in \Gamma$

8. If $(\neg\neg \Rightarrow) \in S$ and $\neg\neg\varphi \in \Gamma$ then $\varphi \in \Gamma$

9. If $(\Rightarrow \neg\neg) \in S$ and $\neg\neg\varphi \in \Delta$ then $\varphi \in \Delta$

10. If $(\neg \supset\Rightarrow)_1 \in S$ and $\neg(\varphi \supset \psi) \in \Gamma$ then $\varphi \in \Gamma$

11. If $(\neg \supset\Rightarrow)_2 \in S$ and $\neg(\varphi \supset \psi) \in \Gamma$ then $\neg\psi \in \Gamma$

12. If $(\Rightarrow \neg \supset) \in S$ and $\neg(\varphi \supset \psi) \in \Delta$ then $\varphi \in \Delta$ or $\neg\psi \in \Delta$

13. If $(\neg\wedge \Rightarrow) \in S$ and $\neg(\varphi \wedge \psi) \in \Gamma$ then $\neg\varphi \in \Gamma$ or $\neg\psi \in \Gamma$

14. If $(\Rightarrow \neg\wedge)_1 \in S$ and $\neg(\varphi \wedge \psi) \in \Delta$ then $\neg\varphi \in \Delta$

15. If $(\Rightarrow \neg\wedge)_2 \in S$ and $\neg(\varphi \wedge \psi) \in \Delta$ then $\neg\psi \in \Delta$

16. If $(\neg\vee \Rightarrow)_1 \in S$ and $\neg(\varphi \vee \psi) \in \Gamma$ then $\neg\varphi \in \Gamma$

17. If $(\neg\vee \Rightarrow)_2 \in S$ and $\neg(\varphi \vee \psi) \in \Gamma$ then $\neg\psi \in \Gamma$

18. If $(\Rightarrow \neg\vee) \in S$ and $\neg(\varphi \vee \psi) \in \Delta$ then $\neg\varphi \in \Delta$ or $\neg\psi \in \Delta$

Assume now that $\Gamma \Rightarrow \Delta$ does not have a cut free proof in $PLK(S)$. It is a standard matter to prove that $\Gamma \Rightarrow \Delta$ can be extended to a saturated sequent $\Gamma^* \Rightarrow \Delta^*$ (where $\Gamma \subseteq \Gamma^*$ and $\Delta \subseteq \Delta^*$) which also has no cut free proof in $PLK(S)$. We may assume therefore that $\Gamma \Rightarrow \Delta$ is already saturated. We now inductively define a refuting valuation v of $\Gamma \Rightarrow \Delta$ as follows:

- $v(\varphi) = f$ iff one of the following conditions is satisfied:

  f1 $\varphi \in \Delta$

  f2 $\varphi = \neg\psi$ and $v(\psi) = t$

  f3 $\varphi = \psi_1 \wedge \psi_2$ and $v(\psi_1) = f$ or $v(\psi_2) = f$

  f4 $\varphi = \psi_1 \vee \psi_2$ and $v(\psi_1) = v(\psi_2) = f$

  f5 $\varphi = \psi_1 \supset \psi_2$ and $v(\psi_1) \neq f$ while $v(\psi_2) = f$

- $v(\varphi) = t$ iff one of the following conditions is satisfied:

  t1 $\neg\varphi \in \Delta$

  t2 $\varphi = \neg\psi$, $(\neg\neg \Rightarrow) \in S$, and $v(\psi) = f$

  t3 $\varphi = \psi_1 \wedge \psi_2$, $(\neg\wedge \Rightarrow) \in S$ and $v(\psi_1) = v(\psi_2) = t$

  t4 $\varphi = \psi_1 \vee \psi_2$, $(\neg\vee \Rightarrow)_1 \in S$ and $v(\psi_1) = t$

  t5 $\varphi = \psi_1 \vee \psi_2$, $(\neg\vee \Rightarrow)_2 \in S$ and $v(\psi_2) = t$

  t6 $\varphi = \psi_1 \supset \psi_2$, $(\neg \supset\Rightarrow)_1 \in S$ and $v(\psi_1) = f$

  t7 $\varphi = \psi_1 \supset \psi_2$, $(\neg \supset\Rightarrow)_2 \in S$ and $v(\psi_2) = t$

- $v(\varphi) = \top$ in any other case.

We prove now by induction on the complexity of φ that $v(\varphi)$ is well defined (i.e.: if one of the conditions f1-f5 is satisfied then none of the conditions t1-t7 is satisfied), that if $\varphi \in \Gamma$ then $v(\varphi) \neq f$, and that if $\neg\varphi \in \Gamma$ then $v(\varphi) \neq t$. First note that since $\Gamma \Rightarrow \Delta$ has no cut-free proof, $\Gamma \cap \Delta = \emptyset$. By the definition of a saturated sequent this entails that if φ satisfies t1 then it cannot satisfy f1 (since if φ satisfies t1 then $\varphi \in \Gamma$). With these two observations in the background, the induction proceeds as follows:

$\varphi = p$ (p atomic): In this case only conditions are f1 and t1 are relevant. Since these conditions cannot be satisfied together, $v(\varphi)$ is well defined. Moreover: if $\varphi \in \Gamma$ then f1 is not satisfied and so $v(\varphi) \neq f$, while if $\neg\varphi \in \Gamma$ then t1 is not satisfied and so $v(\varphi) \neq t$.

$\varphi = \neg\psi$: The relevant conditions in this case are f1, f2, t1, and t2. f2 trivially contradicts t2. We check the remaining cases:

- If φ satisfies f1 then $\psi \in \Gamma$. Hence by induction hypothesis $v(\psi) \neq f$ and so t2 is not satisfied.

- Suppose φ satisfies t1. Then $\varphi \in \Gamma$. Hence $\neg\psi \in \Gamma$, and so $v(\psi) \neq t$ by induction hypothesis. It follows that f2 is not satisfied.

It follows that $v(\varphi)$ is well defined. Moreover, our argument shows that if $\varphi \in \Gamma$ then neither f1 nor f2 are satisfied, and so $v(\varphi) \neq f$. Suppose next that $\neg\varphi = \neg\neg\psi \in \Gamma$. Then t1 is not satisfied. If $(\neg\neg \Rightarrow) \notin S$ then so is t2. Otherwise $\psi \in \Gamma$, and so $v(\psi) \neq f$ by induction hypothesis. Hence again t2 is not satisfied. It follows that $v(\varphi)$ cannot be t in this case.

$\varphi = \psi_1 \wedge \psi_2$: The relevant conditions in this case are f1, f3, t1, and t3. f3 and t3 trivially contradict each other (by the induction hypothesis). We check the remaining two cases:

- If φ satisfies f1 then either $\psi_1 \in \Delta$ or $\psi_2 \in \Delta$ (since $\Gamma \Rightarrow \Delta$ is saturated). Hence by induction hypothesis either $v(\psi_1) = f$ or $v(\psi_2) = f$ (and both are well-defined), and so t3 is not satisfied.

- Suppose φ satisfies t1. Then $\varphi \in \Gamma$. Hence both $\psi_1 \in \Gamma$ and $\psi_2 \in \Gamma$. By induction hypothesis both $v(\psi_1) \neq f$ and $v(\psi_2) \neq f$, and so f3 is not satisfied.

It follows that $v(\varphi)$ is well defined. Moreover, our argument shows that if $\varphi \in \Gamma$ then neither f1 nor f3 are satisfied, and so $v(\varphi) \neq f$. Suppose next that $\neg\varphi \in \Gamma$. Then t1 is not satisfied. If $(\neg\wedge \Rightarrow) \notin S$ then so is t3. Otherwise either $\neg\psi_1 \in \Gamma$ or $\neg\psi_2 \in \Gamma$ (since $\Gamma \Rightarrow \Delta$ is saturated), and so by induction hypothesis either $v(\psi_1) \neq t$ or $v(\psi_2) \neq t$. Hence again t3 is not satisfied. It follows that $v(\varphi)$ cannot be t in this case.

$\varphi = \psi_1 \vee \psi_2$: The relevant conditions in this case are f1, f4, t1, t4, and t5. f4 trivially contradicts both t4 and t5. We check the remaining cases:

- If φ satisfies f1 then both $\psi_1 \in \Delta$ and $\psi_2 \in \Delta$ (since $\Gamma \Rightarrow \Delta$ is saturated). Hence by induction hypothesis both $v(\psi_1) = f$ and $v(\psi_2) = f$ (and both are well-defined). Hence t4 and t5 cannot be satisfied.

- Suppose φ satisfies t1. Then $\varphi \in \Gamma$. Hence either $\psi_1 \in \Gamma$ or $\psi_2 \in \Gamma$. By induction hypothesis either $v(\psi_1) \neq f$ or $v(\psi_2) \neq f$, and so f4 is not satisfied.

It follows that $v(\varphi)$ is well defined. Moreover, our argument shows that if $\varphi \in \Gamma$ then neither f1 nor f4 are satisfied, and so $v(\varphi) \neq f$. Suppose next that $\neg\varphi \in \Gamma$. Then t1 is not satisfied. If $(\neg\vee \Rightarrow)_1 \notin S$

then so is t4. Otherwise $\neg\psi_1 \in \Gamma$, and so by induction hypothesis $v(\psi_1) \neq t$. Hence again t4 is not satisfied. That t5 is not satisfied is shown similarly. It follows that $v(\varphi)$ cannot be t in this case.

$\varphi = \psi_1 \supset \psi_2$: We leave this case to the reader.

We have shown that v is well defined, and that if $\varphi \in \Gamma$ then $v(\varphi) \neq f$ (and so $v(\varphi) \in \mathcal{D}$ in this case). Note also that if $\varphi \in \Delta$ then $v(\varphi) = f$ by definition of v. Hence v refutes $\Gamma \Rightarrow \Delta$, and it only remains to prove that v is a legal valuation. For this we need first to show that v is $\mathcal{M}_P$-legal, and for this it suffices to show that it satisfies the four conditions listed in the Note that follows Definition 8. Now the "if" parts of these conditions directly follow from the definition of v (conditions f2-f5). We show the "only if" parts.

- Assume $v(\neg\psi) = f$. By definition, this is possible only if $\neg\psi$ satisfies either condition f2 or condition f1. If it satisfies f1 then by condition t1 $v(\psi) = t$, and this is of course also true if $\neg\psi$ satisfies condition f2.

- Assume $v(\varphi \wedge \psi) = f$. By definition, this is possible only if $\varphi \wedge \psi$ satisfies either condition f3 or condition f1. If it satisfies f1 then either $\varphi \in \Delta$ or $\psi \in \Delta$ (since $\Gamma \Rightarrow \Delta$ is saturated). Hence in this case either $v(\varphi) = f$ or $v(\psi) = f$, and this is of course true also if $\varphi \wedge \psi$ satisfies condition f3.

- The cases of $\varphi \vee \psi$ and $\varphi \supset \psi$ are similar to the case of $\varphi \wedge \psi$.

We next show that v satisfies the various conditions induced by the rules of S. For the left introduction rules this is immediate from the definition of v (conditions t2-t7). We prove now the conditions induced by the right introduction rules.

$C(\Rightarrow \neg\neg)$: Assume $(\Rightarrow \neg\neg) \in S$, and $v(\psi) = \top$. Then ψ does not satisfy t1, and so $\neg\psi$ does not satisfy f1. Obviously it does not satisfy f2-f5 either, and so $v(\neg\psi) \neq f$. Assume now that $\neg\neg\psi \in \Delta$. Then $\psi \in \Delta$ (since $\Gamma \Rightarrow \Delta$ is saturated and $(\Rightarrow \neg\neg) \in S$). Hence $v(\psi) = f$, contradicting $v(\psi) = \top$. It follows that $\neg\psi$ does not satisfy t1. Obviously it does not satisfy t2-t7 either, and so $v(\neg\psi) \neq t$. The only possibility that remains is that $v(\neg\psi) = \top$.

$C(\Rightarrow \neg \supset)$: Assume $(\Rightarrow \neg \supset) \in S$, and $v(\varphi) \neq f$, $v(\psi) = \top$. Assume further that $\varphi \supset \psi \in \Delta$. Then $\varphi \in \Gamma$ and $\psi \in \Delta$ (since $\Gamma \Rightarrow \Delta$ is saturated). This entails that $v(\psi) = f$, contradicting $v(\psi) = \top$. Hence $\varphi \supset \psi$ does not satisfy f1. Obviously it does not satisfy f2-f5 either, and so $v(\varphi \supset \psi) \neq f$. Assume next that $\neg(\varphi \supset \psi) \in \Delta$. Then $\varphi \in \Delta$ or $\neg\psi \in \Delta$ (since $\Gamma \Rightarrow \Delta$ is saturated and $(\Rightarrow \neg \supset) \in S$). Hence $v(\varphi) = f$ or $v(\psi) = t$, contradicting our assumptions concerning $v(\varphi)$ and $v(\psi)$. It follows that $\varphi \supset \psi$ does not satisfy t1. Obviously it does not satisfy t2-t7 either, and so $v(\varphi \supset \psi) \neq t$. The only possibility that remains is that $v(\varphi \supset \psi) = \top$.

We leave the cases of $C(\Rightarrow \neg\vee)$, $C(\Rightarrow \neg\wedge)_1$, and $C(\Rightarrow \neg\wedge)_2$ to the reader. ∎

COROLLARY 14. *If $S \subseteq NR$ then $PLK(S)$ ($PLK^{ff}(S)$) is sound and complete with respect to $\mathcal{M}_P[S]$ ($\mathcal{M}_P^{ff}[S]$).*

COROLLARY 15. *If $S \subseteq NR$ then $PLK(S)$ and $PLK^{ff}(S)$ admit cut-elimination.*

Note: In particular, it follows from Corollary 15 that the above Gentzen-type systems for C_{min}, PI^* and $LFI1$ admit cut-elimination.

3.3   Some Applications

In this subsection we apply our soundness and completeness results for deriving interesting properties of some of the systems considered above. Our main tool will be the following Definition and simple Lemma, the trivial proof of which we leave to the reader:

DEFINITION 16. Let $\mathcal{L}$ be a propositional language, and let $\mathcal{M} = \langle \mathcal{T}, \mathcal{D}, \mathcal{O} \rangle$ be an Nmatrix for $\mathcal{L}$. A *semivaluation* in $\mathcal{M}$ is any function $v' : \mathcal{F}' \to \mathcal{T}$ such that $\mathcal{F}'$ is a set of formulas of $\mathcal{L}$ which is closed under subformulas, and v' respects $\mathcal{M}$ (in the sense that $\diamond(\psi_1, \ldots, \psi_n) \in \mathcal{F}'$ implies $v'(\diamond(\psi_1, \ldots, \psi_n)) \in \tilde{\diamond}(v(\psi_1), \ldots, v(\psi_n)))$.

LEMMA 17. *Any semivaluation can be extended to a valuation v in $\mathcal{M}$.*

COROLLARY 18. *$PLK(S)$ and $PLK^{ff}(S)$ are decidable for every $S \subseteq NR$.*

Proof. Let $\Gamma \Rightarrow \Delta$ be a sequent of the language of $PLK(S)$. Let $\mathcal{F}'$ be the set of all subformulas of formulas in $\Gamma \Rightarrow \Delta$. To decide whether $\Gamma \Rightarrow \Delta$ is provable in $PLK(S)$, it suffices by Lemma 17 (together with the soundness and completeness of $PLK(S)$ with respect to $\mathcal{M}_P[S]$) to check whether every $v' : \mathcal{F}' \to \{t, f, \top\}$ which is a semivaluation in $\mathcal{M}_P[S]$ has the property that either $v'(\varphi) = f$ for some $\varphi \in \Gamma$, or $v'(\varphi) \neq f$ for some $\varphi \in \Delta$. Since the number of such semivaluations is finite, this is a decision procedure. The proof for $PLK^{ff}(S)$ is similar. ∎

THEOREM 19. *Let S_L be the set of the left introduction rules of NR.*

1. *If φ is not a subformula of ψ then $\neg\psi \nvdash_{PLK(S_L)} \neg\varphi$.*

2. *Two formulas $\neg\varphi$ and $\neg\psi$ can be equivalent in $PLK(S_L)$ (in the sense that $\neg\varphi \vdash_{PLK(S_L)} \neg\psi$ and $\neg\psi \vdash_{PLK(S_L)} \neg\varphi$) only if they are identical.*

Proof. The second part is immediate from the first. To show the first, let $\mathcal{F}$ be the set of subformulas of $\{\neg\varphi, \neg\psi\}$. Define a function v' on $\mathcal{F}$ by assigning t to φ, f to $\neg\varphi$, and $\top$ to all other formulas of $\mathcal{F}$. It is easy to verify that v' is a semivaluation of $PLK(S_L)$. By Lemma 17 v' can be extended to a valuation v in $\mathcal{M}_P[S_L]$. Now v is a model of $\neg\psi$ in $\mathcal{M}_P[S_L]$ which is not a model of $\neg\varphi$. ∎

THEOREM 20. *Let $S_0 = NR - \{(\Rightarrow \neg\neg), (\Rightarrow \neg\wedge)_1, (\Rightarrow \neg\wedge)_2\}$. Then no formula of the form $\neg\varphi$ is provable in $PLK(S_0)$ (or any subsystem of $PLK(S_0)$).*

Proof. We prove first the following Lemma: If Γ is a finite set of formulas of the language of $PLK(S_0)$, then there is a valuation v in $\mathcal{M}_P[S_0]$ which assigns a designated element to all the formulas of Γ, and assigns t to at least one of them.

To prove the lemma it suffices by Lemma 17 to show that if $\mathcal{F}'$ is the set of subformulas of Γ then there exists a semivaluation v' defined on $\mathcal{F}'$ which has the desired property. We construct such v' by an induction on the total sum of connectives in Γ. If this sum is 0 (i.e. if all formulas in Γ are atomic) then the existence of an appropriate v' is trivial. Otherwise pick some $\varphi \in \Gamma$ which is not a subformula of any other formula in Γ, and let $\mathcal{F}'' = \mathcal{F}' - \{\varphi\}$. Then $\mathcal{F}''$ is closed under subformulas. Construct now v' as follows:

$\varphi = \neg\psi$: Let v'' assign $\top$ to all formulas of $\mathcal{F}''$. It is easy to see that v'' is a semivaluation. Extend v'' to v' with domain $\mathcal{F}'$ by letting $v'(\varphi) = t$. Since $(\Rightarrow \neg\neg) \notin S$, v' is a semivaluation in $\mathcal{M}_P[S_0]$, and it obviously has the required properties.

$\varphi = \psi_1 \vee \psi_2$ **or** $\varphi = \psi_1 \wedge \psi_2$: Let $\Gamma' = \Gamma \cup \{\psi_1, \psi_2\} - \varphi$. By induction hypothesis there is a semivaluation v'' in $\mathcal{M}_P[S_0]$ with domain $\mathcal{F}''$ such that v'' assigns a designated element to all the formulas of Γ', and assigns t to at least one of them. Extend v'' to v' with domain $\mathcal{F}'$ by letting $v'(\varphi) = t$ if either $v''(\psi_1) = t$ or $v''(\psi_2) = t$, $v'(\varphi) = \top$ otherwise. Since neither $(\Rightarrow \neg\wedge)_1$ nor $(\Rightarrow \neg\wedge)_2$ is in S_0, v' is a semivaluation in $\mathcal{M}_P[S_0]$, and it obviously has the required properties.

$\varphi = \psi_1 \supset \psi_2$: Let $\Gamma' = \Gamma \cup \{\psi_2\} - \varphi$, and let $\mathcal{F}'''$ be the set of subformulas of Γ'. By induction hypothesis there is a semivaluation v''' in $\mathcal{M}_P[S_0]$ with domain $\mathcal{F}'''$ such that v''' assigns a designated element to all the formulas of Γ', and assigns t to at least one of them. By Lemma 17 v''' can be extended to a semivaluation v'' with domain $\mathcal{F}''$. Extend v'' further to v' with domain $\mathcal{F}'$ by letting $v'(\varphi) = t$ if $v''(\psi_2) = t$, $v'(\varphi) = \top$ otherwise. It is easily seen that v' is a semivaluation in $\mathcal{M}_P[S_0]$ as desired.

The proof of the Theorem itself is now easy. Let φ be a formula of $PLK(S_0)$. By the lemma there is a valuation v in $\mathcal{M}_P[S_0]$ such that $v(\varphi) = t$. Hence $v(\neg\varphi) = f$, and so $\neg\varphi$ is not valid in $\mathcal{M}_P[S_0]$. It follows that $\neg\varphi$ is not provable in $PLK(S_0)$.   ∎

Note: The second part of Theorem 19 and Theorem 20 have been proved in [Carnielli and Marcos, 1999] for the much weaker system C_{min}.

PROPOSITION 21. *Theorem 20 cannot be improved: If no formula of the form $\neg\varphi$ is provable in $PLK(S)$ then $S \subseteq S_0$.*

Proof: This follows from the following (easily established) facts:

- If $(\Rightarrow \neg\neg) \in S$ then $\vdash_{PLK(S)} \neg\neg(\varphi \supset \varphi)$.

- If $(\Rightarrow \neg\wedge)_1 \in S$ then $\vdash_{PLK(S)} \neg(\varphi \wedge \neg\varphi)$.

- If $(\Rightarrow \neg\wedge)_2 \in S$ then $\vdash_{PLK(S)} \neg(\neg\varphi \wedge \varphi)$.

4 The Intuitionistic Case

4.1 The General semantics

The previous section was devoted to paraconsistent extensions of positive classical logics.[3] It seems however that positive *intuitionistic* logic is a better starting point for investigating negation. The valid sentences of this fragment are all intuitively correct. Positive classical logic, in contrast, includes counterintuitive tautologies like $(A \supset B \vee C) \supset (A \supset B) \vee (A \supset C)$ or $((A \supset B) \supset A) \supset A$. Moreover: the classical natural deduction rules for the positive connectives ($\wedge, \vee$ and $\supset$) define the intuitionistic positive logic, not the classical one. It is only with the aid of the classical rules for the classical negation that one can prove the counterintuitive positive tautologies mentioned above.

Now it is well known that it is impossible to conservatively add to intuitionistic positive logic a negation which is both explosive (i.e.: $\neg A, A \vdash B$

---

[3]By this it follows the survey [Carnielli and Marcos, 2002].

for all A, B) and for which LEM is valid. With such an addition we get classical logic. The intuitionists reject indeed LEM, retaining the explosive nature of negation (which is defined using the constant **ff** and implication). In this section we investigate conservative extensions of intuitionistic (positive) logic (with or without **ff**) with a paraconsistent negation for which LEM *is* valid.

As in the case of LK, we start with generalizing the standard, two-valued semantics of LJ^+ (or LJ). Recall that this semantics is provided by the class of all frames of the form $\mathcal{W} = \langle W, \leq, v \rangle$, where $\langle W, \leq \rangle$ is a nonempty partially ordered set (of "worlds"), and $v : W \times \mathcal{F} \to \mathcal{T}$ (where $\mathcal{F}$ is the set of formulas of the language) satisfies the following conditions:

1. If $y \geq x$ and $v(x, \varphi) = t$ then $v(y, \varphi) = t$.[4]

2.
   - $v(x, \varphi \wedge \psi) = f$ iff $v(x, \varphi) = f$ or $v(x, \psi) = f$
   - $v(x, \varphi \vee \psi) = f$ iff $v(x, \varphi) = f$ and $v(x, \psi) = f$
   - $v(x, \textbf{ff}) = f$ (in case **ff** is in the language).

3. $v(x, \varphi \supset \psi) = t$ iff $v(y, \psi) = t$ for every $y \geq x$ such that $v(y, \varphi) = t$

Obviously, if $\mathcal{W} = \langle W, \leq, v \rangle$ is a frame, then for every $x \in W$ the function $\lambda \varphi . v(x, \varphi)$ behaves like an ordinary classical valuation with respect to all the connectives except $\supset$. The treatment of $\supset$ is indeed what distinguishes between (positive) classical and intuitionistic logics. This observation leads to the following nondeterministic generalization of Kripke frames for intuitionistic logic:

DEFINITION 22. Let $\mathcal{L}$ be a propositional language which has $\supset$ as one of its connectives, and let $\mathcal{M} = \langle \mathcal{T}, \mathcal{D}, \mathcal{O} \rangle$ be an Nmatrix for $\mathcal{L}$. Denote by $\mathcal{F}$ be the set of formulas of $\mathcal{L}$. An $\mathcal{M}$-*frame* for $\mathcal{L}$ is a triple $\mathcal{W} = \langle W, \leq, v \rangle$ such that:

1. $\langle W, \leq \rangle$ is a nonempty partially ordered set

2. $v : W \times \mathcal{F} \to \mathcal{T}$ satisfies:

   - The persistence condition: if $y \geq x$ and $v(x, \varphi) \in \mathcal{D}$ then $v(y, \varphi) \in \mathcal{D}$
   - For every $x \in W$, $\lambda \varphi . v(x, \varphi)$ is a legal $\mathcal{M}$-valuation.

---

[4] For the language of LJ it suffices to demand this for atomic formulas only. It is then possible to prove that every formula has this property. This does not remain true for the nondeterministic generalizations we present below.

- $v(x, \varphi \supset \psi) \in \mathcal{D}$ iff $v(y, \psi) \in \mathcal{D}$ for every $y \geq x$ such that $v(y, \varphi) \in \mathcal{D}$

We say that a formula φ is *true* in a world $x \in W$ of a frame $\mathcal{W}$ if $v(x, \varphi) \in \mathcal{D}$. A sequent $\Gamma \Rightarrow \Delta$ is *valid* in $\mathcal{W}$ if for every $x \in W$ there is either $\varphi \in \Gamma$ such that φ is not true in x, or $\psi \in \Delta$ such that ψ is true in x.

Note: Obviously, if $\mathcal{M}_1$ is a refinement of $\mathcal{M}_2$, then any $\mathcal{M}_1$-frame is also an $\mathcal{M}_2$-frame.

DEFINITION 23.

1. Let $\mathcal{M} = \langle \mathcal{T}, \mathcal{D}, \mathcal{O} \rangle$ be an Nmatrix for a language which includes the language of LJ^+. We say that $\mathcal{M}$ is *suitable* for LJ^+ if the following conditions are satisfied:

   - If $a \in \mathcal{D}$ and $b \in \mathcal{D}$ then $a \tilde{\wedge} b \subseteq \mathcal{D}$
   - If $a \notin \mathcal{D}$ then $a \tilde{\wedge} b \subseteq \mathcal{T} - \mathcal{D}$
   - If $b \notin \mathcal{D}$ then $a \tilde{\wedge} b \subseteq \mathcal{T} - \mathcal{D}$

   - If $a \in \mathcal{D}$ then $a \tilde{\vee} b \subseteq \mathcal{D}$
   - If $b \in \mathcal{D}$ then $a \tilde{\vee} b \subseteq \mathcal{D}$
   - If $a \notin \mathcal{D}$ and $b \notin \mathcal{D}$ then $a \tilde{\vee} b \subseteq \mathcal{T} - \mathcal{D}$

   - If $b \in \mathcal{D}$ then $a \tilde{\supset} b \subseteq \mathcal{D}$
   - If $a \in \mathcal{D}$ and $b \notin \mathcal{D}$ then $a \tilde{\supset} b \subseteq \mathcal{T} - \mathcal{D}$

2. Let $\mathcal{M} = \langle \mathcal{T}, \mathcal{D}, \mathcal{O} \rangle$ be an Nmatrix for a language which includes the language of LJ. We say that $\mathcal{M}$ is *suitable* for LJ if it is suitable for LJ^+, and the following condition is satisfied:

   - $\tilde{\mathbf{ff}} \subseteq \mathcal{T} - \mathcal{D}$

Note: An Nmatrix which is suitable for LJ^+ (LJ) is also suitable for LK^+ (LK) iff it satisfies just one more condition: that if $a \notin \mathcal{D}$ then $a \supset b \subseteq \mathcal{D}$.

THEOREM 24. *Assume $\mathcal{W}$ is an $\mathcal{M}$-frame where $\mathcal{M}$ is suitable for LJ^+ (LJ). Then any sequent which is provable in LJ^+ (LJ) is valid in $\mathcal{W}$.*

Proof. Again we leave the easy proof to the reader. ■

Note: A more general formulation of the last theorem is that an adequate model for LJ is a tuple $\langle \mathcal{T}, \mathcal{D}, W, \leq, v \rangle$, where $\emptyset \subset \mathcal{D} \subset \mathcal{T}$, $\langle W, \leq \rangle$ is a nonempty partially ordered set, and $v : W \times \mathcal{F} \to \mathcal{T}$ is a valuation which satisfies the following conditions:

- If $y \geq x$ and $v(x, \varphi) \in \mathcal{D}$ then $v(y, \varphi) \in \mathcal{D}$

- $v(x, \varphi \wedge \psi) \in \mathcal{D}$ iff $v(x, \varphi) \in \mathcal{D}$ and $v(x, \psi) \in \mathcal{D}$

- $v(x, \varphi \vee \psi) \in \mathcal{D}$ iff $v(x, \varphi) \in \mathcal{D}$ or $v(x, \psi) \in \mathcal{D}$

- $v(x, \mathbf{ff}) \notin \mathcal{D}$

- $v(x, \varphi \supset \psi) \in \mathcal{D}$ iff $v(y, \psi) \in \mathcal{D}$ for every $y \geq x$ such that $v(y, \varphi) \in \mathcal{D}$

We turn now to $\mathcal{M}$-frames for paraconsistent negation where $\mathcal{M}$ is an Nmatrix which is based on our basic three truth values.

DEFINITION 25. Let $\mathcal{M}_{IP}$ ($\mathcal{M}_{IP}^{\mathbf{ff}}$) be the unique Nmatrix $\langle \mathcal{T}, \mathcal{D}, \mathcal{O} \rangle$ for the language $\{\neg, \wedge, \vee, \supset\}$ ($\{\neg, \wedge, \vee, \supset, \mathbf{ff}\}$) which satisfies the following conditions:

- $\mathcal{T} = \{t, \top, f\}$

- $\mathcal{D} = \{t, \top\}$

- $\mathcal{M}_{IP}$ ($\mathcal{M}_{IP}^{\mathbf{ff}}$) is suitable for LJ^{+} (LJ)

- $\tilde{\diamond}(x, y) \in \{\mathcal{T}, \mathcal{D}, \{f\}\}$ for $x, y \in \mathcal{T}$ and $\diamond \in \{\wedge, \vee, \supset\}$

- $f \tilde{\supset} f = \mathcal{T}$

- $\tilde{\neg} t = \{f\} \quad \tilde{\neg} f = \neg \tilde{\top} = \mathcal{D}$

Note: The only difference between $\mathcal{M}_{IP}$ and $\mathcal{M}_{P}$ (or $\mathcal{M}_{IP}^{\mathbf{ff}}$ and $\mathcal{M}_{P}^{\mathbf{ff}}$) is in the truth table for $\supset$. For $\mathcal{M}_{IP}$ (and $\mathcal{M}_{IP}^{\mathbf{ff}}$) this table is:

$\supset$	f	$\top$	t
f	$t, \top, f$	$t, \top$	$t, \top$
$\top$	f	$t, \top$	$t, \top$
t	f	$t, \top$	$t, \top$

PROPOSITION 26. *PLJ ($PLJ^{\mathbf{ff}}$) is sound for every $\mathcal{M}_{IP}$-frame ($\mathcal{M}_{IP}^{\mathbf{ff}}$-frame).*

Proof. This follows from Theorem 24 and the fact that $\neg f \subseteq \mathcal{D}$.   ∎

PROPOSITION 27. *In the definition of an $\mathcal{M}_{IP}$-frame (or $\mathcal{M}_{IP}^{\text{ff}}$-frame) the persistence condition (see Definition 22) can be replaced by the following monotonicity condition:*

- *If $x \leq y$ then $v(x, \varphi) \leq_k v(y, \varphi)$*

where $\leq_k$ on $\{t, f, \top\}$ is defined by: $t \leq_k \top$, $f \leq_k \top$ (and $a \leq_k a$).

Proof. The monotonicity condition trivially implies the persistence condition. For the converse we should show that if $x \leq y$ and $v(y, \varphi) \in \{t, f\}$ then $v(x, \varphi) = v(y, \varphi)$. This is immediate from the persistence condition in case $v(y, \varphi) = f$. Suppose that $v(y, \varphi) = t$. Then $v(y, \neg\varphi) = f$ (since v respects $\mathcal{M}_{IP}$). Hence $v(x, \neg\varphi) = f$. Since v respects the operations of $\mathcal{M}_{IP}$, this is possible only if $v(x, \varphi) = t$.   ∎

4.2   The Effects of the Negation Rules

We are interested in this section in extensions of positive intuitionistic logic with a paraconsistent negation. However, one should be cautious when rules for negations are added to positive intuitionistic logic. Thus the addition of both $(\neg \Rightarrow)$ and $(\Rightarrow \neg)$ to LJ^+ is not conservative: the positive fragment of the resulting logic is equivalent to LK^+, not to LJ^+. Accordingly, our first problem is: what combinations of our 11 rules for negation can be added to PLJ (or PLJ^{ff}) so that the resulting system is conservative over LJ^+ (or LJ)? In this section this problem is solved with the help of nondeterministic frames.

We start by following what we have done in the classical case.

DEFINITION 28. *For $S \subseteq (NR)$ let $\mathcal{M}_{IP}[S]$ and $\mathcal{M}_{IP}^{\text{ff}}[S]$ be the weakest refinements of $\mathcal{M}_{IP}$ and $\mathcal{M}_{IP}^{\text{ff}}$ (respectively) in which the conditions in $C(S)$ are satisfied.*

The following soundness theorem can again easily be proved:

PROPOSITION 29. *If $S \subseteq NR$ then $PLJ(S)$ is sound for $\mathcal{M}_{IP}[S]$-frames, and $PLJ^{\text{ff}}(S)$ is sound for $\mathcal{M}_{IP}^{\text{ff}}[S]$-frames.*

For itself proposition 29 does not have much value. Thus it does not guarantee that $PLJ(S)$ is conservative over LJ^+. It is not even constructively usable for showing non-provability in some $PLJ(S)$ (or $PLJ^{\text{ff}}(S)$). The reason is that a valuation is an infinite object, and it is not clear that a partial description would suffice. Let for example $W = \{a, b\}$ with $a < b$, and define $v'(a,p) = v'(a,q) = v'(b,q) = f$, $v'(b,p) = \top$. Then v' respects the monotonicity condition, but if $(\neg \supset \Rightarrow)_1 \in S$ then there is no extension v of v' such that $\langle W, \leq v \rangle$ is an $\mathcal{M}_{IP}[S]$-frame (since $v(a, p \supset q)$ should be f according to the definition of an $\mathcal{M}_{IP}$-frame, while according to $C(\neg \supset \Rightarrow)_1$ it should be t). We next show that $C(\neg \supset \Rightarrow)_1$ is the only rule that causes such problems.

DEFINITION 30. $INR =_{Df} NR - \{(\neg \supset \Rightarrow)_1\}$

DEFINITION 31. Let $\mathcal{M} = \mathcal{M}_{IP}[S]$ (or $\mathcal{M}_{IP}^{ff}[S]$) for some $S \subseteq INR$. An $\mathcal{M}$-semiframe is a triple $\mathcal{W} = \langle W, \leq, v' \rangle$ such that:

1. $\langle W, \leq \rangle$ is a nonempty partially ordered set.

2. $v' : W \times \mathcal{F}' \to \mathcal{T}$ is a partial valuation satisfying:

   - $\mathcal{F}'$ is a subset of $\mathcal{F}$ which is closed under subformulas.
   - The monotonicity condition:
     if $y \geq x$ and $\varphi \in \mathcal{F}'$, then $v(x, \varphi) \leq_k v(y, \varphi)$.
   - v' respects $\mathcal{M}$: If $\diamond$ is an n-ary connective of $\mathcal{L}$, and $\diamond(\psi_1, \ldots, \psi_n) \in \mathcal{F}'$, then $v'(x, \diamond(\psi_1, \ldots, \psi_n)) \in \tilde{\diamond}(v(x, \psi_1), \ldots, v(x, \psi_n))$.
   - If $\varphi \supset \psi \in \mathcal{F}'$ then $v'(x, \varphi \supset \psi) \in \mathcal{D}$ iff $v'(y, \psi) \in \mathcal{D}$ for every $y \geq x$ such that $v'(y, \varphi) \in \mathcal{D}$.

THEOREM 32. Let $S \subseteq INR$, and let $\langle W, \leq, v' \rangle$ (where $v' : W \times \mathcal{F}' \to \mathcal{T}$) be an $\mathcal{M}_{IP}[S]$-semiframe ($\mathcal{M}_{IP}^{ff}[S]$-semiframe). Then there exists an $\mathcal{M}_{IP}[S]$-frame ($\mathcal{M}_{IP}^{ff}[S]$-frame) $\mathcal{W} = \langle W, \leq, v \rangle$ such that v extends v'.

Proof. We imitate as far as possible the construction used in the proof of Theorem 13. The required v is inductively defined as follows:

- $v(x, \varphi) = f$ iff one of the following conditions is satisfied:

  f1 $\varphi \in \mathcal{F}'$, and $v'(x, \varphi) = f$

  f2 $\varphi = \neg \psi$ and $v(x, \psi) = t$

  f3 $\varphi = \psi_1 \wedge \psi_2$ and $v(x, \psi_1) = f$ or $v(x, \psi_2) = f$

  f4 $\varphi = \psi_1 \vee \psi_2$ and $v(x, \psi_1) = v(x, \psi_2) = f$

  f5 $\varphi = \psi_1 \supset \psi_2$ and there exists $y \geq x$ such that $v(y, \psi_1) \neq f$ while $v(y, \psi_2) = f$

- $v(x, \varphi) = t$ iff one of the following conditions is satisfied:

  t1 $\varphi \in \mathcal{F}'$, and $v'(x, \varphi) = t$

t2 $\varphi = \neg\psi$, $(\neg\neg \Rightarrow) \in S$, and $v(x, \psi) = f$

t3 $\varphi = \psi_1 \wedge \psi_2$, $(\neg\wedge \Rightarrow) \in S$ and $v(x, \psi_1) = v(x, \psi_2) = t$

t4 $\varphi = \psi_1 \vee \psi_2$, $(\neg\vee \Rightarrow)_1 \in S$ and $v(x, \psi_1) = t$

t5 $\varphi = \psi_1 \vee \psi_2$, $(\neg\vee \Rightarrow)_2 \in S$ and $v(x, \psi_2) = t$

t6 $\varphi = \psi_1 \supset \psi_2$, $(\neg \supset\Rightarrow)_2 \in S$ and $v(x, \psi_2) = t$

- $v(x, \varphi) = \top$ in any other case.

We prove first by a simultaneous induction on the complexity of φ that if $\varphi \in \mathcal{F}'$ then $v(x, \varphi) = v'(x, \varphi)$, that $\lambda x.v(x, \varphi)$ is $\leq_k$-monotonic, and that $v(x, \varphi)$ is well defined (i.e.: if one of the conditions f1-f5 is satisfied for x and φ then none of the conditions t1-t6 is satisfied for them).

$v(x, \varphi)$ **is well-defined:** Since $\langle W, \leq, v'\rangle$ is an $\mathcal{M}_{\mathcal{IP}}[\mathcal{S}]$-semiframe, it is obvious that if φ and x satisfy f1 then they cannot satisfy any of the conditions t1-t6, and if they satisfy t1 then they cannot satisfy any of the conditions f1-f5. It is also very easy to see that with the exception of the pair f5/t6, no condition from f2-f5 can be in conflict with a condition from t2-t6. Assume therefore that x and φ satisfy t6. Then $\varphi = \psi_1 \supset \psi_2$, and $v(x, \psi_2) = t$. By induction hypothesis on ψ_2, $v(y, \psi_2) = t$ for all $y \geq x$. Hence x and φ do not satisfy $f5$ in this case.[5]

v **is an extension of** v'**:** It is trivial that $v(x, \varphi) = v'(x, \varphi)$ in case $\varphi \in \mathcal{F}'$ and $v'(x, \varphi) = f$ or $v'(x, \varphi) = t$. It is also easy to check that since $\langle W, \leq, v'\rangle$ is an $\mathcal{M}_{\mathcal{IP}}[\mathcal{S}]$-semiframe, the induction hypothesis concerning the subformulas of φ implies that none of the conditions f1-f5, t1-t6 can be satisfied for φ and x in case $\varphi \in \mathcal{F}'$ and $v'(x, \varphi) = \top$. Hence $v(x, \varphi) = \top = v'(x, \varphi)$ in this case as well.

$\lambda x.v(x, \varphi)$ **is** $\leq_k$**-monotonic:** We show that if $y \geq x$ and $v(y, \varphi) = f$ then $v(x, \varphi) = f$, while if $v(y, \varphi) = t$ then $v(x, \varphi) = t$. Assume e.g. that $v(y, \varphi) = f$. Then y and φ satisfy one of the condition f1-f5. Our assumptions concerning v', the transitivity of $\leq$ and the induction hypothesis concerning the subformulas of φ together imply that x and φ satisfy the same condition in case $y \geq x$. Hence $v(x, \varphi) = f$ too. The argument in case $v(y, \varphi) = t$ is similar.

We next show that for every $x \in W$, $v(x, \varphi)$ satisfies the constraints imposed by $\mathcal{M}_{\mathcal{IP}}[\mathcal{S}]$. This follows from our assumptions concerning v' in case $\varphi \in \mathcal{F}'$.

---

[5]Note that the argument breaks down if $(\neg \supset\Rightarrow)_1 \in S$, since $C(\neg \supset\Rightarrow)_1$ may be in conflict with condition f5!

On the other hand the definition of v entails that in case $\varphi \notin \mathcal{F}'$, $v(x, \varphi) = f$ iff this is dictated by $\mathcal{M}_{\mathcal{IP}}[\mathcal{S}]$, and also $v(x, \varphi) = t$ iff this is dictated by $\mathcal{M}_{\mathcal{IP}}[\mathcal{S}]$. This immediately entails that also $v(x, \varphi) = \top$ in this case if (but not only if!) this is dictated by $\mathcal{M}_{\mathcal{IP}}[\mathcal{S}]$.

To complete the proof it remains to show that $v(x, \psi_1 \supset \psi_2) = f$ iff there exists $y \geq x$ such that $v(y, \psi_1) \neq f$ while $v(y, \psi_2) = f$. Well, the "if" part follows from the definition of v (condition f5). The converse follows from our assumptions concerning v' in case $\psi_1 \supset \psi_2 \in \mathcal{F}'$, and from the definition of v in case $\psi_1 \supset \psi_2 \notin \mathcal{F}'$. ∎

The following is an easy corollary of Theorem 32:

THEOREM 33. *If $S \subseteq INR$ then $PLJ(S)$ is a conservative extension of LJ^+ (and $PLJ^{\text{ff}}(S)$ is a conservative extension of LJ).*

Proof. We do the case of LJ^+ (the case of LJ is identical). Let $\mathcal{F}'$ be the set of $\neg$-free formulas of $\mathcal{L}$. Obviously, $\mathcal{F}'$ is closed under subformulas. Let s be a sequent consisting of formulas from $\mathcal{F}'$, and assume that s is not provable in LJ^+. We show that it is also not provable in $PLJ(S)$. By the completeness of LJ^+ relative to ordinary Kripke frames, there is an ordinary Kripke frame $\langle W, \leq, u \rangle$ in which s is not valid. Define a partial valuation $v' : W \times \mathcal{F}' \to \{t, f, \top\}$ by:

$$v'(x, \varphi) = \begin{cases} \top & u(x, \varphi) = t \\ f & u(x, \varphi) = f \end{cases}$$

It is easy to check that $\langle W, \leq, v' \rangle$ is an $\mathcal{M}_{\mathcal{IP}}[\mathcal{S}]$-semiframe (since a formula φ might be forced by $\mathcal{M}_{\mathcal{IP}}[\mathcal{S}]$ to be assigned t only if it either involves $\neg$ or if some subformula of φ is assigned t.[6]). By Theorem 32 there exists an $\mathcal{M}_{\mathcal{IP}}[\mathcal{S}]$-frame $\mathcal{W} = \langle W, \leq, v \rangle$ such that v extends v'. Obviously s is not valid in $\mathcal{W}$, and so $PLJ(S) \not\vdash s$ by Theorem 29. ∎

Our next proposition shows that Theorem 33 cannot be improved:

PROPOSITION 34. *If $(\neg \supset \Rightarrow)_1 \in S$ then $PLJ(S)$ is not conservative over LJ^+. In fact, every valid sequent of LK^+ is provable in it.*

Proof. The sequent $\Rightarrow \varphi \supset \psi, \varphi$ can be proved in $PLJ(\{(\neg \supset \Rightarrow)_1\}$ as follows:

---

[6]Note that again this is not true if $(\neg \supset \Rightarrow)_1 \in S$, as an examination of $C(\neg \supset \Rightarrow)_1$ quickly reveals.

$$\frac{\varphi \supset \psi \Rightarrow \varphi \supset \psi}{\Rightarrow \varphi \supset \psi, \neg(\varphi \supset \psi)} \qquad \frac{\varphi \Rightarrow \varphi}{\neg(\varphi \supset \psi) \Rightarrow \varphi}$$
$$\Rightarrow \varphi \supset \psi, \varphi$$

Now it is well known that this sequent is not provable in LJ^+, and that by adding it to LJ^+ we get a system which is equivalent to LK^+. ∎

From now on we shall concentrate on systems of the form $PLJ(S)$ (or $PLJ^{\text{ff}}(S)$) where $S \subseteq INR$. We shall formulate and prove our results only for the extensions of PLJ. Completely analogous results for extensions of PLJ^{ff} can easily be formulated and proved (using almost identical arguments).

4.3   Completeness and Cut-elimination

In Section 3 we have simultaneously proved the completeness of our various extensions of (positive) classical logic and the cut-elimination theorem for these systems. We are going to do now something similar for our extensions of (positive) intuitionistic logic. However, here things are more complicated because of the following two facts:

The cut-elimination theorem fails: It is easy to see that $\Rightarrow p, q \supset \neg p$ is provable in PLJ using a cut on $\neg p$, but no cut-free proof for this sequent exists even in $PLJ(INR)$.

Even analytic cuts are not always sufficient: A standard, quite satisfactory, substitute for full cut-elimination when the latter fails is to allow only *analytic* cuts. These are cuts in proofs in which the cut-formula is a subformula of the endsequent of the proof. Unfortunately, not for every $S \subseteq INR$ it is the case that non-analytic cuts can be eliminate in $PLJ(S)$. Two rules are problematic from this point of view:

- If $(\Rightarrow \neg \supset) \in S$ then the sequent $\Rightarrow p, q \supset \neg(q \supset p)$ is provable using a non-analytic cut on $\neg p$. However, there is no proof in $PLJ(S)$ of this sequent in which all cuts are analytic. To see this, consider the following sequent s:

$$q \supset p \Rightarrow p, q, \neg(q \supset p), q \supset \neg(q \supset p)$$

  The two sides of s are disjoint, and it is easy to check that in any proof of a (provable) subsequent of s in which all cuts are on subformulas of s, one of the premises of the last inference is also a subsequent of s with disjoint sides. Hence no subsequent of s has a proof of this sort, and so any proof of a subsequent of s includes non-analytic cuts.

- If $(\Rightarrow \neg\lor) \in S$ then the sequent $\Rightarrow p, q \supset (\neg r \supset \neg(p \lor r))$ is provable using a non-analytic cut on $\neg p$. However, there is no proof in $PLJ(S)$ of this sequent in which all cuts are analytic. To see this, consider the following sequent s:

$$r, p \lor r \Rightarrow p, q, \neg r, \neg(p \lor r), \neg r \supset \neg(p \lor r), q \supset (\neg r \supset \neg(p \lor r))$$

Again it is easy to check that in any proof of a subsequent of s in which all cuts are on subformulas of s, one of the premises of the last inference is a subsequent of s with disjoint sides. Hence any proof of a subsequent of s includes non-analytic cuts.

Besides proving the completeness of our various conservative extensions of LJ^+, this subsection has two other main goals. One is to determine exactly in which of these systems one can eliminate non-analytic cuts. The other is to present a satisfactory substitute (and use it to prove completeness of *all* the systems) in case this is impossible. We start with the second goal.

DEFINITION 35. Let s be a sequent. A cut is called s-analytic if the cut formula is a subformula of some formula of s. A proof in a Gentzen-type system is called s-analytic iff all cuts in it are s-analytic. A proof of a sequent s is called analytic if it is s-analytic.

DEFINITION 36.

1. $nsf(\varphi)$ is inductively defined as follows:

   (a) If φ is atomic then $nsf(\varphi) = \{\varphi\}$

   (b) If $\varphi = \neg p$ (p atomic) then $nsf(\varphi) = \{p, \neg p\}$

   (c) If $\varphi = \psi_1 \diamond \psi_2$ ($\diamond \in \{\land, \lor, \supset\}$) then $nsf(\varphi) = \{\varphi\} \cup nsf(\psi_1) \cup nsf(\psi_2)$

   (d) If $\varphi = \neg\neg\psi$ then $nsf(\varphi) = \{\varphi\} \cup nsf(\neg\psi)$

   (e) If $\varphi = \neg(\psi_1 \diamond \psi_2)$ then $nsf(\varphi) = \{\varphi, \psi_1 \diamond \psi_2\} \cup nsf(\neg\psi_1) \cup nsf(\neg\psi_2)$

2. ψ is called an *n-subformula* of φ if $\psi \in nsf(\varphi)$. It is called an *n-subformula* of a sequent s iff it is an n-subformula of some formula in s.

3. A cut in a proof is called *s-n_analytic* if it is done on some n-subformula of *s*. A proof is called *s-n_analytic* if all cuts in it are *s*-n_analytic.

4. A proof of a sequent *s* is called *n_analytic* if it is *s*-n_analytic.

Note that every subformula of a formula or a sequent is an n-subformula of that formula or sequent, and that an n-subformula of an n-subformula of φ is itself an n-subformula of φ. Note also that every n-subformula of a formula φ is either a subformula of φ, or a negation of such a subformula. Moreover: proper n-subformulas of φ are always simpler than φ, and are "internal" to φ (in the sense that if φ is written in Polish notation than any n-subformula of φ is obtained from it by omitting some of the symbols of φ, without changing the order of the remaining ones). Now a crucial observation here is that in a system $PLJ(S)$ even cut-free proofs do not have the usual subformula property, but it is easy to see that they have the *n-subformula property*. This property is kept if *n-analytic* cuts are allowed. Hence in the present context a very satisfactory substitute for full cut-elimination is the possibility to eliminate cuts which are not n-analytic.

THEOREM 37. *Let $S \subseteq INR$ and assume that the sequent s does not have an n_analytic proof in $PLJ(S)$. Then there exists an $\mathcal{M}_{\mathcal{IP}}[S]$-frame in which s is not valid.*

Proof. Let $\mathcal{F}'$ be the set of subformulas of s, $\mathcal{F}''$ — the set of n-subformulas of s, and let W be the set of all sequents which do not have an *s*-n_analytic proof in $PLJ(S)$, and the union of their two sides is $\mathcal{F}''$. Obviously, if $\Gamma \Rightarrow \Delta$ does not have an *s*-n_analytic proof in a Gentzen-type system G, and ψ is an n_subformula of s, then either $\psi, \Gamma \Rightarrow \Delta$ or $\Gamma \Rightarrow \Delta, \psi$ does not have an *s*-n_analytic proof in G. It follows that any sequent which consists of elements of $\mathcal{F}''$ and has no *s*-n_analytic proof in $PLJ(S)$, can be extended to an element of W. In particular s itself is a subsequent of some sequent $\Gamma_0 \Rightarrow \Delta_0 \in W$. Define now a partial order $\leq$ on W as follows: $\Gamma_1 \Rightarrow \Delta_1 \leq \Gamma_2 \Rightarrow \Delta_2$ if $\Gamma_1 \subseteq \Gamma_2$ (iff $\Delta_2 \subseteq \Delta_1$, since $\Gamma_1 \cup \Delta_1 = \Gamma_2 \cup \Delta_2$).

Finally, define inductively $v' : W \times \mathcal{F}' \to \mathcal{T}$ as follows:

- $v'(\Gamma \Rightarrow \Delta, \varphi) = f$ iff $\varphi \in \Delta$.

- $v'(\Gamma \Rightarrow \Delta, \varphi) = t$ iff one of the following conditions is satisfied:

    t1 $\neg\varphi \in \Delta$

    t2 $\varphi = \neg\psi$, $(\neg\neg \Rightarrow) \in S$, and $v'(\Gamma \Rightarrow \Delta, \psi) = f$

    t3 $\varphi = \psi_1 \wedge \psi_2$, $(\neg\wedge \Rightarrow) \in S$ and $v'(\Gamma \Rightarrow \Delta, \psi_1) = v'(\Gamma \Rightarrow \Delta, \psi_2) = t$

t4 $\varphi = \psi_1 \vee \psi_2$, $(\neg\vee \Rightarrow)_1 \in S$ and $v'(\Gamma \Rightarrow \Delta, \psi_1) = t$

t5 $\varphi = \psi_1 \vee \psi_2$, $(\neg\vee \Rightarrow)_2 \in S$ and $v'(\Gamma \Rightarrow \Delta, \psi_2) = t$

t6 $\varphi = \psi_1 \supset \psi_2$, $(\neg \supset\Rightarrow)_2 \in S$ and $v'(\Gamma \Rightarrow \Delta, \psi_2) = t$

- $v'(\Gamma \Rightarrow \Delta, \varphi) = \top$ in any other case.

We prove now by induction on the complexity of φ that v' is well defined (i.e. if φ and $\Gamma \Rightarrow \Delta$ satisfy one of the conditions t1-t6 then $\varphi \notin \Delta$).

t1 : Since $\Gamma \Rightarrow \Delta$ has no s-n_analytic proof, If $\neg\varphi \in \Delta$ then $\varphi \notin \Delta$.

t2 : Suppose $\varphi = \neg\psi$ and $v'(\Gamma \Rightarrow \Delta, \psi) = f$. Then $\psi \in \Delta$, and so $\varphi \notin \Delta$ (since $\Gamma \Rightarrow \Delta$ has no s-n_analytic proof).

t3 : Suppose $\varphi = \psi_1 \wedge \psi_2$ and $v'(\Gamma \Rightarrow \Delta, \psi_1) = v'(\Gamma \Rightarrow \Delta, \psi_2) = t$. By induction hypothesis, $\psi_1 \notin \Delta$ and $\psi_2 \notin \Delta$. Since ψ_1 and ψ_2 are in $\mathcal{F}'$, $\{\psi_1, \psi_2\} \subseteq \Gamma$. Hence $\varphi \notin \Delta$ (otherwise $\Gamma \Rightarrow \Delta$ would have a cut-free proof).

t4-t6 : we leave the proof to the reader.

We next prove by induction that if $v'(\Gamma \Rightarrow \Delta, \varphi) = t$ then $\neg\varphi, \Gamma \Rightarrow \Delta$ has an s-analytic (in fact, cut-free) proof.

φ **and** $\Gamma \Rightarrow \Delta$ **satisfy t1:** This case is trivial.

φ **and** $\Gamma \Rightarrow \Delta$ **satisfy t2:** Then $\psi \in \Delta$. Now $\neg\varphi \Rightarrow \psi$ has a cut-free proof in this case, since $\varphi = \neg\psi$ and $(\neg\neg \Rightarrow) \in S$. Hence $\neg\varphi, \Gamma \Rightarrow \Delta$ also has cut-free proof.

φ **and** $\Gamma \Rightarrow \Delta$ **satisfy t3:** By induction hypothesis, $\neg\psi_1, \Gamma \Rightarrow \Delta$ and $\neg\psi_2, \Gamma \Rightarrow \Delta$ have cut-free proofs. Using $(\neg\wedge \Rightarrow)$ (which is in S in this case), we get a cut-free proof of $\neg\varphi, \Gamma \Rightarrow \Delta$.

We leave the other three cases to the reader.

We show now that $\langle W, \leq, v' \rangle$ is an $\mathcal{M}_{\mathcal{IP}}[S]$-semiframe.

$\mathcal{F}'$ **is a subset of** $\mathcal{F}$ **closed under subformulas:** This is obvious.

The monotonicity condition: Let $\Gamma_1 \Rightarrow \Delta_1 \leq \Gamma_2 \Rightarrow \Delta_2$. Assume that $v'(\Gamma_2 \Rightarrow \Delta_2, \varphi) = f$. Then $\varphi \in \Delta_2$, and so $\varphi \in \Delta_1$ by the definition of $\leq$. It follows that $v'(\Gamma_1 \Rightarrow \Delta_1, \varphi) = f$. The proof that if $v'(\Gamma_2 \Rightarrow \Delta_2, \varphi) = t$ then $v'(\Gamma_1 \Rightarrow \Delta_1, \varphi) = t$ is by induction on the complexity of φ. If $v'(\Gamma_2 \Rightarrow \Delta_2, \varphi) = t$ because φ and $\Gamma_2 \Rightarrow \Delta_2$ satisfy t1, then $v'(\Gamma_1 \Rightarrow \Delta_1, \varphi) = t$ because in this case also φ and $\Gamma_1 \Rightarrow \Delta_1$ satisfy t1 (since $\Delta_2 \subseteq \Delta_1$). If $v'(\Gamma_2 \Rightarrow \Delta_2, \varphi) = t$ because they satisfy one of t2-t6, then the claim easily follows from the induction hypothesis and the already shown fact that if $v'(\Gamma_2 \Rightarrow \Delta_2, \psi) = f$ then $v'(\Gamma_1 \Rightarrow \Delta_1, \psi) = f$.

v' **respects the condition concerning** $\supset$: Let $\varphi \supset \psi \in \mathcal{F}'$ and $\Gamma \Rightarrow \Delta \in W$. Assume first that there exists $\Gamma_1 \Rightarrow \Delta_1 \in W$ such that $\Gamma \Rightarrow \Delta \leq \Gamma_1 \Rightarrow \Delta_1$, and $v'(\Gamma_1 \Rightarrow \Delta_1, \varphi) \neq f$ while $v'(\Gamma_1 \Rightarrow \Delta_1, \psi) = f$. Then $\varphi \in \Gamma_1$ and $\psi \in \Delta_1$. since $\varphi, \varphi \supset \psi \Rightarrow \psi$ has a cut-free proof, $\varphi \supset \psi \notin \Gamma_1$, and so $\varphi \supset \psi \in \Delta_1$. Since $\Delta_1 \subseteq \Delta$, also $\varphi \supset \psi \in \Delta$, and so $v'(\Gamma \Rightarrow \Delta, \varphi \supset \psi) = f$.

For the converse, assume that $v'(\Gamma \Rightarrow \Delta, \varphi \supset \psi) = f$. Then $\varphi \supset \psi \in \Delta$. It follows that $\Gamma, \varphi \Rightarrow \psi$ is a sequent which consists of elements of $\mathcal{F}''$, and does not have an s-n_analytic proof. Hence it can be extended to a sequent $\Gamma_1 \Rightarrow \Delta_1$ in W. Obviously, $\Gamma \Rightarrow \Delta \leq \Gamma_1 \Rightarrow \Delta_1$, and $v'(\Gamma_1 \Rightarrow \Delta_1, \varphi) \neq f$ while $v'(\Gamma_1 \Rightarrow \Delta_1, \psi) = f$.

v' **respects** $\mathcal{M}_{\mathcal{IP}}[S]$: We prove first that if $\varphi \in \mathcal{F}'$ is a complex formula and $\Gamma \Rightarrow \Delta \in W$ then $v'(\Gamma \Rightarrow \Delta, \varphi) = f$ iff this should be the case according to $\mathcal{M}_{\mathcal{IP}}[S]$ (and the definition of a semiframe). This has already been proved in case $\varphi = \psi_1 \supset \psi_2$. We prove the other cases:

- Suppose $\varphi = \neg\psi \in \mathcal{F}'$. Then also $\psi \in \mathcal{F}'$. If $v'(\Gamma \Rightarrow \Delta, \varphi) = f$ then $\varphi \in \Delta$, and so (by condition t1), $v'(\Gamma \Rightarrow \Delta, \psi) = t$. For the converse, assume that $v'(\Gamma \Rightarrow \Delta, \psi) = t$. Then $\varphi, \Gamma \Rightarrow \Delta$ has a cut-free proof (by what we have proved above). Hence $\varphi \notin \Gamma$. Since $\varphi \in \mathcal{F}'$, this implies that $\varphi \in \Delta$, and so $v'(\Gamma \Rightarrow \Delta, \varphi) = f$.

- Suppose $\varphi = \psi_1 \wedge \psi_2 \in \mathcal{F}'$. Assume first that $v'(\Gamma \Rightarrow \Delta, \psi_1) = f$. Then $\psi_1 \in \Delta$. Hence $\varphi \notin \Gamma$ (since $\psi_1 \wedge \psi_2 \Rightarrow \psi_1$ has a cut-free proof), and so $\varphi \in \Delta$. It follows that $v'(\Gamma \Rightarrow \Delta, \varphi) = f$ in this case. The proofs that if $v'(\Gamma \Rightarrow \Delta, \psi_2) = f$ then $v'(\Gamma \Rightarrow \Delta, \varphi) = f$, and that if $v'(\Gamma \Rightarrow \Delta, \psi_1) \neq f$ and $v'(\Gamma \Rightarrow \Delta, \psi_2) \neq f$ then $v'(\Gamma \Rightarrow \Delta, \varphi) \neq f$ are similar.

- The proof that if $\varphi = \psi_1 \vee \psi_2 \in \mathcal{F}'$, then $v'(\Gamma \Rightarrow \Delta, \varphi) = f$ iff either $v'(\Gamma \Rightarrow \Delta, \psi_1) = f$ or $v'(\Gamma \Rightarrow \Delta, \psi_2) = f$, is left to the reader.

We prove now that v' respects the conditions in $C(S)$. This is true for $S \cap \{C(\neg\neg \Rightarrow), C(\neg\wedge \Rightarrow), C(\neg\vee \Rightarrow)_1, C(\neg\vee \Rightarrow)_2, C(\neg \supset \Rightarrow)_2\}$ by definition of v' (conditions t2-t6).

$C(\Rightarrow \neg\neg)$: Assume $(\Rightarrow \neg\neg) \in S$, $v'(\Gamma \Rightarrow \Delta, \psi) = \top$, and $\neg\psi \in \mathcal{F}'$. Then $\psi \in \mathcal{F}'$, and $\psi \notin \Delta$. Hence $\psi \in \Gamma$. Since $\psi \Rightarrow \neg\neg\psi$ has a cut-free proof in case $(\Rightarrow \neg\neg) \in S$, $\neg\neg\psi \notin \Delta$. It follows that $\Gamma \Rightarrow \Delta$ and $\neg\psi$ do not satisfy t1. Obviously they do not satisfy t2-t6 either, and so $v'(\Gamma \Rightarrow \Delta, \neg\psi) \neq t$. That in this case $v'(\Gamma \Rightarrow \Delta, \neg\psi) \neq f$ has already been proved. It follows that $v'(\Gamma \Rightarrow \Delta, \neg\psi) = \top$.

$C(\Rightarrow \neg \supset)$: Assume that $(\Rightarrow \neg \supset) \in S$, $v'(\Gamma \Rightarrow \Delta, \varphi) \neq f$, $v'(\Gamma \Rightarrow \Delta, \psi) = \top$, and that $\varphi \supset \psi \in \mathcal{F}'$. Then $\varphi \in \mathcal{F}'$, $\varphi \notin \Delta$, $\psi \in \mathcal{F}'$, $\psi \notin \Delta$, and $\neg\psi \notin \Delta$. Hence $\varphi \in \Gamma$. Assume that $\neg(\varphi \supset \psi) \in \Delta$. Then $\neg(\varphi \supset \psi) \in \mathcal{F}''$. Hence also $\neg\psi \in \mathcal{F}''$. It followed that $\neg\psi \in \Gamma \cup \Delta$, and since $\neg\psi \notin \Delta$, $\neg\psi \in \Gamma$. But $\varphi, \neg\psi \Rightarrow \neg(\varphi \supset \psi)$ has a cut-free proof in case $(\Rightarrow \neg \supset) \in S$, and so $\Gamma \Rightarrow \Delta$ has such a proof too, contradicting $\Gamma \Rightarrow \Delta \in W$. It follows that $\neg(\varphi \supset \psi) \notin \Delta$ and so $\Gamma \Rightarrow \Delta$ and $\varphi \supset \psi$ do not satisfy t1. Obviously they do not satisfy t2-t6 either, and so $v'(\Gamma \Rightarrow \Delta, \varphi \supset \psi) \neq t$. That in this case $v'(\Gamma \Rightarrow \Delta, \varphi \supset \psi) \neq f$ has already been proved. It follows that $v'(\Gamma \Rightarrow \Delta, \varphi \supset \psi) = \top$.

Again we leave the cases of $C(\Rightarrow \neg\vee)$, $C(\Rightarrow \neg\wedge)_1$, and $C(\Rightarrow \neg\wedge)_2$ to the reader.

We have shown that $\langle W, \leq, v' \rangle$ is an $\mathcal{M}_{\mathcal{IP}}[S]$-semiframe. Therefore by Theorem 32 there exists an $\mathcal{M}_{\mathcal{IP}}[S]$-frame $\mathcal{W} = \langle W, \leq, v \rangle$ such that v extends v'. Now it is easy to see that our original sequent s is false in the world $\Gamma_0 \Rightarrow \Delta_0$ of $\mathcal{W}$ (of which s is a subsequent). Hence s is not valid in the $\mathcal{M}_{\mathcal{IP}}[S]$-frame $\mathcal{W}$. ∎

COROLLARY 38. *If $S \subseteq INR$ then $PLJ(S)$ is sound and complete for $\mathcal{M}_{\mathcal{IP}}[S]$-frames.*

Note: In the special case of da Costa's C_ω Corollary 38 provides illuminating semantics which is much simpler than the Kripke-type semantics given in [Baaz, 1986] and the bivaluations semantics of [Loparić, 1986]. Here is a compact description of this semantics for the reader particularly interested in C_ω: a Kripke-type frame for C_ω is a triple $\langle W, \leq, v \rangle$ such that $\langle W, \leq \rangle$ is a nonempty partially ordered set, and $v : W \times \mathcal{F} \to \{t, f, \top\}$ is a valuation which satisfies the following conditions:

- If $x \leq y$ then $v(x, \varphi) \leq_k v(y, \varphi)$

- $v(x, \varphi \wedge \psi) = f$ iff $v(x, \varphi) = f$ or $v(x, \psi) = f$

- $v(x, \varphi \vee \psi) = f$ iff $v(x, \varphi) = f$ and $v(x, \psi) = f$

- $v(x, \varphi \supset \psi) = f$ iff there exists $y \geq x$ such that $v(y, \varphi) \neq f$ while $v(y, \psi) = f$

- $v(x, \neg\varphi) = f$ iff $v(x, \varphi) = t$

- If $v(x, \varphi) = f$ then $v(x, \neg\varphi) = t$

It is interesting to note that by changing the last clause to an "iff" we get an adequate semantics for the system obtained from C_ω by adding the axiom $\varphi \supset \neg\neg\varphi$ (which corresponds to the rule $(\Rightarrow \neg\neg)$. In both cases a frame is of course a model of a formula φ if $v(x, \varphi) \neq f$ for every $x \in W$.

COROLLARY 39. *If s has a proof in $PLJ(S)$ ($S \subseteq INR$) then s has there a n_analytic proof.*

We show finally that n_analytic cuts can actually be replaced by strictly *analytic* ones in all of the systems $PLJ(S)$ except those which contain at least one of the two problematic rules noted at the beginning of this sub-section.

THEOREM 40. *Let $S \subseteq INR - \{(\Rightarrow \neg \supset), (\Rightarrow \neg\vee)\}$. If s has a proof in $PLJ(S)$ then s has there an analytic proof.*

Proof. Assume that the sequent s does not have an analytic proof in $PLJ(S)$. Again it suffices to show that there exists an $\mathcal{M}_{\mathcal{IP}}[S]$-semiframe in which s is not valid.

Let $\mathcal{F}'$ and $\mathcal{F}''$ be as in the proof of Theorem 37. Call a sequent $\Gamma \Rightarrow \Delta$ *s-acceptable* if $\Gamma, \Delta \subseteq \mathcal{F}''$, and there are sets $\Delta^0, \ldots, \Delta^k$ such that $\Delta = \bigcup_{i=0}^k \Delta^i$, $\Delta^0 \subseteq \mathcal{F}'$, and for every $i > 0$ and φ, $\varphi \in \Delta^i$ iff $\varphi \notin \bigcup_{j<i} \Delta^j$, and it satisfies one of the following conditions:

1. $(\Rightarrow \neg\neg) \in S$ and $\neg\neg\varphi \in \Delta^{i-1}$.

2. $(\Rightarrow \neg\wedge)_1 \in S$ and there exist ψ_1, ψ_2 such that $\varphi = \neg\psi_1$ and $\neg(\psi_1 \wedge \psi_2) \in \Delta^{i-1}$.

3. $(\Rightarrow \neg\wedge)_2 \in S$ and there exist ψ_1, ψ_2 such that $\varphi = \neg\psi_2$ and $\neg(\psi_1 \wedge \psi_2) \in \Delta^{i-1}$.

Let W be the set of all sequents which are s-acceptable, contain all the formulas of $\mathcal{F}'$, and have no s-analytic proof.

LEMMA 41. *Let $s' = \Gamma' \Rightarrow \Delta'$ be a sequent which has no s-analytic proof, and such that $\Gamma' \subseteq \mathcal{F}''$, $\Delta' \subseteq \mathcal{F}'$. Then s' is a subsequent of some sequent in W.*

Proof of Lemma 41: Since s' has no s-analytic proof, one can first add to it in stages all the formulas of $\mathcal{F}'$, so that the resulting sequent $s'' = \Gamma'' \Rightarrow \Delta''$ still has no s-analytic proof. Let $\Delta^0 = \Delta''$. Since only formulas from $\mathcal{F}'$ have been added to s', $\Delta^0 \subseteq \mathcal{F}'$. Define Δ^i for $i > 0$ by letting $\varphi \in \Delta^i$ iff $\varphi \notin \bigcup_{j<i} \Delta^j$, and it satisfies one of the conditions 1-3 above. Then $\Delta_i \subseteq \mathcal{F}''$ for all i. Since $\mathcal{F}''$ is finite, There is k such that $\Delta_k = \emptyset$. For this k $\Gamma'' \Rightarrow \bigcup_{j<k} \Delta^j$ is obviously an element of W with the required properties.

Define now $\leq$ on W by: $\Gamma_1 \Rightarrow \Delta_1 \leq \Gamma_2 \Rightarrow \Delta_2$ if either $\Gamma_1 = \Gamma_2$ and $\Delta_1 = \Delta_2$, or Γ_1 is a proper subset of Γ_2.

LEMMA 42. If $\Gamma_1 \Rightarrow \Delta_1 \leq \Gamma_2 \Rightarrow \Delta_2$ then $\Delta_2 \subseteq \Delta_1$.

Proof of Lemma 42: Let $\Delta_2 = \bigcup_{i=0}^{k} \Delta_2^i$, where $\Delta_2^0, \ldots, \Delta_2^k$ are as in the definition of an s-acceptable sequent. We prove by induction on i that if $\varphi \in \Delta_2^i$ then $\varphi \in \Delta_1$.

For the base case, assume that $\varphi \in \Delta_2^0$. Then $\varphi \in \mathcal{F}'$. Hence $\varphi \in \Gamma_1 \cup \Delta_1$. Suppose that $\varphi \in \Gamma_1$. Then also $\varphi \in \Gamma_2$ (by definition of $\leq$), and so $\varphi \in \Gamma_2 \cap \Delta_2$. This contradicts the fact that $\Gamma_2 \Rightarrow \Delta_2$ has no s-analytic proof. It follows that $\varphi \in \Delta_1$.

For the induction step assume that $\Delta_2^i \subseteq \Delta_1$, and let $\varphi \in \Delta_2^{i+1}$. By definition there are three cases to consider. Assume, e.g., that $(\Rightarrow \neg\wedge)_1 \in S$ and there exist ψ_1, ψ_2 such that $\varphi = \neg\psi_1$ and $\neg(\psi_1 \wedge \psi_2) \in \Delta_2^i$ (the other cases are treated similarly). By induction hypothesis, $\neg(\psi_1 \wedge \psi_2) \in \Delta_1$. Since $\Gamma_1 \Rightarrow \Delta_1$ is s-acceptable and $(\Rightarrow \neg\wedge)_1 \in S$, also $\varphi = \neg\psi_1 \in \Delta_1$.[7]

We now define $v' : W \times \mathcal{F}' \to \mathcal{T}$ exactly as we did in the proof of Theorem 37. Then we prove that v' is well defined, and that if $v'(\Gamma \Rightarrow \Delta, \varphi) = t$ then $\neg\varphi, \Gamma \Rightarrow \Delta$ has an s-analytic proof (the proof of these facts are again identical to those given in the proof of Theorem 37).

We now show that $\langle W, \leq, v' \rangle$ is an $\mathcal{M}_{\mathcal{I}\mathcal{P}}[S]$-semiframe.

The monotonicity condition: The proof is almost identical to that given in in the proof of Theorem 37. The only difference is that Lemma 42 should be used when facts of the form $\Delta_2 \subseteq \Delta_1$ are needed.

v' respects the condition concerning $\supset$: Let $\varphi \supset \psi \in \mathcal{F}'$ and $\Gamma \Rightarrow \Delta \in W$. Then $\psi \in \mathcal{F}'$. Assume that $v'(\Gamma \Rightarrow \Delta, \varphi \supset \psi) = f$. We show that there exists $\Gamma_1 \Rightarrow \Delta_1 \in W$ such that $\Gamma_1 \Rightarrow \Delta_1 \geq \Gamma \Rightarrow \Delta$, and

---

[7]This is the step in the proof which will fail if one of the two problematic rules is added.

$v'(\Gamma_1 \Rightarrow \Delta_1, \varphi) \neq f$ while $v'(\Gamma_1 \Rightarrow \Delta_1, \psi) = f$. From our assumption it follows that $\varphi \supset \psi \in \Delta$. Since $\psi \Rightarrow \varphi \supset \psi$ has a cut-free proof, $\psi \notin \Gamma$. Hence $\psi \in \Delta$ (since $\psi \in \mathcal{F}'$). If $\varphi \in \Gamma$ then $v'(\Gamma \Rightarrow \Delta, \varphi) \neq f$ while $v'(\Gamma \Rightarrow \Delta, \psi) = f$. Hence we can take in this case $\Gamma_1 = \Gamma$ and $\Delta_1 = \Delta$. If $\varphi \notin \Gamma$ consider $\Gamma, \varphi \Rightarrow \psi$. This sequent has no s-analytic proof (otherwise $\Gamma \Rightarrow \Delta$ would have one), contains only formulas from $\mathcal{F}''$, and its left-hand side contains only formulas from $\mathcal{F}'$. By Lemma 41 it can therefore be extended to a sequent $\Gamma_1 \Rightarrow \Delta_1 \in W$. Since $\varphi \notin \Gamma$, Γ is a proper subset of Γ_1, and so $\Gamma \Rightarrow \Delta \leq \Gamma_1 \Rightarrow \Delta_1$. Obviously $v'(\Gamma_1 \Rightarrow \Delta_1, \varphi) \neq f$ while $v'(\Gamma_1 \Rightarrow \Delta_1, \psi) = f$, as required.

The converse of what we have just proven is shown exactly as in the proof of Theorem 37.

v' **respects** $\mathcal{M}_{\mathcal{IP}}[S]$: The proofs that $v'(\Gamma \Rightarrow \Delta, \varphi) = f$ iff it should, and that $v'(\Gamma \Rightarrow \Delta, \varphi) = t$ in any case it should, are exactly as in the proof of Theorem 37. It remains to prove that $v'(\Gamma \Rightarrow \Delta, \varphi) = \top$ whenever it should. By the constraints on S, there are only three cases to consider here:

- Assume $(\Rightarrow \neg\neg) \in S$, $v'(\Gamma \Rightarrow \Delta, \psi) = \top$, and $\neg\psi \in \mathcal{F}'$. The proof that in this case $v'(\Gamma \Rightarrow \Delta, \neg\psi) = \top$ is again identical to the one given in the proof of Theorem 37.

- Assume that $(\Rightarrow \neg\wedge)_1 \in S$, $v'(\Gamma \Rightarrow \Delta, \varphi) = \top$, $v'(\Gamma \Rightarrow \Delta, \psi) \neq f$, and $\varphi \wedge \psi \in \mathcal{F}'$. We should show that $v'(\Gamma \Rightarrow \Delta, \varphi \wedge \psi) = \top$. Well, that $v'(\Gamma \Rightarrow \Delta, \varphi\wedge\psi) \neq f$ in such a case was already proven (since v' assigns the value f only if it should). We show now that $v'(\Gamma \Rightarrow \Delta, \varphi \wedge \psi) \neq t$ as well. Suppose otherwise. Since t2-t6 obviously fail in this case, this can happen only if $\neg(\varphi \wedge \psi) \in \Delta$. But since $(\Rightarrow \neg\wedge)_1 \in S$ and $\Gamma \Rightarrow \Delta$ is s-acceptable, this implies that $\neg\varphi \in \Delta$ too. It follows that $v'(\Gamma \Rightarrow \Delta, \varphi) = t$ by definition of v', contradicting $v'(\Gamma \Rightarrow \Delta, \varphi) = \top$.

- An argument similar to the previous one applies in the case where $(\Rightarrow \neg\wedge)_2 \in S$, $v'(\Gamma \Rightarrow \Delta, \psi) = \top$, $v'(\Gamma \Rightarrow \Delta, \varphi) \neq f$, and $\varphi \wedge \psi \in \mathcal{F}'$.

Together with Lemma 41, the fact that $\langle W, \leq, v' \rangle$ is an $\mathcal{M}_{\mathcal{IP}}[S]$-semiframe easily entails the Theorem (see the proof of Theorem 37). ∎

Acknowledgment

This research was supported by THE ISRAEL SCIENCE FOUNDATION (grant No 33/02-1).

BIBLIOGRAPHY

[Avron and Lev, 2001] Arnon Avron and Iddo Lev. Canonical propositional Gentzen-type systems. In R. Goré, A Leitsch, and T. Nipkow, editors, *Proc. of the 1st International Joint Conference on Automated Reasoning (IJCAR 2001)*, number 2083 in Lecture Notes in AI, pages 529–544. Springer Verlag, 2001.

[Avron and Lev, 2005] Arnon Avron and Iddo Lev. Non-deterministic multiple-valued structures. *Journal of Logic and Computation*, 15:241–261, 2005.

[Avron, 1986] Arnon Avron. On an implication connective of RM. *Notre Dame Journal of Formal Logic*, 27:201–209, 1986.

[Avron, 1991] Arnon Avron. Natural 3-valued logics: Characterization and proof theory. *Journal of Symbolic Logic*, 56(1):276–294, 1991.

[Baaz, 1986] Matthias Baaz. Kripke-type semantics for da Costa's paraconsistent logic C_ω. *Notre Dame Journal of Formal Logic*, 27:523–527, 1986.

[Batens et al., 1999] D. Batens, K. De Clercq, and N. Kurtonina. Embedding and interpolation for some paralogics. The propositional case. *Reports on Mathematical Logic*, 33:29–44, 1999.

[Batens, 1980] Diderik Batens. Paraconsistent extensional propositional logics. *Logique et Analyse*, 90-91:195–234, 1980.

[Batens, 2000] Diderik Batens. A survey of inconsistency-adaptive logics. In Diderik Batens, Chris Mortensen, Graham Priest, and Jan Paul Van Bendegem, editors, *Frontiers of Paraconsistent Logic*, pages 49–73. King's College Publications, Research Studies Press, Baldock, UK, 2000.

[Carnielli and Marcos, 1999] W. A. Carnielli and J. Marcos. Limits for paraconsistent calculi. *Notre Dame Journal of Formal Logic*, 40:375–390, 1999.

[Carnielli and Marcos, 2002] W. A. Carnielli and J. Marcos. A taxonomy of C-systems. In W. A. Carnielli, M. E. Coniglio, and I. L. M. D'Ottaviano, editors, *Paraconsistency — the logical way to the inconsistent*, Lecture notes in pure and applied Mathematics, pages 1–94. Marcell Dekker, 2002.

[Carnielli, 2000] W. A. Carnielli. Possible-translations semantics for paraconsistent logics. In Diderik Batens, Chris Mortensen, Graham Priest, and Jan Paul Van Bendegem, editors, *Frontiers of Paraconsistent Logic*, pages 149–163. King's College Publications, Research Studies Press, Baldock, UK, 2000.

[Crawford and Etherington, 1998] J. M. Crawford and D. W. Etherington. A non-deterministic semantics for tractable inference. In *Proc. of the 15th International Conference on Artificial Intelligence and the 10th Conference on Innovative Applications of Artificial Intelligence*, pages 286–291. MIT Press, Cambridge, 1998.

[da Costa., 1974] Newton C. A. da Costa. On the theory of inconsistent formal systems. *Notre Dame Journal of Formal Logic*, 15:497–510, 1974.

[D'Ottaviano and da Costa, 1970] Itala L. M. D'Ottaviano and Newton C. A. da Costa. Sur un problème de Jaśkowski. *Comptes Rendus de l'Academie de Sciences de Paris (A-B)*, (270):1349–1353, 1970.

[D'Ottaviano, 1985] Itala L. M. D'Ottaviano. The completeness and compactness of a three-valued first-order logic. *Revista Colombiana de Matematicas*, XIX(1-2):31–42, 1985.

[Epstein, 1990] Richard L. Epstein. *The semantic foundation of logic*, volume I: propositional logics, chapter IX. Kluwer Academic Publisher, 1990.

[Gentzen, 1969] Gerhard Gentzen. Investigations into logical deduction. In M. E. Szabo, editor, *The Collected Works of Gerhard Gentzen*, pages 68–131. North Holland, Amsterdam, 1969.

[Loparić, 1986] A. Loparić. A semantical study of some propositional calculi. *The Journal of Non-Classical Logics*, 3:73–95, 1986.

[Raggio, 1968] A.R. Raggio. Propositional sequence-calculi for inconsistent systems. *Notre Dame Journal of Formal Logic*, 9:359–366, 1968.

[Rozoner, 1989] L. I. Rozoner. On interpretation of inconsistent theories. *Information Sciences*, 47:243–266, 1989.

[Schütte, 1960] Kurt Schütte. *Beweistheorie*. Springer, Berlin, 1960.

[Takeuti, 1975] G. Takeuti. *Proof Theory*. American Elsevier Publishing Company, 1975.

Arnon Avron
School of Computer Science, Tel-Aviv University, Israel.
aa@math.tau.ac.il

Possible-Translations Algebraizability

J. Bueno-Soler, M.E. Coniglio and W.A. Carnielli

ABSTRACT.   The interest of investigating combinations of logics
has a philosophical side, a purely logic-theoretical side and promising applicational interests.   This paper concentrates on the logic-
theoretical side and studies some general methods for combination
and decomposition, taking into account that the general idea of combining logics requires methods not only for building new logics, but
also for breaking logics into families of logics with lower semantical
complexity.   The possible-translations semantics, introduced in [8]
and subsequently refined, are a particularly apt tool for analyzing
and providing semantical meaning and algebraic contents to certain
complex logics as paraconsistent logics, and this paper characterizes
the possible-translations semantics (**PTS**) and the concept of algebraizability via **PTS**'s in categorial terms, showing that the product
of finitely-algebraizable (or Blok-Pigozzi algebraizable) propositional
logics is also finitely-algebraizable, under certain very reasonable conditions. Examples and some research directions are discussed.

Keywords: possible-translations semantics; finitely algebraizable logics; categories of logical systems.

1 Algebraizing, splitting and interpreting: the scope of possible-translations semantics

The emergent area of combinations of logics can be approached from three
sides: from a philosophical perspective, from a purely logic-theoretical side
and from the point of view of applications. From the philosophical side, it
is connected with a pluralist view on logic and its consequences; from the
side of applications, many tasks in knowledge representation or software engineering involve combining knowledge, time and spatial reasoning, among
other logic dimensions. But the purely logic-theoretical side offers real chal-
lenges; in particular, the general idea of combining logics involves not only
the construction of new logics starting from simpler components, but also
the concept of breaking logics into families of logics with lower semantical

complexity. In a broad view, the activity of combining logics includes, besides synthesizing logical systems by means of compositional procedures (as, for instance, fibring), also the contrary direction of breaking down a logic in terms of other (less complex) logics. Such a reversing procedure is called *splitting*, as opposite to the process of *splicing* (cf. [10]).

This paper is an expanded version of [4], and our intention here is to present in all details a categorial characterization of the process of splitting logics called *possible-translations semantics* **PTS**'s, (cf. [8] and [9]). Although such semantics are adequate to providing interpretation to several non-standard logics (as to paraconsistent and to many-valued logics, for instance), and were conceived with these aims, they also constitute a widely general method for splitting logics.

On the other hand, by combining algebraic techniques with **PTS**'s one obtains a new notion of algebraizability (cf. [5]) extending the method of finitely algebraizable logics due to W. Blok and D. Pigozzi in [2]. This extended notion of algebraization offers a solution to the question of obtaining an exact algebraic counterpart to certain logics which are not amenable to the method of Blok and Pigozzi, as it is the case of various paraconsistent logics.

The results of this paper are triple: first, to obtain a representation theorem for possible-translations semantics (**PTS**'s); second, to characterize the concept of algebraizability via **PTS**'s in categorial terms, and third to proof a Finiteness Preservation result (Lemma 21), showing that, under certain conditions, the product of finitely-algebraizable (or Blok-Pigozzi algebraizable) propositional logics is also finitely-algebraizable.

Our intention is not only to present such results as if they came out of the blue, but to reconstruct the main steps of our investigations devoted to reach the adequate definitions and the precise constraints under which some basic lemmas hold, aiming to leave a method for the benefit of the reader, so that other topics in the more technical side of logic could be treated in analogous way.

This is achieved by defining the categories **PS** of propositional languages, and **CR** of propositional logics (defined through consequence relations), and showing that they are closed under arbitrary products. This permits to specify the category **ACR** of algebraizable logics as a subcategory of **CR**. Examples and some research directions are also discussed.

2 Propositional languages in a categorial clothing

This long section gives, in a good number of details, several ingredients aimed to characterize propositional languages in general in a categorial clothing. This, on the contrary of making them costly, make propositional languages a very useful resource, as we shall make clear.

DEFINITION 1. A signature is a denumerable family $\Sigma = \{\Sigma_k\}_{k \in \omega}$, where each Σ_k is a set (of connectives of arity k) such that $\Sigma_k \cap \Sigma_n = \emptyset$ if $k \neq n$. The domain of Σ is the set $|\Sigma| = \bigcup_{n \in \omega} \Sigma_n$. We fix a denumerable set $\mathcal{V} = \{p_k : k \in \omega, k \geq 1\}$ of (propositional) variables such that $p_k \neq p_n$ whenever $k \neq n$. The (propositional) language generated by Σ, denoted by $L(\Sigma)$, is the algebra of type Σ freely generated by $\mathcal{V}$. Elements of $L(\Sigma)$ are called formulas. For every $n \geq 0$ let

$$L(\Sigma)[n] = \{\varphi \in L(\Sigma) : \text{the variables occurring in } \varphi \text{ are exactly } p_1, \ldots, p_n\}.$$

We write $\varphi(p_1, \ldots, p_n)$ to indicate that the propositional variables occurring in φ are among $p_1, \ldots, p_n$. The notion of complexity $l(\varphi)$ of a formula φ is defined as usual, stipulating that $l(\varphi) = 1$ whenever $\varphi \in \mathcal{V} \cup \Sigma_0$ and $l(c(\alpha_1, \ldots, \alpha_n)) = 1 + l(\alpha_1) + l(\alpha_2) + \ldots + l(\alpha_n)$, if $c \in \Sigma_n$.

DEFINITION 2. Let Σ be a signature. A substitution on $L(\Sigma)$ is a function $\sigma : \mathcal{V} \to L(\Sigma)$. We denote by $\widehat{\sigma}$ the unique extension of σ to an endomorphism $\widehat{\sigma} : L(\Sigma) \to L(\Sigma)$, such that

(a) $\widehat{\sigma}(p) = \sigma(p)$, if $p \in \mathcal{V}$;

(b) $\widehat{\sigma}(c) = c$, if $c \in \Sigma_0$;

(c) $\widehat{\sigma}(c(\alpha_1, \ldots, \alpha_n)) = c(\widehat{\sigma}(\alpha_1), \ldots, \widehat{\sigma}(\alpha_n))$, if $c \in \Sigma_n$ and $\alpha_1, \ldots, \alpha_n \in L(\Sigma)$.

Given substitutions $\sigma, \sigma' : \mathcal{V} \to L(\Sigma)$ then their product $\sigma'\sigma$ is the substitution $\widehat{\sigma'} \circ \sigma$.

NOTATION 3. Given $\varphi(p_1, \ldots, p_n)$ and σ such that $\sigma(p_i) = \alpha_i$ ($i = 1, \ldots, n$) then $\widehat{\sigma}(\varphi)$ will be denoted by $\varphi(\alpha_1, \ldots, \alpha_n)$

DEFINITION 4. Let Σ and Σ' be signatures. A signature morphism f from Σ to Σ', denoted $\Sigma \xrightarrow{f} \Sigma'$, is a map $f : |\Sigma| \to L(\Sigma')$ such that, if $c \in \Sigma_n$ then $f(c) \in L(\Sigma')[n]$.

Given a signature morphism $\Sigma \xrightarrow{f} \Sigma'$, a map $\widehat{f} : L(\Sigma) \to L(\Sigma')$ can be defined in a natural way:

1. $\widehat{f}(p) = p$ if $p \in \mathcal{V}$;

2. $\widehat{f}(c) = f(c)$ if $c \in \Sigma_0$;

3. $\widehat{f}(c(\alpha_1, \ldots, \alpha_n)) = f(c)(\widehat{f}(\alpha_1), \ldots, \widehat{f}(\alpha_n))$ if $c \in \Sigma_n$ and $\alpha_1, \ldots, \alpha_n \in L(\Sigma)$.

LEMMA 5. *The extension $\widehat{f}$ of f is unique.*

Proof. Suppose that $\widehat{g}$ is another extension of f; we shall show, by induction on complexity of formulas, that $\widehat{f} = \widehat{g}$.

- If φ is a propositional letter p, i.e, $p \in \mathcal{V}$:
$$\widehat{f}(\varphi) \overset{\varphi \,=\, p}{=} \widehat{f}(p) = p = \widehat{g}(p) \overset{\varphi \,=\, p}{=} \widehat{g}(\varphi).$$

- If φ is a constant c, i.e., $c \in \Sigma_0$:
$$\widehat{f}(\varphi) \overset{\varphi \,=\, c}{=} \widehat{f}(c) = f(c) = \widehat{g}(c) \overset{\varphi \,=\, c}{=} \widehat{g}(\varphi).$$

- If φ is a formula of the form $c(\alpha_1, \ldots, \alpha_n)$, for, $c \in \Sigma_n$:

    Suppose, by induction hypothesis, that the result is valid for every formula α, such that $l(\alpha) < l(\varphi)$; then:
$$\widehat{f}(\varphi) = \widehat{f}(c(\alpha_1, \ldots, \alpha_n)) = f(c)(\widehat{f}(\alpha_1), \ldots, \widehat{f}(\alpha_n)) \overset{\text{Ind.Hyp.}}{=}$$
$$f(c)(\widehat{g}(\alpha_1), \ldots, \widehat{g}(\alpha_n)) = \widehat{g}(c(\alpha_1, \ldots, \alpha_n)) = \widehat{g}(\varphi).$$

∎

DEFINITION 6. Let $\Sigma \overset{f}{\to} \Sigma'$ and $\Sigma' \overset{g}{\to} \Sigma''$ be signature morphisms. The *composition* $g \cdot f$ of f and g is the signature morphism $\Sigma \overset{g \cdot f}{\to} \Sigma''$ given by the map $\widehat{g} \circ f : |\Sigma| \to L(\Sigma'')$.

The following technical results will be useful.

LEMMA 7. $\widehat{\sigma' \sigma} = \widehat{\sigma'} \circ \widehat{\sigma}$.

Proof. By induction on the complexity of φ.

- If φ is a variable p, i.e, $p \in \mathcal{V}$:
$$\widehat{\sigma' \sigma}(\varphi) \overset{\varphi \,=\, p}{=} \widehat{\sigma' \sigma}(p) \overset{\text{Def.2(a)}}{=} \sigma' \sigma(p) \overset{\text{Def.2}}{=} (\widehat{\sigma'} \circ \sigma)(p) = \widehat{\sigma'}(\sigma(p)) \overset{\text{Def.2(a)}}{=}$$
$$\widehat{\sigma'}(\widehat{\sigma}(p)) \overset{\text{Def.2}}{=} (\widehat{\sigma'} \circ \widehat{\sigma})(p) \overset{\varphi \,=\, p}{=} (\widehat{\sigma'} \circ \widehat{\sigma})(\varphi).$$

- If φ is a constant c, i.e., $c \in \Sigma_0$:
$$\widehat{\sigma' \sigma}(\varphi) \overset{\varphi \,=\, c}{=} \widehat{\sigma' \sigma}(c) \overset{\text{Def.2(b)}}{=} c \overset{\text{Def.2(b)}}{=} \widehat{\sigma'}(c) \overset{\text{Def.2(b)}}{=} \widehat{\sigma'}(\widehat{\sigma}(c)) = (\widehat{\sigma'} \circ \widehat{\sigma})(c) \overset{\varphi \,=\, c}{=}$$
$$(\widehat{\sigma'} \circ \widehat{\sigma})(\varphi).$$

- If φ is a formula $c(\alpha_1, \ldots, \alpha_n)$, for $c \in \Sigma_n$:
$$\widehat{\sigma' \sigma}(\varphi) = \widehat{\sigma' \sigma}(c(\alpha_1, \ldots, \alpha_n)) \overset{\text{Def.2(c)}}{=} c(\widehat{\sigma' \sigma}(\alpha_1), \ldots, \widehat{\sigma' \sigma}(\alpha_n)) \overset{\text{Ind.Hyp.}}{=}$$
$$c(\widehat{\sigma'}(\widehat{\sigma}(\alpha_1)), \ldots, \widehat{\sigma'}(\widehat{\sigma}(\alpha_n))) \overset{\text{Def.2(c)}}{=} \widehat{\sigma'}(c(\widehat{\sigma}(\alpha_1), \ldots, \widehat{\sigma}(\alpha_n))) \overset{\text{Def.2(c)}}{=}$$
$$\widehat{\sigma'}(\widehat{\sigma}(c(\alpha_1, \ldots, \alpha_n))) = (\widehat{\sigma'} \circ \widehat{\sigma})(c(\alpha_1, \ldots, \alpha_n)) = (\widehat{\sigma'} \circ \widehat{\sigma})(\varphi).$$

■

LEMMA 8. *Let $\varphi(p_1, \ldots, p_n)$ be a formula, and let $\sigma, \sigma' : \mathcal{V} \to L(\Sigma)$ be substitutions such that: $\sigma(p_i) = \alpha_i$ $(i = 1, \ldots, n)$. Then $\widehat{\sigma}'(\varphi(\alpha_1, \ldots, \alpha_n)) = \varphi(\widehat{\sigma}'(\alpha_1), \ldots, \widehat{\sigma}'(\alpha_n))$.*

Proof.

$$
\begin{aligned}
\widehat{\sigma}'(\varphi(\alpha_1, \ldots, \alpha_n)) \quad &\overset{\text{c.f. 3}}{=} \quad \widehat{\sigma}'(\widehat{\sigma}(\varphi(p_1, \ldots, p_n))) \\
&= \quad (\widehat{\sigma}' \circ \widehat{\sigma})(\varphi(p_1, \ldots, p_n)) \\
&\overset{\text{c.f. 7}}{=} \quad \widehat{\sigma'\sigma}(\varphi(p_1, \ldots, p_n)) \\
&\overset{\text{c.f. 3}}{=} \quad \varphi(\widehat{\sigma'\sigma}(p_1), \ldots, \widehat{\sigma'\sigma}(p_n)) \\
&\overset{\text{c.f. 7}}{=} \quad \varphi((\widehat{\sigma}' \circ \widehat{\sigma})(p_1), \ldots, (\widehat{\sigma}' \circ \widehat{\sigma})(p_n)) \\
&= \quad \varphi(\widehat{\sigma}'(\widehat{\sigma}(p_1)), \ldots, \widehat{\sigma}'(\widehat{\sigma}(p_n))) \\
&= \quad \varphi(\widehat{\sigma}'(\sigma(p_1)), \ldots, \widehat{\sigma}'(\sigma(p_n))) \\
&\overset{\sigma(p_i) = \alpha_i}{=} \quad \varphi(\widehat{\sigma}'(\alpha_1), \ldots, \widehat{\sigma}'(\alpha_n)).
\end{aligned}
$$

■

LEMMA 9. *Let $\varphi = \varphi(p_1, \ldots, p_n)$ in $L(\Sigma)$, $\alpha_1, \ldots, \alpha_n \in L(\Sigma)$ and $\Sigma \overset{f}{\to} \Sigma'$ a signature morphism. Then $\widehat{f}(\varphi(\alpha_1, \ldots, \alpha_n)) = \widehat{f}(\varphi)(\widehat{f}(\alpha_1), \ldots, \widehat{f}(\alpha_n))$.*

Proof. By induction on the complexity $l(\varphi)$ of φ. Let us denote the sequence $\alpha_1, \ldots, \alpha_n$ by $\vec{\alpha}$.

- If φ is a propositional variable $p_i \in \mathcal{V}$;

$$
\begin{aligned}
\widehat{f}(\varphi(\vec{\alpha})) \quad &\overset{\varphi = p_i}{=} \quad \widehat{f}(p_i(\vec{\alpha})) \\
&= \quad \widehat{f}(\alpha_i) \\
&= \quad p_i(\widehat{f}(\alpha_1), \ldots, \widehat{f}(\alpha_n)) \\
&= \quad \widehat{f}(p_i)(\widehat{f}(\alpha_1), \ldots, \widehat{f}(\alpha_n)) \\
&\overset{\varphi = p_i}{=} \quad \widehat{f}(\varphi)(\widehat{f}(\alpha_1), \ldots, \widehat{f}(\alpha_n))
\end{aligned}
$$

- If φ is a constant $c \in \Sigma_0$, Then $\varphi(\vec{\alpha}) = c$, hence:

$$
\begin{aligned}
\widehat{f}(\varphi(\vec{\alpha})) \quad &\overset{\varphi = c}{=} \quad \widehat{f}(c(\vec{\alpha})) \\
&= \quad f(c)(\widehat{f}(\alpha_1), \ldots, \widehat{f}(\alpha_n)) \\
&= \quad \widehat{f}(c)(\widehat{f}(\alpha_1), \ldots, \widehat{f}(\alpha_n)) \\
&\overset{\varphi = c}{=} \quad \widehat{f}(\varphi)(\widehat{f}(\alpha_1), \ldots, \widehat{f}(\alpha_n))
\end{aligned}
$$

- If $\varphi = c(\beta_1, \ldots, \beta_k)$, with $\beta_i = \beta_i(p_1, \ldots, p_n)$ $(i = 1, \cdots, k)$, then $\varphi(\vec{\alpha}) = c(\beta_1(\vec{\alpha}), \ldots, \beta_k(\vec{\alpha}))$, hence:

$$
\begin{aligned}
\widehat{f}(\varphi(\vec{\alpha})) \quad &= \quad \widehat{f}(c(\beta_1(\vec{\alpha}), \dots, \beta_k(\vec{\alpha}))) \\
&= \quad f(c)(\widehat{f}(\beta_1(\vec{\alpha})), \dots, \widehat{f}(\beta_k(\vec{\alpha}))) \\
&\overset{\text{Ind.Hyp.}}{=} \quad f(c)(\widehat{f}(\beta_1)(\widehat{f}(\alpha_1), \dots, \widehat{f}(\alpha_n)), \dots, \widehat{f}(\beta_k)(\widehat{f}(\alpha_1), \dots, \widehat{f}(\alpha_n))) \\
&= \quad f(c)(\widehat{f}(\beta_1), \dots, \widehat{f}(\beta_k))(\widehat{f}(\alpha_1), \dots, \widehat{f}(\alpha_n)) \\
&= \quad \widehat{f}(c(\beta_1, \dots, \beta_k))(\widehat{f}(\alpha_1), \dots, \widehat{f}(\alpha_n)) \\
&= \quad \widehat{f}(\varphi)(\widehat{f}(\alpha_1), \dots, \widehat{f}(\alpha_n)).
\end{aligned}
$$

$\blacksquare$

LEMMA 10. *Let $\Sigma \overset{f}{\to} \Sigma'$ and $\Sigma' \overset{g}{\to} \Sigma''$ be signature morphisms. Then $\widehat{g \cdot f} = \widehat{g} \circ \widehat{f}$.*

Proof. By induction on the complexity of $l(\varphi)$, for $\varphi \in L(\Sigma)$, we prove that $\widehat{g \cdot f}(\varphi) = \widehat{g} \circ \widehat{f}(\varphi)$.

- If $\varphi = p$, $p \in \mathcal{V}$, then:

$$
\widehat{g \cdot f}(\varphi) \overset{\varphi \equiv p}{=} \widehat{g \cdot f}(p) = p = \widehat{g}(p) = \widehat{g}(\widehat{f}(p)) = \widehat{g} \circ \widehat{f}(p) \overset{\varphi \equiv p}{=} \widehat{g} \circ \widehat{f}(\varphi).
$$

- If $\varphi = c$, for $c \in \Sigma_0$, then:

$$
\widehat{g \cdot f}(\varphi) \overset{\varphi \equiv c}{=} \widehat{g \cdot f}(c) = g \cdot f(c) \overset{\text{Def.6}}{=} \widehat{g} \circ f(c) = \widehat{g}(f(c)) = \widehat{g}(\widehat{f}(c)) = \widehat{g} \circ \widehat{f}(c) \overset{\varphi \equiv c}{=} \widehat{g} \circ \widehat{f}(\varphi).
$$

- If $\varphi = c(\alpha_1, \dots, \alpha_n)$ and $c \in \Sigma_n$, then:

$$
\begin{aligned}
\widehat{g \cdot f}(\varphi) \quad &= \quad \widehat{g \cdot f}(c(\alpha_1, \dots, \alpha_n)) \\
&= \quad g \cdot f(c)(\widehat{g \cdot f}(\alpha_1), \cdots, \widehat{g \cdot f}(\alpha_n)) \\
&\overset{\text{Def.6}}{=} \quad \widehat{g} \circ f(c)(\widehat{g \cdot f}(\alpha_1), \dots, \widehat{g \cdot f}(\alpha_n)) \\
&= \quad \widehat{g}(f(c))(\widehat{g \cdot f}(\alpha_1), \dots, \widehat{g \cdot f}(\alpha_n)) \\
&\overset{\text{Ind.Hyp.}}{=} \quad \widehat{g}(f(c))(\widehat{g} \circ \widehat{f}(\alpha_1), \cdots, \widehat{g} \circ \widehat{f}(\alpha_n)) \\
&= \quad \widehat{g}(f(c))(\widehat{g}(\widehat{f}(\alpha_1)), \cdots, \widehat{g}(\widehat{f}(\alpha_n))) \\
&\overset{\text{c.f. } 9}{=} \quad \widehat{g}(f(c)(\widehat{f}(\alpha_1), \cdots, \widehat{f}(\alpha_n))) \\
&= \quad \widehat{g}(\widehat{f}(c(\alpha_1, \dots, \alpha_n))) \\
&= \quad \widehat{g}(\widehat{f}(\varphi)) \\
&= \quad \widehat{g} \circ \widehat{f}(\varphi).
\end{aligned}
$$

$\blacksquare$

LEMMA 11. *Let $\Sigma \overset{f}{\to} \Sigma'$ be a signature morphism, and let $\sigma : \mathcal{V} \to L(\Sigma)$ be a substitution over Σ. Then there is a substitution $\sigma' : \mathcal{V} \to L(\Sigma')$ over Σ' such that $\widehat{f} \circ \widehat{\sigma} = \widehat{\sigma'} \circ \widehat{f}$.*

Proof. Define $\sigma'(p) = \widehat{f}(\sigma(p))$ for every $p \in \mathcal{V}$. By induction on the complexity $l(\alpha)$ we can prove that $\widehat{f}(\widehat{\sigma}(\alpha)) = \widehat{\sigma'}(\widehat{f}(\alpha))$ for every α.

- If $\alpha = p \in \mathcal{V}$, then:
  $$\widehat{f}(\widehat{\sigma}(\alpha)) \overset{\alpha\equiv p}{=} \widehat{f}(\widehat{\sigma}(p)) \overset{\text{Def.2(a)}}{=} \widehat{f}(\sigma(p)) = \sigma'(p) \overset{\text{Def.2(a)}}{=} \widehat{\sigma}'(p) \overset{\text{Def.2(a)}}{=} \widehat{\sigma}'(\widehat{f}(p)) \overset{\alpha\equiv p}{=}$$
  $$\widehat{\sigma}'(\widehat{f}(\alpha))$$

- If $\alpha = c \in \Sigma_0$, then:
  $$\widehat{f}(\widehat{\sigma}(\alpha)) \overset{\alpha\equiv c}{=} \widehat{f}(\widehat{\sigma}(c)) \overset{\text{Def.2(b)}}{=} \widehat{f}(c) = f(c) \overset{\text{Def.2(b)}}{=} \widehat{\sigma}'(\widehat{f}(c)) \overset{\alpha\equiv c}{=} \widehat{\sigma}'(\widehat{f}(\alpha)),$$
  because $f(c) \in L(\Sigma')[0]$.

- If $\alpha = c(\alpha_1, \ldots, \alpha_k)$ for $c \in \Sigma_k$, then:
  $$\begin{aligned}
  \widehat{f}(\widehat{\sigma}(\alpha)) \quad &\overset{\text{Sub.}}{=} \quad \widehat{f}(\widehat{\sigma}(c(\alpha_1, \ldots, \alpha_k))) \\
  &\overset{\text{Def.2 (c)}}{=} \quad \widehat{f}(c(\widehat{\sigma}(\alpha_1), \ldots, \widehat{\sigma}(\alpha_k))) \\
  &= \quad f(c)(\widehat{f}(\widehat{\sigma}(\alpha_1)), \ldots, \widehat{f}(\widehat{\sigma}(\alpha_k))) \\
  &\overset{\text{Ind.Hyp.}}{=} \quad f(c)(\widehat{\sigma}'(\widehat{f}(\alpha_1)), \ldots, \widehat{\sigma}'(\widehat{f}(\alpha_k))) \\
  &\overset{\text{c.f. 8}}{=} \quad \widehat{\sigma}'(f(c)(\widehat{f}(\alpha_1), \ldots, \widehat{f}(\alpha_k))) \\
  &= \quad \widehat{\sigma}'(\widehat{f}(c(\alpha_1, \ldots, \alpha_k))) \\
  &= \quad \widehat{\sigma}'(\widehat{f}(\alpha))
  \end{aligned}$$

∎

DEFINITION 12. The category **PS** of (propositional) languages is defined as follows:

- Objects: Propositional signatures (cf. Definition 1);

- Morphisms: Signature morphisms (cf. Definition 4);

- Composition: As in Definition 6;

- Identity morphisms: For every signature Σ the identity morphism $\Sigma \overset{id_\Sigma}{\to} \Sigma$ is defined by:
  $id_\Sigma(c) = c$ (for $c \in \Sigma_0$) and
  $id_\Sigma(c) = c(p_1, \ldots, p_n)$ (for $c \in \Sigma_n$, $n \geq 1$).

PROPOSITION 13. **PS** *is, in fact, a category.*

Proof. Let $\Sigma \overset{f}{\to} \Sigma' \overset{g}{\to} \Sigma'' \overset{h}{\to} \Sigma'''$ be signature morphisms. We must show that the composition . is associative.

$$h \cdot (g \cdot f) \overset{\text{Def.6}}{=} h \cdot (\widehat{g} \circ f) \overset{\text{Def.6}}{=} \widehat{h} \circ (\widehat{g} \circ f) = (\widehat{h} \circ \widehat{g}) \circ f \overset{\text{c.f. 10}}{=} \widehat{\widehat{h} \cdot g} \circ f \overset{\text{Def.6}}{=} (h \cdot g) \cdot f.$$

Finally, it rests to show that the signature morphisms $\Sigma' \overset{f}{\to} \Sigma \overset{g}{\to} \Sigma''$ verify the following identities: $id_\Sigma \cdot f = f$ and $g \cdot id_\Sigma = g$. But it easily follows from Lemma 7, Definition 6 and Lemma 5. ∎

PROPOSITION 14. *The category* **PS** *has products of arbitrary (small, non-empty) families of objects.*

Proof. Let $\mathcal{F} = \{\Sigma^i\}_{i \in I}$ be a family of signatures such that I is a non-empty set. Consider the signature $\Sigma^{\mathcal{F}}$ such that, for every $n \in \omega$,

$$\Sigma_n^{\mathcal{F}} = \{(\varphi_i)_{i \in I} : \varphi_i \in L(\Sigma^i)[n] \text{ for every } i \in I\}.$$

For each $i \in I$, consider the map $\pi_i : |\Sigma^{\mathcal{F}}| \to L(\Sigma^i)$ such that $\pi_i((\varphi_i)_{i \in I}) = \varphi_i$ if $(\varphi_i)_{i \in I} \in \Sigma_n^{\mathcal{F}}$, for $n \in \omega$. Then π_i determines a **PS**-morphism $\Sigma^{\mathcal{F}} \xrightarrow{\pi_i} \Sigma^i$. Consider a signature Σ' together with **PS**-morphisms $\Sigma' \xrightarrow{f_i} \Sigma^i$, for $i \in I$. Let $f : |\Sigma'| \to L(\Sigma^{\mathcal{F}})$ such that $f(c) = (f_i(c))_{i \in I}(p_1, \ldots, p_n)$ if $c \in \Sigma'_n$, for $n \in \omega$. Then f defines a **PS**-morphism $\Sigma' \xrightarrow{f} \Sigma^{\mathcal{F}}$ such that $f_i = \pi_i \centerdot f$ for every $i \in I$. If $\Sigma' \xrightarrow{g} \Sigma^{\mathcal{F}}$ is a morphism such that $f_i = \pi_i \centerdot g$ for every $i \in I$ then clearly $g = f$. This proves that $\langle \Sigma^{\mathcal{F}}, \{\pi_i\}_{i \in I} \rangle$ is the product of the family $\mathcal{F}$ in **PS**. $\blacksquare$

3   Consequence relations, also in a categorial vestment

In this section we introduce the category **CR** of (propositional) logics defined through consequence relations. As much as for the case of propositional languages, this careful treatment makes the family of logics also a very useful collectivity.

DEFINITION 15. A *(propositional) logic* is a pair $\mathcal{L} = \langle \Sigma, \vdash_{\mathcal{L}} \rangle$, where Σ is a signature (cf. Definition 1) and $\vdash_{\mathcal{L}}$ is a subset of $\wp(L(\Sigma)) \times L(\Sigma)$ satisfying the following properties, for every $\Gamma \cup \Theta \cup \{\varphi\} \subseteq L(\Sigma)$:

- If $\varphi \in \Gamma$ then $\Gamma \vdash_{\mathcal{L}} \varphi$ (Extensivity);

- If $\Gamma \vdash_{\mathcal{L}} \varphi$ and $\Theta \vdash_{\mathcal{L}} \psi$ for all $\psi \in \Gamma$ then $\Theta \vdash_{\mathcal{L}} \varphi$ (Transitivity);

- If $\Gamma \vdash_{\mathcal{L}} \varphi$ then $\Delta \vdash_{\mathcal{L}} \varphi$ for some finite set $\Delta \subseteq \Gamma$ (Finitariness);

- If $\Gamma \vdash_{\mathcal{L}} \varphi$ then $\widehat{\sigma}(\Gamma) \vdash_{\mathcal{L}} \widehat{\sigma}(\varphi)$ for every substitution σ (Structurality).

The relation $\vdash_{\mathcal{L}}$ is called the *consequence relation* of $\mathcal{L}$.

Note that, because of Extensivity and Transitivity, the following property is satisfied by any consequence relation $\vdash_{\mathcal{L}}$:

- If $\Gamma \vdash_{\mathcal{L}} \varphi$ and $\Gamma \subseteq \Theta$ then $\Theta \vdash_{\mathcal{L}} \varphi$ (Monotonicity).

DEFINITION 16. Let $\mathcal{L} = \langle \Sigma, \vdash_{\mathcal{L}} \rangle$ and $\mathcal{L}' = \langle \Sigma', \vdash_{\mathcal{L}'} \rangle$ be logics. A *morphism of logics* $\mathcal{L} \xrightarrow{f} \mathcal{L}'$ from $\mathcal{L}$ to $\mathcal{L}'$ is a **PS**-morphism $\Sigma \xrightarrow{f} \Sigma'$ which is a *translation*, that is, it satisfies, for every $\Gamma \cup \{\varphi\} \subseteq L(\Sigma)$:

$$\Gamma \vdash_{\mathcal{L}} \varphi \text{ implies } \widehat{f}(\Gamma) \vdash_{\mathcal{L}'} \widehat{f}(\varphi).$$

By defining composition of morphisms and identity morphisms as in **PS** we then obtain a category of (propositional) logics defined through consequence relations, called **CR**. A fundamental property of **CR** is the following:

PROPOSITION 17. *The category* **CR** *has products of arbitrary (small, non-empty) families of objects.*

Proof. Let $\mathcal{F} = \{\mathcal{L}_i\}_{i \in I}$ be a family of logics, where I is a non-empty set and each $\mathcal{L}_i$ is of form $\langle \Sigma^i, \vdash_{\mathcal{L}_i} \rangle$. Consider the product $\langle \Sigma^{\mathcal{F}}, \{\pi_i\}_{i \in I} \rangle$ of $\{\Sigma^i\}_{i \in I}$ in the category **PS** (cf. Proposition 14). Let $\vdash_{\mathcal{F}} \subseteq \wp(L(\Sigma^{\mathcal{F}})) \times L(\Sigma^{\mathcal{F}})$ be the relation defined as follows:

$\Gamma \vdash_{\mathcal{F}} \varphi$ iff there exists a finite set $\Delta \subseteq \Gamma$ such that $\widehat{\pi}_i(\Delta) \vdash_{\mathcal{L}_i} \widehat{\pi}_i(\varphi)$ for every $i \in I$.

Let $\mathcal{L}^{\mathcal{F}} = \langle \Sigma^{\mathcal{F}}, \vdash_{\mathcal{F}} \rangle$. We will show that the pair $\langle \mathcal{L}^{\mathcal{F}}, \{\pi_i\}_{i \in I} \rangle$ is the product in the category **CR** of the family $\mathcal{F}$. Firstly we will see that $\mathcal{L}^{\mathcal{F}}$ is a logic, that is, the relation $\vdash_{\mathcal{F}}$ is a consequence relation (cf. Definition 15).

(i) $\vdash_{\mathcal{F}}$ is extensional:
Consider $\Gamma \subseteq L(\Sigma^{\mathcal{F}})$. Let $\varphi \in \Gamma$ and $\Delta = \{\varphi\}$. Then $\varphi \in \Delta$ and Δ is a finite subset of Γ. Since $\langle \Sigma^i, \vdash_{\mathcal{L}_i} \rangle$ is a propositional logic then it satisfies Extensivity, and so $\widehat{\pi}_i(\Delta) \vdash_{\mathcal{L}_i} \widehat{\pi}_i(\varphi)$, for every $i \in I$. But this means that $\Gamma \vdash_{\mathcal{F}} \varphi$.

(ii) $\vdash_{\mathcal{F}}$ is transitive:
Suppose that $\Gamma \vdash_{\mathcal{F}} \varphi$ and $\Theta \vdash_{\mathcal{F}} \psi$ for every $\psi \in \Gamma$. Then there exists a finite subset $\Delta = \{\gamma_1, \ldots, \gamma_n\}$ of Γ such that $\widehat{\pi}_i(\Delta) \vdash_{\mathcal{L}_i} \widehat{\pi}_i(\varphi)$, for every $i \in I$. Let $1 \leq j \leq n$. Then $\Theta \vdash_{\mathcal{F}} \gamma_j$ and so there exists a finite subset Δ_j of Θ such that $\widehat{\pi}_i(\Delta_j) \vdash_{\mathcal{L}_i} \widehat{\pi}_i(\gamma_j)$, for every $i \in I$. Let $\Delta' = \bigcup_{j=1}^{n} \Delta_j$. Then Δ' is a finite subset of Θ such that $\widehat{\pi}_i(\Delta') \vdash_{\mathcal{L}_i} \psi$, for every $\psi \in \widehat{\pi}_i(\Delta_j)$, every $j = 1, \ldots n$ and every $i \in I$, because every $\vdash_{\mathcal{L}_i}$ satisfies Extensivity. Since every $\vdash_{\mathcal{L}_i}$ satisfies Transitivity then $\widehat{\pi}_i(\Delta') \vdash_{\mathcal{L}_i} \widehat{\pi}_i(\gamma_j)$, for every $j = 1, \ldots n$ and every $i \in I$. Using again the Transitivity of $\vdash_{\mathcal{L}_i}$ we infer that $\widehat{\pi}_i(\Delta') \vdash_{\mathcal{L}_i} \widehat{\pi}_i(\varphi)$, for every $i \in I$. Therefore $\Theta \vdash_{\mathcal{F}} \varphi$.

(iii) $\vdash_{\mathcal{F}}$ is finitary by the very definition.

(iv) $\vdash_{\mathcal{F}}$ is structural:
Consider a set $\Gamma \cup \{\varphi\} \subseteq L(\Sigma^{\mathcal{F}})$ such that $\Gamma \vdash_{\mathcal{F}} \varphi$. Then, there is a finite set $\Delta \subseteq \Gamma$ such that $\widehat{\pi}_i(\Delta) \vdash_{\mathcal{L}_i} \widehat{\pi}_i(\varphi)$ for every $i \in I$. Let $\sigma : V \to L(\Sigma^{\mathcal{F}})$

be a substitution over $\Sigma^{\mathcal{F}}$. Since every π_i is a **PS**-morphism then, for every $i \in I$, there exists a substitution $\sigma_i : \mathcal{V} \to L(\Sigma^i)$ over Σ^i such that $\widehat{\pi}_i \circ \widehat{\sigma} = \widehat{\sigma}_i \circ \widehat{\pi}_i$, by Lemma 11. Since each $\mathcal{L}_i$ satisfies Structurality then $\widehat{\sigma}_i(\widehat{\pi}_i(\Delta)) \vdash_{\mathcal{L}_i} \widehat{\sigma}_i(\widehat{\pi}_i(\varphi))$, i.e. $\widehat{\pi}_i(\widehat{\sigma}(\Delta)) \vdash_{\mathcal{L}_i} \widehat{\pi}_i(\widehat{\sigma}(\varphi))$ for every $i \in I$, where $\widehat{\sigma}(\Delta)$ is a finite subset of $\widehat{\sigma}(\Gamma)$. Therefore $\widehat{\sigma}(\Gamma) \vdash_{\mathcal{F}} \widehat{\sigma}(\varphi)$ and so $\vdash_{\mathcal{F}}$ satisfies Structurality.

This shows that $\mathcal{L}^{\mathcal{F}}$ is a logic. By the very definition of $\mathcal{L}^{\mathcal{F}}$, each π_i is a **CR**-morphism $\mathcal{L}^{\mathcal{F}} \overset{\pi_i}{\to} \mathcal{L}_i$. Suppose that $\mathcal{L}' = \langle \Sigma', \vdash_{\mathcal{L}'} \rangle$ is a logic and $\mathcal{L}' \overset{f_i}{\to} \mathcal{L}_i$ is a **CR**-morphism, for every $i \in I$. Then there exists a unique **PS**-morphism $\Sigma' \overset{f}{\to} \Sigma^{\mathcal{F}}$ such that, in **PS**, $\pi_i \centerdot f = f_i$, for every $i \in I$, because $\langle \Sigma^{\mathcal{F}}, \{\pi_i\}_{i \in I} \rangle$ is the product of $\{\Sigma^i\}_{i \in I}$ in the category **PS**. Suppose that $\Gamma \cup \{\varphi\} \subseteq L(\Sigma')$ is such that $\Gamma \vdash_{\mathcal{L}'} \varphi$. Since $\mathcal{L}'$ satisfies Finitariness, there exists a finite set $\Delta \subseteq \Gamma$ such that $\Delta \vdash_{\mathcal{L}'} \varphi$. Since each f_i is a **CR**-morphism then $\widehat{f}_i(\Delta) \vdash_{\mathcal{L}_i} \widehat{f}_i(\varphi)$, for every $i \in I$, hence $\widehat{\pi_i \centerdot f}(\Delta) \vdash_{\mathcal{L}_i} \widehat{\pi_i \centerdot f}(\varphi)$. Using the Lemma 10 we have that $\widehat{\pi}_i \circ \widehat{f}(\Delta) \vdash_{\mathcal{L}_i} \widehat{\pi}_i \circ \widehat{f}(\varphi)$, i.e. $\widehat{\pi}_i(\widehat{f}(\Delta)) \vdash_{\mathcal{L}_i} \widehat{\pi}_i(\widehat{f}(\varphi))$, for every $i \in I$, where $\widehat{f}(\Delta)$ is a finite subset of $\widehat{f}(\Gamma)$. Therefore, by definition of $\vdash_{\mathcal{F}}$ we have that $\widehat{f}(\Gamma) \vdash_{\mathcal{F}} \widehat{f}(\varphi)$ and then f is a **CR**-morphism $\mathcal{L}' \overset{f}{\to} \mathcal{L}^{\mathcal{F}}$ such that, in **CR**, $\pi_i \centerdot f = f_i$, for every $i \in I$. The unicity of f is a consequence of the universal property in the category **PS** of the product $\langle \Sigma^{\mathcal{F}}, \{\pi_i\}_{i \in I} \rangle$. This show that $\langle \mathcal{L}^{\mathcal{F}}, \{\pi_i\}_{i \in I} \rangle$ is the product in the category **CR** of the family $\mathcal{F}$. ∎

4 Products of algebraizable logics

In this section we prove that, given a (small and non-empty) family $\mathcal{F}$ of finitely algebraizable logics (in the sense of Blok-Pigozzi, cf. [2]) satisfying a finite bound condition, then the product of $\mathcal{F}$ in **CR** is also an algebraizable logic. This will we used in Section 6.

We begin by briefly recalling the basic definitions of [2].

DEFINITION 18. A propositional logic $\mathcal{L} = \langle \Sigma, \vdash_{\mathcal{L}} \rangle$ is *algebraizable* (in the sense of Blok-Pigozzi) if there exists a finite set $\Delta = \{\Delta^i(p_1, p_2) : 1 \le i \le n\}$ of formulas in $L(\Sigma)[2]$, and a finite set $\langle \varepsilon, \delta \rangle = \{\langle \varepsilon^i(p_1), \delta^i(p_1) \rangle : 1 \le i \le m\}$ contained in $L(\Sigma)[1] \times L(\Sigma)[1]$ such that, for every $\varphi, \psi, \gamma \in L_{\Sigma}$:

1. $\vdash_{\mathcal{L}} \varphi \Delta \varphi$;

2. $\varphi \Delta \psi \vdash_{\mathcal{L}} \psi \Delta \varphi$;

3. $\varphi \Delta \psi, \psi \Delta \gamma \vdash_{\mathcal{L}} \varphi \Delta \gamma$;

4. $\varphi_1 \Delta \psi_1, \ldots, \varphi_k \Delta \psi_k \vdash_{\mathcal{L}} c(\varphi_1, \ldots, \varphi_k) \Delta c(\psi_1, \ldots, \psi_k)$ for every $c \in \Sigma_k$ and every $\varphi_1, \ldots, \varphi_k, \psi_1, \ldots, \psi_k$ in $L(\Sigma)$;

5. $\varphi \vdash_{\mathcal{L}} \varepsilon(\varphi)\Delta\delta(\varphi)$, and $\varepsilon(\varphi)\Delta\delta(\varphi) \vdash_{\mathcal{L}} \varphi$.

We say that $\langle \Delta, \langle \varepsilon, \delta \rangle \rangle$ is an *algebraizator* for $\mathcal{L}$.

Some remarks on the notation adopted in Definition 18: For any $\varphi, \psi \in L(\Sigma)$ then $\varphi\Delta\psi$ denotes the set of formulas $\{\Delta^i(\varphi, \psi) : 1 \leq i \leq n\}$, and $\varepsilon(\varphi)\Delta\delta(\varphi)$ denotes the set $\{\Delta^j(\varepsilon^i(\varphi), \delta^i(\varphi)) : 1 \leq j \leq n$ and $1 \leq i \leq m\}$. And given sets Γ, Θ of formulas then $\Gamma \vdash_{\mathcal{L}} \Theta$ means that $\Gamma \vdash_{\mathcal{L}} \varphi$ for every $\varphi \in \Theta$. Following [15] and [16] we define the category **ACR** of algebraizable logics.

DEFINITION 19. The category **ACR** of *algebraizable logics* is the subcategory of **CR** defined as follows:

- Objects: propositional logics $\mathcal{L} = \langle \Sigma, \vdash_{\mathcal{L}} \rangle$ which are algebraizable (cf. Definitions 15 and 18);

- Morphisms: a morphism $\mathcal{L} \xrightarrow{f} \mathcal{L}'$ is a **CR**-morphism $\mathcal{L} \xrightarrow{f} \mathcal{L}'$ such that, if $\langle \Delta, \langle \varepsilon, \delta \rangle \rangle$ and $\langle \Delta', \langle \varepsilon', \delta' \rangle \rangle$ are algebraizators for $\mathcal{L}$ and $\mathcal{L}'$, respectively, then $p_1\widehat{f}(\Delta)p_2 \vdash_{\mathcal{L}'} p_1\Delta'p_2$ and $p_1\Delta'p_2 \vdash_{\mathcal{L}'} p_1\widehat{f}(\Delta)p_2$, where $p_1\widehat{f}(\Delta)p_2$ denotes the set of formulas $\{\widehat{f}(\Delta^i)(p_1, p_2) : 1 \leq i \leq n\}$;

- Composition and identity morphisms: inherited from **CR**.

REMARK 20. From [2] we obtain the following: let $\langle \Delta, \langle \varepsilon, \delta \rangle \rangle$ and $\langle \Delta', \langle \varepsilon', \delta' \rangle \rangle$ be two algebraizators for a logic $\mathcal{L}$. Then $p_1\Delta'p_2 \vdash_{\mathcal{L}} p_1\Delta p_2$ and $p_1\Delta p_2 \vdash_{\mathcal{L}} p_1\Delta'p_2$. Therefore, a **CR**-morphism $\mathcal{L} \xrightarrow{f} \mathcal{L}'$ is a **ACR**-morphism iff there are algebraizators $\langle \Delta, \langle \varepsilon, \delta \rangle \rangle$ and $\langle \Delta', \langle \varepsilon', \delta' \rangle \rangle$ for $\mathcal{L}$ and $\mathcal{L}'$, respectively, such that $p_1\widehat{f}(\Delta)p_2 \vdash_{\mathcal{L}'} p_1\Delta'p_2$ and $p_1\Delta'p_2 \vdash_{\mathcal{L}'} p_1\widehat{f}(\Delta)p_2$ (cf. [15] and [16]).

Now we prove that the product of a family of algebraizable logics satisfying a finitely bounding condition is algebraizable.

THEOREM 21. *(Finiteness Preservation) Let $\mathcal{F} = \{\mathcal{L}_i\}_{i \in I}$ be a family of algebraizable logics, where $\mathcal{L}_i = \langle \Sigma^i, \vdash_{\mathcal{L}_i} \rangle$ for every $i \in I$, and I is a nonempty set. Assume that $\mathcal{F}$ has the following property: there are natural numbers n and m such that, for every $i \in I$, there is an algebraizator $\langle \Delta_i, \langle \varepsilon_i, \delta_i \rangle \rangle$ for $\mathcal{L}_i$ such that Δ_i has at most n elements, and $\langle \varepsilon_i, \delta_i \rangle$ has at most m elements. Then, there exists the product in **ACR** of $\mathcal{F}$.*

Proof. By hypothesis we can take, for any $i \in I$, finite sequences

- $\Delta_i^1(p_1, p_2) \cdots \Delta_i^n(p_1, p_2)$

- $\langle \varepsilon_i^1(p_1), \delta_i^1(p_1) \rangle \cdots \langle \varepsilon_i^m(p_1), \delta_i^m(p_1) \rangle$

such that $\langle \Delta_i, \langle \varepsilon_i, \delta_i \rangle \rangle$ is an algebraizator for $\mathcal{L}_i$, where

- $\Delta_i = \{\Delta_i^1(p_1, p_2), \dots, \Delta_i^n(p_1, p_2)\}$

- $\langle \varepsilon_i, \delta_i \rangle = \{\langle \varepsilon_i^1(p_1), \delta_i^1(p_1) \rangle, \dots, \langle \varepsilon_i^m(p_1), \delta_i^m(p_1) \rangle\}$, for every $i \in I$.

In fact, it is enough to consider, for every $i \in I$, an algebraizator with at most n elements in Δ_i and at most m elements in $\langle \varepsilon_i, \delta_i \rangle$ and list their elements, repeating, if necessary, some elements, in order to define sequences of length n and m, respectively. Now, consider the product $\langle \mathcal{L}^{\mathcal{F}}, \{\pi_i\}_{i \in I} \rangle$ in **CR** of the family $\mathcal{F}$ (recall the proof of Proposition 17), and define the following formulas in $L(\Sigma^{\mathcal{F}})$:

- $\Delta_{\mathcal{F}}^j(p_1, p_2) = (\Delta_i^j(p_1, p_2))_{i \in I}(p_1, p_2)$ for $1 \le j \le n$;

- $\varepsilon_{\mathcal{F}}^j(p_1) = (\varepsilon_i^j(p_1))_{i \in I}(p_1)$ for $1 \le j \le m$;

- $\delta_{\mathcal{F}}^j(p_1) = (\delta_i^j(p_1))_{i \in I}(p_1)$ for $1 \le j \le m$.[1]

Finally, let:

- $\Delta_{\mathcal{F}} = \{\Delta_{\mathcal{F}}^j(p_1, p_2) : 1 \le j \le n\}$

- $\langle \varepsilon_{\mathcal{F}}, \delta_{\mathcal{F}} \rangle = \{\langle \varepsilon_{\mathcal{F}}^j(p_1), \delta_{\mathcal{F}}^j(p_1) \rangle : 1 \le j \le m\}$.

We will show that $\langle \Delta_{\mathcal{F}}, \langle \varepsilon_{\mathcal{F}}, \delta_{\mathcal{F}} \rangle \rangle$ is a algebraizator for $\mathcal{L}^{\mathcal{F}}$ (cf. Definition 18).

(1.) $\vdash_{\mathcal{F}} \varphi \Delta_{\mathcal{F}} \varphi$.
Since $\langle \Delta_i, \langle \varepsilon_i, \delta_i \rangle \rangle$ is a algebraizator for $\mathcal{L}_i$ then $\vdash_{\mathcal{L}_i} \widehat{\pi}_i(\varphi) \Delta_i \widehat{\pi}_i(\varphi)$ for every $i \in I$. By definition of $\Delta_{\mathcal{F}}$ we have that $\vdash_{\mathcal{L}_i} \widehat{\pi}_i(\varphi \Delta_{\mathcal{F}} \varphi)$ for every $i \in I$ and hence $\vdash_{\mathcal{F}} \varphi \Delta_{\mathcal{F}} \varphi$.

(2.) $\varphi \Delta_{\mathcal{F}} \psi \vdash_{\mathcal{F}} \psi \Delta_{\mathcal{F}} \varphi$.
Since, for every $i \in I$, $\langle \Delta_i, \langle \varepsilon_i, \delta_i \rangle \rangle$ is a algebraizator for $\mathcal{L}_i$ then we know that $\widehat{\pi}_i(\varphi) \Delta_i \widehat{\pi}_i(\psi) \vdash_{\mathcal{L}_i} \widehat{\pi}_i(\psi) \Delta_i \widehat{\pi}_i(\varphi)$. By definition of $\Delta_{\mathcal{F}}$ we have that $\widehat{\pi}_i(\varphi \Delta_{\mathcal{F}} \psi) \vdash_{\mathcal{L}_i} \widehat{\pi}_i(\psi \Delta_{\mathcal{F}} \varphi)$, for every $i \in I$, and hence $\varphi \Delta_{\mathcal{F}} \psi \vdash_{\mathcal{F}} \psi \Delta_{\mathcal{F}} \varphi$.

(3.) $\varphi \Delta_{\mathcal{F}} \psi, \psi \Delta_{\mathcal{F}} \gamma \vdash_{\mathcal{F}} \varphi \Delta_{\mathcal{F}} \gamma$.
Similar to case (ii).

---

[1] Recall that the family of formulas $(\Delta_i^j(p_1, p_2))_{i \in I}$ is a binary connective of signature $\Sigma^{\mathcal{F}}$. Analogously, the families $(\varepsilon_i^j(p_1))_{i \in I}$ and $(\delta_i^j(p_1))_{i \in I}$ are unary connectives of $\Sigma^{\mathcal{F}}$. This justify the apparently redundant notation employed here.

(4.) Since $\mathcal{L}^{\mathcal{F}}$ is structural, it is enough to prove the following:
$p_1\Delta_{\mathcal{F}}p_{k+1},\ldots,p_k\Delta_{\mathcal{F}}p_{2k} \vdash_{\mathcal{F}} c(p_1,\ldots,p_k)\Delta_{\mathcal{F}}c(p_{k+1},\ldots,p_{2k})$ for every $c \in \Sigma_k^{\mathcal{F}}$. Let $c = (\varphi_i)_{i\in I} \in \Sigma_k^{\mathcal{F}}$.
If $k = 0$ the result follows from clause (i) of Definition 18.
Suppose that $k > 0$. By induction on the length of $\varphi = \varphi(p_1,\ldots,p_k)$ it is easy to show that, if $\langle\Delta', \langle\varepsilon', \delta'\rangle\rangle$ is a algebraizator for a logic $\mathcal{L}'$, then

$$\varphi_1\Delta'\psi_1,\ldots,\varphi_k\Delta'\psi_k \vdash_{\mathcal{L}'} \varphi(\varphi_1,\ldots,\varphi_k)\Delta'\varphi(\psi_1,\ldots,\psi_k)$$

for every $\varphi_1,\ldots,\varphi_k$, $\psi_1,\ldots,\psi_k$ in $L(\Sigma')$. In particular, for every $i \in I$ it holds that $p_1\Delta_i p_{k+1},\ldots,p_k\Delta_i p_{2k} \vdash_{\mathcal{L}_i} \varphi_i(p_1,\ldots,p_k)\Delta_i\varphi_i(p_{k+1},\ldots,p_{2k})$, that is,

$$\widehat{\pi}_i(p_1\Delta_{\mathcal{F}}p_{k+1}),\ldots,\widehat{\pi}_i(p_k\Delta_{\mathcal{F}}p_{2k}) \vdash_{\mathcal{L}_i} \widehat{\pi}_i(c(p_1,\ldots,p_k)\Delta_{\mathcal{F}}c(p_{k+1},\ldots,p_{2k})).$$

Then $p_1\Delta_{\mathcal{F}}p_{k+1},\ldots,p_k\Delta_{\mathcal{F}}p_{2k} \vdash_{\mathcal{F}} c(p_1,\ldots,p_k)\Delta_{\mathcal{F}}c(p_{k+1},\ldots,p_{2k})$.

(5.) $\varphi \dashv\vdash_{\mathcal{F}} \varepsilon_{\mathcal{F}}(\varphi)\Delta_{\mathcal{F}}\delta_{\mathcal{F}}(\varphi)$.[2]
Since, for every $i \in I$, $\langle\Delta_i, \langle\varepsilon_i, \delta_i\rangle\rangle$ is a algebraizator for $\mathcal{L}_i$ then $\widehat{\pi}_i(\varphi) \dashv\vdash_{\mathcal{L}_i} \varepsilon_i(\widehat{\pi}_i(\varphi))\Delta_i\delta_i(\widehat{\pi}_i(\varphi))$. By definition of $\langle\Delta_{\mathcal{F}}, \langle\varepsilon_{\mathcal{F}}, \delta_{\mathcal{F}}\rangle\rangle$ we have that $\widehat{\pi}_i(\varphi) \dashv\vdash_{\mathcal{L}_i} \widehat{\pi}_i(\varepsilon_{\mathcal{F}}(\varphi)\Delta_{\mathcal{F}}\delta_{\mathcal{F}}(\varphi))$, for every $i \in I$. From this, we conclude that

$$\varphi \dashv\vdash_{\mathcal{F}} \varepsilon_{\mathcal{F}}(\varphi)\Delta_{\mathcal{F}}\delta_{\mathcal{F}}(\varphi).$$

This shows that $\langle\Delta, \langle\varepsilon, \delta\rangle\rangle$ is an algebraizator for $\mathcal{L}^{\mathcal{F}}$.

Finally, we must prove that $\langle\mathcal{L}^{\mathcal{F}}, \{\pi_i\}_{i\in I}\rangle$ is the product in **ACR** of the family $\mathcal{F}$. Using Remark 20, it is clear that every projection π_i is a **ACR**-morphism. Suppose that $\mathcal{L}'$ is an algebraizable logic having a **ACR**-morphism $\mathcal{L}'\xrightarrow{f_i}\mathcal{L}_i$, for every $i \in I$, and let $\langle\Delta', \langle\varepsilon', \delta'\rangle\rangle$ be an algebraizator for $\mathcal{L}'$. Using the universal property of $\langle\mathcal{L}^{\mathcal{F}}, \{\pi_i\}_{i\in I}\rangle$ in **CR** we obtain a **CR**-morphism $\mathcal{L}'\xrightarrow{f}\mathcal{L}^{\mathcal{F}}$ such that, in **CR**, $f_i = \pi_i \cdot f$ for every $i \in I$. Since $p_1\widehat{f_i}(\Delta')p_2 \vdash_{\mathcal{L}_i} p_1\Delta_i p_2$ for every $i \in I$ then $\widehat{\pi}_i(p_1\widehat{f}(\Delta')p_2) \vdash_{\mathcal{F}} \widehat{\pi}_i(p_1\Delta_{\mathcal{F}}p_2)$ for every $i \in I$, by Lemma 10, thus $p_1\widehat{f}(\Delta')p_2 \vdash_{\mathcal{F}} p_1\Delta_{\mathcal{F}}p_2$. Analogously we prove that $p_1\Delta_{\mathcal{F}}p_2 \vdash_{\mathcal{F}} p_1\widehat{f}(\Delta')p_2$. Using Remark 20, this shows that f is a **ACR**-morphism such that, in **ACR**, $f_i = \pi_i \cdot f$ for every $i \in I$. The uniqueness of f follows from the universal property of $\langle\mathcal{L}^{\mathcal{F}}, \{\pi_i\}_{i\in I}\rangle$ in **CR**. ∎

5 Possible-translations semantics, and when they can be replaced by just one translation

Combining and factoring logics are two sides of the same coin. Besides the interest on the synthesis of given logics by means of a combination procedure

---

[2] Here $\Delta \dashv\vdash_{\mathcal{L}} \Gamma$ denotes that $\Delta \vdash_{\mathcal{L}} \Gamma$ and $\Gamma \vdash_{\mathcal{L}} \Delta$.

in order to obtain a new logic (as is the case, for instance, when fibring logics), it is also interesting to split a logic into a family of other (hopefully simpler) logics. This kind of 'reverse' technique is what is called *splitting logics*, as opposite to the process of *splicing logics* (cf. [10]; see also [11]). In this section we provide a categorial characterization of the process of splitting logics called *possible-translations semantics* (cf. [9]). Possible-translations semantics are a tool for assigning semantical interpretations to logics in general, but this interpretation is done in such a way that a factoring or splitting of the logic in terms of simpler logics is obtained.

We begin by adapting the original definitions of [9] to our formalism.

DEFINITION 22. Let $\mathcal{L} = \langle \Sigma, \vdash_{\mathcal{L}} \rangle$ be a logic, and let $\{\mathcal{L}_i\}_{i \in I}$ be a family of logics such that I is a non-empty set and $\mathcal{L}_i = \langle \Sigma^i, \vdash_{\mathcal{L}_i} \rangle$ for every $i \in I$. Let $\mathcal{L} \overset{f_i}{\to} \mathcal{L}^i$ be a **CR**-morphism for every $i \in I$. Then $P = \langle \{\mathcal{L}_i\}_{i \in I}, \{f_i\}_{i \in I} \rangle$ is a *possible-translations semantics for* $\mathcal{L}$ (in short, a **PTS**) if, for every $\Gamma \cup \{\varphi\} \subseteq L(\Sigma)$,

$$\Gamma \vdash_{\mathcal{L}} \varphi \text{ iff there is a finite set } \Delta \subseteq \Gamma \text{ such that } \widehat{f_i}(\Delta) \vdash_{\mathcal{L}_i} \widehat{f_i}(\varphi) \text{ for every } i \in I.$$

The meaning of having a **PTS** for a logic $\mathcal{L}$ is that $\mathcal{L}$ splits into the family $\{\mathcal{L}_i\}_{i \in I}$ through the translations $\{f_i\}_{i \in I}$.

Inspired by [13] we say that a **CR**-morphism $\mathcal{L} \overset{f}{\to} \mathcal{L}'$ is a *conservative translation* if, for every $\Gamma \cup \{\varphi\} \subseteq L(\Sigma)$,

$$\Gamma \vdash_{\mathcal{L}} \varphi \text{ iff } \widehat{f}(\Gamma) \vdash_{\mathcal{L}'} \widehat{f}(\varphi).$$

Each logic $\mathcal{L}_i$ is called a *traduct* of $\mathcal{L}$.

Using the results stated in the previous sections we can characterize **PTS**'s in categorial terms.

We show below that possible-translations semantics for a logic induce conservative translations over products of families of logics in **CR**, and vice-versa. Such (induced) conservative translations turn out to be apt to extend the method of finite algebraizability with interesting applications, as shown in Section 6.

THEOREM 23. *Given a possible-translations semantics for a logic $\mathcal{L}$ there exists a conservative translation $\mathcal{L} \overset{f}{\to} \mathcal{L}'$, where $\mathcal{L}'$ is a product in **CR** of some family of logics indexed by a non-empty set. Conversely, every conservative translation $\mathcal{L} \overset{f}{\to} \mathcal{L}'$, where $\mathcal{L}'$ is a product in **CR** of some family of logics indexed by a non-empty set, induces a possible-translations semantics for $\mathcal{L}$.*

Proof. Let $P = \langle\{\mathcal{L}_i\}_{i\in I}, \{f_i\}_{i\in I}\rangle$ be a possible-translations semantics for $\mathcal{L}$, and consider the product $\langle\mathcal{L}^{\mathcal{F}}, \{\pi_i\}_{i\in I}\rangle$ of the family $\mathcal{F} = \{\mathcal{L}_i\}_{i\in I}$ in the category **CR** (cf. Proposition 17). By the universal property of products, there exists a unique **CR**-morphism $\mathcal{L} \xrightarrow{\mathbf{t}(P)} \mathcal{L}^{\mathcal{F}}$ such that $f_i = \pi_i \cdot \mathbf{t}(P)$ for every $i \in I$. We claim that $\mathbf{t}(P)$ is a conservative translation that encodes P. Since $\mathbf{t}(P)$ is a translation (because it is a **CR**-morphism), it is enough to show that, for every $\Gamma \cup \{\varphi\} \subseteq L(\Sigma)$,

$$\widehat{\mathbf{t}(P)}(\Gamma) \vdash_{\mathcal{F}} \widehat{\mathbf{t}(P)}(\varphi) \text{ implies that } \Gamma \vdash_{\mathcal{L}} \varphi.$$

Assume that $\widehat{\mathbf{t}(P)}(\Gamma) \vdash_{\mathcal{F}} \widehat{\mathbf{t}(P)}(\varphi)$. Then, by definition of $\mathcal{L}^{\mathcal{F}}$, there exists a finite set $\Delta \subseteq \Gamma$ such that

$$\widehat{\pi_i}(\widehat{\mathbf{t}(P)}(\Delta)) \vdash_{\mathcal{L}_i} \widehat{\pi_i}(\widehat{\mathbf{t}(P)}(\varphi))$$

for every $i \in I$ and, then, using Lemma 10 and the fact that $f_i = \pi_i \cdot \mathbf{t}(P)$ for every $i \in I$, we have that $\widehat{f_i}(\Delta) \vdash_{\mathcal{L}_i} \widehat{f_i}(\varphi)$ for every $i \in I$. Since P is a possible-translations semantics for $\mathcal{L}$, we obtain that $\Gamma \vdash_{\mathcal{L}} \varphi$ and then $\mathbf{t}(P)$ is a conservative translation. Clearly, $\mathbf{t}(P)$ together with its codomain $\mathbf{L}(P) = \mathcal{L}^{\mathcal{F}}$ codify P: every logic $\mathcal{L}_i$ is obtained as the codomain of π_i, and every translation f_i is obtained as $f_i = \pi_i \cdot \mathbf{t}(P)$.

Conversely, let $\mathcal{L} \xrightarrow{f} \mathcal{L}'$ be a conservative translation, where $\mathcal{L}'$ is a product in **CR** of a family $\{\mathcal{L}_i\}_{i\in I}$ of logics with projections π_i for every $i \in I$, such that I is a non-empty set. Let $f_i = \pi_i \cdot f$ for every $i \in I$, and define $\mathbf{PT}(f) = \langle\{\mathcal{L}_i\}_{i\in I}, \{f_i\}_{i\in I}\rangle$. We will show that $\mathbf{PT}(f)$ is a **PTS** for $\mathcal{L}$ such that $\mathbf{t}(\mathbf{PT}(f)) = f$. Thus, let $\Gamma \cup \{\varphi\} \subseteq L(\Sigma)$. Since $\mathcal{L}$ satisfies Finitariness and every f_i is a translation, $\Gamma \vdash_{\mathcal{L}} \varphi$ implies that there is a finite set $\Delta \subseteq \Gamma$ such that $\widehat{f_i}(\Delta) \vdash_{\mathcal{L}_i} \widehat{f_i}(\varphi)$ for every $i \in I$. On the other hand, suppose that there exists a finite set $\Delta \subseteq \Gamma$ such that $\widehat{f_i}(\Delta) \vdash_{\mathcal{L}_i} \widehat{f_i}(\varphi)$ for every $i \in I$. By Lemma 10,

$$\widehat{\pi_i}(\widehat{f}(\Delta)) \vdash_{\mathcal{L}_i} \widehat{\pi_i}(\widehat{f}(\varphi))$$

for every $i \in I$, where $\widehat{f}(\Delta) \subseteq \widehat{f}(\Gamma)$ is finite. Then $\widehat{f}(\Gamma) \vdash_{\mathcal{F}} \widehat{f}(\varphi)$ and so $\Gamma \vdash_{\mathcal{L}} \varphi$, by the fact that f is conservative. This shows that $\mathbf{PT}(f)$ is a **PTS** for $\mathcal{L}$. Clearly, we recover the information about f and $\mathcal{L}'$ from $\mathbf{PT}(f)$: in fact, $f = \mathbf{t}(\mathbf{PT}(f))$ and $\mathcal{L}'$ is the product of a family of logics of $\mathbf{PT}(f)$. Finally, it is clear that, if P is a **PTS** for $\mathcal{L}$ then $\mathbf{PT}(\mathbf{t}(P)) = P$. This concludes the proof. ■

6 Algebraizing logics via possible-translations semantics

The results above give support to a method for algebraizing logics using **PTS**'s which extends the well-known method of finite algebraizability of

[2].

DEFINITION 24. A *possible-translations algebraic semantics* (in short, a **PTAS**) for a propositional logic $\mathcal{L}$ is a triple $\mathbf{PA} = \langle \{\mathcal{L}_i\}_{i \in I}, \{\mathcal{A}_i\}_{i \in I}, \{f_i\}_{i \in I} \rangle$ such that:
(i) $P = \langle \{\mathcal{L}_i\}_{i \in I}, \{f_i\}_{i \in I} \rangle$ is a possible-translations semantics for $\mathcal{L}$;
(ii) For each $i \in I$, $\mathcal{A}_i = \langle \Delta_i, \langle \varepsilon_i, \delta_i \rangle \rangle$ is an algebraizator for $\mathcal{L}_i$ (cf. Definition 18).

The idea, proposed and studied in [5] and [4], is the following: consider a propositional logic $\mathcal{L}$, and let $P = \langle \{\mathcal{L}_i\}_{i \in I}, \{f_i\}_{i \in I} \rangle$ be a **PTS** for $\mathcal{L}$. Suppose that every $\mathcal{L}_i$ is algebraizable, and assume that the family $\mathcal{F} = \{\mathcal{L}_i\}_{i \in I}$ satisfies the condition of Theorem 21. Then, by Theorem 23, the product $\langle \mathcal{L}^{\mathcal{F}}, \{\pi_i\}_{i \in I} \rangle$ of the family $\mathcal{F}$ in **ACR** encodes P. Moreover, it is possible to build an algebraizator for $\mathcal{L}^{\mathcal{F}}$ from a bounded (in the sense of Theorem 21) family of algebraizators for $\mathcal{F}$. This shows that there exists a conservative translation $\mathcal{L} \xrightarrow{f} \mathcal{L}^{\mathcal{F}}$, where $\mathcal{L}^{\mathcal{F}}$ is an algebraizable logic. The conservative translation f is a link between $\mathcal{L}$ and $\mathcal{L}^{\mathcal{F}}$ that preserves derivability and so, using the algebraization for $\mathcal{L}^{\mathcal{F}}$, one obtains a kind of 'remote' algebraization for $\mathcal{L}$: in order to algebraically analyze $\mathcal{L}$, it is sufficient to conservative translate $\mathcal{L}$ into $\mathcal{L}^{\mathcal{F}}$ and then analyze the result using the algebraic resources of $\mathcal{L}^{\mathcal{F}}$.

A natural generalization of **PTAS**'s is the following:

DEFINITION 25. Let $\mathcal{L} = \langle \Sigma, \vdash_{\mathcal{L}} \rangle$ be a propositional logic and let

$$P = \langle \{\mathcal{L}_i\}_{i \in I}, \{f_i\}_{i \in I} \rangle$$

be a **PTS** for $\mathcal{L}$. We define recursively the following:
(i) P is a **PTAS** of level 0 if every traduct $\mathcal{L}_i$ admits an algebraizator $\mathcal{A}_i$;
(ii) P is a **PTAS** of level $n + 1$ if every $\mathcal{L}_i$ admits a **PTAS** of level n, for some $n \in \omega$.

If $P = \langle \{\mathcal{L}_i\}_{i \in I}, \{f_i\}_{i \in I} \rangle$ is a **PTAS** for $\mathcal{L}$ of level n, then each $\mathcal{L}_i$ is a traduct of level 0; the traducts of each $\mathcal{L}_i$ are traducts of level 1, and so on. Therefore, just the traducts of level n are algebraizable in the sense of Blok-Pigozzi. Thus, the algebraizators of P are the algebraizators of the traducts of level n. Note that a **PTAS** of level 0 for $\mathcal{L}$ is equivalent to a **PTAS** for $\mathcal{L}$, in the sense of Definition 24. Thus, Definition 25 generalizes Definition 24. However, the next result shows that this generalization is innocuous, under certain hypothesis.

THEOREM 26. *Let* $P = \langle \{\mathcal{L}_i\}_{i \in I}, \{f_i\}_{i \in I} \rangle$ *a* **PTAS** *of level* $n > 1$ *for a propositional logic* $\mathcal{L}$ *such that the algebraizators (of the traducts of level* n*) are globally bounded in the sense of Theorem 21. Then* P *can be transformed* *in a* **PTAS** *for* $\mathcal{L}$ *of level* 0 *and bounded algebraizators.*

Proof. Clearly, it is enough to show how to obtain a **PTAS** of level 0 from a given **PTAS** of level 1. Thus, let $P = \langle\{\mathcal{L}_i\}_{i\in I}, \{f_i\}_{i\in I}\rangle$ be a **PTAS** of level 1 for a propositional logic $\mathcal{L}$. Then P is of the form displayed below (for simplicity, in the figure below we consider finite sets of indices everywhere).

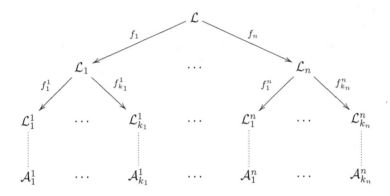

In the general case, for each $i \in I$ the logic $\mathcal{L}_i$ admits a **PTAS**, say

$$\mathbf{PA}_i = \langle\{\mathcal{L}^i_j\}_{j\in J^i}, \{\mathcal{A}^i_j\}_{j\in J^i}, \{f^i_j\}_{j\in J^i}\rangle$$

such that the cardinals of the algebraizators involved are globally bounded, in the sense of Theorem 21; that is, there are bounds r and m for the cardinal of every algebraizator $\mathcal{A}^i_j$. For each $i \in I$ let $P_i = \langle\{\mathcal{L}^i_j\}_{j\in J^i}, \{f^i_j\}_{j\in J^i}\rangle$ be the **PTS** for $\mathcal{L}_i$ obtained from $\mathbf{PA}_i$, and let $\mathcal{F}_i = \{\mathcal{L}^i_j\}_{j\in J^i}$. By the proof of Theorem 21, if $\langle\mathcal{L}^{\mathcal{F}_i}, \{\pi^i_j\}_{j\in J^i}\rangle$ is the product in the category **CR** of the family $\mathcal{F}_i$ then it is, in fact, the product in the category **ACR** of the family $\mathcal{F}_i$, and thus $\mathcal{L}^{\mathcal{F}_i}$ admits an algebraizator $\mathcal{A}^{\mathcal{F}_i} = \langle\Delta_{\mathcal{F}_i}, \langle\varepsilon_{\mathcal{F}_i}, \delta_{\mathcal{F}_i}\rangle\rangle$ bounded by r and m. For each $i \in I$ let $\mathcal{L}_i \overset{\mathbf{t}(P_i)}{\to} \mathcal{L}^{\mathcal{F}_i}$ be the conservative translation such that $f^i_j = \pi^i_j \cdot \mathbf{t}(P_i)$ for every $j \in J^i$, guaranteed by the proof of Theorem 23. In the finite case, we obtain the following figure:

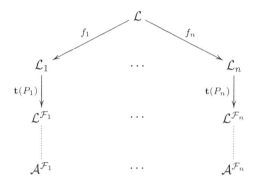

Now, for every $i \in I$ let $g_i = \mathbf{t}(P_i) \cdot f_i$. Then $\mathcal{L} \xrightarrow{g_i} \mathcal{L}^{\mathcal{F}_i}$ is a morphism in $\mathbf{CR}$, for every $i \in I$. In the finite case, the new situation is displayed below.

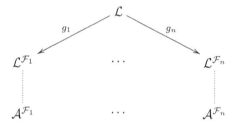

Thus, $\bar{P} = \langle \{\mathcal{L}^{\mathcal{F}_i}\}_{i \in I}, \{g_i\}_{i \in I} \rangle$ is a $\mathbf{PTAS}$ of level 0 for $\mathcal{L}$, because $\bar{P}$ is a $\mathbf{PTS}$ for $\mathcal{L}$ such that every traduct $\mathcal{L}^{\mathcal{F}_i}$ is algebraizable. Moreover, the algebraizators of $\bar{P}$ are bounded by r and m in the sense of Theorem 21. ■

The next result is an application of Theorem 26 that shows that the logic C_{Lim}, introduced in [6]) has a possible-translations algebraic semantics. In that article, the authors show that C_{Lim} is characterized by a possible-translations semantics whose traducts are the paraconsistent logics C_n. Since these logics are characterized by possible-translations semantics where the traducts are all coincident with the logic $\mathbf{LFI1}$ (see [5]) , the result will be obtained by compounding these facts.

THEOREM 27. *The paraconsistent logic C_{Lim} has a $\mathbf{PTAS}$.*

Proof. Theorem 26 together with the facts above. ■

The last result is a concrete application of Theorem 26. Possible-translations semantics are used (cf. [9] and [18]) to obtain new semantics for the paraconsistent systems C_n introduced in [12]. Such semantics permit to characterize the logics C_n in terms of a family $\mathcal{F}_n$ of copies of the three-valued logic **LFI1** (cf. [7]). The latter is equivalent to the three-valued paraconsistent logic J_3 (introduced in [14]).[3] It turns out that J_3, **LFI1** and the three-valued Łukasiewicz logic L_3 are all finitely algebraizable with the same equivalent quasivariety, to wit, the quasivariety of the three-valued Moisil algebras, as shown in [3], p. 43.

It is clear then that every traduct J_3 (or **LFI1**) in the family $\mathcal{F}_n$ is finitely algebraizable and so the algebraizators of the family $\mathcal{F}_n$ obviously satisfy the bounded conditions of Theorem 21. As a consequence of this, the product $\mathcal{L}^{\mathcal{F}_n}$ of the family $\mathcal{F}_n$ is also algebraizable, as argued in [5].

Furthermore, as a consequence of Theorem 23, there exists a conservative translation from each C_n into $\mathcal{L}^{\mathcal{F}_n}$.

This shows that our categorial characterization extends the concept of finite algebraizability in very adequate terms, offering a non *ad hoc* solution to the question of algebraizing logics in general, as it amply extends the method of [2]. Other interesting questions, as the characterization of logics having the Craig interpolation property for their consequence relations, can be recast as a challenging problem in our setting: indeed, it is known (see [17] page 43 for a discussion) that a logic enjoying the deduction-detachment theorem has the Craig interpolation property iff its algebraization has the amalgamation property. Since the amalgamation property can be seen as a universal construction in our (sub)category **CR** of algebraizable logics, it remains as an open problem to know whether this would correspond to any form of Craig interpolation property.

Acknowledgements

This research was financed by FAPESP (Brazil), Thematic Project ConsRel 2004/1407-2. The first author was also supported by a grant from CAPES (Brazil), and the second and third authors were also supported by individual research grants from The National Council for Scientific and Technological Development (CNPq), Brazil.

BIBLIOGRAPHY

[1] D. Batens. Dynamic dialectical logics. In G. Priest, R. Routley, and J. Norman, editors, *Paraconsistent Logic. Essays on the Inconsistent*, pages 187–217, München, 1989. Philosophia Verlag.

[2] W. Blok and D. Pigozzi. *Algebraizable Logics*, volume 77 (396) of *Memoirs of the American Mathematical Society*. AMS, Providence, Rhode Island, 1989.

[3] W. J. Blok and D. Pigozzi. Abstract algebraic logic and the deduction theorem. To appear.

---

[3]It is noticeable that J_3 and **LFI1** are also equivalent in terms of consequence relations with the system **CLuNs** (cf. [1]) and with the system Φ_v introduced in [19].

[4] J. Bueno, M.E. Coniglio, and W.A. Carnielli. Finite algebraizability via possible-translations semantics. In W.A. Carnielli, F.M. Dionísio, and P. Mateus, editors, *Proceedings of CombLog'04 - Workshop on Combination of Logics: Theory and Applications*, pages 79–86, Lisboa, Portugal, 2004. Departamento de Matemática, Instituto Superior Técnico.

[5] J. Bueno-Soler and W.A. Carnielli. Possible-translations algebraization for paraconsistent logics. *Bulletin of The Section of Logic*, 34(2):77–92, 2005.

[6] W. A. Carnielli and J. Marcos. Limits for paraconsistent calculi. *Notre Dame Journal of Formal Logic*, 40(3):375–390, 1999.

[7] W. A. Carnielli, J. Marcos, and S. de Amo. Formal inconsistency and evolutionary databases. *Logic and Logical Philosophy*, 8:115–152, 2000.

[8] W.A. Carnielli. Many-valued logics and plausible reasoning. In *Proceedings of the XX International Congress on Many-Valued Logics*, pages 328–335, University of Charlotte, North Carolina, U.S.A, 1990. IEEE Computer Society.

[9] W.A. Carnielli. Possible-Translations Semantics for Paraconsistent Logics. In D. Batens, C. Mortensen, G. Priest, and J. P. Van Bendegem, editors, *Frontiers of Paraconsistent Logic: Proceedings of the I World Congress on Paraconsistency*, Logic and Computation Series, pages 149–163. Baldock: Research Studies Press, King's College Publications, 2000.

[10] W.A. Carnielli and M.E. Coniglio. A categorial approach to the combination of logics. *Manuscrito*, 22(2):69–94, 1999.

[11] W.A. Carnielli and M.E. Coniglio. Splitting logics. In A. Garcez S. Artemov, H. Barringer and L. Lamb, editors, *We Will Show Them! Essays in Honour of Dov Gabbay*, volume 1, pages 389–414. College Publications, 2005.

[12] N. C. A. da Costa. *Inconsistent Formal Systems* (in Portuguese). PhD thesis, Federal University of Parana, Curitiba, Brazil, 1963. Edited by Editora UFPR, Curitiba, 1993.

[13] J. J. da Silva, I. M. L. D'Ottaviano, and A. M. Sette. Translations between logics. In X. Caicedo and C. H. Montenegro, editors, *Models, Algebras and Proofs: Selected Papers of the X Latin American Symposium on Mathematical Logic Held in Bogota*, pages 435–448. Marcel Dekker, 1999.

[14] I.M.L. D'Ottaviano and N. C. A. da Costa. Sur un problème de Jaśkowski. *Comptes Rendus de l'Academie de Sciences de Paris (A-B)*, 270:1349–1353, 1970.

[15] V.L. Fernández and M.E. Coniglio. Fibring algebraizable consequence systems. In W.A. Carnielli, F.M. Dionísio, and P. Mateus, editors, *Proceedings of CombLog'04 - Workshop on Combination of Logics: Theory and Applications*, pages 93–98, Lisboa, Portugal, 2004. Departamento de Matemática, Instituto Superior Técnico.

[16] V. L. Fernández and M. E. Coniglio. Fibring in the Leibniz Hierarchy. *Logic Journal of the IGPL*, to appear, 2007.

[17] J. M. Font, R. Jansana, and D. Pigozzi. A survey of abstract algebraic logic. *Studia Logica*, 74:13–97, 2003.

[18] J. Marcos. Semânticas de Traduções Possíveis (Possible Translations Semantics, in Portuguese). Master's thesis, IFCH-UNICAMP, Campinas, Brazil, 1999. URL = http://www.cle.unicamp.br/pub/thesis/J.Marcos/.

[19] K. Schütte. *Beweistheorie*. Springer-Verlag, 1960.

J. Bueno-Soler, M. E. Coniglio and W. A. Carnielli

GTAL - Department of Philosophy and Centre for Logic, Epistemology and the History of Science (CLE), State University of Campinas - P.O. Box 6133, 13083-970 - Campinas, SP, Brazil.

{juliana.bueno,coniglio, carniell}@cle.unicamp.br

Ineffable inconsistencies

João Marcos

ABSTRACT. For any given consistent tarskian logic it is possible to find another non-trivial logic that allows for an inconsistent model yet completely coincides with the initial given logic from the point of view of their associated single-conclusion consequence relations.

A paradox? This short note shows you how to do it.

This may be read as the description of an expedition into unexplored regions of abstract logic, the theory of valuations and paraconsistency.

1 Inconsistent classical logic

> Plus on voit ce monde, et plus on le voit plein de contradictions et d'inconséquences.
> —Voltaire, *Dictionnaire Philosophique*, XVIII century.

Take your preferred presentation of classical propositional logic. More concretely, take some denumerable set At of atomic sentences and some non-empty functionally complete set of logical constants C. As usual, the set $\mathcal{S}$ of classical formulas will be inductively built as the free algebra generated by C over At. Let $\mathcal{V}$ be a set of truth-values, $\mathcal{D} \subseteq \mathcal{V}$ a set of designated values (shades of truth) and $\mathcal{U} \subseteq \mathcal{V}$ a set of undesignated values (shades of falsehood), where $\mathcal{D} \cup \mathcal{U} = \mathcal{V}$ and $\mathcal{D} \cap \mathcal{U} = \varnothing$. Semantically, a classical state of the world will be simulated by an assignment Asg : At $\to \mathcal{V}$, where both $\mathcal{D}$ and $\mathcal{U}$ are required to be non-empty —usually, they are taken to be singletons, symbolizing 'the true' and 'the false', if you like. Yes, if you have a boolean mind, you will probably be expecting each such assignment Asg to be uniquely extendable into a valuation $\S : \mathcal{S} \to \mathcal{V}$, according to the truth-functional interpretation of each connective in C. Indeed, say you are talking about disjunction and negation, $\vee$ and $\sim$. In that case you are probably expecting their semantical interpretations to be induced by the set Sem of all valuations $\S : \mathcal{S} \to \mathcal{V}$ such that:

$$\S(\alpha \vee \beta) \in \mathcal{D} \quad \text{iff} \quad \S(\alpha) \in \mathcal{D} \text{ or } \S(\beta) \in \mathcal{D}$$
$$\S(\sim\alpha) \in \mathcal{D} \quad \text{iff} \quad \S(\alpha) \in \mathcal{U}$$

Because classical logic has a truth-functional semantics and because this semantics was formulated above in order to display the dependence of each complex classical formula on its immediate subformulas, and only on them, each of the §-clauses regulating the set Sem could be written with an 'iff' and have a very specific format, indicating the similarity between the algebra of classical formulas and the classical (boolean) algebra of truth-values.

The canonical single-conclusion tarskian consequence relation induced by Sem, denoted by $\models^s_{Sem} \subseteq Pow(S) \times S$, is defined by:

(Ent^s) $\Gamma \models^s_{Sem} \varphi$ iff $\S(\Gamma) \not\subseteq \mathcal{D}$ or $\S(\varphi) \notin \mathcal{U}$, for every $\S \in$ Sem,

where $\Gamma \cup \{\varphi\} \subseteq S$.

Now, given any other set of formulas S and any set of truth-values $\mathcal{V}$, one can take Sem* as any set of valuations $\S : S \to \mathcal{V}$, and the definition of $\models^s_{Sem^\star}$ as in (Ent^s) will still make perfect sense, and it will define the consequence relation of *some* tarskian logic. With that idea in mind, theorists of valuations (cf. [9, 7, 6]) come and ask you to simply forget about the structure of the set of truth-values and concentrate on the set of valuations itself, whichever way it might be introduced. And that is precisely what we shall be doing from now on.

Let $\S^d : S \to \mathcal{V}$ be an arbitrary mapping such that $\S^d(\varphi) \in \mathcal{D}$, for any $\varphi \in S$. A valuation like this plays the role of an inconsistent model, making everything 'true' at once. Suppose you now build a set Sem^d by just adjoining $\S^d$ to the classical set of valuations Sem. Is the new associated single-conclusion consequence relation, $\models^s_{Sem^d}$, any different from the original consequence relation from classical logic? Surprising as it might seem, the answer is 'NO'. Indeed, suppose $\Gamma \models^s_{Sem} \varphi$, for some formulas $\Gamma \cup \{\varphi\} \subseteq S$. In that case, $\S(\varphi) \in \mathcal{D}$ whenever $\S(\Gamma) \subseteq \mathcal{D}$, for any $\S \in$ Sem, by definition. This obviously still holds good for $\S^d$. Conversely, suppose $\Gamma \models^s_{Sem^d} \varphi$. Then, $\S(\varphi) \in \mathcal{D}$ whenever $\S(\Gamma) \subseteq \mathcal{D}$, for any $\S \in$ Sem^d. So, in particular, this holds good for every $\S \in$ Sem. Thus, Sem^d provides an alternative sound and complete semantics for classical logic in a single-conclusion formulation!

The literature on paraconsistent logics is prolific on vague definitions of the very phenomenon of paraconsistency, at all levels. It is not without some disquietness that we find in the paraconsistent jungle definitions such as:

- "Paraconsistent logics are non-trivial logics which can accomodate contradictory theories."

- "Paraconsistent logics are non-explosive logics."

- "Paraconsistent logics are logics having some inconsistent models."

From a semantical perspective, all such definitions tend to say, when properly formalized,[1] that the above alternative formulation of classical logic is paraconsistent. Yet it is characterized by the very same single-conclusion consequence relation of the first and more usual formulation of classical logic!

What's wrong, if anything?

2 The general recipe

> My desire and wish is that the things I start with should be so obvious that you wonder why I spend my time stating them. This is what I aim at because the point of philosophy is to start with something so simple as not to seem worth stating, and to end with something so paradoxical that no one will believe it.
> —Bertrand Russell, *The Philosophy of Logical Atomism*, 1918.

Again, take some set $\mathcal{S}$ of formulas built from the logical constants in some set C over the atomic sentences in At. As soon as we need below to talk about negation, we will simply suppose that there are schemas of the form $\sim\!\varphi$, where $\sim\, \in \mathsf{C}$, available for us. Next, take some set $\mathcal{D}$ of designated truth-values and some disjoint set $\mathcal{U}$ of undesignated truth-values. As usual, $\mathcal{V} = \mathcal{D} \cup \mathcal{U}$. Any set $\Gamma \subseteq \mathcal{S}$ will here be called a *theory*. In the last section we talked about single-conclusion consequence relations. Given some set Sem of valuations $\S : \mathcal{S} \to \mathcal{V}$, you can also define the associated *canonical* MULTIPLE-CONCLUSION consequence relation $\models^m_{\mathsf{Sem}} \subseteq \mathsf{Pow}(\mathcal{S}) \times \mathsf{Pow}(\mathcal{S})$ (cf. [15]), by simply setting:

(Ent^m) $\Gamma \models^m_{\mathsf{Sem}} \Delta$ iff $\S(\Gamma) \not\subseteq \mathcal{D}$ or $\S(\Delta) \not\subseteq \mathcal{U}$, for every $\S \in \mathsf{Sem}$,

where $\Gamma \cup \Delta \subseteq \mathcal{S}$. Taking commas as unions and omitting curly braces, from an abstract viewpoint any *tarskian* consequence relation $\models$ defined as above will be characterized by the following universal axioms, where $\mathsf{Ptn}(\Sigma)$ denotes the set of all partitions of the set Σ:

(C1) $(\Gamma, \varphi \models \varphi, \Delta)$ (overlap)
(C2) $(\forall \langle \Sigma_1, \Sigma_2 \rangle \in \mathsf{Ptn}(\Sigma))(\Gamma, \Sigma_1 \models \Sigma_2, \Delta)\ /\ (\Gamma \models \Delta)$ (cut)
(C3) $(\Gamma \models \Delta)\ /\ (\Gamma', \Gamma \models \Delta, \Delta')$ (dilution)

To be sure, a strong adequacy theorem connects canonical consequence relations and tarskian consequence relations: A consequence relation is characterizable by (Ent^m) if and only if it satisfies axioms (C1), (C2) and (C3) (check [15]). Having said that, I will from this point on assume that every logic has an associated consequence relation, though not necessarily a

---

[1] For such a formalization from the perspective of abstract logics, check for instance [4].

canonical / tarskian one. Each particular consequence relation is intended to embody some specific notion of inference, a collection of directives about what-follows-from-what.[2]

There are of course some dumb examples of tarskian logics that you might prefer to avoid, for the sake of 'minimal enlightenment'. Given $\mathcal{S}$, $\mathcal{D}$ and $\mathcal{U}$, collect in $\mathsf{Sem}(\mathcal{D}) = \{\S : \S(\mathcal{S}) \subseteq \mathcal{D}\}$ all the valuations that are 'biased towards truth', and collect in $\mathsf{Sem}(\mathcal{U}) = \{\S : \S(\mathcal{S}) \subseteq \mathcal{U}\}$ all the valuations that are 'biased towards falsehood'. Any valuation $\S^d \in \mathsf{Sem}(\mathcal{D})$ will from now on be said to constitute a *dadaistic model*, and any valuation $\S^n \in \mathsf{Sem}(\mathcal{U})$ will be said to constitute a *nihilistic model*. Let Dada denote some non-empty subset of $\mathsf{Sem}(\mathcal{D})$, and let Nihil denote some non-empty subset of $\mathsf{Sem}(\mathcal{U})$. Obviously, in a logic having a non-empty set of designated values and a consequence relation characterized by Dada, every formula is a tautology, a thesis, a top particle; in a logic having a non-empty set of undesignated values and characterized by Nihil, every formula is an antilogy, an antithesis, a bottom particle; in a logic having both designated and undesignated values and characterized by models which are either dadaistic or nihilistic, any given formula follows from any other given formula; in a logic with no models, any given theory follows from any other given theory. We will call *overcomplete* any of the above four logics. If you have not seen this before, the surprising bit is that, while the distinctions are clearly visible if you use a multiple-conclusion abstract framework, the four paths to overcompleteness lead to only two different logics in a single-conclusion abstract framework.

For a quick summary, here are the names we will give to each of the above four kinds of overcompleteness, and the way they are characterized:

---

[2] An authoritative referee has called my attention to the alleged 'mistake' of calling 'tarskian' the class of logics whose consequence relation is *multiple-conclusion* and is axiomatized through clauses (C1)–(C3). He claimed that this is "what the literature calls 'Scott Consequence Relations'", and advised me to "see e.g. Gabbay's book on intuitionistic logic". Well, if there is a mistake involved in my decision, it is certainly not *my* mistake, and maybe not even of Gabbay's book (which book?). Dana Scott has indeed been one of the foremost authors to propose the study of multiple-conclusion versions of the preceding tarskian axioms, initially formulated only in terms single-conclusion consequence relations, or rather, equivalently, in terms of consequence operators (check [21]). Typically, [12, 11, 13] are the papers published by Scott that are cited by those who claim that 'multiple-conclusion logics are scottian'. I know that too well —I have, elsewhere, made that confusion myself. Nonetheless, axiom (C2) is never to be found in those papers; at best one can find, in its place, the strictly weaker version of (C2) where Σ is a singleton. Scott's approach in the aforementioned papers, in fact, always seems quite tentative, and it shows no hint of a deep underlying semantic motivation. Not surprisingly, nowhere has Scott an adequacy theorem to offer about the weaker notion of consequence relation that he proposes. My own approach here, thus, cannot be 'scottian'. It is based instead on the work of Shoesmith & Smiley (cf. [14, 15, 22]).

(1)	(2)	(3)	(4)
dadaistic logic	**nihilistic logic**	**semitrivial logic**	**trivial logic**

Semantical conditions:

$\mathcal{D}_1 \neq \varnothing$	$\mathcal{U}_2 \neq \varnothing$	$\mathcal{D}_3 \neq \varnothing$ and $\mathcal{U}_3 \neq \varnothing$	—
$\mathrm{Sem}_1 = \mathrm{Dada}$	$\mathrm{Sem}_2 = \mathrm{Nihil}$	$\mathrm{Sem}_3 = \mathrm{Dada} \cup \mathrm{Nihil}$	$\mathrm{Sem}_4 = \mathrm{Dada} \cap \mathrm{Nihil}$

Single-conclusion abstract characterizations:

$(\forall \beta \Gamma)$	$(\forall \alpha \beta \Gamma)$	$(\forall \alpha \beta \Gamma)$	$(\forall \beta \Gamma)$
$\Gamma \models_1^s \beta$	$\Gamma, \alpha \models_2^s \beta$	$\Gamma, \alpha \models_3^s \beta$	$\Gamma \models_4^s \beta$

Multiple-conclusion abstract characterizations:

$(\forall \beta \Gamma \Delta)$	$(\forall \alpha \Gamma \Delta)$	$(\forall \alpha \beta \Gamma \Delta)$	$(\forall \Gamma \Delta)$
$\Gamma \models_1^m \beta, \Delta$	$\Gamma, \alpha \models_2^m \Delta$	$\Gamma, \alpha \models_3^m \beta, \Delta$	$\Gamma \models_4^m \Delta$

All four overcomplete logics are obviously tarskian (you can check, as an exercise, that they respect (C1), (C2) and (C3)). Moreover, if a logic is trivial then it is both dadaistic and nihilistic, and being either dadaistic or nihilistic a logic will also be semitrivial. As you should notice, $\models_1^s = \models_4^s$, so the single-conclusion framework cannot *see* the difference between the situation in which all models satisfy all formulas and the situation in which the logic has no models. Even worse, $\models_2^s = \models_3^s$, so single-conclusion consequence relations for which all formulas are always false are identical to consequence relations for which all formulas are either all false or all true. But perhaps we should agree that truth-blindness is a serious variety of blindness?

Single-conclusion truth-blindness and the upgraded multiple-conclusion consequence relation can help sorting out the paradox from the last section. Say that we have a *consistent logic* in case (i) the logic is non-dadaistic but (ii) every theory of the logic is derivable from the set of all of its formulas, that is, $\mathcal{S} \models^m \Delta$, for every $\Delta \subseteq \mathcal{S}$. Clause (i) might be read as regulating the number of tautologies of our logic and clause (ii), stronger than (i), says that the set of all formulas cannot be compatibly sustained, all at once. This seems to meet our intuitions according to which varieties of inconsistency appear when 'too many things' are allowed to be true, in a logic. If one can count on dilution, (C3), the above definition implies that there is some β such that $\not\models^m \beta$, and at the same time $\mathcal{S} \models^m$, that is, $\mathcal{S} \models^m \varnothing$. In that case, the addition of a dadaistic model to a consistent logic, as it was done in the last section, clearly gives place to inconsistency, once it occasions $\mathcal{S} \not\models^m$. But in the single-conclusion case, given (C1), both semantics will give you just the same: $\mathcal{S} \models^s \varphi$, for every $\varphi \in \mathcal{S}$.

The situation gets particularly spiky when you think of a logic having a negation symbol $\sim$. Say that we have a *$\sim$-contradictory context* $\langle \Gamma, \Delta \rangle$, where $\Gamma \cup \Delta \subseteq \mathcal{S}$, in case there is some formula $\varphi \in \mathcal{S}$ such that both

$\Gamma \models \varphi, \Delta$ and $\Gamma \models \sim\varphi, \Delta$; say that we have a $\sim$-*inconsistent model* $\S \in$ Sem in case there is some formula $\varphi \in \mathcal{S}$ such that both $\S(\varphi) \in \mathcal{D}$ and $\S(\sim\varphi) \in \mathcal{D}$. Given (C1) and a logic with a negation symbol, contradictory theories are unavoidable. The same does not happen, though, with inconsistent models —the usual set of models for classical logic and for other usual consistent logics does indeed avoid such anomalous models. Consistency of a logic $\mathcal{L}$ should of course be a presupposition for its $\sim$-*consistency*. Now, one further presupposition for a particular formulation of a logic $\mathcal{L}$ in terms of a semantics Sem to be called $\sim$-*inconsistent* is that Sem should have a $\sim$-inconsistent model.

Let's explain this again. Consider the following classical universal rules:

(R1) $(\Gamma, \alpha, \sim\alpha \models \Delta)$ (*pseudo-scotus*, or *explosion*)
(R2) $(\Gamma, \alpha, \sim\alpha \models \beta, \Delta)$ (*ex contradictione sequitur quodlibet*)

Obviously, (R1) implies (R2). Now, while the failure of *pseudo-scotus* corresponds to the existence of some $\sim$-inconsistent model (such as the dadaistic one), the failure of *ex contradictione* corresponds, more specifically, to the existence of some non-dadaistic $\sim$-inconsistent model (which is much more interesting). Yet the two rules will look exactly the same (as (R1) collapses into (R2)) inside a single-conclusion environment.

Summing up the above observations, we will from now on say that we are talking about a $\sim$-*consistent logic* in case this logic is non-dadaistic but it still does respect *pseudo-scotus* —thus, in case it has a canonical semantics, it will admit of no $\sim$-inconsistent model. Here then is the **Paradox of Ineffable Inconsistencies**:

> Let $\mathcal{L}$ be any fixed non-overcomplete tarskian logic.
> Then it is always possible to find an inconsistent logic $\mathcal{IL}$ such that:
>
> $\Gamma \models^{\mathrm{m}}_{\mathcal{IL}} \beta, \Delta$ iff $\Gamma \models^{\mathrm{m}}_{\mathcal{L}} \beta, \Delta$ (and, in particular, $\Gamma \models^{\mathrm{s}}_{\mathcal{IL}} \beta$ iff $\Gamma \models^{\mathrm{s}}_{\mathcal{L}} \beta$),
>
> yet:
> $$\mathcal{S} \not\models^{\mathrm{m}}_{\mathcal{IL}} \quad \text{(while, by definition, } \mathcal{S} \models^{\mathrm{m}}_{\mathcal{L}} \Delta \text{, for every } \Delta\text{).}$$
> In case $\mathcal{L}$ has a symbol $\sim$for negation and is $\sim$-consistent, then
>
> $$\alpha, \sim\alpha \not\models^{\mathrm{m}}_{\mathcal{IL}} \quad \text{(while, by definition, } \Gamma, \alpha, \sim\alpha \models^{\mathrm{m}}_{\mathcal{L}} \Delta \text{, for every } \Gamma \text{ and } \Delta\text{).}$$

You already know the simple strategy to make the above trick work: Just add to Sem$_{\mathcal{L}}$ some dadaistic valuation. We will call the logic $\mathcal{IL}$ thus obtained the *inconsistent counterpart of* $\mathcal{L}$. As in the case of classical logic, in the last section, the inconsistent counterpart of a consistent logic is always identical to the original formulation of the logic from a single-conclusion perspective. But now we know that while the inconsistent counterpart of classical logic still validates rules such as *ex contradictione*, it does NOT validate *pseudo-scotus* any longer. Note that the paradox does not subsist if you add a nihilistic valuation instead of a dadaistic one. In that case you would need a single-premise multiple-conclusion framework for it to make sense.

The only conundrum we are left with is the following. Logics such as $\mathcal{IL}$ are very naturally obtained from their consistent counterparts, and they happen to be neither consistent nor, in general, overcomplete (at least $\mathcal{IL}$ is not overcomplete if the original logic $\mathcal{L}$ was not overcomplete either). Are we willing to call them *paraconsistent*?

3 Paraconsistency is not enough

> To make advice agreeable, try paradox or rhyme.
> —Mason Cooley, *City Aphorisms*, 14th Selection, 1994.

Universal logicians (cf. [2, 3, 1]) believe that logic should be seen a mother-structure (in the sense of Bourbaki) based on some given set of formulas and a consequence relation defined over it. They do not require in general these formulas and relation to bring any further built-in structure (say, an algebraic structure over the set of formulas). But in practical cases, of course, it is often interesting to fix for instance some set of axioms or another over the consequence relation. I have indeed presented above a multiple-conclusion version (cf. [15]) of the customary tarskian axioms (cf. [19]) and immediately after that I exhibited some trivial examples of tarskian logics: the overcomplete ones. Should we modify the given axioms in order to rule out these examples as illegitimate? One could surely do that, and it has indeed been done here and there in the literature, but I am not convinced that this is a very wise manoeuvre. First of all, the overcomplete logics fit very naturally both in the abstract and the semantical frameworks. Besides, I am only talking about 'overcomplete logics' once I had decided that they should be called 'logics', to start with. *Ad hoc* modifications of the definition of logic in order to avoid the above mentioned unpleasant examples do not seem to carry much persuasive power —for one thing, a good question is: Where will they stop?

Imagine the following conversation overheard between two philosophers:

    ($\forall$belard) 'I bought an arm chair today.'
      ($\exists$loise) 'How nice.'

(∀belard)   'It has flatulence filter seat cushion.'
  (∃loise)   'Good.'
(∀belard)   'It has a purple upholstered back.'
  (∃loise)   'Hmmm...'
(∀belard)   'It has 42 slender chippendale legs.'
  (∃loise)   'Wait a moment. I wouldn't call a 'chair' any object
         having more than 4 legs!'

Now, was ∀belard wrong in using the word 'chair' from the very start? Maybe ∃loise has a sound intuition, and this anomalous object will turn out to be impractical as a chair —its many legs are too difficult to clean, the ensemble is too heavy to carry, or something. Suppose the philosophers will some day agree about the essential properties of a chair, including its maximal number of legs. Will post-modernist designers still have a job? If they will, then what will be the next development to *trivialize* the notion of 'chair'?

Going back to logic, consider the *minimal* tarskian logic defined over some fixed set of formulas $\mathcal{S}$. This logic is characterized by:

$$\Gamma \models^m \Delta \quad \text{iff} \quad \Gamma \cap \Delta \neq \varnothing,$$

where $\Gamma \cup \Delta \subseteq \mathcal{S}$. Clearly, this is the minimal logic respecting (C1), and it is easy to check that both (C2) and (C3) are also respected. ∃loise, again, finds this construction quite 'trivial' and dull. Should we then add a further restriction to the definition of logic so as to please her?

It does seem hopeless, and even counterproductive, to expect logicians to reach a final agreement about the answers to fundamental questions such as 'what is logic?', 'what is negation?' (or conjunction, or some other connective), 'what is paraconsistency?' and so on. This does not mean, however, that 'anything goes'. It often seems more realistic and reasonable to look for properties that we do *not* want to allow 'interesting' logics, negations, conjunctions, paraconsistent logics etc to have. This principle that combines a strong wish both for economy and for significance was made transparent as a sort of motto for paraconsistency since its infancy (cf. [5]): 'From the syntactico-semantical standpoint, every mathematical theory is admissible, unless it is trivial' (notice however that the author did not clear up what 'theory' or 'trivial' were supposed to mean). Investing on that idea, clarifying and updating it, the paper [10] shows one way of implementing this *negative* approach to general abstract nonsense. For the purposes of the present paper, it will be sufficient to require non-overcompleteness for the definition of a so-called *minimally decent logic*. From an abstract viewpoint, that can be done by saying for instance that a minimally decent tarskian logic should also respect a further negative axiom, denying the very possibility of semitriviality. From a semantical viewpoint the thing gets a bit more complicated. It is not enough for theorists of valuations to add the requirement that both the set of designated values and the set of undesignated values should be non-empty. One needs also to directly constrain the set of all valuations of an intended semantics —or else collections of dadaistic and nihilistic models might reappear. There is no need to go into details of that here. At any rate, other necessary conditions for minimal decency might of course still impose themselves at some future moment, according to the interest and experience of logic-designers.

Now, at least two lessons may be drawn from the paradox explored in the previous sections. The first lesson is about the usefulness of a multiple-conclusion environment when doing logic in general, and paraconsistent logic in particular (I recommend again checking [10], where this framework was extensively used for the study of *negation*, its more usual positive properties and some negative properties that make it 'minimally decent'). Obviously, as any other formalism, multiple-conclusion will also have its limitations, and the adequacy of its use will depend on the phenomenon that needs to be seized at the time. On the positive side, however, there are several arguments pro multiple-conclusion. Many of them are well-known, or quite obvious, and I will not try to survey them here (for the interested reader, it might be a good idea to check [15]).[3] I will mention only one further particularly interesting advantage of that formalism, as connected to paraconsistency.

Even after the wide acknowledgment of the inferential character of logic, philosophy continues to suffer from a certain 'bias towards truth'. Arguably, because a compact tarskian logic sees no difference between inferences with a finite or an infinite set of premises, because the single-conclusion notation derived from the notion of a closure operator cannot mark the difference between constructive and non-constructive sets of theses, and because of persisting positivistic influences, the logico-philosophical community ended up accommodating with a lot of inertia around the notion of theoremhood, as opposed to the notion of inference from a set of premises. Even nowadays, the study of 'logics as sets of theorems' or 'logics as sets of truths' is very

---

[3]A particularly attractive advantage of the multiple-conclusion framework, from a semantical viewpoint, is the so-called *categoricity* of the class of models of each given tarskian logic (check [8]): Any two adequate 2-valued canonical semantics for a given logic $\mathcal{L}$ must be isomorphic. The result is always true for multiple-conclusion logics, and never true for non-overcomplete single-conclusion logics. That explains why each variety of overcompleteness is clear distinguished from the other varieties in a multiple-conclusion framework, and why the so-called Paradox of Ineffable Inconsistencies is immediately disclosed when one moves from single- to multiple-conclusion consequence relations.

likely to find more practitioners than the more inferential-related approach. Besides, even proposals as interesting as those of Łukasiewicz in axiomatizing his modal many-valued logics using the notion of rejected propositions, alongside with accepted propositions, were soon to fall into almost complete disregard (the papers [18, 17, 16] are among the few interesting modern exceptions to that trend). But why should truth be privileged over falsehood? Why should acceptance be privileged over rejection?

The multiple-conclusion approach allows not only the inferential character of logic to be taken into proper account but its full symmetry also allows truth and falsehood to be put on equal footing. Playing with the right-left symmetry of the consequence relation turnstile symbol one can very naturally talk for instance about the notion of *duality* of logics, of connectives, and of rules. Given a consequence relation $\triangleright$, its dual $\blacktriangleright$ is such that

$$(\Gamma \blacktriangleright \Delta) \text{ iff } (\Delta \triangleright \Gamma).$$

Similarly, given any rule of a connective in the first consequence relation, $\triangleright$, one can immediately look for the corresponding rule of the dual connective in the second consequence relation, $\blacktriangleright$, just reading the rule the other way around. This way an introduction rule for classical conjunction can be characterized as dual to an elimination rule for classical disjunction, implication can be characterized as dual to right residuation, negation as consistency (explosion) as dual to negation as completeness or determinedness (excluded middle). Any definition involving paraconsistency can immediately be converted into a definition involving its dual, paracompleteness. In semantical terms, given a two-valued interpretation of a tarskian logic, its dual is obtained by uniformly substituting 'true' for 'false', and vice-versa.

It is about time for the 'single-conclusion bias' to be defeated once and for all. If not just for the hidden prejudice against multiple-conclusion, or plain sluggishness of many logicians, the only extra reason I see for no version of the above paradox to have been explicitly reported before (as far as I know) is because there seems not to have been much interest in exploring single-premise inferences (that idea has been taken forward, though, in papers such as [20]). Notice, at any rate, that at the single-premise-single-conclusion case the interpretation of the entailment sign, $\models$, confuses itself with the interpretation of the (classical) material conditional.

If you recall the definition of a $\sim$-consistent logic proposed in the last section, you will see that an $\sim$-inconsistent logic will either be overcomplete or it will disrespect *pseudo-scotus*. Once a *paraconsistent logic* is A LOGIC before anything else, it should be a minimally decent logic, hence it should be $\sim$-inconsistent but not overcomplete (notice that the failure of *pseudo-scotus* is perfectly compatible with dadaism). It is sad to recognize that, several decades after its initial developments, paraconsistent logic remains by and large a terrain wide open for adventurers and for intellectual impostures. The general inability demonstrated by the paraconsistent community so far in having constructive conversations attests to the great lack of coordination in the field. These last grumpy (yet justified) comments of mine might help explaining, at least partially, the serious lack of foundational papers which would help in finally setting some necessary conditions for minimal decency in paraconsistent logic. The second lesson of the present paper intends to be a contribution to that. Instead of proposing changes to the very definition of paraconsistency —say, to those hazy definitions recorded in section 1— my sole suggestion here is that a *minimally decent paraconsistent logic* should, in a multiple-conclusion abstract environment, avoid not only *pseudo-scotus* but also *ex contradictione* —as it has generally been done in the single-conclusion environment, where the two rules are indistinguishable. Semantically, as observed in section 2, this amounts to requiring the semantics of minimally decent (tarskian) paraconsistent logics not just to allow for ∼-inconsistent models, but, more specifically, to allow for non-dadaistic ∼-inconsistent models.

Acknowledgements

This study was partially supported by FCT (Portugal) and FEDER (European Union), namely, via the Project FibLog POCTI / MAT / 37239 / 2001 of the Center for Logic and Computation (IST, Portugal) and the FCT grant SFRH / BD / 8825 / 2002. I take the chance to thank Carlos Caleiro, Walter Carnielli and two anonymous referees for their many helpful comments on an earlier version of this paper.

BIBLIOGRAPHY

[1] J.-Y. Béziau, editor. *Logica Universalis*. Birkhäuser Verlag, Basel, Switzerland, 2005.

[2] Jean-Yves Béziau. Universal Logic. In T. Childers and O. Majers, editors, Logica'94, *Proceedings of the* VIII *International Symposium*, pages 73–93. Czech Academy of Science, Prague, CZ, 1994.

[3] Jean-Yves Béziau. Non-truth-functional many-valuedness. In J.-Y. Béziau, A. Costa-Leite, and A. Facchini, editors, *Aspects of Universal Logic*, pages 199–218. Centre de Recherches Sémiologiques, Neuchâtel, 2004.

[4] Walter A. Carnielli, Marcelo E. Coniglio, and João Marcos. Logics of Formal Inconsistency. In D. Gabbay and F. Guenthner, editors, *Handbook of Philosophical Logic*, volume 14, pp. 1–94. Springer, 2nd edition, 2007. In print.

[5] Newton C. A. da Costa. Observações sobre o conceito de existência em matemática. *Anuário da Sociedade Paranaense de Matemática*, 2:16–19, 1959.

[6] Newton C. A. da Costa and Jean-Yves Béziau. La théorie de la valuation en question. In M. Abad, editor, *Proceedings of the* IX *Latin American Symposium on Mathematical Logic, Part 2* (Bahía Blanca, 1992), pages 95–104. Bahía Blanca: Universidad Nacional del Sur, 1994.

[7] Newton C. A. da Costa and Jean-Yves Béziau. Théorie de la valuation. *Logique et Analyse (N.S.)*, 37(146):95–117, 1994.

[8] Gary M. Hardegree. Completeness and super-valuations. *Journal of Philosophical Logic*, 34(1):81–95, 2005.

[9] Andrea Loparić and Newton C. A. da Costa. Paraconsistency, paracompleteness, and valuations. *Logique et Analyse (N.S.)*, 27(106):119–131, 1984.

[10] João Marcos. On negation: Pure local rules. *Journal of Applied Logic*, 3(1):185–219, 2005. Preprint available at:
http://www.cle.unicamp.br/e-prints/revised-version-vol_4,n_4,2004.html.

[11] Dana Scott. On engendering an illusion of understanding. *Journal of Philosophy*, 68:787–807, 1971.

[12] Dana Scott. Completeness and axiomatizability in many-valued logic. In *Proceedings of the Tarski Symposium* (Proc. Sympos. Pure Math., Vol. XXV, held at UC Berkeley, 1971), pages 411–435, Providence / RI, 1974. American Mathematical Society.

[13] Dana Scott. Rules and derived rules. In S. Stenlund, editor, *Logical Theory and Semantical Analysis*, pages 147–161. D. Reidel, Dordrecht, 1974.

[14] D. J. Shoesmith and Timothy J. Smiley. Deducibility and many-valuedness. *The Journal of Symbolic Logic*, 36(4):610–622, 1971.

[15] D. J. Shoesmith and Timothy J. Smiley. *Multiple-Conclusion Logic*. Cambridge University Press, Cambridge / MA, 1978.

[16] Tomasz Skura. Aspects of refutation procedures in the intuitionistic logic and related modal systems. *Logika*, 20, 2000.

[17] Timothy Smiley. Rejection. *Analysis (Oxford)*, 56(1):1–9, 1996.

[18] Allard M. Tamminga. Logics of rejection: Two systems of natural deduction. *Logique et Analyse (N.S.)*, 146:169–208, 1994.

[19] Alfred Tarski. On the concept of following logically. *History and Philosophy of Logic*, 23:155–196, 2002. Translated by M. Stroińska and D. Hitchcock. Original versions published in Polish under the title 'O pojciu wynikania logicznego', in *Przegląd Filozoficzny*, 39:58–68, 1936, and then in German under the title 'Über den Begriff der logischen Folgerung', in *Actes du Congrès International de Philosophie Scientifique*, 7:1–11, 1936.

[20] Igor Urbas. Dual-intuitionistic logic. *Notre Dame Journal of Formal Logic*, 37(3):440–451, 1996.

[21] Ryszard Wójcicki. *Theory of Logical Calculi*. Kluwer, Dordrecht, 1988.

[22] Jan Zygmunt. *An Essay in Matrix Semantics for Consequence Relations*. Wydawnictwo Uniwersytetu Wrocławskiego, Wrocław, 1984.

Joã Marcos
Department of Informatics and Applied Mathematics (DIMAp), UFRN, Natal, RN, Brazil and Security and Quantum Information Group (SQIG), TU-Lisbon, Portugal.

Part IV
Results about
Paraconsistent Logic

Natural Deduction and Weak Normalization for the Paraconsistent Logic of Epistemic Inconsistency

ANA TERESA MARTINS, LÍLIA RAMALHO MARTINS AND
FELIPE FERREIRA DE MORAIS

ABSTRACT. Proof theory for paraconsistent logics through natural deduction systems has not been widely investigated. For da Costa's hierarchy of calculi Cn, $1 \leq n \leq \omega$, we may mention works in natural deduction systems and sequent calculus [23, 1, 2, 17], but normalization theorems are only proved for $C\omega$ [16], the logic in the limit of this hierarchy with an extremely weak negation. Here we will present such results for another paraconsistent logic, the Logic of Epistemic Inconsistency (LEI) [19, 12, 11, 7]. LEI was specially designed to be the monotonic basis of a default logic, allowing us to tolerate contradictions that could arise from reasoning under partial knowledge represented through formulas suffixed by plausible operators: ? and !. α? is to be true when someone asserts that α is plausible based on sensible reasons, and if everyone agrees with the plausibility of α, based also on sensible reasons, then α! is true. Sensible reasons are introduced by default rules. Paraconsistency is achieved by a combination of LEI negation $\neg$ and these plausible operators. Hence, it is possible to tolerate, in LEI, both α? and $\neg(\alpha?)$ without leading us to an explosion of theorems. In order to design a natural deduction system that is weakly normalizable, an investigation of rules and related problems of substitution and reductions is presented. The weak normalization theorem will be proved here for the whole LEI language, including all quantifiers, connectives and plausible operators as primitive signs. Main results are concerned with the use of proof techniques to rules, and in reduction steps related to the paraconsistent negation and the plausible operators.

1 Introduction

The Logic of Epistemic Inconsistency (LEI) is a paraconsistent logic designed to tolerate contradictions that arise from reasoning under partial knowledge [19, 12, 11, 7]. It was originally proposed to be the monotonic basis of a default logic, the Inconsistent Default Logic IDL [18, 7, 8, 9], but it may also be used by itself to formalize reasoning in situations where multiple and cooperative agents are involved.

LEI has two signs for plausibility: ? that represents a credulous one, and ! for a skeptical one. α? means that α is plausible if at least one agent believes, based on sensible reasons, that it is true, and α! represents that α is plausible whenever all agents believe, based also in sensible reasons, that it is true. LEI paraconsistent negation $\neg$ allows one to deal with conflicting conclusions between plausible formulas without explosion of theorems.

Sensible reasons are introduced by IDL default rules. The usual example for representing that 'a bird typically flies, unless it is a penguin' will be represented as an IDL default rule as: $\dfrac{bird(x) \; : \; fly(x) \; ; \; \neg penguin(x)}{fly(x)?}$.

Thence, if Tweety is a bird, '$bird(tweety)$)', and it is consistent to assume that it flies, '$fly(tweety)$', and it is not a penguin, '$\neg penguin(tweety)$', so it is plausible that it flies, '$fly(tweety)?$'. A second example, also from the literature, is the following: suppose that we want to represent that 'if someone is a republican, then he is plausibly a non-pacifist' but 'if he is a quaker, he is plausibly a pacifist'. It will be represented using IDL default rules as: $\dfrac{republican(x) \; : \; \neg pacifist(x)}{(\neg pacifist(x))?}$ and $\dfrac{quaker(x) \; : \; pacifist(x)}{pacifist(x)?}$. Now, suppose that Nixon is a republican and a quaker. So from '$quaker(nixon)$' we get '$pacifist(nixon)?$' and from '$republican(nixon)$' we infer '$(\neg pacifist(nixon))?$'. Since from one point of view, expressed by the first default rule, there are sensible reasons to belive that '$(\neg pacifist(nixon))?$', from this same point of view it indicates that '$\neg (pacifist(nixon)\ ?)$'. In fact, $(\neg \alpha)? \leftrightarrow \neg(\alpha?)$ is a valid formula in LEI (see section 2). We then reach a contradiction between plausible formulas equally assumed. How can it be solved? As there is insufficient knowledge at hand to decide in favour of one of them, we just tolerate such plausible conflict without contaminating safe deductions until fresh knowledge allows us to solve it.

Since LEI was specially designed to be the monotonic basis for IDL, the meaning of the plausibility operator '?' (and its dual '!') can only be completely understood within the IDL&LEI system. A default theory [24] is usually presented as a set of facts and default rules. An extension is the set of formulas that are monotonically derived from the set of facts and conclusions of applicable defaults, defaults that are not blocked by the consistency test (as shown above). In an IDL theory, all formulas in the set of facts are ?-free (without '?' or '!'), the credulous plausibility operator '?' is only introduced as a suffix in the default conclusion, and the skeptical plausibility operator '!' appears only in formulas that belongs to an extension. An IDL extension is the set of formulas that are derived, by using the paraconsistent logic LEI, from the set of facts and conclusions of applicable IDL defaults. Since we are, at present time, just interested in proof theory for LEI, we will focus only on LEI semantics in section 2. Semantics for the whole IDL&LEI system is detailed in [8, 7].

Aiming to better understand the deductive power of LEI signs individually, a natural deduction system is introduced in section 3. This kind of system was originally proposed by Gentzen [6] for classical and intuitionistic logics. Within natural deduction systems, the deductive role of each logical constant is detached. The inferences are broken into atomic steps, namely introduction I* and elimination E* rules, where * stands for a logical symbol. They reflect the compositionality principle, by Frege, where the denotation (truth value) of a sentence is a function of the denotation (truth value) of its subformulas. The compositionality principle leads us to the subformula principle which states that, in a canonical proof, called normal proof, all formulas are subformulas of the assumptions not discharged or of the formula to be proved, except for assumptions discharged by applications of the classical absurdity rule and for some occurrences of ⊥ (the absurdity sign). The system introduced in section 3 is also closed under substitution. Substitution can be considered as the operation of composition, if proofs are regarded as functions [30]. Some examples will be presented to clarify how the notion of connection, used in the definition of plausible rules, are central to prove such property.

The individual analysis of each logical constant, in a natural deduction system, allows us to see the existence of a close correspondence between the constructive meaning of a logical constant and its introduction rule [21]. Moreover, the symmetry between atomic inferences indicates how introductions and eliminations are the inverse of each other. It will be emphasized by the rewriting process that involves the reduction steps of the normalization proofs allowing us to identify the inversion principle. This principle says that the conclusion obtained by an elimination rule does not state anything else than what must have already been obtained if the major premise (the one where the connective in focus appears) of the elimination was inferred by an introduction rule. That is, the proof of the conclusion of an elimination is already contained in the proofs of the premises when the major premise is inferred by introduction. The reduction steps for LEI constants will be presented in section 4.

Besides being used to build proofs and to better understand the meaning of the logical constants, a natural deduction system may also be used to study metalogical properties under the perspective of proof theory. By a deep analysis of the structure of natural deduction proofs, one can identify that they should be written in a normal form which could give a simpler understanding of the notion of proof. In his PhD thesis [22], Prawitz proved normalization theorems for classical, intuitionistic and modal logics that assures the existence of a normal form for natural deduction proofs. Following Prawitz's steps, we will prove the weak normalization theorem for LEI natural deduction system and the subformula property in section 5.

It is important to note that weak normalization for classical logic was originally proved, by Prawitz [22], just for a language without disjunction and existential quantifier, and with restrictions for the application of the classical absurdity rule ($\bot \sim$) to atomic formulas. Reduction steps involving disjunction, existential quantifier and classical negation brings about a lot of problems which are not easy to solve. The first proof of weak normalization for the whole classical language is presented by Statman in [28] but it is not a direct proof. The weak normalization theorem was directly proved for the whole language, without restrictions to the classical absurdity rule, rather recently [27, 25, 13, 20]. Probably this is the reason why proof theory for paraconsistent logics through natural deduction presentations has not been widely investigated yet. For Cn, $1 \leq n \leq \omega$, da Costa's family of logics [5, 4], we may mention proof-theoretical works by Alves [1], Raggio [23], Béziau [2] and Moura [16], for instance, in both natural deduction and sequent calculus. However, normalization results for natural deduction calculus was only achieved for $C\omega$ [17], the logic in the limit of this hierarchy with an extremely weak negation. In the future, we want to investigate the possibility of applying our absurdity reductions, here considered with disjuction and existential quantifier analyses, in the proof of weak normalization for other logics in Cn hierarchy with stronger paraconsistent negations.

2 LEI Semantics

Before introducing LEI semantics, we have to define the set of LEI formulas. Formulas in LEI are written in the language $\mathcal{L}_?$. The alphabet $\mathcal{A}_?$ of LEI contains $\wedge$ (conjunction), $\vee$ (disjunction), $\rightarrow$ (implication), $\forall$ (universal quantifier), $\exists$ (existential quantifier), $\neg$ (negation), ? (credulous plausibility) and ! (skeptical plausibility) as primitive logical constants. $\mathcal{L}_?$ extends a classical first-order language $\mathcal{L}$, defined as usual [22], adding the following clause in the inductive definition: if α is a formula, then so are $(\alpha?)$ and $(\alpha!)$. The absurdity symbol $\bot$ is defined as $\bot \equiv (A \wedge \neg A)$, where A is a formula without ? and !, and without free variables. As in [22], $\bot$ is considered an atomic formula. The classical negation $\sim$ is introduced as a derived symbol: $(\sim \alpha) \equiv (\alpha \rightarrow \bot)$. The negation symbol $\neg$ behaves as a paraconsistent negation.

Formulas in $\mathcal{L}_?$ are called ?-formulas, or simply formulas. In ?-formulas, ? or ! do not necessarily appear. We will use small Greek letters α, β, γ, ... as meta-variables for ?-formulas and capital Roman letters $A, B, C, ...$ for ?-free formulas, formulas without ? and !. For sets of ?-formulas we will use capital Greek letters Γ, Δ, Λ, ..., but Π and Σ will be used for a deduction and a sequence, possibly empty, of deductions, respectively (see section 3). Whenever it does not lead to ambiguities, we will omit parentheses in formulas following the same conventions stated in [22].

LEI semantics will be introduced here as in [8, 7]. It is an alternative version of its original semantics [19]. Whereas in the original semantics multiple observers are used to represent possible distinct points of view of the state of affairs, the idea of a plurality of views is replaced here by a Kripke-like plurality of worlds, called *plausible worlds*.

A set of plausible worlds could be intuitively thought as a subset of a set of possible worlds in an S5 structure, for instance. In comparison with possible operators $\Box$ and $\Diamond$, we can say that what is plausible is possible (that is, if α? then $\Diamond\alpha$), although not the converse. As an example, we can think that it is possible to snow near the Equator (if the Sun dies and the Earth turns cold), but it is not (credulously) plausible, i.e., there are no sensible reasons to think like that now. Since '!' (the skeptical plausibility) is defined as dual to '?', then what is necessary is skeptically plausible (that is, if $\Box\alpha$ then α!), but not the converse.

LEI (modal) semantics will be defined through a set $\mathcal{W}_{LEI}$ of plausible worlds where each plausible world w within this set is intended to index the set V, $V = \{v_w \mid w \in \mathcal{W}_{LEI}\}$, of valuation functions for atomic formulae. The set V of valuation functions v_w may be thought as representing points of view of different observers or different points of view of the same observer.

A plausible formula 'α?' may be analysed from two angles: a maximal and a minimal one. Under the maximal view, 'α?' is true if 'α' is plausible for some observer (i.e., true under v_w, for some plausible world w in $\mathcal{W}_{LEI}$). Under the minimal perspective, 'α' must be plausible to all observers (i.e., true under v_w, for all plausible worlds w in $\mathcal{W}_{LEI}$) in order to affirm that 'α?' is true. In definition 4, we will see that the maximal view, if compared with the minimal view, is taken as the preferential meaning of '?'. Semantics for the plausible operator '!' is taken as dual to '?'.

DEFINITION 1 (Plausible World Structure $\mathcal{M}$). $\mathcal{M} =<\mid \mathcal{A} \mid, \sigma, \mathcal{W}_{LEI}, V >$ is a plausible world structure for $\mathcal{L}_?$ where:

1. $\mid \mathcal{A} \mid$ is a nonempty universe of discourse;

2. σ is an interpretation mapping for terms which assigns to each constant symbol c a member $c^{\mathcal{A}}$ of the universe $\mid \mathcal{A} \mid$ and to each n-place

function symbol f an n-place operation $f^{\mathcal{A}}$ on $|\mathcal{A}|$, i.e., $f^{\mathcal{A}} : |\mathcal{A}|^n \to |\mathcal{A}|$;

3. $\mathcal{W}_{LEI}$ is a non-empty set (possibly finite) of plausible worlds w;

4. V is an indexed set of valuation functions $\{v_w \mid w \in \mathcal{W}_{LEI}\}$. Any valuation function $v_w \in V$ assigns to each n-place predicate symbol P its interpretation $P_w^{\mathcal{A}}$ on $|\mathcal{A}|$, that is, $P_w^{\mathcal{A}} \subseteq |\mathcal{A}|^n$. $P_w^{\mathcal{A}}$ is intended to represent a relation whose elements are n-tuples of members of the universe $|\mathcal{A}|$.

In the following definitions, let $\mathcal{M} = <|\mathcal{A}|, \sigma, \mathcal{W}_{LEI}, V>$ be a Plausible World Structure for $\mathcal{L}_?$ under consideration.

DEFINITION 2 (Assignment $\bar{s}$ for Terms in $\mathcal{L}_?$). :
Let $s : \mathcal{V} \to |\mathcal{A}|$ be an assignment function for variables in $\mathcal{L}_?$. $\bar{s} : Terms \to |\mathcal{A}|$ is defined as usual:

1. for each variable x, $\bar{s}(x) = s(x)$;

2. for each constant symbol c, $\bar{s}(c) = \sigma(c) = c^{\mathcal{A}}$;

3. if $t_1, \ldots, t_n$ are terms and f is an n-place function symbol, then $\bar{s}(f(t_1, \ldots, t_n)) = \sigma(f)(\bar{s}(t_1), \ldots, \bar{s}(t_n)) = f^{\mathcal{A}}(\bar{s}(t_1), \ldots, \bar{s}(t_n))$.

As it was said before, in order to define the satisfaction of a formula α in a structure $\mathcal{M}$, two auxiliary notions of maximal and minimal satisfaction $\mathcal{M} \models_{max}^w$ and $\mathcal{M} \models_{min}^w$, respectively, are defined. They are intended to capture the behaviour of the paraconsistent negation and its relation to the other connectives. The function s is also used as the assignment function for variables.

DEFINITION 3 ($\mathcal{M} \models_{max}^w \alpha \ [s]$ and $\mathcal{M} \models_{min}^w \alpha \ [s]$). : Let $w \in \mathcal{W}_{LEI}$. Then:

1. $\mathcal{M} \models_{max}^w P(t_1, \ldots, t_n) \ [s]$ iff $\mathcal{M} \models_{min}^w P(t_1, \ldots, t_n) \ [s]$ iff $< \bar{s}(t_1), \ldots, \bar{s}(t_n) > \in v_w(P) = P_w^{\mathcal{A}}$ where $P(t_1, \ldots, t_n)$ is an atomic formula in $\mathcal{L}_?$. For other formulae in $\mathcal{L}_?$, the definition is the following:

2. $\mathcal{M} \models_{max}^w \alpha? \ [s]$ iff for some $w' \in \mathcal{W}_{LEI}$, $\mathcal{M} \models_{max}^{w'} \alpha \ [s]$;

3. $\mathcal{M} \models_{min}^w \alpha? \ [s]$ iff for all $w' \in \mathcal{W}_{LEI}$, $\mathcal{M} \models_{min}^{w'} \alpha \ [s]$;

4. $\mathcal{M} \models_{max}^w \alpha! \ [s]$ iff for all $w' \in \mathcal{W}_{LEI}$, $\mathcal{M} \models_{max}^{w'} \alpha \ [s]$;

5. $\mathcal{M} \models_{min}^w \alpha! \ [s]$ iff for all $w' \in \mathcal{W}_{LEI}$, $\mathcal{M} \models_{max}^{w'} \alpha \ [s]$;

6. $\mathcal{M} \models_{max}^{w} \neg\alpha$ $[s]$ iff $\mathcal{M} \not\models_{min}^{w} \alpha$ $[s]$;

7. $\mathcal{M} \models_{min}^{w} \neg\alpha$ $[s]$ iff $\mathcal{M} \not\models_{max}^{w} \alpha$ $[s]$;

8. $\mathcal{M} \models_{max}^{w} \alpha \to \beta$ $[s]$ iff $\mathcal{M} \not\models_{max}^{w} \alpha$ $[s]$ or $\mathcal{M} \models_{max}^{w} \beta$ $[s]$;

9. $\mathcal{M} \models_{min}^{w} \alpha \to \beta$ $[s]$ iff $\mathcal{M} \not\models_{max}^{w} \alpha$ $[s]$ or $\mathcal{M} \models_{min}^{w} \beta$ $[s]$;

10. $\mathcal{M} \models_{max}^{w} \alpha \wedge \beta$ $[s]$ iff $\mathcal{M} \models_{max}^{w} \alpha$ $[s]$ and $\mathcal{M} \models_{max}^{w} \beta$ $[s]$;

11. $\mathcal{M} \models_{min}^{w} \alpha \wedge \beta$ $[s]$ iff $\mathcal{M} \models_{min}^{w} \alpha$ $[s]$ and $\mathcal{M} \models_{min}^{w} \beta$ $[s]$;

12. $\mathcal{M} \models_{max}^{w} \alpha \vee \beta$ $[s]$ iff $\mathcal{M} \models_{max}^{w} \alpha$ $[s]$ or $\mathcal{M} \models_{max}^{w} \beta$ $[s]$;

13. $\mathcal{M} \models_{min}^{w} \alpha \vee \beta$ $[s]$ iff $\mathcal{M} \models_{min}^{w} \alpha$ $[s]$ or $\mathcal{M} \models_{min}^{w} \beta$ $[s]$;

14. $\mathcal{M} \models_{max}^{w} \forall x\alpha$ $[s]$ iff $\mathcal{M} \models_{max}^{w} \alpha$ $[s(x \mid d)]$, for every $d \in\mid \mathcal{A} \mid$;

15. $\mathcal{M} \models_{min}^{w} \forall x\alpha$ $[s]$ iff $\mathcal{M} \models_{min}^{w} \alpha$ $[s(x \mid d)]$, for every $d \in\mid \mathcal{A} \mid$;

16. $\mathcal{M} \models_{max}^{w} \exists x\alpha$ $[s]$ iff $\mathcal{M} \models_{max}^{w} \alpha$ $[s(x \mid d)]$, for some $d \in\mid \mathcal{A} \mid$;

17. $\mathcal{M} \models_{min}^{w} \exists x\alpha$ $[s]$ iff $\mathcal{M} \models_{min}^{w} \alpha$ $[s(x \mid d)]$, for some $d \in\mid \mathcal{A} \mid$;

where $s(x \mid d) = s(y)$, if $y \neq x$ or $s(x \mid d) = d$, if $y = x$.
$\mathcal{M}$ **satisfies** α **at** w **with** s if $\mathcal{M} \models_{max}^{w} \alpha$ $[s]$.

In order to capture the intended meaning of the paraconsistent negation and its relation with implication, $\mathcal{M} \models_{max}^{w} \alpha \to \beta$ and $\mathcal{M} \models_{min}^{w} \alpha \to \beta$ are not dually defined. As a consequence $(\alpha \to \beta) \to (\neg\alpha \vee \beta)$, but not the converse, is a valid formula (see definition 6).

DEFINITION 4 ($\mathcal{M} \models \alpha$). [1] α is satisfied in a structure $\mathcal{M}$ or $\mathcal{M}$ is a **Plausible World Model** of α, represented by $\mathcal{M} \models \alpha$, iff for all worlds $w \in \mathcal{W}_{LEI}$ and all functions s, $\mathcal{M} \models_{max}^{w} \alpha$ $[s]$.

DEFINITION 5 ($\mathcal{M} \models \Gamma$). $\mathcal{M}$ is a plausible world model of Γ iff it is a plausible world model of each formula in Γ.

DEFINITION 6 ($\Gamma \models \alpha$). : $\Gamma \models \alpha$ iff every plausible world model $\mathcal{M}$ of Γ is also a plausible world model of α. If the set Γ of formulae is empty, then α is said to be a **valid formula**.

Some examples of valid formulae are:

---

[1] In [3], it is presented the paracomplete logic LSR, the *Logic of Skeptical Reasoning*, dual to LEI. Its satisfaction relation is based on the auxiliary notion of minimal satisfaction.

$\alpha \to \alpha?$	$\alpha?? \to \alpha?$
$(\alpha \to B) \to ((\alpha \to \neg B) \to \neg \alpha)$	$\alpha! \to \alpha?$
$\alpha! \to \alpha$	$(\alpha!)? \to (\alpha!)$
$(\alpha \lor \beta)? \to \alpha? \lor \beta?$	$(\exists x \alpha)? \to \exists x(\alpha?)$
$\neg(\alpha \to \beta) \leftrightarrow (\alpha \land \neg \beta)$	$\neg(\alpha \land \beta) \leftrightarrow (\neg \alpha \lor \neg \beta)$
$\neg(\alpha \lor \beta) \leftrightarrow (\neg \alpha \land \neg \beta)$	$\neg\neg\alpha \leftrightarrow \alpha$
$\neg \forall x \alpha \leftrightarrow \exists x \neg \alpha$	$\neg \exists x \alpha \leftrightarrow \forall x \neg \alpha$
$\neg(\alpha \land \neg \alpha)$	$(\neg \alpha)? \leftrightarrow \neg(\alpha?)$

The paraconsistent negation $\neg$ behaves classically for ?-free formulas.

THEOREM 7. *[19, 3] Let A be a ?-free formula, Γ be a set of ?-free formulae and let $\models_C$ be the classical consequence relation. Then, $\Gamma \models A$ iff $\Gamma \models_C A$.*

As we said in the first section, the meaning of the plausible operators are only properly understood in the whole IDL&LEI system. Plausible formulas are there first introduced by default rules which give sensible reasons to assert them.

3   LEI Natural Deduction System - ND_{LEI}

In this section, we will present a natural deduction system for LEI, ND_{LEI}, that is normalizable. Recall, from section 2, that Π and Σ will be used for a deduction and a sequence, possibly empty, of deductions, respectively. Π and Σ may appear with indexes. Although $\sim$ and $\bot$ are derived symbols, they are considered here as primitive ones in order to simplify the presentation of rules and reductions.

DEFINITION 8 (?-closed formula). Let α and β be any formula in $\mathcal{L}_?$. The notion of ?-closed formula is inductively defined by the following cases:

1. $\bot$, $\alpha?$ and $\alpha!$ are ?-closed;

2. If α is ?-closed, so are $\neg \alpha$, $\forall x \alpha$, $\exists x \alpha$;

3. If α and β are ?-closed, so are $\alpha \land \beta$, $\alpha \lor \beta$, $\alpha \to \beta$.

The definition of ?-closed formula is similar to the definition, by Prawitz [22], of essentially modal formula for S5 modal logic.

THEOREM 9. *[3, 7] If α is ?-closed, then $\alpha? \to \alpha$ and $\alpha \to \alpha!$.*

Proof. We can easily check this result from LEI semantics since the interpretation of ? and ! involve a quantification over plausible worlds. If α is ?-closed, but is not $\bot$, the interpretation of the outer ? (or !) in α?

(or α!) is a sort of a vacuous quantification. If α is $\perp$, the semantical proof that $\perp? \rightarrow \perp$ (or $\perp \rightarrow \perp$!) is based on the assumption that for all atomic formula α, α is true or false under a LEI valuation.   ∎

Definition 8 and theorem 9 will be essential in the design of a natural deduction proof system for LEI that is able to be normalized.

Our notion of deduction differs slightly from the usual ones first presented by Gentzen [6] and Prawitz [22]. By successive applications of the inference rules, we can generate a formula-tree, in which the leaves are the assumptions (top-formulas) and the root is the conclusion (end-formula). Such formula-tree will be called a deduction Π, following all conventions presented in [22], but the Complete Discharge Convention (see [29], section 1.3), that is, in our system, we will use a tag associated to a top-formula α, and this tag will be placed on the right side of the application r of the rule that discharges it. If δ is a premise of r, we synonymously say that α is discharged at δ. Furthermore, top-formulas occurrences that are not of the same shape α or are discharged by other applications of inference rules different from r cannot be indexed by the same tag. This notion of deduction is required to allow a better control, through the normalization procedure, over the set of assumptions discharged. Without it, we cannot guarantee strong normalization[2] (see [29], section 6.6).

The notions of *formula occurrence*, *shape of a formula*, *side-connected formulas*, *a formula that occurs (immediately) above/below another*, *major premise* and *minor premise* are taken for granted [22].

DEFINITION 10 (Thread). [22] A thread is a sequence of formulas $\alpha_1, \alpha_2, ... \alpha_n$ in a deduction Π such that α_i occurs immediately above α_{i+1}, $1 \leq i < n$, α_1 is a top-formula, and α_n is the end-formula of Π. We say that a thread is determined by α_1.

DEFINITION 11 (Dependency between a formula β and a top-formula α). : β is said to *depend on* the assumption α if β belongs to some thread in a deduction Π that begins with α and α is not discharged at a formula occurrence δ that precedes β.

Note that β may depend on α and α be discharged at β. For example:

$$\frac{\begin{array}{c} [\alpha]^u \\ \Pi \\ \beta \end{array}}{\alpha \rightarrow \beta} \ I\rightarrow \ u$$

is an application of I→ that discharges a set of occurrences of top-formulas α, the assumption class denoted by $[\alpha]$ as in [22, 29], on which β depends.

---

[2] A unique normal form from a finite number of reductions.

The tag u on the right side of the inference application defines the place where assumptions α, tagged by u, are discharged. Recall that, in [22], notation:

$$\begin{array}{c} \Gamma \\ \Pi \\ \alpha \end{array}$$

means that Γ is a set of top-formulas in Π (it is neither necessary to exhibit all top-formulas nor exhibit any of them) that α depends on and α is the end-formula of Π.

All inference rules for ND_{LEI} will be divided in: introduction rules (I?, I!, $I\wedge$, $I\vee_1$, $I\vee_2$, $I\rightarrow$, $I\forall$, $I\exists$, $I\neg$, $I\sim$), elimination rules (E?, E!, $E\wedge_1$, $E\wedge_1$, $E\vee$, $E\rightarrow$, $E\forall$, $E\exists$ $E\neg$, $E\sim$), classical absurdity ($\perp\sim$), paraconsistent absurdity ($\perp\neg$) and special rules ($I_{\neg n}$, $E_{\neg n}$, $n = 1..7$). All rules, but the special ones presented in the second table, are presented in the first table. The introduction and elimination rules for conjunction $\wedge$, disjunction $\vee$, implication $\rightarrow$, universal quantifier $\forall$, existential quantifier $\exists$, classical negation $\sim$, and classical absurdity ($\perp\sim$) are exactly as the ones in classical logic [22]. For paraconsistent negation, beyond the paraconsistent absurdity rule ($\perp\neg$), we have an introduction rule $I\neg$ and an elimination rule $E\neg$, but $E\neg$ deals only with ?-free formulas. In LEI, $\alpha \wedge \neg\alpha$ does not really represent a contradiction able to trivialize a theory. Contradictions able to do that are $A \wedge \neg A$ (or $\alpha\wedge \sim \alpha$), and they will be denoted by $\perp$.

The following definitions will be essential in the formulation of normalizable rules I! and E?.

DEFINITION 12 (Connection). By a connection in a deduction Π between two formula occurrences α and β, we understand a sequence $\alpha_1, \ldots, \alpha_n$ of formula occurrences in Π such that $\alpha_1 = \alpha, \alpha_n = \beta$, and one of the following conditions holds for each $i \leq n$;

1. α_i is not the major premise of an application of $E\vee$, $E\exists$ and $E?$, and α_{i+1} stands immediately below α_i; or vice versa;

2. α_i is a premise of an application of $E\rightarrow$, $E\neg$ or $E\sim$, and α_{i+1} is side connected with α_i;

3. α_i is the major premise of an application of $E\vee$, $E\exists$ and $E?$, and α_{i+1} is a hypothesis discharged by this application; or vice versa;

4. α_i is a consequence of an application of $I\rightarrow$, $I\neg$, $I\sim$, $\perp\neg$, $\perp\sim$, and α_{i+1} is a hypothesis discharged by this application; or vice versa;

DEFINITION 13 (Modally Independent). Two formula occurrences α and β are said to be modally independent in a deduction Π iff every connection in Π between α and β contains an occurrence of a ?-closed formula.

The first and second table present the whole ND_{LEI} system. Restrictions below are for plausible rules I! and E?.

REMARK 14 (Restrictions over $I!$). : Let β be a top-formula such that α depends on. Then α must be modally independent in Π with β.

REMARK 15 (Restrictions over $E?$). : Let Γ be the set of assumptions, other than α^u, on which β depends. Then each occurrence of α^u must be modally independent in Π with each formula in Γ as well as β.

$I!\quad \dfrac{\begin{array}{c}\Pi\\ \alpha\end{array}}{\alpha!}$	$E!\quad \dfrac{\alpha!}{\alpha}$
$I?\quad \dfrac{\alpha}{\alpha?}$	$E?\quad \dfrac{\alpha? \qquad \begin{array}{c}[\alpha]^u\\ \Pi\\ \beta\end{array}}{\beta}\,u$
$I_1\vee\quad \dfrac{\alpha}{\alpha\vee\beta}\qquad \dfrac{\beta}{\alpha\vee\beta}\quad I_2\vee$	$E\vee\quad \dfrac{\alpha\vee\beta \qquad \begin{array}{c}[\alpha]^u\\ \Pi_1\\ \gamma\end{array} \qquad \begin{array}{c}[\beta]^v\\ \Pi_2\\ \gamma\end{array}}{\gamma}\,u,v$
$I\wedge\quad \dfrac{\alpha\ \ \beta}{\alpha\wedge\beta}$	$E_1\wedge\quad \dfrac{\alpha\wedge\beta}{\alpha}\qquad \dfrac{\alpha\wedge\beta}{\beta}\quad E_2\wedge$
$I\rightarrow\quad \dfrac{\begin{array}{c}[\alpha]^u\\ \Pi\\ \beta\end{array}}{\alpha\rightarrow\beta}\,u$	$E\rightarrow\quad \dfrac{\alpha\ \ \alpha\rightarrow\beta}{\beta}$
$I\neg\quad \dfrac{\begin{array}{c}[\alpha]^u\\ \Pi\\ \bot\end{array}}{\neg\alpha}\,u$	$E\neg\quad \dfrac{A\ \ \neg A}{\bot}$
$I\sim\quad \dfrac{\begin{array}{c}[\alpha]^u\\ \Pi\\ \bot\end{array}}{\sim\alpha}\,u$	$E\sim\quad \dfrac{\alpha\ \ \sim\alpha}{\bot}$
$\bot\sim\quad \dfrac{\begin{array}{c}[\sim\alpha]^u\\ \Pi\\ \bot\!\!=_u\end{array}}{\alpha}$	$\bot\neg\quad \dfrac{\begin{array}{c}[\neg\alpha]^u\\ \Pi\\ \bot\!\!=_u\end{array}}{\alpha}$
$I\forall\quad \dfrac{\alpha(x)}{\forall x\alpha(x)}$ the free variable x must not occur free in any assumptions that $\alpha(x)$ depends on	$E\forall\quad \dfrac{\forall x\alpha(x)}{\alpha(t)}$ t is free for x in $\alpha(x)$
$I\exists\quad \dfrac{\alpha(t)}{\exists x\alpha(x)}$ t is free for x in $\alpha(x)$	$E\exists\quad \dfrac{\exists x\alpha(x) \qquad \begin{array}{c}[\alpha(x)]^u\\ \Pi\\ \gamma\end{array}}{\gamma}\,u$ the free variable x must neither occur free in γ nor in any assumptions that γ depends on other than $\alpha(x)$

$I_{\neg 1}$	$\dfrac{\alpha \wedge \neg\beta}{\neg(\alpha \to \beta)}$	$E_{\neg 1}$	$\dfrac{\neg(\alpha \to \beta)}{\alpha \wedge \neg\beta}$
$I_{\neg 2}$	$\dfrac{\neg\alpha \vee \neg\beta}{\neg(\alpha \wedge \beta)}$	$E_{\neg 2}$	$\dfrac{\neg(\alpha \wedge \beta)}{\neg\alpha \vee \neg\beta}$
$I_{\neg 3}$	$\dfrac{\neg\alpha \wedge \neg\beta}{\neg(\alpha \vee \beta)}$	$E_{\neg 3}$	$\dfrac{\neg(\alpha \vee \beta)}{\neg\alpha \wedge \neg\beta}$
$I_{\neg 4}$	$\dfrac{\alpha}{\neg\neg\alpha}$	$E_{\neg 4}$	$\dfrac{\neg\neg\alpha}{\alpha}$
$I_{\neg 5}$	$\dfrac{\exists x \neg\alpha}{\neg\forall x \alpha}$	$E_{\neg 5}$	$\dfrac{\neg\forall x \alpha}{\exists x \neg\alpha}$
$I_{\neg 6}$	$\dfrac{\forall x \neg\alpha}{\neg\exists x \alpha}$	$E_{\neg 6}$	$\dfrac{\neg\exists x \alpha}{\forall x \neg\alpha}$
$I_{\neg 7}$	$\dfrac{(\neg\alpha)?}{\neg(\alpha?)}$	$E_{\neg 7}$	$\dfrac{\neg(\alpha?)}{(\neg\alpha)?}$

Rule I? is intuitively acceptable: if we have a deduction of a formula α, then we have a deduction of its plausibility $\alpha?$. For skeptical plausibility !, the elimination rule E! has no restrictions.

Rules I! and E? are not so natural. The use of connection was inspired in the related notion of connection proposed by Prawitz for defining the rule of Introduction of $\square$ in his S5 natural deduction system [22]. Our notion is different from Prawitz's since, instead of $\square$, we will consider the plausible operator ! and also ?.

The notion of connection combined with the notion of two formulas be modally independent are essential to preserve correctness after the rewriting process related to the reduction steps (see section 4) in the normalization procedure.

In principle, one could think that the notion of connection could be simply replaced by the notion of thread. For instance, instead of restriction 14, we could just consider that in each thread determined by the premise of an application of I! and some assumption on which the premise α depends on, a ?-closed formula occurs such that this occurrence does not depend on any assumptions that the premise does not depend on. To check that the use of connection, instead of thread, is really essential, we will show the following example where α and β are not ?-closed. References for maximum segment/formula, operational/permutative/absurdity reductions and reduction steps will be explained in Section 4.

EXAMPLE 16.

$$\dfrac{\dfrac{\neg(\alpha! \vee \alpha!)^1 \quad \neg(\alpha! \vee \alpha!) \to \beta}{\beta} \; \text{E}{\to} \quad \sim\beta}{\dfrac{\dfrac{\bot}{\alpha! \vee \alpha!} \; \bot\neg \; 1 \qquad \dfrac{\alpha!^2}{\alpha} \; \text{E!} \quad \dfrac{\alpha!^3}{\alpha} \; \text{E!}}{\dfrac{\alpha}{\alpha!} \; \text{I!}} \; \text{EV} \; 2,3} \; \text{E}\sim}$$

Since $\alpha! \vee \alpha!$ is ?-closed, both threads determined by α and assumptions $\sim\beta/\neg(\alpha! \vee \alpha!) \to \beta$ satisfy the weak restriction over I!, the one that uses the notion of thread, instead of connection. However, $\alpha! \vee \alpha!$ is a maximum formula and we have to apply an absurdity reduction in the normalization procedure. Thus, we obtain the following deduction:

$$\dfrac{\dfrac{\alpha! \vee \alpha!^4 \quad \dfrac{\alpha!^2}{\alpha} \; \text{E!} \quad \dfrac{\alpha!^3}{\alpha} \; \text{E!}}{\dfrac{\alpha}{\neg(\alpha! \vee \alpha!)} \; \text{I}\neg \; 4} \; \text{EV} \; 2,3 \quad \sim\alpha^5}{\dfrac{\dfrac{\bot}{\neg(\alpha! \vee \alpha!)} \; \text{I}\neg \; 4 \qquad \neg(\alpha! \vee \alpha!) \to \beta}{\beta} \; \text{E}{\to} \quad \sim\beta}{\dfrac{\bot}{\dfrac{\alpha}{\alpha!}} \; \text{I!} \; \sim \; 5} \; \text{E}\sim}$$

In the thread determined by α and the assumption $\sim\beta$ that α depends on, there is a ?-closed formula $\bot$, but it depends on assumption 5 that α does not depend on. Hence, the application of I! becomes wrong after this reduction if we use the notion of thread, instead of connection.

The use of connection allows us also to guarantee that the system is closed under substitution [30]. For instance, consider the following example where we replace the assumption β by its deduction Π_1:

EXAMPLE 17.

$$\dfrac{\Pi_1 \quad \dfrac{\alpha? \quad \dfrac{\beta^w[\alpha]^k}{\Pi_2} \; \gamma}{\dfrac{\gamma}{\beta \to \tau} \; \text{I}{\to} \; w} \; \text{E?} \; k}{\tau} \; \text{E}{\to} \qquad \rhd\rhd \qquad \dfrac{\alpha? \quad \dfrac{\Pi_1 \quad \dfrac{\beta[\alpha]^k}{\Pi_2}}{\gamma}}{\dfrac{\gamma}{\Pi_3}} \; \text{E?} \; k$$

After the operational reduction used to remove $\beta \to \tau$, even if γ depends on ?-free top-formulas that occurs in Π_1, the notion of connection and of modally independent formulas assure a correct application of E?.

Rules for $\neg$ (I$\neg$, E$\neg$ and $\bot\neg$) state that $\neg$ behaves paraconsistently for ?-formulas and classically for ?-free formulas [3].

The special rules (I$\neg n$, E$\neg n$, $n = 1..7$) do not follow the same pattern of other rules since the special rules involve more than one logical constant. They are introduced in order to recover the deductive power of the paraconsistent negation $\neg$, lost by restrictions imposed on the application of E$\neg$ just to ?-free formulas. In fact, they are a sort of definition and could be introduced as axioms, instead[3]. The subformula property will be then proved with a new restriction: formulas in a deduction can also be a transformation, by special rules, of (subformulas of) top-formulas and conclusion. A related problem is found in system G4ip, in [29].

The correctness and completeness proof of this system can be found in http://www.lia.ufc.br/~ana/NDLEIappendix.ps.

4   Reductions

Reductions are the central part of the proof of the weak normalization procedure for ND_{LEI}. They will be defined for each kind of maximum segment described below aiming to eliminate them. Reductions will be classified as operational, permutative and those related to the absurdity sign. Since operational reductions for $\to, \vee, \wedge, \exists, \forall$, and permutative reductions for $\vee, \exists$ are as in [22], we will omit them. For the same reason, we will also omit reductions that involve the absurdity rules for $\sim$ (see [13, 20] for details). Our focus will be on reductions for plausibilities ? and !, and for the paraconsistent negation $\neg$, that is, we will just present reductions that are related to rules: I?, E?, I!, E!, I$\neg$, E$\neg$, $\bot\neg$, I$\neg n$, E$\neg n$, n=1..7. Auxiliary permutative reductions that involve $\vee$ and $\exists$ were just presented because $\bot$ appears as a result of E$\neg$ (or E$\sim$). What is important to emphasize is that what lead us to define *exactly* these additional reductions is the definition 19 of Maximum Segment.

Before presenting all reductions, the notion of Normal Form and other required definitions are introduced.

---

[3]In a sequent calculus formulation [6], for instance, if $\alpha \leftrightarrow \beta$ is an axiom of the logical system not proved by the sequent rules, then it could be represented as the following Cut rule: $\dfrac{\Gamma\vdash\Delta,\alpha \quad \beta,\Gamma'\vdash\Delta'}{\Gamma,\Gamma'\vdash\Delta,\Delta'}$ (Cut$_{\alpha\leftrightarrow\beta}$).

In a system with this new kind of cut rule, the cut elimination theorem could now be rephased as: instead of eliminating all cut occurrences, we will only eliminate cut applications that match the usual pattern of cut rule $\dfrac{\Gamma\vdash\Delta,\alpha \quad \alpha,\Gamma'\vdash\Delta'}{\Gamma,\Gamma'\vdash\Delta,\Delta'}$ (Cut). A normal form that keeps cut rule applications such as Cut$_{\alpha\leftrightarrow\beta}$ is non-standard. It may be used to explicit the definitional character of the axioms.

Both in our LEI sequent calculus [7, 12, 11] and in ND_{LEI}, we choose a different presentation: we addopted special rules (instead of axioms) and proved subformula property with restrictions, but we kept the usual notion of normal proof.

4.1   Definitions

DEFINITION 18 (Segment). A segment in a deduction Π is a sequence $\sigma = \alpha_1, \alpha_2, ..., \alpha_n$, with length n, of consecutive occurrences of formulas in a subthread of Π, such that:

1. α_1 is not a conclusion of an application of $E?$, $E\vee$ or $E\exists$;

2. $\alpha_i, (i < n)$, is a minor premise of an application of $E?$, $E\vee$ or $E\exists$;

3. α_n is not a minor premise of an application of $E?$, $E\vee$ or $E\exists$.

We will extend the usual definition of maximum segment (see [13, 20]) as follows.

DEFINITION 19 (Maximum Segment). Let $\sigma = \alpha_1, \alpha_2, ..., \alpha_n$ be a segment. We can classify it, depending on α_n, as:

1. α_n is a conclusion of an application of an introduction rule, or of an application of intuitionistic absurdity[4], or of an application of $E?$, $E\vee$ or $E\exists$, and it is, at the same time, a major premise of an elimination rule;

2. α_n is a conclusion of an application of the classical absurdity ($\perp \sim$), or of paraconsistent absurdity ($\perp\neg$), and it is, at the same time, a major premise of an elimination rule;

3. α_n is a conclusion of an application of the classical absurdity ($\perp \sim$), or of paraconsistent absurdity ($\perp\neg$), or of an application of $E?$, $E\vee$ or $E\exists$, and it is, at the same time, a minor premise of $E\sim$ or $E\neg$, where the major premise is a top-formula;

A maximum segment that is a conclusion of an introduction rule and it is, at the same time, major premise of an elimination rule is also called *Maximum Formula* [22].

Segments that follow the third pattern are non-usual. They must be removed to prevent the appearance of maximum segments of the first and second patterns that really disturb the normalization procedure. Hence, they will not contribute to the degree of a deduction.

DEFINITION 20 (Degree of a Formula). The degree of a formula α, $d(\alpha)$, is the number of connectives in α.

---

[4]Intuitionistic absurdity is an application of $\perp\neg$ or $\perp \sim$ that does not discharge any assumptions.

DEFINITION 21 (Degree of a Segment). The degree of a segment σ, $d(\sigma)$, is the degree of the formula that occurs in it.

DEFINITION 22 (Degree of a Deduction). The degree of a deduction Π, $d(\Pi)$, is defined as $d(\Pi) = \max \{\, d(\sigma) \colon \sigma$ is a maximum segment in Π of kind 1 or 2 in definition 19 $\}$.

Since strong normalization will not be proved in this paper, for simplicity, we will define the notion of normal form of a deduction without being worried about redundant applications of E?, E$\vee$ or E$\exists$.

DEFINITION 23 (Normal Form of a Deduction). A deduction in ND_{LEI} that do not contain any occurrence of maximum segments is said to be Normal, or to be in a Normal Form.

4.2 Reductions

In all reductions defined below, the deduction on the right is an *immediate reduction* (notation $\rhd_i\rhd$) of the reduction on the left. The reducibility relation, denoted by $\rhd\rhd$, is the transitive closure of $\rhd_i\rhd$. These reductions are based on the ones presented in [15, 14]. Differences are in reductions that involve rules I! and E?, and in the necessity of checking if restrictions over connections are respected after reductions. Absurdity reductions for paraconsistent negations are strongly inspired on the ones for classical negation by [13, 20].

1. Operational Reductions

   (a) Operational Reduction for ?

$$
\begin{array}{c}
\Pi_1 \quad [\alpha]^j \\
\dfrac{\alpha \quad \Pi_2}{\alpha? \quad \beta} \\
\dfrac{}{\beta} \; \text{E?}_j \\
\Pi_3
\end{array}
\qquad \underset{i}{\rhd\rhd} \qquad
\begin{array}{c}
\Pi_1 \\
[\alpha] \\
\Pi_2 \\
\beta \\
\Pi_3
\end{array}
$$

   REMARK 24. Note that new connections may appear after the substitution of the top-formula α by its proof Π_1, but all restrictions over applications of I! and E?, if any, are preserved since Π_1 is placed above $[\alpha]$, in the right deduction, and α? is not connected with β, in the left deduction.

   (b) Operational Reduction for !

$$
\begin{array}{c}
\Pi_1 \\
\dfrac{\alpha}{\alpha!} \; \text{I!} \\
\alpha \\
\Pi_2
\end{array}
\qquad \underset{i}{\rhd\rhd} \qquad
\begin{array}{c}
\Pi_1 \\
\alpha \\
\Pi_2
\end{array}
$$

REMARK 25. Note that there could exist an application of I! and E? in Π_2 and α! be used as a ?-closed formula in connections. However, restriction over I! also assures the existence of other ?-closed formulas in these connections, beyond α!. Hence, this reduction preserve correctness.

(c) Operational Reduction for Special Rules

  i. $I_{\neg n}$ followed by $E_{\neg n}$, $1 \leq n \leq 7$.

$$
\begin{array}{c}
\Pi_1 \\
\alpha \\
I \; \dfrac{\beta}{\;} \\
E \; \dfrac{\alpha}{\;} \\
\Pi_2
\end{array}
\quad \underset{i}{\rhd\rhd} \quad
\begin{array}{c}
\Pi_1 \\
\alpha \\
\Pi_2
\end{array}
$$

REMARK 26. In this step, letter I denotes an application of $I_{\neg n}$, and letter E denotes an application of rule $E_{\neg n}$, $1 \leq n \leq 7$. Note that, even if β was the ?-closed referred in a connection of any possible application of I! and E? in the left deduction, α can play the same role since if a conclusion of a special rule is ?-closed, obviously the premise is too, and vice-versa.

 ii. $I_{\neg 1}$ followed by $E_{\neg}$
    For all cases below related to $I_{\neg n}$, $1 \leq n \leq 6$, the unique possible elimination is $E_{\neg}$, and formulas must be ?-free. A conclusion of an application of $I_{\neg 7}$ cannot be major premise of $E_{\neg}$. Hence this case will be not considered.

$$
\dfrac{\Pi_1 \quad \dfrac{\Pi_2 \quad A \wedge \neg B}{\neg(A \to B)} I_{\neg 1}}{\bot} E_{\neg}
\qquad \underset{i}{\rhd\rhd} \qquad
\dfrac{\dfrac{\dfrac{\Pi_2}{A \wedge \neg B}}{A} E1\wedge \quad \Pi_1}{B} E_{\to} \qquad \dfrac{\dfrac{\Pi_2}{A \wedge \neg B}}{\neg B} E2\wedge}{\bot} E_{\neg}
$$

iii. $I_{\neg 2}$ followed by $E_{\neg}$

$$
\dfrac{\Pi_1 \quad \dfrac{\Pi_2 \quad \neg A \vee \neg B}{\neg(A \wedge B)} I_{\neg 2}}{\bot} E_{\neg}
\qquad \underset{i}{\rhd\rhd} \qquad
\dfrac{\dfrac{\dfrac{\Pi_1}{A \wedge B}}{A} E1\wedge \quad \neg A^j}{\bot} E_{\neg} \quad \dfrac{\dfrac{\Pi_1}{A \wedge B}}{B} E2\wedge \quad \neg B^k}{\bot} E \;\; EV_{j,k}
$$

iv. $I_{\neg 3}$ followed by $E_{\neg}$

$$
\dfrac{\Pi_1 \quad \dfrac{\Pi_2 \quad \neg A \wedge \neg B}{\neg(A \vee B)} I_{\neg 3}}{\bot} E_{\neg}
\qquad \underset{i}{\rhd\rhd} \qquad
\dfrac{\Pi_1}{A \vee B} \quad \dfrac{A^j \quad \dfrac{\dfrac{\Pi_2}{\neg A \wedge \neg B}}{\neg A} E1\wedge}{\bot} E_{\neg} \quad B^k \quad \dfrac{\dfrac{\Pi_2}{\neg A \wedge \neg B}}{\neg B} E2\wedge}{\bot} E_{\neg} \;\; EV_{j,k}
$$

 v. $I_{\neg 4}$ followed by $E_{\neg}$

$$\cfrac{\Pi_1 \quad \cfrac{\cfrac{\Pi_2}{A} \quad }{\neg\neg A} \ I_{\neg}4}{\cfrac{\neg A}{\bot} \ E\neg} \qquad \vartriangleright_i\vartriangleright \qquad \cfrac{\Pi_2 \quad \Pi_1}{\cfrac{A \quad \neg A}{\bot}} \ E\neg$$

vi. $I_{\neg}5$ followed by $E\neg$

$$\cfrac{\Pi_1 \quad \cfrac{\Pi_2}{\exists x\neg A} \ I_{\neg}5}{\cfrac{\forall xA \quad \neg\forall xA}{\bot} \ E\neg} \qquad \vartriangleright_i\vartriangleright \qquad \cfrac{\Pi_2}{\cfrac{\exists x\neg A \quad \cfrac{\cfrac{\Pi_1}{\forall xA}}{\cfrac{A(b)}{\bot}} \ E\forall \quad \neg A(b)^j}{\bot} \ E\exists \ j}$$

Such that b is a new proper parameter.

vii. $I_{\neg}6$ followed by $E\neg$

$$\cfrac{\Pi_1 \quad \cfrac{\Pi_2}{\forall x\neg A} \ I_{\neg}6}{\cfrac{\exists xA \quad \neg\exists xA}{\bot} \ E\neg} \qquad \vartriangleright_i\vartriangleright \qquad \cfrac{\cfrac{\Pi_1}{\exists xA} \quad \cfrac{A(b)^j \quad \cfrac{\cfrac{\Pi_2}{\forall x\neg A}}{\neg A(b)} \ E\forall}{\bot} \ E\neg}{\bot} \ E\exists \ j$$

Such that b is a new proper parameter.

viii. $I\neg$ followed by $E_{\neg}1$

$$\cfrac{\cfrac{[\alpha \to \beta]^j}{\Pi_1}}{\cfrac{\cfrac{\bot}{\neg(\alpha \to \beta)} \ I\neg \ j}{\alpha \wedge \neg\beta} \ E_{\neg}1} \qquad \vartriangleright_i\vartriangleright \qquad \cfrac{\cfrac{\cfrac{\cfrac{\alpha^k \quad \sim \alpha^j}{\bot} \ E\sim}{\beta} \ \bot\neg \quad \bot\neg}{\cfrac{}{[\alpha \to \beta]} \ I\to \ k}{\Pi_1}}{\cfrac{\cfrac{\bot}{\alpha} \ \bot\sim \ j \qquad \cfrac{\cfrac{\beta^l}{[\alpha \to \beta]} \ I\to}{\Pi_1}}{\alpha \wedge \neg\beta}} \ I\wedge$$

ix. $I\neg$ followed by $E_{\neg}2$

$$\cfrac{\cfrac{[\alpha \wedge \beta]^j}{\Pi_1}}{\cfrac{\cfrac{\bot}{\neg(\alpha \wedge \beta)} \ I\neg \ j}{\neg\alpha \vee \neg\beta} \ E_{\neg}2} \qquad \vartriangleright_i\vartriangleright$$

ix continued:

$$\cfrac{\cfrac{\cfrac{\cfrac{\neg\alpha^k}{\neg\alpha \vee \neg\beta} \ I1\vee \quad \sim(\neg\alpha \vee \neg\beta)^j}{\cfrac{\bot}{\alpha} \ \bot\neg \ k} \ E\sim \qquad \cfrac{\cfrac{\neg\beta^l}{\neg\alpha \vee \neg\beta} \ I2\vee \quad \sim(\neg\alpha \vee \neg\beta)^j}{\cfrac{\bot}{\beta} \ \bot\neg \ l} \ E\sim}{[\alpha \wedge \beta]} \ I\wedge}{\cfrac{\Pi_1}{\cfrac{\bot}{\neg\alpha \vee \neg\beta} \ \bot\sim \ j}}$$

x. $I\neg$ followed by $E_{\neg}3$

$$\cfrac{\cfrac{[\alpha \vee \beta]^j}{\Pi_1}}{\cfrac{\cfrac{\bot}{\neg(\alpha \vee \beta)} \ I\neg \ j}{\neg\alpha \wedge \neg\beta} \ E_{\neg}3} \qquad \vartriangleright_i\vartriangleright \qquad \cfrac{\cfrac{\cfrac{\alpha^j}{[\alpha \vee \beta]} \ I1\vee}{\Pi_1}}{\cfrac{\bot}{\neg\alpha} \ I\neg \ j} \qquad \cfrac{\cfrac{\beta^k}{[\alpha \vee \beta]} \ I2\vee}{\cfrac{\Pi_1}{\cfrac{\bot}{\neg\beta}} \ I\neg \ k}}{\neg\alpha \wedge \neg\beta} \ I\wedge$$

xi. $I\neg$ followed by $E_{\neg}4$

$$\cfrac{\cfrac{[\neg\alpha]^j}{\Pi_1}}{\cfrac{\cfrac{\bot}{\neg\neg\alpha} \ I\neg \ j}{\alpha} \ E_{\neg}4} \qquad \vartriangleright_i\vartriangleright \qquad \cfrac{\cfrac{[\neg\alpha]^j}{\Pi_1}}{\cfrac{\bot}{\alpha} \ \bot\neg \ j}$$

xii. $I\neg$ followed by $E_{\neg}5$

$$
\begin{array}{c}
[\forall x\alpha]^j \\
\Pi_1 \\
\dfrac{\bot}{\neg\forall x\alpha}\ \text{I}\neg\ j \\[2pt]
\dfrac{}{\exists x\neg\alpha}\ \text{E}_{\neg 5}
\end{array}
\qquad \underset{i}{\triangleright\triangleright} \qquad
\begin{array}{c}
\dfrac{\neg\alpha(b)^k}{\exists x\neg\alpha}\ \text{I}\exists \qquad \sim\exists x\neg\alpha^j \\[4pt]
\dfrac{\bot}{\alpha(b)}\ \text{E}\sim \\[2pt]
\quad\ \bot\neg\ k \\
\dfrac{}{[\forall x\alpha]}\ \text{I}\forall \\
\Pi_1 \\
\dfrac{\bot}{\exists x\neg\alpha}\ \bot\sim\ j
\end{array}
$$

Such that b is a new proper parameter.

xiii. I¬ followed by $\text{E}_{\neg 6}$

$$
\begin{array}{c}
[\exists x\alpha]^j \\
\Pi_1 \\
\dfrac{\bot}{\neg\exists x\alpha}\ \text{I}\neg\ j \\[2pt]
\dfrac{}{\forall x\neg\alpha}\ \text{E}_{\neg 6}
\end{array}
\qquad \underset{i}{\triangleright\triangleright} \qquad
\begin{array}{c}
\dfrac{\alpha(b)^j}{[\exists x\alpha]}\ \text{I}\exists \\
\Pi_1 \\
\dfrac{\bot}{\neg\alpha(b)}\ \text{I}\neg\ j \\[2pt]
\dfrac{}{\forall x\neg\alpha}\ \text{I}\forall
\end{array}
$$

Such that b is a new proper parameter.

xiv. I¬ followed by $\text{E}_{\neg 7}$

$$
\begin{array}{c}
[\alpha?]^j \\
\Pi_1 \\
\dfrac{\bot}{\neg(\alpha?)}\ \text{I}\neg\ j \\[2pt]
\dfrac{}{(\neg\alpha)?}\ \text{E}_{\neg 7}
\end{array}
\qquad \underset{i}{\triangleright\triangleright} \qquad
\begin{array}{c}
\dfrac{\alpha^j}{[\alpha?]}\ \text{I?} \\
\Pi_1 \\
\dfrac{\bot}{\neg\alpha}\ \text{I}\neg\ j \\[2pt]
\dfrac{}{(\neg\alpha)?}\ \text{I?}
\end{array}
$$

2. Permutative Reductions

In all permutative reductions, restrictions over applications of I! and E?, if any, are respected since connections are preserved.

(a) Permutative Reduction for ?

$$
\begin{array}{c}
\qquad\ [\alpha]^j \\
\Pi_1 \quad \Pi_2 \\
\dfrac{\alpha? \quad \beta}{\dfrac{\beta}{\tau}}\begin{array}{l}\text{E?}_j\\ \ \\ \text{R}\end{array}\ \Sigma_3
\end{array}
\qquad \underset{i}{\triangleright\triangleright} \qquad
\begin{array}{c}
\qquad [\alpha]^j \\
\Pi_2 \\
\dfrac{\alpha?\quad \dfrac{\beta\quad \Sigma_3}{\tau}\ \text{R}}{\tau}\ \text{E?}_j \\
\Pi_1
\end{array}
$$

REMARK 27. β is the major premise of the elimination rule R and Σ_3 may occur on the left of β.

(b) Auxiliary Permutative Reduction

$$
\dfrac{\begin{array}{cc}\Pi_1 & \dfrac{[\alpha]^j}{\begin{array}{c}\Pi_2\\\beta\end{array}}\\[2pt]\dfrac{\alpha?\quad\beta}{\beta}\ \mathrm{E?}_j & \end{array}\quad \delta^w}{\bot}\ \mathrm{R}
\qquad \vartriangleright\vartriangleright_i \qquad
\dfrac{\Pi_1\quad \dfrac{\dfrac{[\alpha]^j}{\begin{array}{c}\Pi_2\\\beta\end{array}}\quad \delta^w}{\bot}\ \mathrm{R}}{\dfrac{\alpha?}{\bot}\ \mathrm{E?}_j}
$$

REMARK 28. β is the minor premise of the elimination rule R and δ is a top-formula of shape $\sim\beta$ or $\neg\beta$ and, in the last case, β must be ?-free.

(c) Auxiliary Permutative Reduction

$$
\dfrac{\dfrac{\begin{array}{ccc}\Pi_1 & \dfrac{[\alpha]^i}{\begin{array}{c}\Pi_2\\\gamma\end{array}} & \dfrac{[\beta]^j}{\begin{array}{c}\Pi_3\\\gamma\end{array}}\\[2pt]\alpha\vee\beta & \gamma & \gamma\end{array}}{\gamma}\ \mathrm{EV}_{i,j}\quad \delta^w}{\bot}\ \mathrm{R}
\qquad \vartriangleright\vartriangleright_i \qquad
\dfrac{\dfrac{\Pi_1}{\alpha}\quad \dfrac{\dfrac{[\alpha]^i}{\begin{array}{c}\Pi_2\\\gamma\end{array}}\quad \delta^w}{\bot}\ \mathrm{R}\quad \dfrac{\dfrac{[\beta]^j}{\begin{array}{c}\Pi_3\\\gamma\end{array}}\quad \delta^w}{\bot}\ \mathrm{R}}{\bot}\ \mathrm{EV}_{i,j}
$$

REMARK 29. γ is the minor premise of the elimination rule R and δ is a top-formula of shape $\sim\gamma$ or $\neg\gamma$ and, in the last case, γ must be ?-free.

(d) Auxiliary Permutative Reduction

$$
\dfrac{\dfrac{\Pi_1\quad \dfrac{[\alpha_a^x]^i}{\begin{array}{c}\Pi_2\\\gamma\end{array}}}{\dfrac{\exists x\alpha\quad \gamma}{\gamma}\ \mathrm{E\exists}_i}\quad \delta^w}{\bot}\ \mathrm{R}
\qquad \vartriangleright\vartriangleright_i \qquad
\dfrac{\Pi_1\quad \dfrac{\dfrac{[\alpha_a^x]^i}{\begin{array}{c}\Pi_2\\\gamma\end{array}}\quad \delta^w}{\bot}\ \mathrm{R}}{\dfrac{\exists x\alpha}{\bot}\ \mathrm{E\exists}_i}
$$

REMARK 30. γ is the minor premise of the elimination rule R and δ is a top-formula of shape $\sim\gamma$ or $\neg\gamma$ and, in the last case, γ must be ?-free.

3. Absurdity Reductions

(a) Absurdity Reductions for $\neg$

$$
\dfrac{\dfrac{[\neg\alpha]^k}{\begin{array}{c}\Pi_1\\\frac{\bot}{\alpha}\quad \bot\neg k\end{array}}\quad \Sigma_2}{\gamma}\ \mathrm{R}
\qquad \vartriangleright\vartriangleright_i \qquad
\dfrac{\dfrac{\dfrac{\alpha^j\quad \Sigma_2}{\gamma}\ \mathrm{R}\quad \sim\gamma^i}{\dfrac{\bot}{[\neg\alpha]}\ j}}{\begin{array}{c}\Pi_1^*\\\frac{\bot}{\gamma}\ i\end{array}}
$$

Π_1^* is obtained from Π_1 by substitution of all deductions of shape

$$\dfrac{\overset{\Pi_n}{\alpha} \quad \neg\alpha^k}{\bot}$$

by

$$\dfrac{\dfrac{\overset{\Pi_n}{\alpha} \quad \Sigma_2}{\gamma} \quad \sim\gamma^i}{\bot}$$

REMARK 31. α is the major premise of the elimination rule R. The same reduction step is used to reduce maximum segments of this kind resulted from an application of $\bot\sim$ (as stated in the definition of maximum segment). Σ_2 can occur at the left of α. Note that all new connections pass through $\neg\alpha$ and restrictions on any applications of I! and E? in Π_1^*, if any, are respected. The same reasoning about connections is valid for the following absurdity reductions.

(b) Auxiliary Absurdity Reductions for $\neg$

$$\dfrac{\dfrac{[\neg\alpha]^k}{\overset{\Pi_1}{\dfrac{\bot}{\alpha}} \; \bot\neg k} \quad (\sim\alpha)^w}{\bot} \; R \qquad\qquad \underset{i}{\triangleright\triangleright} \qquad\qquad \dfrac{\dfrac{\dfrac{\alpha^j \quad (\sim\alpha)^w}{\bot} \, R}{[\neg\alpha]} \, j}{\dfrac{\Pi_1^*}{\bot}}$$

Π_1^* is obtained from Π_1 by substitution of all deductions of shape

$$\dfrac{\overset{\Pi_n}{\alpha} \quad (\neg\alpha)^k}{\bot}$$

by

$$\dfrac{\overset{\Pi_n}{\alpha} \quad (\sim\alpha)^w}{\bot}$$

REMARK 32. The same reduction step is used to reduce maximum segments of this kind resulted from an application of $\bot\sim$ which are minor premise of E$\neg$ (as stated in the definition of maximum segment).

(c) Auxiliary Absurdity Reductions for $\neg$

$$
\begin{array}{c}
[\sim A]^k \\
\Pi_1 \\
\dfrac{\perp}{A} \ {\perp\sim k} \quad (\neg A)^w \\
\hline
\perp
\end{array}
\qquad \underset{i}{\triangleright\triangleright} \qquad
\begin{array}{c}
\dfrac{A^j \quad (\neg A)^w}{\perp} \\
\dfrac{}{[\sim A]} \ j \\
\Pi_1^* \\
\perp
\end{array}
$$

Π_1^* is obtained from Π_1 by substitution of all deductions of shape

$$
\begin{array}{c}
\Pi_n \\
\dfrac{A \quad (\sim A)^k}{\perp}
\end{array}
$$

by

$$
\begin{array}{c}
\Pi_n \\
\dfrac{A \quad (\neg A)^w}{\perp}
\end{array}
$$

(d) Auxiliary Absurdity Reductions for $\neg$

$$
\begin{array}{c}
[\neg A]^j \\
\Pi_1 \\
\dfrac{\perp}{A} \ {\perp\neg j} \quad (\neg A)^k \\
\hline
\perp
\end{array}
\qquad \underset{i}{\triangleright\triangleright} \qquad
\begin{array}{c}
[\neg A]^k \\
\Pi_1 \\
\perp
\end{array}
$$

REMARK 33. The same reduction step is used to reduce maximum segments of this kind resulted from an application of $\perp\sim$ which are minor premise of E$\sim$ (as stated in the definition of maximum segment).

5 Weak Normalization for ND_{LEI}

The proof of weak normalization theorem for LEI is the same as the procedure presented in [13, 20] for full Classical Logic (including $\vee$ and $\exists$ as primitive signs), *mutatis mutandis*.

DEFINITION 34 (Length of a Deduction). The length of a deduction Π, $l(\Pi)$, is the number of formula occurrences in Π.

DEFINITION 35 (Last Inference of a Deduction). The last inference of a deduction Π will be denoted by $r(\Pi)$.

LEMMA 36. *Let Π be a deduction in ND_{LEI} of α from Γ with $d(\Pi) = n$ such that each maximum segment σ with degree n that contributes for the* *degree of* Π *is a conclusion of an application of an introduction rule, or of an application of intuitionistic absurdity, and it is, at the same time, a major premise of an elimination rule. Then,* Π *reduces to a deduction* Π' *of* α *from* Δ *(*$\Delta \subseteq \Gamma$*) such that* $d(\Pi') < d(\Pi)$*.*

Proof. The proof is analogous as the one by [22], *mutatis mutandis*.   ∎

LEMMA 37. *Every deduction* Π *such that* $d(\Pi) = 0$ *reduces to a normal deduction* Π'.

Proof. If Π is normal, $\Pi' = \Pi$. If Π is not normal, then the result is obtained by induction on K, where $K =$ the sum of the lengths of all maximum segments in Π. Choose a maximum segment σ in Π, such that there is neither other segment above it nor above (or that contains) a formula side-connected to the last formula of σ. Let Π_1 be the reduction of Π that eliminates the maximum segment σ. In reduction steps 2(b), 2(c), 2(d), 3(b), 3(c) and 3(d), the induction value of Π_1 is smaller than the value of Π. The result follows immediately.   ∎

LEMMA 38 (Critical Deduction). *Let* Π *be a deduction in* ND_{LEI} *of* β *from* Γ *such that:*

1. β *is the conclusion of an elimination rule, in which the major premise is the last formula occurrence of the unique maximum segment* σ *in* Π;

2. $d(\Pi) > 0$.

Then, Π *reduces to a deduction* Π' *of* β *from* Δ *(*$\Delta \subseteq \Gamma$*), such that* $d(\Pi') < d(\Pi)$*.*

Proof. By induction on $l(\Pi)$ (see definition 34).   ∎

THEOREM 39 (Weak Normalization). *Every deduction in* ND_{LEI} *reduces to a normal form in an effective way, by using the reduction steps.*

Proof. The proof is by induction on the pair (m, l), as in [13], such that m is the degree of Π, $d(\Pi)$, and l is the length of Π, $l(\Pi)$. All possible cases are listed below:

- If $r(\Pi)$ (see definition 35) is an application of an introduction rule or of any absurdity rule, then the result follows from the inductive hypothesis.

- If $r(\Pi)$ is an elimination rule, then Π is of form:

$$\Pi \equiv \frac{\begin{array}{ccc} \Pi_1 & & \Pi_n \\ \alpha_1 & \cdots & \alpha_n \end{array}}{\beta}$$

By using the inductive hypothesis, each Π_i, $(1 \leq i \leq n)$, reduces to a normal deduction Π'_i. Consider the following deduction Σ:

$$\Sigma \equiv \frac{\begin{array}{ccc} \Pi'_1 & & \Pi'_n \\ \alpha_1 & \cdots & \alpha_n \end{array}}{\beta}$$

If Σ is normal, then $\Pi' \equiv \Sigma$. If Σ is not normal, then Π' is obtained from Σ by using lemma 37 when $d(\Sigma) = 0$, or by lemma 38 when $d(\Sigma) > 0$. The result follows by the inductive hypothesis.

∎

The subformula property easily follows by an analysis of the form of normal deductions:

COROLLARY 40 (Subformula Property). *Every formula occurrence in a normal deduction in ND_{LEI} of α from Γ has the shape of a subformula (or of a transformation, by the special rules, of subformulas) of α or of some formula of Γ, except for assumptions discharged by applications of absurdity rules and for occurrences of $\perp$ that stand immediately below such assumptions.*

Proof. The proof is similar to the one, by [22], for intuitionistic logic. We have to first define the notion of path and then prove that all normal deductions follow a specific pattern where all elimination rules (E-part) come before introductions (I-part). Subformula property follows from this result with a difference: since any conclusion of a special rule is not a subformula of its premise, we have to consider that formulas in Π can also be a transformation, by special rules, of subformulas of $\Gamma \cup \alpha$[5]. ∎

6   Conclusion and Further Works

The Logic of Epistemic Inconsistency is a paraconsistent logic specially designed to support contradictions that come from reasoning under plausible knowledge introduced by default rules. LEI has two signs for plausibility, ? and !, which represents credulous and skeptical plausibility, respectively.

---

[5]see system G4ip, in [29], chapter 4, for comparisons.

In a situation of multiple and cooperative agents, α? is true if at least one agent believes in α, based on sensible reasons represented by default rules, and α! is true if all agents believe in α, also based on sensible reasons. LEI paraconsistent negation $\neg$ allows one to tolerate conflicting conclusions between plausible formulas α? and $\neg(\alpha?)$ without explosion of theorems.

Aiming to better understand the deductive power of these signs individually, a natural deduction system was introduced here. However, in order to design a natural deduction system that is weakly normalizable, a deep investigation of rules and related problems of substitution and reductions was done. Main results are concerned with the use of proof techniques to rules, and in the definition of reduction steps related to the paraconsistent negation and the plausible operators. The *weak normalization* theorem was presented for full LEI language, including $\vee, \exists, ?$ and the paraconsistent negation $\neg$. *Subformula property* was stated as corollary. Since normalization theorems through natural deduction system for da Costa's family of paraconsistent logics Cn, $1 \leq n < \omega$ were not yet investigated, we want to further analize if our absurdity reductions steps can be applied in order to prove weak normalization for the whole Cn family.

It was proved, in [10, 7], that the plausibility operators in LEI are closely related to modal operators in S5 modal logic. In fact, if we consider all modal operators $\Box, \Diamond, !, ?$ we could say that what is (credulously) plausible is possible, although not the inverse, and what is necessary is (skeptically) plausible, although not the inverse. Normalization for S5 logic was obtained, in Prawitz [22], not for the entire language: the possibility operator was defined as a dual operator to the necessity one, disjunction and existential quantifier were not considered, and the absurdity rule was applied with restrictions. Further works, as in [26], tried to normalize intuitionistic modal systems by presenting natural deduction rules with labels attached to formulas. Throughout the connection between credulous plausibility in LEI and possibility in S5, we want propose new rules for S5 modal operators, and to prove weak normalization for full S5, that is, for S5 modal logic using all connectives, quantifiers and modal operators as primitive symbols and without restrictions in the applications of the classical absurdity rule.

Acknowledgements

This research is partially supported by PADCT/CNPq and PRONEX/FUNCAP.

BIBLIOGRAPHY

[1] E.H. Alves. *Lógica e Inconsistência: Um Estudo dos Cálculos Cn, $1 \leq n \leq \omega$.* Master's thesis, Universidade Estadual de Campinas, 1976. preprint by Institute of Mathematics.

[2] J.-Y. Beziau. La Logique Paraconsistante C1 de Newton da Costa. 1990. Paris.

[3] A.R.Buchsbaum. *Lógicas da Inconsistência e Incompletude: Semântica, Axioma-tização e Automatização*. PhD thesis, Departamento de Informática, Pontifícia Universidade Católica do Rio de Janeiro, 1995.

[4] N.C.A. da Costa. On the Theory of Inconsistent Formal Systems. *Notre Dame Journal of Formal Logic*, 15. 1974. pp.497-510.

[5] N.C.A. da Costa. *Sistemas Formais Inconsistentes*. PhD thesis, Universidade Federal do Paraná, Curitiba, 1963.

[6] G. Gentzen. Investigations into Logical Deductions. In M.E.Szabo, editor, *The Collected Papers of Gerhard Gentzen*, pages 68-131, North-Holland, Amsterdam, 1969.

[7] A.T. Martins. *A Syntactical and Semantical Uniform Treatment for the IDL & LEI Nonmonotonic System*. PhD Thesis, Departamento de Informática, Universidade Federal de Pernambuco, Recife, 1997. www.lia.ufc.br/~ana/tese.ps

[8] A.T.Martins, M. Pequeno and T. Pequeno. Multiple Worlds Semantics to a Paraconsistent Nonmonotonic Logic. In W. A. Carnielli, M. E. Coniglio e I. M. L. D' Ottaviano (eds.), *Paraconsistency: The Logical Way To The Inconsistency*, Marcel Dekker,Inc., pgs.187-211, 2002.

[9] A.T.Martins, M. Pequeno and T. Pequeno. Well-Behaved IDL Theories. In: Proceedings of the *13th Brazilian Symposium on Artificial Intelligence*. Curitiba, Brazil, Oct., 1996. *Lecture Notes in Artificial Intelligence*. 1159:11-20. Springer. 1996.

[10] A.T. Martins and T. Pequeno. Paraconsistency and Plausibility in the Logic of Epistemic Inconsistency. In *Proceedings of the 1st World Congress on Paraconsistency*, University of Ghent, Belgium, Jul 31, Ago, 1-2 1997. Abstract.

[11] A.T.Martins and T. Pequeno. A Sequent Calculus of the Logic of Epistemic Inconsistency. In: Proceedings of the *11th Brazilian Symposium on Artificial Intelligence*. Fortaleza, Brazil, Oct. 17-20, 1994.

[12] A.T. Martins and T. Pequeno. Proof-theoretical considerations about the Logic of Epistemic Inconsistency. *Logique et Analise*, 143-144:245-60, 1993.

[13] C.D.B. Massi. *Provas de Normalização para a Lógica Clássica*, PhD Thesis, Departamento de Filosofia, Universidade Estadual de Campinas, Campinas, 1990.

[14] F.F. Morais. *Teoremas de Normalização para a Lógica da Inconsistência Epistêmica*, Master's thesis, Departamento de Computação, Universidade Federal do Ceará, Fortaleza, 2004. www.lia.ufc.br/~ana/felipe.pdf

[15] F.F. Morais. and A.T.Martins. Weak Normalization for the Logic of Epistemic Inconsistency. *Proceedings of the III world Congress on Paraconsistency*. IRIT, France, Jul 28-31, 2003. page 53.

[16] J.E.A. Moura. *Um Estudo de C_w em Cálculo de Seqüentes e Dedução Natural*. PhD. thesis, Departamento de Filosofia, Universidade Estadual de Campinas, Campinas, 2001.

[17] J.E.A. Moura, I.M.L. D'Ottaviano. NCG_ω: A Paraconsistent Sequent Calculus. In W.A. Carnielli, M.E. Coniglio and I.M.L D'Ottaviano, editors, *Paraconsistency, the Logical Way to the Inconsistent*. Lecture Notes in Pure and Applied Mathematics, Marcel Dekker, Inc, N.Y., 2002. pages 227-40.

[18] T. Pequeno. *A Logic for Inconsistent Nonmonotonic Reasoning*, Technical Report 90/6, Department of Computing, Imperial college, London, 1990.

[19] T. Pequeno and A. Buchsbaum. The Logic of Epistemic Inconsistency, In Proceedings of the *2nd International Conference on the Principles of Knowledge Representation and Reasoning*, J.A.Allen, R.Fikes and E.Sandewall, editors, pp. 453-60. San Mateo, CA, Apr. 1991. Morgan Kaufmann Publishers Inc.

[20] L.C.Pereira and C.D.B. Massi. Normalização para a Lógica Clássica, In O que nos faz pensar. *Cadernos de Filosofia da PUC-RJ*, n.2, pp.49-53, 1990.

[21] D. Prawitz. Ideas and Results in Proof Theory. In J.E.Fenstad, editor, *Proceedings of the Second Scandinavian Logic Symposium*, pages 235-250, North-Holland, Amsterdam, 1971.

[22] D. Prawitz. Natural Deduction: A Proof-theoretical Study. In *Stockholm Studies in Philosophy 3*, Almqvist and Wiksell, Stockholm, 1965.

[23] A.R. Raggio. Propositional Sequent-Calculi for Inconsistent Systems. *Notre Dame Journal of Formal Logic*, IX(4):359-66, 1968.

[24] R. Reiter. A Logic of Default Reasoning. *Artificial Intelligence*, 13:81-132, 1980.

[25] J. Seldin. Normalization and Excluded Middle I, *Studia Logica 48*, pages 193-217, 1989.

[26] A. K. Simpson. *The Proof Theory and Semantics of Intuitionistic Modal Logic.* Ph.D. thesis, University of Edinburgh, 1994.

[27] G. Stalmark. Normalizations Theorems for Full First Order Classical Natural Deduction. *The Journal of Symbolic Logic.* 56, 1991.

[28] R. Statman. *Structural Complexity of Proofs.* PhD Thesis, Stanford. 1974.

[29] A.S. Troelstra and H. Schwichtenberg. *Basic Proof Theory.* Cambridge University Press, Cambridge, 1996.

[30] A.M. Ungar. *Normalization, Cut-Elimination and the Theory of Proofs.* CSLI Lecture Notes, 28, 1992.

Ana Teresa Martins, Lília Ramalho Martins and Felipe Ferreira de Morais
Department of Computation, Federal University of Ceará, Phone +55 85 40089847 P.O.Box 12166 Fortaleza Ce Brazil 60455-760, Brazil.
{ana,lilia,felipe}@lia.ufc.br

A Non-standard graphical decision procedure and some praconsistent theorems for the Vasil'evian logic IL2

Alessio Moretti

ABSTRACT. There is a growing, deserved interest on the thought of N.A. Vasil'ev (1880-1940), a Russian visionary, philosopher and logician, considered together with A. Meinong a forerunner of contemporary paraconsistent logics. But there also seems to be a "Vasil'evian problem": while he foresaw, prophetically enough, many issues of non standard logics (which he conceived as "imaginary logics"), we still seem to lack a formal *faithful* representation of Vasil'evian logics. Effectively, due to the indirectness of the non-Russian-speaking approaches, the existing logical formalizations of his thought (outside Russia), although logically interesting *per se*, are most of the time highly unsatisfactory: not faithful enough, in fact very distant from the real Vasil'ev, most of the time truly ignored. While Vasil'ev's authority is called, *his* precise views are not formalized. The Russian logical school (at least from V.A. Smirnov onwards), on the contrary, has provided the most interesting bunch of works on this topic (but in Russian ...), developing logical axiomatizations very close to the texts. In particular, in 1999, V.I. Markin and D.V. Zaitzev have proposed the logical system **IL2** for capturing a conceptualist version of imaginary logic put forward by Vasil'ev himself. This recent work, very close to the letter and spirit of Vasil'ev, has nevertheless two drawbacks: (1) it is unrepresentable in terms of the standard logical diagrams (they do fail to represent this deviant conceptualist logic); (2) it does not yet show interesting enough "paraconsistent" results, while Vasil'ev explicitly advocated several, which he called "indifferent judgements and syllogisms". The first drawback seems normal (and temporary): Vasil'ev's logics were meant to be totally non-standard. The second drawback is curious: Vasil'ev's logics were *definitely* meant to be paraconsistent, i.e. able to assume non trivial inconsistency. In this paper we show that facing the first drawback by elaborating a non standard graphical decision procedure (GDP) for

IL2, very elegant and intuitive, suggests that the second drawback relies precisely on the lack of such an intuitive decision procedure. Our GDP expresses graphically all known axioms and theorems of **IL2**, plus several very Vasil'evian new paraconsistent theorems (still of **IL2**). We claim that this GDP, serving **IL2**, helps solving the "Vasil'evian problem" in two senses: (a) it explains most of what Vasil'ev saw clearly; (b) it throws light on several things that Vasil'ev had problems in seeing clearly. Our GDP seems thus to be a fruitful basis for exploring the Vasil'evian logical space **IL2**

1   Vasil'ev's non-earthly "imaginary logic" and its intensional "earthly" interpretation as a "logic of concepts" (1912)

The "triangle of contrarieties" and the "principle of the excluded fourth": Vasil'ev's route to the discovery of non standard logic. In 1910, the Russian logician N.A. Vasil'ev (1880-1940)[1], a doctor at the beginning of his professional life and then a professor of philosophy at the university of Kazan (the university of Lobachevski), enters the realm of logical investigation by way of a study (cf. [30]) on "some", a notion traditionally named (in Aristotelian, syllogistic terms) "particular judgement". He remarks that this concept is highly ambiguous, meaning alternatively "some, and perhaps all" (or: "at least some") and "only some" (or: "some, but not all" - these last are also called "accidental" judgements). While natural language (even in science) sticks almost always to the second meaning, the logically good one according to Vasil'ev, logical research until his time has quite often undertaken a use of "some" in the first meaning, which, according to Vasil'ev, is more psychological, heuristic and dynamical than logical[2]. Vasil'ev shows that this use is conceptually dangerous, in so much it loses the fact that, logically speaking, not one but *two* heterogeneous logical ways are possible here. The explanation for this relies, according to him, on the fact that fundamentally there are *two models of logical "opposition"*: a square one, the "square of oppositions", which gives a logic of the excluded third (that is standard Aristotelian logic); and a triangular one, the "triangle of contrarieties", which gives a logic of the excluded fourth (this logic is unknown at that time)[3]. Vasil'ev claims that the first of these

---

[1] For details on the life of Vasil'ev, cf. [2]. In English one can see [35].

[2] This meaning is though the one it has now in contemporary logic and mathematics: ∃.

[3] A fundamental deepening of this profound remark of Vasil'ev was made in 1953 by the French logician R. Blanché, who discovered the notion of "logical hexagon"; then by J.-Y. Béziau (2003), who proved the existence of a multiplicity of such modal hexagons and questioned their mutual ordering in the 3-dimensional space; and by ourself (2004),

logics is the logic of the "judgements on facts", while the second one is the logic of the "judgements on concepts". This strange point will turn out very important.

Starting from the observation that no scientific proof exists that logic is one (and one only), and in order to clarify what resembles to him, after his study on "some", more and more an exciting new start in logic, he will have the (bright) idea of understanding such strange new non-Aristotelian possibilities of logic as closely analogous to the paradoxical non-Euclidean possibilities discovered less than a century before (around 1826), in geometry, by Gauss, Lobachevski and Bolyai. Vasil'ev's logical-geometrical analogy will be very serious - not just a metaphor -, taking three progressive shapes: an analogy of form, of content and of interpretation.

1. The analogy of form. The philosophical Vasil'evian project of an "imaginary, non-Aristotelian logic".

The analogy of form says that in the same way as there are, in geometry, new possibilities beyond Euclid's geometry, there are in logics new possibilities beyond Aristotle's logic. Relying on the formal model of non-Euclidean geometries (mainly the one of Lobachevski, his compatriot), Vasil'ev tried, yet remaining inside the syllogistic pre-Frege-Russell frame[4], to break through Aristotelian logic - a Logic bound, among others, to the principles of identity, of non contradiction and of the excluded third. Such logics (the "imaginary logics", or, later, "n-dimensional logics") were meant to portray realms more complex than the "real world" we inhabit (a world ruled, Vasil'ev recognizes it, by Aristotelian logic). Such are, among others, the worlds of imagination, dream, and fiction in general (and there are more)[5]. Again, remarking that, as was the case for geometry, there is no decent proof showing that our logic is the only one, Vasil'ev thinks the imaginary logic he explores as being irreducibly (i.e. axiomatically) incompatible with our ordinary, normal, logic. More generally, he thinks that from the point of view of each of the yet unknown but possible logical worlds, the other ones must be false (not real), but that logically speaking each of these points of view could be endorsed (in a suited "imaginary world")[6]. So, considering the principle of non contradiction as

---

with the general notion of "logical bi-simplex of dimension n". Nowadays there is a formal science of all this, both (modal) logical and geometrical, the "theory of n-opposition" (for all the references, and for the backbone of the theory, cf. [19]; cf. also [20] and [23]; for an overview on the most recent results, cf. [22]).

[4] Which of course doomed Vasil'ev, unfairly enough, to be almost forgotten by history as a serious logician.

[5] This view shares interesting parallels with I. Matte Blanco's theory of the Freudian unconscious in terms of a "bi-logic" based on mental n-dimensional spaces (cf. [16], [17], [18]).

[6] This sort of "indexical relativism" reminds of D. Lewis' "modal realism" ("in every possible world - there are many! - people think that their own world is the only real

being the logical analogue to the Euclidean geometrical postulate of the parallels (i.e. an arbitrary presupposition, axiomatically independent from the other fundamental geometrical axioms and thus avoidable), he builds, in 1912 (cf. [31]), an "imaginary (non Aristotelian) logic", a logic ruled - among others - by the "principle of the excluded fourth" (as he discovered it possible in his previous study on "some") and in which the principle of non-contradiction is not any more a universally valid axiom or law. Later he will claim that this is further generalizable to a series of logical systems of the excluded n-th, the "imaginary n-dimensional logics" (cf. [27] and [8]). But let us see now more precisely the starting version of imaginary logic.

2. The analogy of content. "Indifferent" judgements and syllogisms.
The analogy of content says that, exactly in the same way as, from Euclidean to non-Euclidean geometry, one passes from a geometry of two possible kinds of lines (parallel or secant to a given line) to a geometry of three possible kinds of lines (parallel, secant, or not parallel *but* not secant to a given line), from Aristotelian to non-Aristotelian (i.e. imaginary) logics one passes from a logic of two possible judgements (affirmative and negative) to a logic of three possible judgements (affirmative, negative and "indifferent")[7]. One of the most impressive features, from our point of view, of Imaginary logic is thus that it is a logic which is also able, in some sense, to cope logically with real contradictions (in some sense, thus, speaking a tongue unknown to Vasil'ev, a "paraconsistent" logic). In Vasil'ev's mind, building syllogistic systems of imaginary logics is not only a strange possibility, but rather a scientifically *necessary* and exciting move in order to determine what can be changed in classical logic, without loss of "logicity": this unchangeable residual part will the so-called meta-logic. In order to seize this invariant part of logic, Vasil'ev preconises infinite variations on the accepted axioms. The result will be constituted by some positive kind of logic (a logic without negations, i.e. ruled by the "principle of the excluded second") as well as by a (small) hypothetically invariant part of Aristotelian logic (a part of the logic of the excluded third), which is unchangeable (as "Absolute Geometry" is with respect to all possible geometries)[8]. For these reasons, and especially for having connected the idea of exploring the "log-

---

one, the others being only possible in fiction"). More than this, we agree on this point ("non-normal worlds") with G. Priest ([24], p. 144), according to which Vasil'ev could be seen as a true precursor of the nowadays very popular idea that there are "impossible possible worlds" (an issue much debated between D. Lewis, who rejects such worlds, and some paraconsistent philosophers and logicians, cf. *Notre Dame Journal of Formal Logic* 38, n. 4, Special Issue on Impossible Worlds).

[7]Indifferent judgements are also said, sometimes, "contradictory" judgements.

[8]This point, what is Vasil'evian metalogic, is debated. Cf. [27].

ically unchangeable" by means of developing systematically the "logically non standard", Vasil'ev is rightly considered as one of the very first fore-runners of "deviant logics" in general, and, possibly (the question remains open), of paraconsistent and multivalued logics in particular (cf. [1], [4], [11], [25] and [26])[9].

Remark, however, that if Vasil'ev's achievement has been to reject suc-cessfully (i.e. without loss of logicity) in his imaginary logic the "principle of non-contradiction" - thus proving the logical relativity of such a "principle" - (logic can speak about *contradictory objects*) he keeps, as metalogical (i.e. absolute) the "principle of non *self*-contradiction" (logic can't formulate judgements in a *contradictory way*)[10].

All this leaves open, for the working logician, the problem of having an intuitive representation of such logical coherent but strange possibilities.

3. The "analogy of interpretation" as a further refinement of the project: the Vasil'evian experimental 3 self-interpretations of his own non-standard logic in terms of old standard logic.
As we saw, Vasil'ev achieved doing, for syllogistic, what Gauss, Lobachevski and Bolyai did for geometry. He showed that, paradoxically, some logical "fundamental" laws can be changed without loss of "logicity"[11], and that this fact opens the doors of realms quite strange. Now, going back to the geometrical source of revolutionary inspiration, because Non-Euclidean ge-ometry can be expressed, by a suited translation, inside Euclidean geometry itself, as did Beltrami (who relied Lobachevski's geometry with the "pseudo-sphere") and Poincaré (who relied Riemann's geometry with the "Euclidean sphere") (cf. [31], p. 81; [28], pp. 487-501), Vasil'ev thinks there should be as well, in the logical plan, a way to express non-Aristotelian imaginary

---

[9]In some sense Vasil'evian metalogic could be seen as a kind of precursor of the approaches like the one of J.-Y. Béziau's, "Universal Logic", although UL is something much more pluralistic, much more dynamic (cf. [3]). But the spirit, the attraction for the non-standard motivated as being an access to the more profound is very akeen to UL. Vasil'ev discusses such a point while criticizing Husserl's *Logische Untersuchungen* (cf. [31] p. 58n, [32] p. 94; p. 329 of the English translation).

[10]This can be seen as reflecting somehow the fact that in contemporary logics, and paradigmatically in the da Costa's systems C_n, $1 \leq n \leq \omega$, paraconsistency is often obtained only at the object level, whereas at the meta-level of the paraconsistent system one still uses classical non-paraconsistent logic. In [1] Arruda discusses the possible relations of Vasil'ev's thought with the different kinds of (then) existing paraconsistent systems (namely those of N.C.A. da Costa, R. Routley an R.K. Meyer, etc.)

[11]This fact was rather unknown at that time (and until recent date many logicians used to laugh ignorantly at those speaking of non standard features like paraconsistency). Only Lukasiewicz in someway had demonstrated the same thing (i.e. the fact that the principle of non contradiction is, logically speaking, an independent axiom) in his study of 1910 *O zasadzie sprzecznosci u Arystotelesa* (*On the principle of contradiction by Aristotle*).

logic inside Aristotelian logic[12]. And effectively, by (a third successful) analogy with the geometrical interpretations just mentioned of non-Euclidean
geometry in terms of Euclidean geometry, Vasil'ev gives *successfully* three
interpretations of his non-Aristotelian "non-earthly" logic in terms of normal (Aristotelian) "earthly" logic (cf. [31], p.81-88).

The main idea of this is that, while Aristotelian logic is the logic of things,
imaginary logic, in our world, is embodied in the logic of concepts, which is
different from the logic of facts (or things), as revealed by his first study on
"some" ([30]). A concrete fact spatio-temporal fact in the world can only be
true or false (third excluded); differently, a concept can only be necessary,
impossible *or contingent* (excluded fourth). Of this paradoxical embodiment
(imaginary logic plunged into the real world!) Vasil'ev proposes three forms,
all related to the conceptual abstract plan. The three "earthly" (or realist)
interpretations Vasil'ev gives of his non-earthly (or imaginary) logic pertain
to: (1) modal logic (cf. [31], pp. 81-87), (2) the logic of absolute resemblance
and difference (cf. [31], p. 87) and (3) the general logic of concepts, of
which he gives very interesting precise semantical elements (cf. [31], pp.
87-88). Of the two last logics Vasil'ev remarks that their axiomatic system
is in fact very similar to, but slightly different from, that, canonical, of the
fundamental abstract version of imaginary logic. This remark will turn out
important, as we will see. In what follows we will solely deal with the third
Vasil'evian self-interpretation of imaginary logic (as a logic of concepts).

However, before going on, one must understand that while the "earthly"
interpretations of imaginary logic will give (as we will see) the impression
that we remain inside old standard logic (just changing the object of application, now less abstract, but keeping the old logic at the metalevel), this
impression is highly misleading: axiomatically speaking imaginary logic is
and remains *highly counter-intuitive* and plainly non-standard, exactly as
much as non-Euclidean geometry remains non-standard despite Beltrami's
and Poincaré's interpretations. Vasil'ev's shocking (but coherent) axioms
and theorems can be used as such, without any "earthly" make-up. But
sofar we remained informal. What about the formal counterpart of all this?

**Contemporary formalizations of Vasil'ev's logical ideas often end
up in friendly betrayals.** Vasil'ev's own formalization of imaginary logic
deals more with words and concepts than with formulas. In a sense, he can
do this because he can evoke syllogistic quite complex conceptual logical pro-

---

[12]Needless to say, such translations are very important. They where related, as it
appeared later, to consistency-proofs and completeness considerations premonitory of
some future tricky proof strategies, such as Gödel 1931's one (on all this cf. the beautiful
study [28]).

cedures by means of technically codified names[13]. But logic has changed, we generally can't think anymore in terms of syllogistic, we simply forgot that primitive but coherent (yet restricted) language of logic. So there have been several useful attempts thereafter in formalizing his ideas[14]. But most of them (at least those done outside Russia, of course) have been done without any knowledge of Russian. And this linguistical point - usually harmless - happens, sadly enough, to be here a very critical one. Because of the rarity of the existing translations until recent date, these attempts rely most of the time on a few English speaking second-hand sources, namely[15]: (1) a paper by G. Kline on the importance of Vasil'ev for logic (1965), (2) a review by D.D. Comey of a 1962 Russian paper by V.A. Smirnov on Vasil'ev's logical ideas (cf. [27]); (3) A. Arruda's 1977 ([1]), 1979 and 1984 papers on Vasil'ev's logic, the second including a partial translation of Vasil'ev into English (through Portuguese!) by E. and I. Braga; and finally - and dramatically - (4) all those who combine such secondary or incomplete sources. For instance, in [26] (1988) L.Z. Puga and N.C.A. da Costa stress the fact that their logical study is neither historical nor exegetical, and when it comes to the point of explaining Vasil'ev's ideas they quote extensively another author's views on the subject, that is D.D. Comey's review of the 1962 paper of V.A. Smirnov on Vasil'ev! In order to confirm their "non-historical" (and non-exegetical) interpretation, in 4 clauses, of that "third-hand lecture" of Vasil'ev, they refer to the authority of Arruda's interpretation. In 1989, G. Priest and R. Routley ([25], p.29-34) try to sum up in a few pages Vasil'ev's forerunning contribution to the development of paraconsistent thought, still relying on Arruda's 1977 presentation of it. In 1996, in a beautiful study on the history of paraconsistent logics ([4]) A. Bobenrieth necessarily deals with the important question of the place of Vasil'ev's ideas in it. He, as well, refers to G.L. Kline, Comey and Arruda. In 2000, in a nice paper (cf. [24]), G. Priest tries to judge logically and philosophically imaginary logic. The attempt is very interesting, he expressly (and wisely) avoids the so lethal pitfall of formalizing Vasil'ev in a totally anachronistic way, recognizing overtly the poverty of the means available to the non-Russian speaking person he is, and decides to stay close to syllogistic notions; but he

---

[13]Syllogistic symbolism was more encoded with words - such as "conversion", "obversion", "n^{th} figure of the mth mood, "camestres", etc. - than with letters (variables, functions, arguments, relations, etc.), as become progressively necessary after G. Boole's translation of logic into algebra (1854) and the subsequent refinements of the mathematization of logics.

[14]One of the first studies on Vasil'ev, besides some early critical reviews (sometimes *very* critical) in Vasil'ev's life time, was: P.V. Kornin, "O logicheskikh vozzrenijakh N.A. Vasil'eva (Iz istorii russkoi logiki)" ("On N.A. Vasil'ev's way of seeing logic (From the story of Russian logics)"), *Trudy Tomskogo gos. universiteta*, T. 112, 1950.

[15]In [4], p. 28n, A. Bobenrieth, while justifying himself, explains this point nicely.

relies only on the English translation of [32], the third of the three papers constituting Vasil'es's main work, ignoring totally Vasil'ev's first two papers (especially the first one); which prevents him to understand correctly the whole. Now, the point is that most of the non Russian *logical* interpretations (i.e. formalizations, as in [1] and [26]) - at least those we know of - of Vasil'ev's ideas keep a very loose relation to them, and proceed in this respect in a too quick manner, thus missing the complexity and refinement of Vasil'ev's original complex and structured views. Worse than this, they rely fully and exclusively on propositional logics, which is heresy with respect to Vasil'ev's very syllogistical thought[16].

The Russian school of studies on Vasil'ev's logical thought. Its quite unknown powerful and fertile research framework. Its technical results But such a failure is not everybody's doom. It is neither the case, for instance, of Smirnov's studies of 1962 and 1989 (cf. [27])[17], nor of the ones of the Russian logicians who followed him up to now. Effectively, at the end of the nineties a group of fine Russian logicians built up a coherent (and official) research project of "Reconstruction of the logical systems of N.A. Vasil'ev"[18]. Inside this remarkable framework, very productive and well-structured, at least four directions are systematically being explored: (1) Vasil'ev's assertoric syllogistic [19], (2) Imaginary logic under its fundamental version[20], (3) imaginary logic under the three "self-interpretations" provided by Vasil'ev himself[21] and (4) n-dimensional logics[22]. These Rus-

---

[16]Arruda planned to develop, after her propositional $V1$, $V2$ and $V3$ "Vasil'evian" systems, new systems of predicate calculus. But her premature death put an end to this promising project.

[17]For an overview on Smirnov's fundamental and impulse-giving contribution to the Vasil'evian studies, cf. [15].

[18]The official members of this research project are V.I. Markin, D.V. Zaitzev, O.J. Karpinskaja and T.P. Kostjuk. Their papers as well as their editorial line are detailed on their web page, http://www.logic.ru/Russian/RusLogic/Vasiliev/index.html.

[19]This issue deals with Vasil'ev's thought of [30], before the invention/discovery of imaginary logics. The main papers here are [9], [6], [7], and [5].

[20]This issue deals with the fundamental (or canonical) version of imaginary logic. There also have been works thematically investigating the relations of it to 3-valued logic. The main papers here are [10], [14] and [11].

[21]Up to now this issue deals with the modal and the conceptualist paradoxical self-interpretations of imaginary logic suggested by Vasil'ev himself. A third self-interpretation (the first in Vasil'ev's list) remains unexplored. The relevant papers here are [13], [36], [38] and [37]. Our present paper belongs to this category.

[22]This issue, perhaps one of the most interesting, seems to be, until now, the less explored one. Precious elements are in Smirnov's paper of 1989 (cf. [27]). Another study is [8]. Unless we are much mistaken the future of this question will necessarily be related to the nowadays emerging and here already mentioned n-opposition theory (cf. [19], [20], [22] and [23]), a branch of contemporary abstract modal logic, of which the Vasil'evian n-dimensional logics seem, *a posteriori*, to be a particular (and simple, syllogistic) but

sian authors (with others outside the four mentioned in the previous foot-note, such as V.V. Anosov, V.A. Bazhanov and V.L. Vasjukov, to quote some) build a logical, careful and exhaustive reconstruction of Vasil'ev's ideas on the "logically possible" that respects and understands Vasil'ev's written doctrine (dealing, for instance, with the syllogistical peculiarities of it - thus following a methodology pioneered by Lukasiewicz's work on Aris-totle's syllogistic). The result is the consensual elaboration of a standard of formal "Vasil'evian systems". According to this reasonable standard, until now Vasil'ev's assertoric syllogistic is expressed at its best by system **C4B** (Kostjuk and Markin, this system is sometimes written **C4V**). Imaginary logic under its fundamental form is expressed by system **IL** (Kostjuk and Markin), while the conceptualist version of it is expressed by the system **IL2** (Markin and Zaitzev). Other systems, for the other parts of the Vasil'evian logical space, are being elaborated.

In this paper we will deal with the third of the four research issues we mentioned. So we turn now to one of the contributions relative to this topic, the most recent and the most complete one.

2   Markin and Zaitzev's recent faithful formalization (1999) of Vasil'ev's (conceptualist) third self-interpretation (1912): the logic IL2

One of the self-interpretations Vasil'ev gave of his own imaginary logic, the one in terms of an intensional logic of concepts (inspired by Leibniz)[23], has been formalized very nicely (very faithfully) as system "**IL2**" in recent years by Markin and Zaitzev (cf. [37]). **IL2** is a monadic fragment of first-order predicate calculus. This system, which has the merit of allowing the non trivial (paraconsistent) expression of contradictory judgements of the form "some A (at the same time) is and is not B", or "every A (at the same time) is and is not B" (the "indifferent judgements", particular and universal), is intensional, i.e. conceptualist, in a sense - partly explicitly sketched by Vasil'ev - which we are now going to explain.

To construct **IL2**, following precise informal instructions of Vasil'ev, Markin and Zaitzev do as follows. We shall consider a non empty set **L** of *letters* (also said *literals*, or *atoms*) - some positive and negative signs - $\{p_1, \neg p_1, p_2, \neg p_2, \ldots\}$ (couples of 2 by 2 mutually contradictory atoms). We will call "concept" every non empty and consistent subset of **L**, in other words a set $\alpha \subseteq \mathbf{L}$, satisfying the following conditions:

   (i) $\alpha \neq \emptyset$;
   (ii) there exists no p_i such that: $p_i \in \alpha$ and $\neg p_i \in \alpha$.

---

interesting case avant la lettre (cf. [21]).

[23] On the Leibnizian inspiration for this kind of conceptualism cf. [12].

Let **M** be then the set of all concepts. We will fix on **M** the operation '
* ', assigning to every concept α the concept α* which is its inverse:
$p_i \in \alpha$* $\Leftrightarrow \neg p_i \in \alpha$ and
$\neg p_i \in \alpha$* $\Leftrightarrow p_i \in \alpha$.
The operation * has the three following properties (remark the third one):
(a) $\alpha \cap \alpha$* $= \emptyset$, (b) α** $= \alpha$, (c) $\alpha \subseteq \beta \Rightarrow \alpha$* $\subseteq \beta$*.
Finally, an *interpreting function* " d " is the function attributing some
linguistically intuitive meanings to the syllogistic terms, attaching to every
term certain concepts : $d(P) \in$ **M** (with, for instance, $d(H)=$ "human").
We find by Vasil'ev six atomic "judgements" (by Aristotle there are four).
A judgement can, accordingly, be:
   1) affirmative ($X_1 SP$), negative ($X_2 SP$) or indifferent ($X_3 SP$), (with X
$= A$ or I); and
   2) universal ($A_n SP$) or particular ($I_n SP$) (with $n = 1$, 2 or 3). expressed
with words, the six possible judgements are:

$A_1 SP =$ "every S is P",
$A_2 SP =$ "no S are P",
$A_3 SP =$ "every S is and is not P",
$I_1 SP =$ "some S are P",
$I_2 SP =$ "some S are not P",
$I_3 SP =$ "some S are and are not P".

In conformity with Vasil'ev's intuition (inspired by Leibniz), such a logic
is *conceptualist*: a concept, such as, for instance, "gold", is not here defined extensionally by the set of all the individuals which are instances of
it (i.e. the golden objects), but rather by the set of the sub-properties
characterizing (and forming) the *concept* "gold" (e.g.: "metal", "yellow",
"heavy", "precious", etc.). As for semantic definitions, Vasil'ev only gives
non-formalized ones for universal judgements (no direct hint is given for
particular judgements). Which gives, once formalized by Markin and Zaitzev in terms of sets of letters:

$|A_1 SP|^d = 1$ iff $d(S) \supseteq d(P)$,
$|A_2 SP|^d = 1$ iff $d(S) \supseteq d(P)$*,
$|A_3 SP|^d = 1$ iff $d(S) \cap d(P) \neq \emptyset$ and $d(S) \cap d(P)$* $\neq \emptyset$.

The two Russian logicians show the equivalence of these definitions of
the universal judgements with the following, more easily workable quantified ones:

$|A_1SP|^d = 1$ iff $\forall \alpha \in \mathbf{M}$ $[d(S) \subseteq \alpha \Rightarrow d(P) \subseteq \alpha]$,
$|A_2SP|^d = 1$ iff $\forall \alpha \in \mathbf{M}$ $[d(S) \subseteq \alpha \Rightarrow d(P)^* \subseteq \alpha]$,
$|A_3SP|^d = 1$ iff $\forall \alpha \in \mathbf{M}$ $[d(S) \subseteq \alpha \Rightarrow d(P) \cap \alpha \neq \emptyset$ and $d(P)^* \cap \alpha \neq \emptyset]$.

(universality is related to universal quantification over the set $\mathbf{M}$ of concepts)

Hence they can easily construct the following analogous definitions for particular judgements (always in terms of quantification over the set $\mathbf{M}$):

$|I_1SP|^d = 1$ iff $\exists \alpha \in \mathbf{M}$ $[d(S) \subseteq \alpha$ and $d(P) \subseteq \alpha]$,
$|I_2SP|^d = 1$ iff $\exists \alpha \in \mathbf{M}$ $[d(S) \subseteq \alpha$ and $d(P)^* \subseteq \alpha]$,
$|I_3SP|^d = 1$ iff $\exists \alpha \in \mathbf{M}$ $[d(S) \subseteq \alpha$ and $d(P) \cap \alpha \neq \emptyset$ and $d(P)^* \cap \alpha \neq \emptyset]$.

(particularity is related to existential quantification over the set $\mathbf{M}$ of concepts)

From there they then prove the following useful definitions for particular judgements (equivalent to the ones just seen), which complete Vasil'ev's semantic definitions as reconstructed by Markin and Zaitzev:

$|I_1SP|^d = 1$ iff $d(P)^* \cap d(S) = \emptyset$,
$|I_2SP|^d = 1$ iff $d(P) \cap d(S) = \emptyset$,
$|I_3SP|^d = 1$ iff $d(P) \backslash d(S) \neq \emptyset$ and $d(P)^* \backslash d(S) \neq \emptyset$.

By hypothesis the six "atomic judgements" can be combined in the usual way by means of the propositional unary and binary connectives ($\neg$, $\vee$, $\wedge$, $\rightarrow$, $\leftrightarrow$). *Validity* is defined as the fact, for a given formula, of taking the value " 1 " under any possible interpretation d. The class of such valid formulae can then be axiomatized as " system **IL2** ", whose axioms are all the classic tautologies of standard propositional logic plus the 17 following axioms given by Markin and Zaitsev:

A1. $(A_1MP \wedge A_1SM) \supset A_1SP$ (A1 to A8: imaginary syllogisms for concepts),
A2. $(A_1MP \wedge A_2SM) \supset A_2SP$,
A3. $(A_2MP \wedge A_1SM) \supset A_2SP$,
A4. $(A_2MP \wedge A_2SM) \supset A_1SP$,
A5. $(A_1MP \wedge I_1SM) \supset I_1SP$,
A6. $(A_1MP \wedge I_2SM) \supset I_2SP$,
A7. $(A_2MP \wedge I_1SM) \supset I_2SP$,
A8. $(A_2MP \wedge I_2SM) \supset I_1SP$,
A9. A_1SS,
A10. $\neg(A_1SP \wedge I_2SP)$ (A10 and A11: contrariety),

A11. $\neg(A_2 SP \land I_1 SP)$,

A12. $I_1 SP \supset I_1 PS$ (A12 and A13: conversion of the particulars),

A13. $I_2 SP \supset I_2 PS$,

A14. $A_1 SP \supset I_1 SP$ (A14 and A15: subalternation),

A15. $A_2 SP \supset I_2 SP$,

A16. $A_3 SP \equiv \neg I_1 SP \land \neg I_2 SP$ (A16 and A17: indifferent judgements),

A17. $I_3 SP \equiv \neg A_1 SP \land \neg A_2 SP$.

From these axioms they deduce 26 theorems (T1-T26)

T1. $\neg I_2 SS$,

T2. $\neg(A_2 SP \land A_1 SP)$ (T2: incompatibility of the contraries)

T3. $\neg(A_1 MP \land A_1 MS \land \neg I_1 SP)$,

T4. $\neg(A_2 MP \land A_2 MS \land \neg I_1 SP)$,

T5. $\neg(A_2 MP \land A_1 MS \land \neg I_2 SP)$,

T6. $\neg(A_1 MP \land A_2 MS \land \neg I_2 SP)$,

T7. $\neg(A_1 PM \land A_2 SM \land I_1 SP)$,

T8. $\neg(A_2 PM \land A_1 SM \land I_1 SP)$,

T9. $\neg(A_1 PM \land A_1 SM \land I_2 SP)$,

T10. $\neg(A_2 PM \land A_2 SM \land I_2 SP)$,

T11. $\neg(\neg I_1 QR \land A_1 PQ \land A_1 SR \land I_1 SP)$,

T12. $\neg(\neg I_1 QR \land A_1 PR \land A_1 SQ \land I_1 SP)$,

T13. $\neg(\neg I_1 QR \land A_2 PQ \land A_2 SR \land I_1 SP)$,

T14. $\neg(\neg I_1 QR \land A_2 PR \land A_1 SQ \land I_1 SP)$,

T15. $\neg(\neg I_2 QR \land A_2 PQ \land A_1 SR \land I_1 SP)$,

T16. $\neg(\neg I_2 QR \land A_2 PR \land A_1 SQ \land I_1 SP)$,

T17. $\neg(\neg I_2 QR \land A_1 PQ \land A_2 SR \land I_1 SP)$,

T18. $\neg(\neg I_2 QR \land A_1 PR \land A_2 SQ \land I_1 SP)$,

T19. $\neg(\neg I_2 QR \land A_1 PQ \land A_1 SR \land I_2 SP)$,

T20. $\neg(\neg I_2 QR \land A_1 PR \land A_1 SQ \land I_2 SP)$,

T21. $\neg(\neg I_2 QR \land A_2 PQ \land A_2 SR \land I_2 SP)$,

T22. $\neg(\neg I_2 QR \land A_2 PR \land A_2 SQ \land I_2 SP)$,

T23. $\neg(\neg I_1 QR \land A_1 PQ \land A_2 SR \land I_2 SP)$,

T24. $\neg(\neg I_1 QR \land A_1 PR \land A_2 SQ \land I_2 SP)$,

T25. $\neg(\neg I_1 QR \land A_2 PQ \land A_1 SR \land I_2 SP)$,

T26. $\neg(\neg I_1 QR \land A_2 PR \land A_1 SQ \land I_2 SP)$,

and, through them, two meta-theorems: (1) a proof of the *soundness* and (2) a proof of the *completeness* of the axiomatic system with respect to the given semantics. The only deduction rule of **IL2** is the modus ponens (i.e.: "if A and $A \supset B$, then B").

Markin and Zaitzev's formalization of Vasil'ev's ideas is a very precious one. As said, it respects the fundamental philosophical intuitions of the Russian logician whereas most of the rival formalization, despite their intrinsic logical interest, do not respect *at all* the original Vasil'evian style of thinking. Nevertheless there seem to be two problems with it: (1) it can be shown that the standard graphical decision procedures do not work with **IL2**; (2) moreover, when left as an axiomatic system, **IL2** remains very poor when we try to speak about paraconsistent judgements, whereas it is not when it has to speak about normal judgements, and this is unfair with respect to Vasil'ev's revolutionary ideas. In what follows we will handle both problems. And here begins our personal contribution.

3   First problem: what kind of graphic decision procedure (GDP) for the intensional logic IL2?

The normal (Aristotelian) syllogistic admits a two-step graphic decision procedure (Euler's and Venn's diagrams, and the like): (1) one falsifies a pseudo-valid formula by the construction of a graphic counter-example; (2) one demonstrates a formula's validity by showing that it is impossible to construct any graphic counter-model of it. In this view *predicates* are generally represented by circles (sets). The *negation* of a set (i.e. of a predicate's extension) is its complement, the other side of its frontier (the exterior of the circle). *Conjunction* (logical product) is expressed as graphical intersection, while *disjunction* (logical sum) is expressed as graphical superposition. *Implication* is expressed as graphical inclusion(P implies S iff P is graphically included into S). *Absurdity* (to which proofs quite often reduce) is a set's non connectivity with respect to its frontier, i.e. the fact that an element lays at both sides of a given closed frontier (cf. Figure 1).

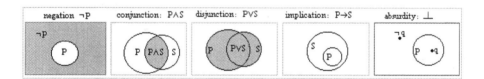

Figure 1. The standard GDP: expressing the propositional connectives

But the case of **IL2** seems to be quite different. It happens that, being *intensional* (because conceptualist, not extensional, think of the concept of gold), the standard graphic representations do fail with it, despite the fact that axiomatically speaking **IL2** works perfectly. There are at least two big problems: one for representing paraconsistency (i.e. A_3 and I_3

judgements) and, more fundamentally, one for representing negation. This can be shown through **IL2**'s axioms and theorems: four axioms (the A2, A4, A6, A8 above) and 18 theorems (the T4, T6, T9, T10, T13-T26 above, as proven deductively by Markin and Zaitzev) are graphically inexpressible (absurd) in the sense that their standard graphic representation does not confirm their status of axioms or of theorems inside **IL2**. Let us see, for instance, the case of two axioms of **IL2**, the second of which is graphically very problematic (cf. figure 2).

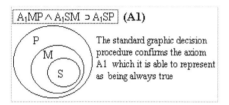 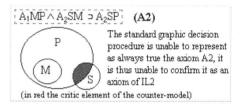

Figure 2. Two axioms of **IL2** analyzed according to the standard GDP: graphically speaking one works, not the other (so the GDP fails for **IL2**!)

(transitivity for inclusion proves very simply A1, but nothing supports graphically the fact that A2 is an axiom: on the contrary, it is possible to draw a graphical counter-example to it)

As already said, the same thing can in fact be shown for many theorems of **IL2** (18 over the 26 known), as for instance T10 (cf. Figure 3, where we show that the standard GDP proves T7 but not T10).

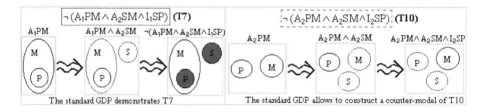

Figure 3. Theorems T7 and T10 of **IL2** according to the standard GDP: graphically speaking one works, but not the other. So the standard GDP fails for **IL2**!

(by the waving arrow we mean the passing from one step to another, inside the graphic construction procedure) Must one give up the idea of a graphical understanding of Vasil'ev's conceptualist logic? No.

4   Our solution of the first problem: IL2-related non-standard GDP ("negation as symmetry")

So we propose a suited non standard graphic decision procedure (GDP), working for **IL2** at least in the sense that it allows to demonstrate graphically *all* the axioms and theorems of **IL2**. The problem of finding the *exact* GDP of **IL2** (i.e. with metatheoretic proofs of adequacy and completeness) is not as simple as it could seem and would deserve a further study, but our present attempt offers a first step in this direction (again, we will first concentrate in finding a GDP which produces at least *all* the conclusions of **IL2**).

To keep prudent, we will consider this non standard GDP as relative to a yet undetermined logical system **IL2X**, the question remaining open as to know if simply **IL2X = IL2**. But **IL2X**, as similar as possible to **IL2**, will anyway consist of at least all the axioms and rules of **IL2** (rooted in **IL2**'s semantics).

The **general rule** (still very classic) of our GDP will be the following: (1) in order to *demonstrate* the validity of a proposition, one has to test its negation, drawing one by one the atomic judgements of the negation of the given proposition; if it is impossible to draw in this way a counter-example, the formula is valid, otherwise it is not. (2) In order to *falsify* a proposition one has to succeed in constructing a model of its negation (the interpretation of the propositional binary connectives is intuitively rather easy), otherwise the proposition is, once again, valid (falsifying is usually quicker than proving). (3) To *draw* a formula is to draw the elements that constitute the translated negation-free expression of the formula. In other terms, to draw a formula's graph one needs, before, to eliminate the negation signs from it, using for this purpose the classical propositional laws (De Morgan and the like) and the mutual definitions of the six atomic judgements. Effectively, as we will see, each one is defined as the negation of the disjunction of two others, which thing allows us to do such a complete elimination of the negation signs (each negated atomic judgement $\neg X$ will be expressed as the disjunction $Y \vee W$ of the two terms Y and W whose conjunction of the respective negations $\neg Y \wedge \neg W$ expresses the starting atomic judgement X). The six atomic judgements (among which two express affirmations, two express negations and two express contradictions), as we will see, are all graphically expressible by our GDP.

We shall see now what is new in our method.

Graphical definitions of conceptualist "negation" and "absurdity". An atom (or a set of atoms) and its *negation* are mutually symmetric with respect to a vertical axis σ_X (*symmetry axis* or *mirror*). We will draw such an axis with a *vertical* scattered line, an horizontal fixed segment (*opposition bar*) being (eventually) reserved to the expression of the link existing between any element and its negation (i.e. its symmetric image with respect to σ). There is an exponentially growing number of such symmetry mirrors σ_X in the logical space of **IL2X**, one σ_X for each concept X, i.e. one for each finite subset of the set of the atoms - it is easy to prove that the relevant formula here is in fact

$\text{Card}(\Sigma) = 2^{(Card(L)/2)-1}$, with $\Sigma =$ the set of all distinguishable mirrors σ.

So we can define, for any concept P, an axis σ_P dividing P from P^*. Now, once drawn one such symmetry axis (say: σ_P) it is possible to draw all remaining concepts (say: Q, R, S, ...) with respect to that axis (the only constraints: no concept can be empty, no concept can contain both an atom and its negation - which is its symmetric image, at the other side of the mirror). In other words: for demonstration's sake one axis (any) will always suffice. For reasons to appear later, we will call "σ_Q-well-behaved" all graphical expressions of a concept P such that the concept's image does not cross the given σ_Q (by definition every concept X is σ_X-well-behaved). We will call "σ_Q-transversal" a concept's P image such that it crosses the given axis σ_Q (cf. Figure 4).

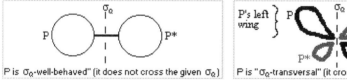

P is σ_Q-well-behaved" (it does not cross the given σ_Q)

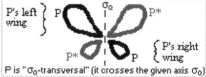

P is "σ_Q-transversal" (it crosses the given axis σ_Q)

Figure 4.   Representing a concept P: σ-well-behavedness and σ-transversality

It is easy to see that a concept containing (i.e. surrounding) a σ_P-transversal concept is itself σ_P-transversal, whereas a concept containing a σ_P-well-behaved concept need not be itself σ_P-well-behaved (it generally will be σ_P-transversal). Because σ_P-well-behavedness of a concept S with respect to another concept P's symmetry axis is an exceptional case (it is rare that two concepts, especially if complex - i.e. constituted of many atoms -, share the same symmetry axis) it will be important graphically speaking to draw the concepts so that they always have two "wings", a left one and a right one (except, of course, when they are allowed to be σ-well-behaved by construction).

So, according to our non standard GDP, given two unrelated concepts (say P and R) there are two possibilities in order to represent them in terms of the symmetry axis of one of them: (1) they can happen to share the same axis ($\sigma_P = \sigma_R$); or (2) they do not (in which case, once fixed σ_P, R is σ_P-transversal and *vice versa* if the same 2 concepts are represented with respect to the axis σ_R). In Figure 5 we show the two ways in which the representation of a concept R can be added to that of a concept P (according to the fact that the starting concept P is depicted with respect to its own symmetry axis σ_P - left side of the Figure - or with respect to any other one σ - right side of the Figure).

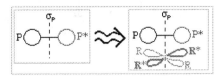 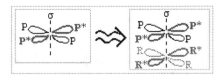

Figure 5. How to add the representation of a concept R to that of a concept P

As by Markin and Zaitzev (1999), the atoms' (or letters') negations will be expressed by the usual negation sign ¬, while a set's (i.e. a concept's) negation will be marked by a star *. *Absurdity* (to which proofs revert) will be defined as the presence of an atom (or set of atoms) and its negation at the same side of the considered symmetry axis, or (equivalently) as the presence of a same element (in its entirety) at both sides of some symmetry axis, or (equivalently) as a non-empty intersection of an element (a literal, an atom) with the axis σ. The graphical definitions of negation (i.e. negation as symmetry of the contradictories) and of absurdity (i.e. absurdity as symmetry disruption) are given in Figure 6.

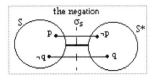 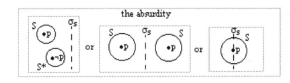

Figure 6. Our non-standard graphical definitions of 'negation' and 'absurdity'

This strange treatment of negation (by graphical symmetry) is due to property (c) of the operation * (cf. *supra*). Remark that in our graphics we will simplify the notation, writing "*P*" instead of "d(*P*)", "*P**" instead of "d(*P*)*", etc.; further, the use of colors will be pedagogical (sorry for color-blind people or black and white printers or Elsevier publishers …), but without any once-for-all fixed rule : at every step, in general, colors will underline what is going on at that moment with respect to the previous step (more generally colors will help to distinguish the different concepts).

Graphical definitions of "universal" affirmative and negative judgements. According to the previous semantic definitions, universality is bound to some form of inclusion (cf. Figure 7).

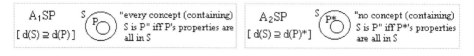

Figure 7. Our non-standard graphical definitions of affirmative and negative universal judgements (A_1SP and A_2SP)

One will however remark the paradoxical reversal: in order to make "Every S is P" true, P must be included in S (and not the other way round). Intuitively this means that every concept containing S will contain P, "S is P" meaning here "The concept S admits P as one of its sub-concepts" (or properties).

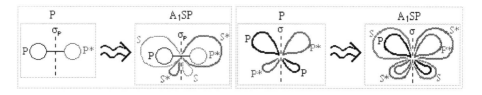

Figure 8. How to build the representations of universal affirmative judgements

In Figure 8 we show how to construct graphically the expressions of A_1 judgements (assuming P σ-well behaved or σ-transversal with respect to the starting symmetry axis). Analogously, in Figure 9 we show how to construct graphically the expressions of A_2 judgements (with P σ-well behaved or σ-transversal).

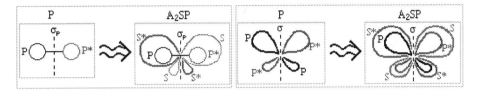

Figure 9. How to build the representations of universal negative judgements

Graphical definitions of "particular" affirmative and negative judgements. "Some S are P" means, from a conceptualist point of view, that for some concept α containing S (as a sub-concept) α also contains P (i.e. is "protected" against containing P^*). This means, from the graphical point of view, that there is some concept α (we represent it with a scattered line) such that α contains S and P, which is the same as S and P^* being unable (because of symmetry constraints) to fit together into this concept α. Reciprocally, "Some S is not P" means that some concept α containing S (as a sub-concept) also contains P's negation (which is P^*). This means that this concept α is such that it contains S and P^* (which is tantamount as saying that S and P are unable to fit together into this concept α) (cf. Figure 10).

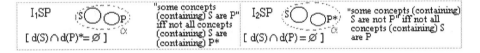

Figure 10. Our non-standard graphical definitions of affirmative and negative particular judgements (I_1SP and I_2SP)

It is very important to remark that one can at any moment draw the negation of any given concept P (by the mirror, drawing such negation as the symmetric image of the given concept), without meaning by that that the concept is negated overall: the concept's complement (negation) is here graphically most of the time implicit (one has to draw the mirror and use it), whereas in the standard GDP it is immediate (for every circle it is its outside). Such an act of drawing ("build the symmetric figure") may suggest some *implicit* properties (useful, maybe necessary ones) to the logician working with **IL2X** and can therefore help seriously on some reasoning during a proof.

In Figures 11 and 12 we show how to construct graphically the expres-

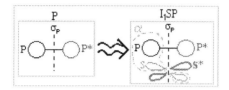 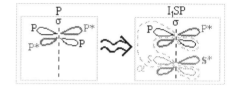

Figure 11. How to build the representations of particular affirmative judgements

sions of I_1 and I_2 judgements (assuming P σ-well behaved or σ-transversal with respect to the starting symmetry axis)

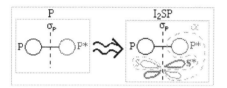 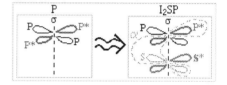

Figure 12. How to build the representations of particular negative judgements

Graphical definition of the "paraconsistent" atomic judgements (indifferent judgements). The non-standard graphical representation of indifference judgements ($A_3 SP$ and $I_3 SP$) differs from the previous ones in so much it is now necessary to represent at least one concept (say S) out of three (say P, P^* and S) as being not connected (i.e. σ_P-transversal, having parts laying at both sides of the axis σ_P). Such S nevertheless still respects, as any concept does, the three previously stated conditions of non-absurdity, which are: (1) the fact that S and S^* do not both lay *entirely* at the same side of σ_P (S^* is itself disconnected, symmetrically with respect to S), (2) the emptiness of the intersection of S and S^* (no element of S can lay simultaneously at both sides of the axis σ_P) and (3) the empty intersection with the axis σ_P (the left and right parts of S, i.e. its 2 asymmetric "wings", are disconnected).

Thus, a *universal* indifferent judgement ($A_3 SP$) affirms that *every* concept (α containing) S contains, with respect to a given concept P, some of its characteristics (left wing of α at the top left side of σ_P) as well as the negation of some others (right wing of α at the bottom right side of σ_P) (cf. Figure 13).

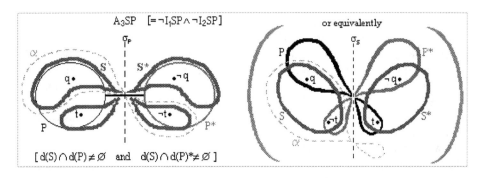

Figure 13. Our non-standard graphical definitions of "indifferent" (i.e. para-consistent) universal judgements (A_3SP)

A *particular* indifferent judgement (I_3SP) affirms that *some* (but not necessarily all) concepts (α containing) S contain, with respect to a given concept P, some of its characteristics (left wing of α) as well as the negation of some others (right wing of α) (cf. Figure 14).

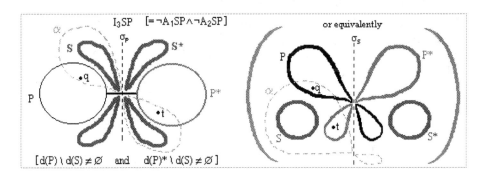

Figure 14. Our non-standard graphical definitions of "indifferent" (i.e. para-consistent) particular judgements (I_3SP)

So, "Every S is and is not P" (A_3SP) means that every concept containing S (as one of its own sub-concepts) must, at the same time, contain and not contain a given concept P (partly be P and partly be not P). S is P because it contains some P (it contains some of P's properties), it is not P because it contains some $P*$ (it contains some of the negations of P's properties). Remark that each A_3SP judgement presupposes (implies) that S as well as P contain each at least two atoms (this remark will turn out important later).

Similarly, the particular indifferent judgement (I_3SP) affirms that some concepts (α containing) S can, at the same time, contain and not contain a given concept P. This means that it is impossible to put them entirely at the same side (of the axis σ_P) than P. Remark that each I_3SP judgement presupposes that P contains at least two atoms, but makes no presupposition on S, which could contain only one atom (this remark will be very important later).

From the point of view of the conceptualist interpretation, one can therefore remark that this (seemingly feeble) "paraconsistency" à la Vasil'ev, even if it allows some kind of non-trivial contradictions (as well as some kind of included third - or excluded fourth - for an indifferent judgement is as much a "neither... nor..." than a "this... and not this...")[24], is constructed on top of a ban of the contradiction at a meta-level (which is the role here of property (a) of the operation *, as well as the role of condition (ii) of the concept formation rules of the Vasil'evan semantics, cf. *supra*). But does this non-standard GDP really work? Yes.

5   All of IL2's known axioms and theorems are regained by our non standard GDP for "IL2X"

All the known axioms and theorems of IL2 can be demonstrated again graphically by this method, which was not possible using the standard graphic method. For instance, we shall first demonstrate graphically an axiom which was unproblematic: A1 (cf. Figure 15).

Figure 15. Graphical justification of **IL2**'s axiom A1

(A_1MP's graphical construction followed by that of A_1SM necessarily gives - by transitivity of the inclusion relation - that of A_1SP, which gives the axiom A1). In the same way we can also demonstrate graphically the

---

[24]In this sense, being at the same time (feebly) paraconsistent (refusing noncontradiction) and paracomplete (i.e. refusing the principle of the excluded third), this logic seems to be (feebly) "non-alethic".

axiom A2 which, on the contrary, was intractable with the standard GDP (cf. Figure 16).

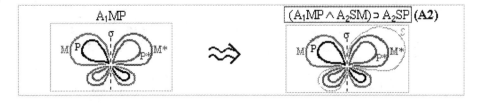

Figure 16. Graphical justification of **IL2**'s axiom A2

(A_1MP's construction followed by that of A_2SM necessarily gives - by transitivity of the inclusion relation - that of A_2SP, establishing the evidence of the axiom A2 as such; this result is conform to the actual axiomatization of **IL2**).

We will see now some examples of theorems of **IL2**. First we will demonstrate graphically the unproblematic T7. Three possible graphical constructions of an hypothetical negation of T7 must be checked (three possible choices of two starting conjuncts over three)[25]. We begin by examining the first over the three hypothetical possible constructions of a negation of T7 (cf. Figure 17).

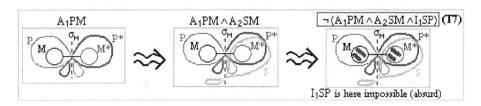

Figure 17. Graphical proof of **IL2**'s theorem T7, first possible path

(We have that the simultaneous construction of A_1PM, A_2SM and I_1SP reveals itself impossible, because the hypothetical α containing S and P should then be contradictory, given that S contains some P^*). Analogously with the second hypothetical possible construction of a negation of T7 (cf.

---

[25] As shown in this (T7) and in the following (T1O) example, in cases like these ones (the negative axioms A10-A11 and the negative theorems T2-T26) there can be several constructive ways, all to be explored (one has to check all the different ways of building a complex conjunction) unless some commutativity property is proven (and here we do not - but it should be easy).

Figure 18).

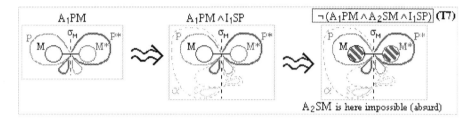

Figure 18. Graphical proof of **IL2**'s theorem T7, second possible path

(We see that the simultaneous construction of A_1PM, I_1SP and A_2SM reveals itself impossible, because S should then include M^*, which is impossible given the existence of an α containing S and P^*: α should then be contradictory, containing M and M^*. Which negates the second possible negation of the theorem). Analogously with the third and last hypothetical possible construction of a negation of T7 (cf. Figure 19).

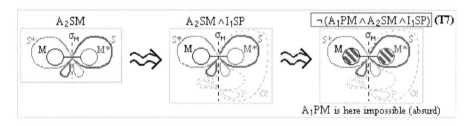

Figure 19. Graphical proof of **IL2**'s theorem T7, third possible path

(As one sees, the simultaneous construction of A_2SM, I_1SP and A_1PM reveals itself impossible, because P should then include M^*, and therefore P^* should include M, which is impossible given the existence of an α containing S and P^*: α should then be contradictory, containing M and M^*. Which negates the third and last possible hypothetical negation of the theorem T7, thus proven).

More interestingly, we will now demonstrate T10, which was graphically problematic with the standard GDP. We begin by examining the first over the three hypothetical possible constructions of a negation of T10 (cf. Figure 20)

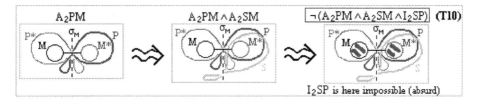

Figure 20. Graphical proof of **IL2**'s theorem T10, first possible path

(The simultaneous construction of A_2PM, A_2SM and I_2SP reveals itself impossible, because the hypothetical α containing S and P^* should then be contradictory, given that S contains some P. Which negates the first possible negation of the theorem) Analogously with the second hypothetical possible construction of a negation of T10 (cf. Figure 21).

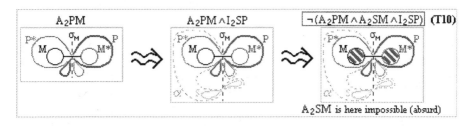

Figure 21. Graphical proof of **IL2**'s theorem T10, second possible path

(The simultaneous construction of A_2PM, I_2SP and A_2SM reveals itself impossible, because S should then include M^*, which is impossible given the existence of an α containing S and P^*: α should then be contradictory, containing M and M^*. Which negates the second possible negation of the theorem). Analogously with the third and last hypothetical possible construction of a negation of T10 (cf. Figure 22).

(The simultaneous construction of A_2SM, I_2SP and A_2PM reveals itself impossible, because P should then include M^*, and therefore P^* should include M, which is impossible given the existence of an α containing S and P^*: α should then be contradictory, containing M and M^*. Which negates the third and last possible hypothetical negation of the theorem T10, thus proved).

Thus, waiting for a logical demonstration of its soundness and completeness (metatheorems which we will not afford here), our non standard GDP

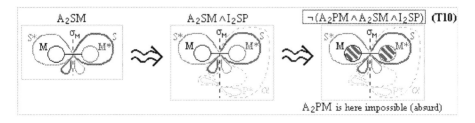

Figure 22. Graphical proof of **IL2**'s theorem T10, third possible path

for **IL2X** seems at least to be intuitively effective (for **IL2**): one can at least check Markin and Zaitzev's axioms and theorems of **IL2** one by one, all are graphically provable by now[26]. The first of the two problems of **IL2** is thus solved.

6 Second problem: which "paraconsistent" theorems of IL2? Well, "IL2X" shows some nice ones ...

The problem. It is worth noting that, while "paraconsistent" features are, from the point of view of contemporary logical research, one of the most exciting parts of the real Vasil'evian systems (i.e. of the systems he had in mind)[27], Markin and Zaitsev do not give any "paraconsistent" theorem of **IL2**: axioms A16 and A17, the only ones which deal with paraconsistency à la Vasil'ev, play no effective role in their study other than supporting the demonstration of the two metatheorems, the two authors seemingly exhibit just the theorems they will need in order to build the proofs of semantical consistency and completeness of **IL2**'s axiomatisation[28]. We would like to "breathe" some more paraconsistency[29].

---

[26] Of course, privately we did it (and this could provide funny pen-and-pencil "homework" for diligent sceptical readers), but exhibiting here all the graphical proofs would be too long.

[27] A famous joke, seemingly attributed to Dana Scott, says that "paraconsistency is to logic what pornography is to love", by which one can presumably understand: something at the same time scandalous and exciting.

[28] Only exception: they *mention*, with a proof just sketched, the paraconsistent universal conversion $A_3 SP \supset A_3 PS$ and, without any proof, the non validity in **IL2** of the other paraconsistent conversion, $I_3 SP \supset I_3 PS$.

[29] Vasil'ev himself gives two imaginary syllogisms (i.e. paraconsistent theorems), "Mindalin" and "Kindirinp" (renewing the syllogistic terminology), in [31] and [32]. And in [33], p. 158, he relates that in one of his unpublished typescripts, one of a lenght of 8 pages, he has already demonstrated rigorously and successfully (but with effort) some such unusual theorems laying at the basis of imaginary logic in order to see if this "logic of contradiction" is viable, (i.e. non contradictory in itself). Unfortunately this typescript has never been found until now. Another reference made by Vasil'ev to unpublished

An obstacle to go further in this nearing investigation is that while we have some hints as to what imaginary logic could be (some theorems - syllogisms - explicitly given by Vasil'ev) the conceptualist, self-interpreted version of imaginary-logic (sketched in [31]) is said by him to bear some differences, relative to the possible theorems, with respect to the canonical version of it (cf. [31], p.88)[30]. The difficulty while looking for a faithful formalization is that (1) we have a model to follow (imaginary logic), but also (2) a warning that this model could be misleading sometimes (there is no strict identity between the two logics, nevertheless very close). The best solution seems thus to be to follow the semantics laid by Vasil'ev at the basis of his intuition (it is by this semantics that he can spot out the diverging conceptualist syllogism).

How can we go further? Up to this point the non standard GDP (for **IL2X**) had just allowed us to represent the *already known* results about **IL2** (and, in particular, a graphical representation of the conceptualist "paraconsistency" à la Vasil'ev, cf. Figures 13-14). But, more than this, this graphic method allows us now to demonstrate *new* theorems: paraconsistent ones.

Two questions arise here. Firstly, are they new theorems of **IL2** (i.e. reachable within a simple use of the axioms and rules given by Markin and Zaitzev) or theorems (and/or axioms) of a new Vasil'evian logical system, similar to but stronger than **IL2**? They seem to belong all to **IL2**. The second question is: are all such theorems really paraconsistent? That is: do they commit to contradiction? Do they escape triviality? It seems they are, and it seems they do.

We propose 8 theorems, among which 2 syllogistical theorems (conversion and subalternation) and 6 syllogisms (only 2 of them were known until now, as long as we know), namely[31]:

T. 27.* $A_3SP \supset A_3PS$ (indifferent universal conversion),

T. 28.* $A_3SP \supset I_3SP$ (indifferent subalternation),

T. 29.* $A_1SP \wedge A_3PR \supset A_3RS$ (T29*-T34*: paraconsistent syllogisms),

T. 30.* $A_2SP \wedge A_3PR \supset A_3RS$.

T. 31.* $I_1SP \wedge A_3PR \supset I_3SR$.

---

results (theorems) of his own - which, he says, "will be published soon" (in fact they never were) - on non trivial imaginary syllogistic is in [32], p.110. So the point is that probably Vasil'ev knew more imaginary theorems than we know by his writings. Are we to rediscover them?

[30]Vasil'ev gives, for instance, the example of a syllogism valid in the conceptualist version but not in the canonical version of imaginary logic, which corresponds to the axiom A2 of Markin and Zaitzev's **IL2**.

[31]We name such theorems with stars (*) to stress that their homing logical system is still undetermined at this point.

T. 32.* $I_2SP \wedge A_3PR \supset I_3SR.$

T. 33.* $A_3SP \wedge I_1PR \supset I_3RS.$

T. 34.* $A_3SP \wedge I_2PR \supset I_3RS.$

The theorems on paraconsistent conversion and subalternation.
We saw by Markin and Zaitzev that conversion obtains in **IL2** for I_1SP
and I_2SP (cf. axioms A12 and A13 above), but not for A_1SP and A_2SP.
What about A_3SP and I_3SP judgements? The two Russian logicians do
not say much about that (except for a footnote). The non-standard GDP
shows in fact, by T27*, that conversion does obtain (in "**IL2X**") for paraconsistent universal judgements (cf. Figure 23), but not for the particular.

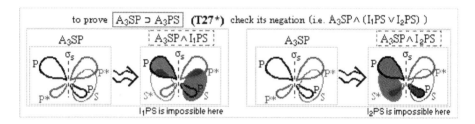

Figure 23. Demonstration of T27*, paraconsistent conversion inside **IL2**

(the negation of $A_3SP \supset A_3PS$ - namely $A_3SP \wedge (I_1SP \vee I_2SP)$ -
is graphically impossible) This theorem can be easily proven inside **IL2**
(Markin and Zaitzev mention the proof in a footnote): by use of axioms
A16, A12 and A13. Remark that Vasil'ev explicitly denies the existence of
negative indifferent conversions in the canonical version of imaginary logic
(cf. [31], p. 109). So this is another point where the normal and the conceptualist versions of imaginary logic diverge.

Paraconsistent subalternation, i.e. implication from the paraconsistent
universal to the paraconsistent particular, can be demonstrated as well, as
it happens in T28* (cf. Figure 24).

(the negation of $A_3SP \supset I_3SP$ - namely $A_3SP \wedge (A_1SP \vee A_2SP)$ - is
graphically impossible: on the left, A_3SP is impossible because then each
of its α should contain P and some P^*; on the right, A_3SP is impossible
because then each of its α should contain P^* and some P. Other proof: if
each α containing S contains some P and some P^*, then for every subset β
of α there is some γ containing β such that γ contains some P and some P^*,
both graphical proofs are intuitively straightforward) This theorem can also

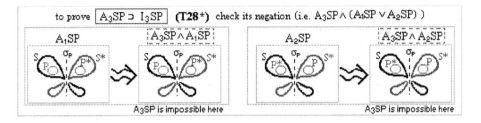

Figure 24. Demonstration of T28*, paraconsistent universal conversion

be proven axiomatically inside **IL2**. Markin and Zaitzev do not mention it, but one can prove it rather easily by use of A16, A14, A15 and A17. It is a theorem of **IL2**.

The next six new theorems, very important, are paraconsistent syllogisms. They are distributed in 3 couples[32].

The first couple of paraconsistent syllogisms. The first couple is constituted by syllogisms with only universal terms (A_n judgements, with $n \in \{1,2,3\}$). $A_1 SP \wedge A_3 PR \supset A_3 RS$, the first syllogism of this first couple, T29*, has an affirmative premise (cf. Figure 25).

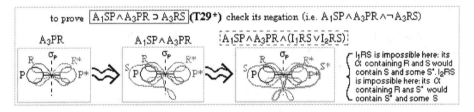

Figure 25. Demonstration of T29*, a new paraconsistent theorem of **IL2**

(if P contains some R and some R^*, then each S containing P contains itself some R and some R^*) Although very intuitive indeed, this theorem is a little bit less easy to prove axiomatically inside **IL2**. In order to make one feel the differences of the three different ways of proving theorems inside **IL2** (the two official ways, semantical and axiomatical, and the new candidate, the graphical approach) I give now the semantical and the axiomatical proofs of this theorem.

---

[32]We will write each syllogism in an order different from the traditional one. Wishing to put into evidence transitivity, we will write $SP \wedge PR \supset RS$ (or SR) instead of $PR \wedge SP \supset RS$ (or SR).

Here is the **semantical proof** of $A_1SP \wedge A_3PR \supset A_3SR$ inside **IL2**.
Suppose $A_1SP \wedge A_3PR$ is true:
$|A_1SP|^d = 1$ iff $\forall \alpha \in \mathbf{M} \ [\mathrm{d}(S) \subseteq \alpha \Rightarrow \mathrm{d}(P) \subseteq \alpha]$,
$|A_3PR|^d = 1$ iff $\forall \alpha \in \mathbf{M} \ [\mathrm{d}(P) \subseteq \alpha \Rightarrow \mathrm{d}(R) \cap \alpha \neq \emptyset$ and $\mathrm{d}(R)^* \cap \alpha \neq \emptyset]$.
And suppose A_3SR is false:
$\exists \beta \in \mathbf{M} \ [\mathrm{d}(S) \subseteq \beta$ and $(\mathrm{d}(R) \cap \beta = \emptyset$ or $\mathrm{d}(R)^* \cap \beta = \emptyset)]$.
This gives two cases to examine (one for each disjunct of the previous line). Let us examine the first:
$\exists \beta \in \mathbf{M} \ [\mathrm{d}(S) \subseteq \beta$ and $\mathrm{d}(R) \cap \beta = \emptyset]$, which gives (using the first line):
$\mathrm{d}(S) \subseteq \beta \Rightarrow \mathrm{d}(P) \subseteq \beta$, which, combined with the previous line (by Modus Ponens) gives:
$\mathrm{d}(P) \subseteq \beta$. This, combined with the second line (Modus Ponens), gives:
$\mathrm{d}(R) \cap \beta \neq \emptyset$ and $\mathrm{d}(R)^* \cap \beta \neq \emptyset$, which leads to contradiction with the hypothesis expressed by the first disjunct.
Let us examine then the second disjunct:
$\exists \beta \in \mathbf{M} \ [\mathrm{d}(S) \subseteq \beta$ and $\mathrm{d}(R^*) \cap \beta = \emptyset]$, which gives (using the first line):
$\mathrm{d}(S) \subseteq \beta \Rightarrow \mathrm{d}(P) \subseteq \beta$, which, combined with the previous line (by Modus Ponens) gives:
$\mathrm{d}(P) \subseteq \beta$. This, combined with the second line (Modus Ponens), gives:
$\mathrm{d}(R) \cap \beta \neq \emptyset$ and $\mathrm{d}(R)^* \cap \beta \neq \emptyset$, which leads to contradiction with the hypothesis expressed by the second disjunct. The theorem is thus proven[33].

Here is the **axiomatic proof** of $A_1SP \wedge A_3PR \supset A_3RS$ inside **IL2**.
Suppose $A_1SP \wedge A_3PR$ is true and A_3RS is false.
From $A_1SP \wedge A_3PR \wedge \neg A_3RS$ we have (A16 and A17):
$A_1SP \wedge \neg I_1PR \wedge \neg I_2PR \wedge (I_1RS \vee I_2RS)$, whence:
$(A_1SP \wedge \neg I_1PR \wedge \neg I_2PR \wedge I_1RS) \vee (A_1SP \wedge \neg I_1PR \wedge \neg I_2PR \wedge I_2RS)$.
To falsify this we have to falsify each of the two disjuncts. Let us try to falsify the first one:
$A_1SP \wedge \neg I_1PR \wedge \neg I_2PR \wedge I_1RS$ gives (A5):
$I_1PR \wedge \neg I_1PR \wedge \neg I_2PR$, which is contradictory.
Let us try now to falsify the second disjunct:
$A_1SP \wedge \neg I_1PR \wedge \neg I_2PR \wedge I_2RS$ gives (A6):
$I_2PR \wedge \neg I_2PR \wedge \neg I_1PR$, which is contradictory as well.
So the initial condition leads to contradiction, which means that the theorem is proven.

---

[33] In this and in the following proof one sees that Vasil'ev's logic, while claiming to cope logically with real contradictions, does not accept them in its metalevel.

One sees that the graphical proof goes a little bit faster with respect to the semantical and the axiomatic ones.

Philosophically speaking this theorem is important because it appears clearly in crucial places of Vasil'ev's writings (cf. [31] p.74, [32] p.109), where he calls it "Mindalin" ("in" meaning, in the acronym Min-Da-Lin, "universal indifferent judgement") as being very peculiar of imaginary logic. The reason for this is probably that this theorem is, among the ones involving indifferent judgements (i.e. paraconsistency), the most easy to think of intuitively (it relies ont the transitivity of the inclusion relation)

Note that this theorem is not trivial, in the sense that *not all* conjunctions of A_1 and A_3 judgements imply A_3 judgements, the order of the predicates S, P and R does matter: it doesn't suffice to "put-in" an indifferent judgement in the premises to get an indifferent judgement in the conclusions. As an example, $A_3 SP$ and $A_1 PR$, as premises, do not imply $A_3 RS$ as conclusion (cf. Figure 26)

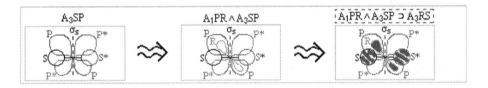

Figure 26. Changing the order of the predicates changes all!

(it is graphically possible to join the two judgements of the premise without necessary involving the judgement of the conclusion) This fact, that was known by Vasil'ev, is rather important and must be stressed: it shows that in some sense - provided one accepts indifferent judgements as being contradictory judgements - it is really *paraconsistency* (i.e. "non trivial inconsistency") we are talking about. Some theorems state inconsistent claims (indifferent judgements), but *not all* theorems. And not all couples of premises containing a universal indifferent judgement do imply a universal indifferent conclusion.

The second syllogism, $A_2 SP \wedge A_3 PR \supset A_3 RS$, with universal conjunctive premises, negative and indifferent, the "negative twin" of the previous one, is T30* (cf. Figure 27).

(if P contains some R and some R^*, *so that similarly does its negation* P^*, then all S containing P^* contains itself - by transitivity of the inclusion - some R and some R^*) Here as well, although very intuitive once stated in

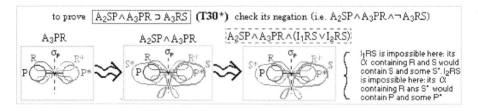

Figure 27. Demonstration of T30*, a new paraconsistent theorem of **IL2**

terms of symmetric negation, it is a bit harder to prove it axiomatically (and semantically) inside **IL2**, whereas graphically speaking it is straightforward.

Here is the **axiomatic proof** that T30* belongs to **IL2**.
Suppose that $A_2SP \land A_3PR \land \neg A_3RS$, then (by A16):
$A_2SP \land \neg I_1PR \land (I_1RS \lor I_2RS)$, hence:
$(A_2SP \land I_1RS \land \neg I_1PR \land \neg I_2PR) \lor (A_2SP \land I_2RS \land \neg I_1PR \land \neg I_2PR)$.
To falsify this we have to falsify each of the two disjuncts. Let us try to falsify the first one:
$A_2SP \land I_1RS \land \neg I_1PR \land \neg I_2PR$ gives (by A7):
$I_2RP \land \neg I_1PR \land \neg I_2PR$, which gives (by A13):
$I_2PR \land \neg I_2PR \land \neg I_1PR$, which is contradictory.
Let us now falsify the second disjunct:
$A_2SP \land I_2RS \land \neg I_1PR \land \neg I_2PR$ gives (by A8):
$I_1RP \land \neg I_1PR \land \neg I_2PR$, which gives (by A13):
$I_1PR \land \neg I_1PR \land \neg I_2PR$, which is contradictory. This ends the proof.

Now, this syllogism, although very similar to the Mindalin because of symmetry considerations (considerations very plausible given the conceptualist semantics we *are* in - where negation truly works as symmetry), does not appear - this time - in Vasil'ev's papers known by us. The reason for that could be that from the extensionalist point of view this theorem is crazy (one must remember or imagine that Vasil'ev could fear to see his intuition going too far, becoming ludicrous). This strongly suggests that, due to the lack of an intuitive method (such as our GDP) for dealing with conceptualist imaginary logic, it could be the case that many important virtual Vasil'evian features could have remained hidden to Vasil'ev himself (of this, of course, we can't be sure at the present). This would confirm the heuristical fruitfulness of our non-standard GDP.

The same remark as previously can be done here: because in T30* the order of the predicates (i.e. concepts) does matter, this theorem is not trivial as such: for instance A_3SP and A_2PR do not imply A_3RS, although altogether - abstraction done with respect to the varaibles for concepts (P, R, S) - the disposition of these three judgements is the same as in T30* (cf. Figure 28).

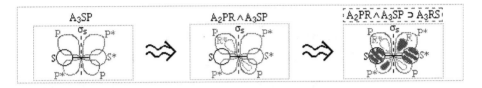

Figure 28. Changing the order of the predicates changes all!

(it is possible graphically speaking to construct A_3SP and A_2PR without implying the construction of A_3RS, which effectively is not the case in the third panel of Figure 28) Remark, additionally, that, even respecting the good order of the predicates inside the three judgements, not all kinds of couples of universal judgements including a paraconsistent one imply a universal paraconsistent judgement: for instance, A_3SP and A_3PR together do not imply A_3RS (cf. Figure 29).

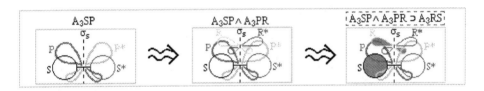

Figure 29. Changing the order of the predicates changes all!

(it is possible graphically speaking to construct A_3SP and A_3PR without implying the construction of A_3RS, which effectively is not the case in the third panel of Figure 29) Once again, this shows that, at this level (the level of universal indifferent judgements), it is really paraconsistency we are dealing with: there are contradictions (universal indifferent judgements), but neither everywhere nor every how; one can deal with them logically (i.e. non trivially).

The next two couples of paraconsistent syllogisms deal with reasonings involving I_n and A_m judgements in the premises, and I_3 judgements in the conclusion. In the second couple the premise is of the form $I_nSP \wedge A_3PR$, while in the third couple of syllogisms the premise is of the form $A_3SP \wedge$

$I_n PR.$

The second couple of paraconsistent syllogisms. The first syllogism
of this couple, $A_3MP \wedge I_1SM \supset I_3SP$, has a particular affirmative judgement among its premisses. Its graphical demonstration is developed in the
two Figures 30 and 31.

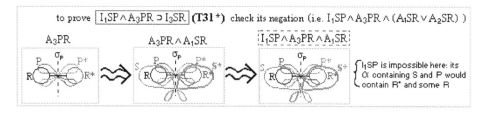

Figure 30. Demonstration of T31* (1), a new paraconsistent syllogism of
IL2

(if each P contains some R and some R^*, and if each S contains R, then
I_1SP is impossible because any concept α containing S and P should contain R^* and some R)

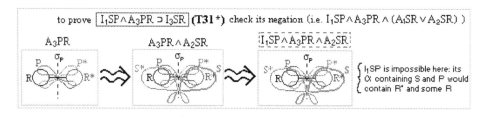

Figure 31. Demonstration of T31* (2), a new paraconsistent syllogism of
IL2

(if each P contains some R and some R^*, and if every S contains R^*,
then I_1SP is impossible because any concept α containing S and P should
contain R^* and some R) This theorem happens to have been known by
Vasil'ev, but inside the canonical version of imaginary logic. It had been
named Kindirinp by him[34]. Consequently, from a Vasil'evian point of view
this theorem T31* proves this until now missing syllogism "Kindirinp" (cf.
[31] p.74, [32] p.108), is valid not only in the canonical version, but also in

---

[34] "Inp", different from "in", means "particular indifferent".

the conceptualist version of imaginary logic.

We give here an **axiomatic proof** of $I_1SP \wedge A_3PR \supset I_3SR$ inside **IL2**.
Suppose $I_1SP \wedge A_3PR$ is true and I_3SR is false, that is $I_1SP \wedge A_3PR \wedge \neg I_3SR$. This gives (by A16 and A17):
$I_1SP \wedge \neg I_1PR \wedge \neg I_2PR \wedge (A_1SR \vee A_2SR)$, which gives:
$(A_1SR \wedge I_1SP \wedge \neg I_1PR \wedge \neg I_2PR) \vee (A_2SR \wedge I_1SP \wedge \neg I_1PR \wedge \neg I_2PR)$.
We must check the two disjuncts. Let us look at the first one:
$A_1SR \wedge I_1PS \wedge \neg I_1PR \wedge \neg I_2PR$ (by A12), which gives (by A5):
$I_1PR \wedge \neg I_1PR \wedge \neg I_2PR$, which is contradictory.
Let us look now for the second disjunct:
$A_2SR \wedge I_1PS \wedge \neg I_1PR \wedge \neg I_2PR$ (by A12), which gives (by A7):
$I_2PR \wedge \neg I_1PR \wedge \neg I_2PR$, which is contradictory.
The theorem is thus proven.

The "negative twin" of Kindirinp, the syllogism $I_2SP \wedge A_3PR \supset I_3SR$, which is our theorem T32*, has a particular negative premise, whose graphical proof is, as previously, distributed in two schemas (cf. Figures 32 and 33).

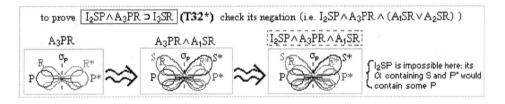

Figure 32. Demonstration of T32* (1), a new paraconsistent syllogism of **IL2**

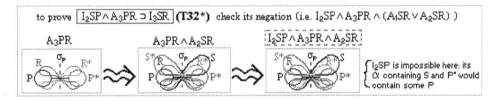

Figure 33. Demonstration of T32* (2), a new paraconsistent syllogism of **IL2**

Here is its **axiomatic proof.** Suppose
$I_2SP \wedge A_3PR \wedge \neg I_3SR$; then (by A16 and A17):
$I_2SP \wedge \neg I_1PR \wedge \neg I_2PR \wedge (A_1SR \vee A_2SR)$, which gives:
$(A_1SR \wedge I_2SP \wedge \neg I_1PR \wedge \neg I_2PR) \vee (A_2SR \wedge I_2SP \wedge \neg I_1PR \wedge \neg I_2PR)$.

We have to check both these disjuncts. Let us check the first one:
$A_1SR \wedge I_2PS \wedge \neg I_1PR \wedge \neg I_2PR$ (by A13), and (by A6):
$I_2PR \wedge \neg I_1PR \wedge \neg I_2PR$, which is contradictory.
Now for the second disjunct. We have (by A13):
$A_2SR \wedge I_2PS \wedge \neg I_1PR \wedge \neg I_2PR$, and (by A8):
$I_1PR \wedge \neg I_1PR \wedge \neg I_2PR$, which is contradictory.
Hence the theorem (its negations can never hold).

This "negative twin" of Kindirinp does not seem to be mentioned by
Vasil'ev. But it holds in **IL2**. Remark that, in a way similar to what seen
previously, syllogisms quite similar to this one (but slightly different with
respect to the disposition of R and S in the conclusions), like $I_1SP \wedge A_3PR$
$\supset I_3RS$ and $I_2SP \wedge A_3PR \supset I_3RS$, are not valid. Neither is valid I_3SP
$\wedge A_3PR \supset I_3SR$, the "indifferent twin" of the syllogisms of theorems T31*
and T32*.

The third and last couple of paraconsistent syllogisms. The last
two paraconsistent Vasil'evian new syllogisms reverse, with respect to the
two previous ones, the order of their premises. The premises now have
the form $A_3SP \wedge I_nPR$ (with $n \in \{1,2\}$). The first of these two syllogisms,
T33*, one has a particular affirmative premise (cf. Figures 34 and 35).

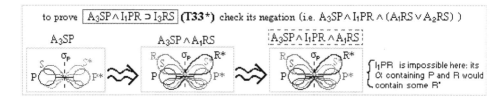

Figure 34. Demonstration of T33* (1), a new paraconsistent syllogism of
IL2

(if each S contains some P and some P^*, and if each R contains S, then
I_1PR is impossible, because any concept α containing P and R would contain some R^* as well)

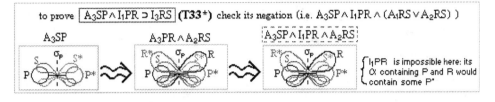

Figure 35. Demonstration of T33* (2), a new paraconsistent syllogism of **IL2**

(if each S contains some P and some P^* - so that each S^* contains some P^* and some P -, and if each R contains S^*, then $I_1 PR$ is impossible because any concept α containing P and R should contain some P^* as well). As in the previous cases, the graphical proof takes two steps (one for each of the two disjuncts of the negation of the starting formula (cf. Figures 34 and 35)

Here is the **axiomatic proof** of T33* inside **IL2**. Suppose
$A_3 SP \wedge I_1 PR \wedge \neg I_3 RS$; then (by A16 and A17):
$\neg I_1 SP \wedge \neg I_2 SP \wedge I_1 PR \wedge (A_1 RS \vee A_2 RS)$, which gives:
$(A_1 RS \wedge I_1 PR \wedge \neg I_1 SP \wedge \neg I_2 SP) \vee (A_2 RS \wedge I_1 PR \wedge \neg I_1 SP \wedge \neg I_2 SP)$.
We have to check both these disjuncts. Let us look at the first one:
$I_1 PS \wedge \neg I_1 SP \wedge \neg I_2 SP$ (by A5), which gives (by A12):
$I_1 SP \wedge \neg I_1 SP \wedge \neg I_2 SP$, which is contradictory.
Let us now look at the second disjunct. We have (by A7):
$I_2 PS \wedge \neg I_1 SP \wedge \neg I_2 SP$, and (by A13):
$I_2 SP \wedge \neg I_1 SP \wedge \neg I_2 SP$, which is contradictory.
Hence the theorem (its negation can never hold).

T34*, the syllogism $A_3 SP \wedge I_2 PR \supset I_3 RS$, second member of this third couple, has a particular negative in the premise (cf. Figures 36 and 37).

(if each S contains some P and some P^* - so that each S^* contains some P^* and some P -, and if each R contains S - so that each R^* contains S^* -, then $I_2 PR$ is not possible, because any concept α containing P and R^* should also contain some P^*)

(if each S contains some P and some P^*, and if each R contains S^* - so that each R^* contains S -, then $I_2 PR$ is impossible, because any concept α containing P and R^* should also contain some P^*)

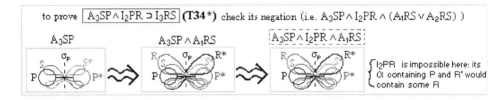

Figure 36. Demonstration of T34* (1), a new paraconsistent syllogism of **IL2**

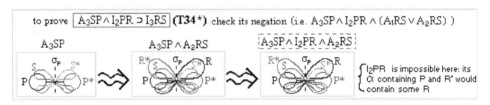

Figure 37. Demonstration of T34* (2), a new paraconsistent syllogism of **IL2**

Here is the **axiomatic proof** of T34* inside **IL2**. Suppose
$A_3SP \wedge I_2PR \wedge \neg I_3RS$; then (by A16 and A17):
$\neg I_1SP \wedge \neg I_2SP \wedge I_2PR \wedge (A_1RS$ or $A_2RS)$, which gives:
$(A_1RS \wedge I_2PR \wedge \neg I_1SP \wedge \neg I_2SP)$ or $(A_2RS \wedge I_2PR \wedge \neg I_1SP \wedge \neg I_2SP)$.

As usual, we have to check both these disjuncts. Let us look at the first one:
$I_2PS \wedge \neg I_1SP \wedge \neg I_2SP$ (by A6), which gives (by A13):
$I_2SP \wedge \neg I_1SP \wedge \neg I_2SP$, which is contradictory.
Let us now look at the second disjunct. We have (by A8):
$I_1PS \wedge \neg I_1SP \wedge \neg I_2SP$, and (by A12):
$I_1SP \wedge \neg I_1SP \wedge \neg I_2SP$, which is contradictory.
Hence the theorem (its negation can never hold).

The moral about the new paraconsistent theorems. So how can we answer the two questions of the beginning of the chapter? The answer to the first ("is **IL2X** equivalent to **IL2**?"), as said, is that all the new theorems T27*-T34* of **IL2X** are in fact theorems T27-T34 of **IL2** (down with the stars!). So **IL2** and our non standard GDP based on it (=**IL2X**), *seem* reasonably to be the same, which wouldn't be so strange, given we built it relying on the formal semantics of **IL2**. But a serious proof of this shall be given in terms of completeness and consistency metatheorems - we won't do it here. As for the answer to the second question ("irrelevance, triviality or paraconsistency?"), it says that such theorems are clearly not trivial at all, which is shawn by the fact that of the many, many, possible combinations, into an imaginary syllogism, of three judgements with at least a paraconsistent one (in the premises and/or in the conclusion), only six are valid (plus a few redundant forms of these 6, reducible to these). As it seems, all paraconsistent simple syllogisms of **IL2** are such that they have an indifferent judgement both in the premises *and* in the conclusion of the syllogism (this indifferent judgement needs not to have the same quantity, universal or particular, in both places of a given paraconsistent syllogism).

It deserves being remarked that even if the paraconsistent theorems we proposed could have been found using only the semantical or the axiomatic means of **IL2**, our graphical method is nevertheless very useful and powerful indeed for handling the system **IL2**: a single drawing can suffice to falsify a candidate proposition claiming to be a theorem of this system (we gave many examples of this), while the same by axiomatic means is very tedious (it is less by semantic means - but again: with drawings one single drawing well constructed can suffice to discard an insufficient pretender).

Remark also that the "golden rule" (not the golden object) for judging graphically a concept's seriousness in this graphical counterpart of **IL2** (i.e. understanding if it is a contradictory one, in the sense Vasil'ev judges unbearable) is very intuitive and very handy, it consists in seeing if the concept's shape is such that it is intersection-free with respect to it negation, i.e. its symmetric image with respect to the given symmetry axis (for instance any concept symmetric with respect of the symmetry axis coincides with its own negation and therefore can't be a concept).

7   Conclusion and perspectives.

As said, in his known published few works, Vasil'ev mentions some unpublished materials of him, containing paraconsistent (in his words "imaginary") theorems and demonstrations, and announces some others to come in future studies. But we know of no real such remaining texts of him (such texts probably existed, but are lost for us, at least until now). Two among the best existing logical formalizations of his thought, **IL** and **IL2**, produced by the Russian school, have not yet stressed the scope of the paraconsistent theorems reachable (and seemingly reached already - at least partly - by Vasil'ev himself in his own times).

In this paper we proposed a non-standard graphical decision procedure

(GDP) and hence some paraconsistent theorems for **IL2**, the faithful formalization by Markin and Zaitzev of Vasil'ev's outstanding ideas on the conceptualist variant of imaginary logic. We showed how this GDP, based on the expression of negation as graphical symmetry (instead as set-theoretical complementarity), seemingly expresses all the known features of **IL2** and allows to discover some new, less visible ones, coherently with the existing axiomatization of it. By doing this we offered a visual, faithful interpretation of this slightly counter-intuitive non-standard logic.

An open problem to be handled in the future is that this non standard GDP still has to be furnished with metatheorems, proving its adequacy with **IL2**. Despite this (hopefully temporary) proof-theoretical defect (or gap) our GDP shows nice features and allows interesting results, more difficult to reach without the help of a graphical device like ours. This suggests that many (far fetched) Vasil'evian theorems could have escaped Vasil'ev's eye, because of his lack of an intuitive visual method. And this suggests the interest of making further investigations into the Vasil'evian logical spaces. In particular, a very natural question to be investigated systematically will be the one of the differences between Vasil'ev's systems **IL** and **IL2**. Again, our method should prove useful in handling this comparison (in this paper we started showing rhapsodically some such differences).

Another open question closely related to our present paper is also that of knowing if similar graphical methods can be found for the other versions of imaginary logic (the canonical, the modal and the n-dimensional ones, for instance). Finding them could greatly increase our understanding of the frontiers of Vasil'ev's amazing and pioneering paraconsitent thought.

One last point deserves to be mentioned here. Vasil'ev's logics seem to us to be much more important - with respect to contemporary issues - than it generally seem to the commentators (as for instance in [24], with whom we disagree strongly on this question). In particular, one crucial point to be explored in the future is the question of the geometrical restructuration that Vasil'ev imposes to the classical "logical square". The relations holding between the six atomic judgements of **IL2** (and of **IL** as well) show clearly that imaginary logic changes effectively the notion of contradictory negation: a point's contradiction is no more another point, but a segment, interpreted logically as the (inclusive) disjunction of its two vertexes (cf. Figure 38).

This bright, until now unremarked (as long as we know), idea of the logician of Kazan is open to infinite generalizations (the contradictory negation of a point shifting from points to segments, to 3-dimensional solids, ..., to n-dimensional solids). The Russian Vasil'evian school explores presently

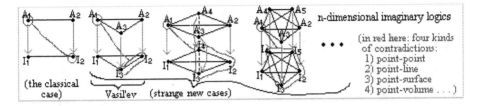

Figure 38. The Vasilevian *virtual* series of the *n*-dimensional logics

the possible faithful axiomatizations of *n*-dimensional logics. We claim that these can (and shall) be expressed geometrically as a series of couples of subalternated "simplexes of dimension *n*-1" (cf. Figure 38). But recent results on abstract contemporary mathematical logic, namely those already cited on "*n*-opposition theory" show a very promising way to understand the Vasil'evian series as being just one small (but very interesting) part of a much more complex series, the one of the "logical bi-simplexes of dimension *n*" (cf. Figure 39).

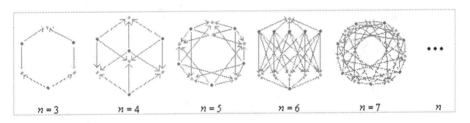

Figure 39. The oppositional series of the bi-simplexes of dimension *n*

Which is a different story, though closely related: the logical bi-simplexes are not linear subalternations of simplexes, as by Vasil'ev, but entanglements of dual simplexes, with many more implicative arrows, with respect to which Vasil'ev's series is isomorph to a modal decorating technique, the series of the modal "*n*(2)-graphs", a classical fragment of modal *n*(*m*)-graphs (we really quit Aristotle's, and fall into a geometrical complexity much more akin to Plato's views). Forerunner of this "still on the move" new logical approach, Vasil'ev has the fantastic interest for us of having developed a whole *philosophy* entangled with his strange deep logical attempts. This philosophy we shall study and meditate, and discuss *presently*, in order to build/discover things we still ignore.

With such issues in mind, it should become clear that, far from being an historical puzzle, Vasil'ev's ideas are much more visions whose sharp originality seems to be still far from exhaustion.

Acknowledgements

I wish to thank here W.A. Carnielli, who first encouraged me in 2002 to expose my results, followed by P. Gochet. Then S. Odintsov and V.L. Vasyukov, who helped me in getting Russian materials, and J.-Y. Béziau for personal constant intellectual support and counseling. I also thank F. Schang for some helpful remarks and discussions.

BIBLIOGRAPHY

[1] Arruda, A., "On the imaginary logic of Vasil'ev", in: Arruda, da Costa, Chuaqui (eds.), *Non-Classical Logics, Model Theory, and Conputability*, 1976.

[2] Bazhanov, V.A., "Nikolai Aleksandrovich Vasil'ev: zhizn' i tvorchestvo" ("Nikolai Alexandrovich Vasil'ev: life and creation"), in [29].

[3] Béziau, J.-Y., "La Logique Universelle en 13 questions", to appear in: *Noésis*, n.10.

[4] Bobenrieth, A., *Inconsistencias ¿por qué no? Un estudio filosófico sobre la lógica paraconsistente*, Colcultura, Bogota, 1996 (ch. II, "La lógica imaginaria de Vasil'ev").

[5] Karpinskaja O.J. and Markin V.I., "K voprosy ob adekvatnoi rekonstrukzii assertoricheskoi sillogistiki N.A. Vasil'eva" ("On the question of adequately reconstructing the assertoric syllogistic of N.A. Vasil'ev"), *Vtoroi Rossiskii Filosofskii Kongress "XXI vek: buduschee Rossii v filosofskom izmerenii"*, Tom 1. Ontologia, gnoseologia i metodologia nauki, logika. Chast' 1., Ekaterinburg, 1999.

[6] Kostjuk, T.P., "Rekonstrukzija assertoricheskoi sillogistiki N.A. Vasil'eva" ("The reconstruction of the assertoric syllogistic of N.A. Vasil'ev"), *Pervyi Rossiskii Filosofskii Kongress "Chelovek, filosofia, gumanism"*, Tom 3. Ontologia, gnoseologia, logika i analiticheskaia filosofia, Isd.-vo S-Pb. universiteta, Sankt-Peterburg, 1997, pp. 185-197.

[7] Kostjuk, T.P., "Pozitivnye sillogistiki vasil'evskogo tipa" ("The Vasil'evian type of positive syllogistics"), *Logicheskie isledovania*, Vyr.6, ROSSPEN, Moskva, 1999, pp. 259-267.

[8] Kostjuk, T.P., "Rekonstrukzija logiki n izmerenii N.A. Vasil'eva" ("The reconstruction of N.A. Vasil'ev's n-dimensional logic"), *Vtoraja mezhdunarodnaja konferenzija "Smirnovskie chtenija"*, IF RAN, Moskva, 1997.

[9] Kostjuk T.P. and Markin V.I., "Problema rekonstrukzii assertoricheskoi sillogistiki N.A. Vasil'eva" ("The problem of reconstructing the assertoric syllogistic of N.A. Vasil'ev"), *Mezhdunarodnaja konferenzija "Smirnovskie chtenija"*, IF RAN, Moskva, 1997.

[10] Kostjuk T.P. and Markin V.I., "Formal'naja rekonstrukzija voobrazhaemoi logiki N.A. Vasil'eva" ("Formal reconstruction of N.A. Vasil'ev's imaginary logic"), *Sovremennaja logika: problemy teorii, istorii i primenenija v nauke (V Obshcherossiskaja naychnaja konferenzija)*, Isd.-vo S-Pb. universiteta, Sankt-Peterburg, 1998.

[11] Markin, V.I., "Pogruzhenie voobrazhaemoi logiki N.A. Vasil'eva v kvantornuju trekhznachnuju logiku" ("Plounging Vasil'ev's imaginary logic into 3-valued logic"), Online Journal "Logical Studies", No.2(1999), http://www.logic.ru/LogStud/index.html.

[12] Markin, V.I., "Intensional'naja semantika traditzionnoi sillogistiki" ("The intensional semantics of the traditional syllogistic"), in: *Online Journal "Logical Studies"*, N.6(2001), http://www.logic.ru/LogStud/index.html.

[13] Markin, V.I., "Modal'naja interpretazija voobrazhaemoi logiki N.A. Vasil'eva" ("The modal interpretation of N.A. Vasil'ev's imaginary logic"), *Sovremennaja logika:*

problemy teorii, istorii i primenenija v nauke (V Obshcherossiskaja nauchnaja konferenzija), Isd.-vo S-Pb. universiteta, Sankt-Peterburg, 1998.

[14] Markin, V.I., "Voobrazhaemaja logika N.A. Vasil'eva i kvantornaja trekhznachnaja logika" ("N.A. Vasil'ev's imaginary logic and 3-valued logic"), *Vtoraja mezhdunarodnaja konferenzija "Smirnovskie chtenija"*, IF RAN, Moskva, 1999, pp.126-128.

[15] Markin, "Kommentarii k rabotam V.A. Smirnova, posvjashchennym logicheskomu naslediju N.A. Vasil'eva" ("Remarks on V.A. Smirnov's works on the logical heritage of N.A. Vasil'ev"), unpublished.

[16] Matte Blanco, I., *The Unconscious as Infinite Sets. An Essay in Bi-logic*, Duckwort, London, 1975.

[17] Moretti, A., "Trois approches de l'irrationnel: Davidson, Matte Blanco et da Costa", in: *Noésis*, No. 5, Vol. 2, 2003.

[18] Moretti, A., "Deux spatialisations convergentes: I. Matte Blanco et P. Gärdenfors", in: M. Sobieszczanski and C. Lacroix (eds.), *Spatialisations en art et en sciences humaines*, Peeters France, Paris, 2004.

[19] Moretti, A., "Geometry for Modalities? Yes: through 'n-Opposition Theory'", in Béziau, Costa-Leite and Facchini (eds.), *Aspects of Universal Logic*, University of Neuchâtel, Neuchâtel, December 2004.

[20] Moretti, A., "Logical 'Hyper-Flowers'. The 'oppositional Cuboctahedron' Belongs to an infinite (fractal) series of n-dimensional solids" (2006 ?), (proceedings of the UNILOG 2005).

[21] Moretti, A., "Imaginary logics or n-opposition? Russian school and/or french school", (to be submitted).

[22] Moretti, A., "Le carré logique et son évolution contemporaine. De la théorie de l'opposition à la thorie de la n-opposition", to appear in *Noésis*, n.10.

[23] Pellissier, R., " 'Setting' n-opposition", (2006 ?), (proceedings of the UNILOG 2005).

[24] Priest, G., "Vasil'ev and Imaginary Logic", in: History and Philosophy of Logics, N.21, 2000, pp. 135-146.

[25] Priest G., Routley R. and Norman J. (eds.), *Paraconsistent Logic. Essays on the Inconsistent*, Philosophia Verlag, Muenchen Hamden Wien, 1989.

[26] Puga L.Z. and da Costa N.C.A., "On the imaginary logic of N.A. Vasiliev", *Zeitschrift für mathematische Logik und Grundlagen der Mathematik*, Bd.34 p.205-211 (1988).

[27] Smirnov, V.A., "Logicheskie idei N.A. Vasil'eva i sovremennaja logika" ("Vasil'ev's logical ideas and contemporary logic") (1989) (in [29]) (english translation in: J.E. Fenstad et alii (eds.), *Logic, Methodology and Philosophy of Science VIII*, Elsevier, 1989, pp. 625-640) (french translation by A. Moretti to appear in *Les Cahiers de l'ATP*, http://alemore.club.fr/CahiersATP.htm).

[28] Toth, I., *Aristotele e i fondamenti assiomatici della geometria. Prolegomeni alla comprensione dei frammenti non-euclidei nel "Corpus Aristotelicum"*, Vita e Pensiero, Milano, 1997.

[29] Vasil'ev, N.A., *Voobrazhaemaja logika. Izbrannye trudy (Imaginary logic. Selected works)*, edited by V.A. Smirnov, Nauka, Moskva, 1989.

[30] Vasil'ev, N.A., "O chastnykh suzhdenijakh, o treugol'nike protivopolozhnostei, o zakone iskljuchennogo cetvertogo" (1910) ("On particular judgements, on the triangle of oppositions, on the law of the excluded fourth"), in [29].

[31] Vasil'ev, N.A., "Voobrazhaemaja (nearistoteleva) logika" (1912) ("Imaginary (non-Aristotelian) logic"), in [29] (an english translation by R. Vergauwen is in [34]).

[32] Vasil'ev, N.A., "Logika i metalogika" (1912-1913) ("Logic and metalogic"), in [29]. (an english translation by V.L. Vasjukov is in: *Axiomathes*, IV, No.3, December 1993).

[33] Vasil'ev, N.A., "Otchet privat-dozenta po kafedre filosofii imperatorskogo kazanskogo universiteta N.A. Vasil'eva o khode ego nauchnykh zanjatii za vremja s 1 ijulja 1911 g. po 1 ijulja 1912 g." ("Report of N.A. Vasil'ev, professor of the chair for philosophy of the Imperial University of Kazan, on the progression of his own scientifical

activities for the time comprised between july the 1st 1911 and july the 1st 1912"), in [29].

[34] Vergauwen R. and Zaytsev E.A. (eds), *Special issue on* Imaginary logic *by N.A. Vasil'ev, Logique et Analyse*, vol.46 No. 182 (2003).

[35] Vergauwen R. and Zaytsev E.A., "The Worlds of Logic and the Logic of Worlds", in: [34].

[36] Zaitzev, D.V., "Interpretazija voobrazhaemoi logiki: rekonstrukzija ideii N.A. Vasil'eva" ("An interpretation of imaginary logics: reconstruction of the ideas of N.A. Vasil'ev"), *Sovremennaja logika: problemy teorii, istorii i primenenija v nauke (V Obshcherossiskaja nauchnaja konferenzija)*, Isd.-vo S-Pb. universiteta, Sankt-Peterburg, 1998, pp. 113-117.

[37] Zaitzev D.V. and Markin V.I., "Voobrazhaemaja logika-2 : rekonstruktzija odnogo iz variantov znamienitoi logicheskoi sistemy N.A. Vasil'eva" ("Imaginary logic 2: a reconstruction of one variant of Vasil'ev's well known logical system"), *Online Journal "Logical Studies"*, N.2(1999), http://www.logic.ru/LogStud/index.html (an english version of it is in: *Logique et Analyse*, 177-178, march-june 2002, pp. 39-54.) (french translation of the Russian original by A. Moretti to appear in *Les Cahiers de l'ATP*, http://alemore.club.fr/CahiersATP.htm).

[38] Zaitzev D.V. and Markin V.I., "Nezamechennaja logicheskaja sistema N.A. Vasil'eva: voobrazhaemaja logika-2 ili logika ponjatii" ("An unnoticed logical system of N.A. Vasil'ev: the imaginary logic-2 or the logic of concepts"), *Mezhdunarodnaja konferenzija "Smirnovskie chtenija"*, IF RAN, Moskva, 1999.

Alessio Moretti

alemore@club-internet.fr

A New *Real* Axiomatization of the Discursive Logic D_2

JANUSZ CIUCIURA

1 Introduction

It is one of the logical truisms to say that the vagueness of a term can lead to *seeming contradictions*. Since we cannot safely assume that imprecise terms and seeming contradictions will never occur among participants in the discussion our formal language should allow the contradictions to be present. Classical and almost all of the non-classical logics Jaśkowski considered in [Jaśkowski, 1948] are not too helpful to gain a real insight into their nature. It partly explains why he decided to apply a translation procedure instead of introducing a set of the axiom schemata or describing a semantics for D_2.

DEFINITION 1. Let *var* be a non-empty denumerable set of all propositional variables $\{p_1, p_2, p_3, ...\}$. For_{D2} is defined to be the smallest set for which the following holds:

(i) $\alpha \in var \Rightarrow \alpha \in For_{D2}$

(ii) $\alpha \in For_{D2} \Rightarrow \sim \alpha \in For_{D2}$

(iii) $\alpha \in For_{D2}$ and $\beta \in For_{D2} \Rightarrow \alpha \bullet \beta \in For_{D2}$, where $\bullet \in \{\vee, \wedge_d, \rightarrow_d\}$.

The symbols $\sim$, $\vee$, $\wedge_d$, $\rightarrow_d$ denote negation, disjunction, discursive conjunction and discursive implication, respectively. It is quite remarkable that not only has some of the classical connectives disappeared, but they were successfully replaced with their discursive counterparts. Hence the language of D_2 is not best thought of as an extension of the language of CPC.

There is an important feature of D_2 that we have hardly touched upon so far. To clarify things, let us define a translation function of the language of D_2 into the language of S_5 of Lewis, $\tau : For_{D2} \Rightarrow For_{S5}$, as follows:

(i) $\tau(p_i) = p_i$ if $p_i \in var$ and $i \in N$

(ii) $\tau(\sim \alpha) = \sim \tau(\alpha)$

(iii) $\tau(\alpha \vee \beta) = \tau(\alpha) \vee \tau(\beta)$

(iv) $\tau(\alpha \wedge_d \beta) = \tau(\alpha) \wedge \Diamond\tau(\beta)$

(v) $\tau(\alpha \rightarrow_d \beta) = \Diamond\tau(\alpha) \rightarrow \tau(\beta)$

and additionally:

(vi) $\forall \alpha \in For_{D2} : \alpha \in D_2 \Leftrightarrow \Diamond\tau(\alpha) \in S_5$.

A little calculation reveals that most of the *notorious* formulas being in a *real* sense paraconsistent are not valid in D_2, e.g.

1. $p \rightarrow_d (\sim p \rightarrow_d q)$

2. $p \rightarrow_d (\sim p \rightarrow_d \sim q)$

3. $(p \rightarrow_d q) \rightarrow_d (\sim q \rightarrow_d \sim p)$

4. $(\sim p \rightarrow_d \sim q) \rightarrow_d (q \rightarrow_d p)$

5. $p \rightarrow_d (\sim p \rightarrow_d (\sim\sim p \rightarrow_d q))$

6. $(p \wedge_d \sim p) \rightarrow_d q$.

On the other hand, it is easy task to prove that the following propositions hold:

(i)   Let α be a formula that does not include constant symbols other than $\sim$ and $\vee$. If α is valid in the classical propositional calculus, then (1) α and (2) $\sim \alpha \rightarrow_d \beta$ are also valid in D_2, where $\beta \in For_{D2}$.

(ii) Let α contain, besides variables, at most the connectives $\wedge$, $\rightarrow$, $\leftrightarrow$ and $\vee$. If α is valid in CPC, then α_d is valid in D_2, where α_d is obtained from α by replacing $\wedge$, $\rightarrow$, $\leftrightarrow$, $\vee$ with $\wedge_d$, $\rightarrow_d$, $\leftrightarrow_d$, $\vee$, respectively.

(iii) Let α include, besides variables, at most the functors $\wedge_d$, $\rightarrow_d$, $\leftrightarrow_d$ and $\vee$. If α is valid in D_2, then α_{cpc} is valid in CPC, where α_{cpc} is obtained from α by replacing $\wedge_d$, $\rightarrow_d$, $\leftrightarrow_d$, $\vee$ with $\wedge$, $\rightarrow$, $\leftrightarrow$, $\vee$, respectively.[1]

In practice, as seen in the item of (ii) and (iii), we may read off the validity of α directly from a classical true-value analysis, of course, so long as α is a negation-free formula. What is really interesting is that if we agree to accept $\tau(\alpha \wedge \beta) = \tau(\alpha) \wedge \tau(\beta)$ in lieu of $\tau(\alpha \wedge_d \beta) = \tau(\alpha) \wedge \Diamond\tau(\beta)$ we will obtain a non-adjunctive system in which the proposition (ii) does not hold plus $(p \wedge \sim p) \rightarrow_d q$ as a thesis.[2] From pure philosophical point of

---

[1] Cf. [Jaśkowski, 1948, pp. 45–59]; [Jaśkowski, 1949, p. 57]; [Ciuciura, 1999, pp. 9–10]; and [Kotas, 1975, pp. 152–153]. To be precise, it should be added that we use here the English translations of Jaśkowski's papers that appeared in 1999. See References for details.

[2] Cf. [Jaśkowski, 1948, p. 47].

view, the thesis imitates the situation when a participant in a discussion has simultaneously stated that p and not-p, i.e. expressed different views on the same topic. It is fairly dangerous since the discussion may lose coherency.

Though we depicted how to transform any discursive formula into its modal counterpart, the procedure was slightly inconvenient to deal with. Before presenting a direct *Kripke-type* semantics for D_2 that solves the problem[3] we introduce an axiomatization given by da Costa, Dubikajtis and Kotas. The axiomatization consists of the following axiom schemata and rules:

(A_1) $\alpha \rightarrow_d (\beta \rightarrow_d \alpha)$

(A_2) $(\alpha \rightarrow_d (\beta \rightarrow_d \gamma)) \rightarrow_d ((\alpha \rightarrow_d \beta) \rightarrow_d (\alpha \rightarrow_d \gamma))$

(A_3) $((\alpha \rightarrow_d \beta) \rightarrow_d \alpha) \rightarrow_d \alpha$

(A_4) $(\alpha \wedge_d \beta) \rightarrow_d \alpha$

(A_5) $(\alpha \wedge_d \beta) \rightarrow_d \beta$

(A_6) $\alpha \rightarrow_d (\beta \rightarrow_d (\alpha \wedge_d \beta))$

(A_7) $\alpha \rightarrow_d (\alpha \vee \beta)$

(A_8) $\beta \rightarrow_d (\alpha \vee \beta)$

(A_9) $(\alpha \rightarrow_d \gamma) \rightarrow_d ((\beta \rightarrow_d \gamma) \rightarrow_d ((\alpha \vee \beta) \rightarrow_d \gamma))$

(A_{10}) $\alpha \rightarrow_d \sim\sim \alpha$

(A_{11}) $\sim\sim \alpha \rightarrow_d \alpha$

(A_{12}) $\sim (\alpha \vee \sim \alpha) \rightarrow_d \beta$

(A_{13}) $\sim (\alpha \vee \beta) \rightarrow_d \sim (\beta \vee \alpha)$

(A_{14}) $\sim (\alpha \vee \beta) \rightarrow_d (\sim \beta \wedge_d \sim \alpha)$

(A_{15}) $\sim (\sim\sim \alpha \vee \beta) \rightarrow_d \sim (\alpha \vee \beta)$

(A_{16}) $(\sim (\alpha \vee \beta) \rightarrow_d \gamma) \rightarrow_d ((\sim \alpha \rightarrow_d \beta) \vee \gamma)$

(A_{17}) $\sim ((\alpha \vee \beta) \vee \gamma) \rightarrow_d \sim (\alpha \vee (\beta \vee \gamma))$

(A_{18}) $\sim ((\alpha \rightarrow_d \beta) \vee \gamma) \rightarrow_d (\alpha \wedge_d \sim (\beta \vee \gamma))$

(A_{19}) $\sim ((\alpha \wedge_d \beta) \vee \gamma) \rightarrow_d (\alpha \rightarrow_d \sim (\beta \vee \gamma))$

(A_{20}) $\sim (\sim (\alpha \vee \beta) \vee \gamma) \rightarrow_d (\sim (\sim \alpha \vee \gamma) \vee \sim (\sim \beta \vee \gamma))$

(A_{21}) $\sim (\sim (\alpha \rightarrow_d \beta) \vee \gamma) \rightarrow_d (\alpha \rightarrow_d \sim (\sim \beta \vee \gamma))$

(A_{22}) $\sim (\sim (\alpha \wedge_d \beta) \vee \gamma) \rightarrow_d (\alpha \wedge_d \sim (\sim \beta \vee \gamma))$

$(MP)^*$ $\alpha, \alpha \rightarrow_d \beta \ / \ \beta$

$(R_d 1)$ $\alpha \leftrightarrow_d \beta = (\alpha \rightarrow_d \beta) \wedge_d (\beta \rightarrow_d \alpha)$

$(R_d 2)$ $\alpha \rightarrow \beta = (\sim \alpha \vee \beta)$

$(R_d 3)$ $O\alpha = \sim (\alpha \vee \sim \alpha)$

$(R_d 4)$ $\Box \alpha = (\sim \alpha \rightarrow_d O\alpha)$

$(R_d 5)$ $\Diamond \alpha = \sim \Box \sim \alpha$

$(R_d 6)$ $\alpha \wedge \beta = \sim (\sim \alpha \vee \sim \beta)$

$(R_d 7)$ $\alpha \leftrightarrow \beta = (\alpha \rightarrow \beta) \wedge (\beta \rightarrow \alpha).$[4]

---

[3]In [Ciuciura, 2004], we gave another solution to the problem.

[4]Cf. [Achtelik *et al.*, 1981; Da Costa and Dubikajtis, 1977; Kotas and da Costa, 1977].

It is surprising that the construction has been widely regarded as a real axiomatization of D_2. To illustrate the point, take the axiom schema:

$(A_{19}) \sim ((\alpha \wedge_d \beta) \vee \gamma) \rightarrow_d (\alpha \rightarrow_d \sim (\beta \vee \gamma))$

apply the translation procedure to get:

$\Diamond(\Diamond \sim ((\alpha \wedge \Diamond\beta) \vee \gamma) \rightarrow (\Diamond\alpha \rightarrow \sim (\beta \vee \gamma)))$

and check if the translated formula is valid in S_5 of Lewis.

COROLLARY 2. *The formula is not valid in S_5 of Lewis (for every $\alpha, \beta, \gamma \in For_{S5}$).*

In fact, da Costa, Dubikajtis and Kotas built a system that differs from D_2 in several respects. Whereas Jaśkowski introduced the notion of the *right* discursive conjunction they coped with the *left* one, viz. $\tau(\alpha \wedge_d \beta) = \Diamond\tau(\alpha) \wedge \tau(\beta)$. This may seem, at first sight, superficial since $(p \wedge_d q) \rightarrow_d (q \wedge_d p)$ is valid in both of the systems, but it is a little deceptive. Let us have a closer look at the formulas:

1. $\sim (p \wedge_d q) \rightarrow_d \sim (q \wedge_d p)$

2. $p \rightarrow_d \sim (\sim p \wedge_d \sim q)$

3. $p \rightarrow_d \sim (\sim q \wedge_d \sim p)$.

Observe that the formula (1) is neither valid in D_2 nor da Costa, Dubikajtis and Kotas' system (D_2^* for short). The formula (2) is valid in D_2, but is not in D_2^* while (3) is valid in D_2^*, but not in Jaśkowski's D_2.[5] On the other hand, it was not by chance or without good reason that Jaśkowski rejected the *left-right* discursive conjunction, i.e. $\tau(\alpha \wedge_d \beta) = \Diamond\tau(\alpha) \wedge \Diamond\tau(\beta)$. A simple calculation shows that if we accepted the conjunction we should also accept $(p \wedge_d q) \rightarrow_d (\sim (p \wedge_d q) \rightarrow_d r)$ as a thesis.

2   *Kripke-type* Semantics for D_2

A frame (D_2-frame) is a pair $\langle W, R \rangle$, where W is a non-empty set of points and R is the equivalence relation on W. By a model (D_2-model) we mean a triple $\langle W, R, v \rangle$, where v is a mapping from propositional variables to sets of worlds, $v : var \Rightarrow 2^W$. The satisfaction relation $\models_m$ is inductively defined:

(var) $x \models_m p_i \Leftrightarrow x \in v(p_i)$ and $i \in N$
$(\sim)$ $x \models_m \sim \alpha \Leftrightarrow x \not\models_m \alpha$
$(\vee)$ $x \models_m \alpha \vee \beta \Leftrightarrow x \models_m \alpha$ or $x \models_m \beta$
$(\wedge_d)$ $x \models_m \alpha \wedge_d \beta \Leftrightarrow x \models_m \alpha$ and $\exists y \in W (xRy$ and $y \models_m \beta)$

---

[5] For more details, see [Ciuciura, 2005].

$(\rightarrow_d)$ $x \models_m \alpha \rightarrow_d \beta \Leftrightarrow$ if $\exists y \in W(xRy$ and $y \models_m \alpha)$ then $x \models_m \beta$.

A formula α is valid in D_2, $\models \alpha$ for short, iff for any model $\langle W, R, v \rangle$, for every $x \in W$, there exists $y \in W$ such that: xRy and $y \models_m \alpha$.

The discursive equivalence is introduced as an abbreviation

$$\alpha \leftrightarrow_d \beta = (\alpha \rightarrow_d \beta) \wedge_d (\beta \rightarrow_d \alpha).$$

Additionally, we can eliminate the discursive implication from the system:

$$\alpha \rightarrow_d \beta = \sim ((p_1 \vee \sim p_1) \wedge_d \alpha) \vee \beta.$$

COROLLARY 3. $\forall \alpha \in For_{D2}$: $\models \alpha \Leftrightarrow \alpha \in D_2$ $(\Leftrightarrow \Diamond \tau(\alpha) \in S_5)$.

Proof. By induction. ■

Notice that Corollary 3 enables us to establish a link between the translation rules and our semantics. The translation procedure became redundant and we succeeded in constructing a new (*direct*) semantics for D_2.

The accessibility relation defined on D_2-frame is reflexive, symmetric and transitive. Any world is accessible from any other. This implies that we may treat a model (D_2-model) as a pair $\langle W, v \rangle$, where W is a non-empty set (of points, possible worlds, etc.) and v is a function that each pair consisting of a formula and a point assigns an element of $\{1, 0\}$, $v : For_{D2} \times W \Rightarrow \{1, 0\}$, defined as follows:

$(\sim)$ $v(\sim \alpha, x) = 1 \Leftrightarrow v(\alpha, x) = 0$
$(\vee)$ $v(\alpha \vee \beta, x) = 1 \Leftrightarrow v(\alpha, x) = 1$ or $v(\beta, x) = 1$
$(\wedge_d)$ $v(\alpha \wedge_d \beta, x) = 1 \Leftrightarrow v(\alpha, x) = 1$ and $\exists y \in W(v(\beta, y) = 1)$
$(\rightarrow_d)$ $v(\alpha \rightarrow_d \beta, x) = 1 \Leftrightarrow \forall y \in W(v(\alpha, y) = 0)$ or $v(\beta, x) = 1$.

Now we can simplify the notion of a valid sentence:

$\models \alpha \Leftrightarrow$ in any model $\langle W, v \rangle$, $\exists y \in W(v(\alpha, y) = 1)$.

3 New Axiomatization of D_2

In this section, we present a new axiomatization of D_2 using the discursive connectives which *directly* occur in a set of axiom schemata.

(A_1) $\alpha \rightarrow_d (\beta \rightarrow_d \alpha)$
(A_2) $(\alpha \rightarrow_d (\beta \rightarrow_d \gamma)) \rightarrow_d ((\alpha \rightarrow_d \beta) \rightarrow_d (\alpha \rightarrow_d \gamma))$
(A_3) $(\alpha \wedge_d \beta) \rightarrow_d \alpha$
(A_4) $(\alpha \wedge_d \beta) \rightarrow_d \beta$
(A_5) $(\alpha \rightarrow_d \beta) \rightarrow_d ((\alpha \rightarrow_d \gamma) \rightarrow_d (\alpha \rightarrow_d (\beta \wedge_d \gamma)))$
(A_6) $\alpha \rightarrow_d (\alpha \vee \beta)$
(A_7) $\beta \rightarrow_d (\alpha \vee \beta)$

(A_8) $(\alpha \rightarrow_d \gamma) \rightarrow_d ((\beta \rightarrow_d \gamma) \rightarrow_d ((\alpha \vee \beta) \rightarrow_d \gamma))$
(A_9) $\sim (\alpha \wedge_d \sim \alpha)$
(A_{10}) $(\alpha \vee \sim \alpha) \rightarrow_d \sim (\beta \wedge_d \sim (\alpha \vee \sim \alpha) \wedge_d \sim (\beta \vee \alpha \vee \sim \alpha))$
(A_{11}) $\sim (\sim \alpha \wedge_d \sim \beta \wedge_d \sim (\alpha \vee \beta)) \rightarrow_d \sim (\sim \alpha \wedge_d \sim \beta \wedge_d \sim \gamma \wedge_d \sim (\alpha \vee \beta \vee \gamma))$
(A_{12}) $\sim (\sim \alpha \wedge_d \sim \beta \wedge_d \sim \gamma \wedge_d \sim (\alpha \vee \beta \vee \gamma)) \rightarrow_d$
$\qquad \sim (\sim \alpha \wedge_d \sim \gamma \wedge_d \sim \beta \wedge_d \sim (\alpha \vee \gamma \vee \beta))$
(A_{13}) $\sim (\sim \alpha \wedge_d \sim \beta \wedge_d \sim (\alpha \vee \beta)) \rightarrow_d ((\alpha \vee \sim \beta) \rightarrow_d \alpha)$
(A_{14}) $\sim (\sim \alpha \wedge_d \sim \beta \wedge_d \sim \gamma \wedge_d \sim (\alpha \vee \beta \vee \gamma)) \rightarrow_d ((\alpha \vee \beta \vee \sim \gamma) \rightarrow_d (\alpha \vee \beta))$
(A_{15}) $\sim (\alpha \wedge_d \beta) \rightarrow_d (\sim \alpha \vee \sim \beta)$
(A_{16}) $(\alpha \vee \sim\sim \beta) \rightarrow_d (\alpha \vee \beta)$
(A_{17}) $(\alpha \vee \beta) \rightarrow_d (\alpha \vee \sim\sim \beta)$.

The sole rule of inference is *Detachment*

$(MP)^*$ α, $\alpha \rightarrow_d \beta$ / β

The consequence relation $\vdash_{D2}$ is determined by the set of axioms and $(MP)^*$.

THEOREM 4. $\Phi \vdash_{D2} \alpha \rightarrow_d \beta \Leftrightarrow \Phi \cup \{\alpha\} \vdash_{D2} \beta$, where $\alpha, \beta \in For_{D2}$, $\Phi \subseteq For_{D2}$.

Proof. Standard.   ∎

COROLLARY 5. *The formulas listed below are provable in* D_2:

(T_1) $(\alpha \vee \alpha) \leftrightarrow_d \alpha$
(T_2) $(\alpha \vee \beta) \leftrightarrow_d (\beta \vee \alpha)$
(T_3) $((\alpha \vee \beta) \vee \gamma) \leftrightarrow_d (\alpha \vee (\beta \vee \gamma))$
(T_4) $(\beta \vee \alpha \vee \beta) \leftrightarrow_d (\alpha \vee \beta)$
(T_5) $(\alpha \vee (\beta \wedge_d \gamma)) \leftrightarrow_d ((\alpha \vee \beta) \wedge_d (\alpha \vee \gamma))$
(T_6) $\sim (\sim \alpha \wedge_d \sim \beta) \rightarrow_d (\alpha \vee \beta)$
(T_7) $\alpha \vee \sim \alpha$
(T_8) $\sim (\sim \alpha \wedge_d \sim \beta \wedge_d \sim (\alpha \vee \beta)) \rightarrow_d (\sim (\sim \alpha \wedge_d \sim\sim \beta \wedge_d \sim (\alpha \vee \sim \beta)) \rightarrow_d \alpha)$

and the set of $\{\alpha : \vdash_{D2} \alpha\}$ *is closed under the rules:*

(R_1) α, β / $\alpha \wedge_d \beta$
(R_2) $\alpha \wedge_d \beta$ / α (β)
(R_3) α (β) / $\alpha \vee \beta$.

Proof. We prove (T_1) - (T_5) in much the same way as it is in the (positive) classical case. In view of the deduction theorems (T_6) and (T_8) are established by:

1. $\sim (\sim \alpha \wedge_d \sim \beta)$ premise
2. $\sim\sim \alpha \vee \sim\sim \beta$ (A_{15}), 1 and $(MP)^*$
3. $\sim\sim \alpha \vee \beta$ (A_{16}), 2 and $(MP)^*$
4. $\beta \vee \sim\sim \alpha$ (T_2), 3 and $(MP)^*$
5. $\beta \vee \alpha$ (A_{16}), 4 and $(MP)^*$
6. $\alpha \vee \beta$ (T_2), 5 and $(MP)^*$.

1. $\sim (\sim \alpha \wedge_d \sim \beta \wedge_d \sim (\alpha \vee \beta))$ premise
2. $\sim (\sim \alpha \wedge_d \sim\sim \beta \wedge_d \sim (\alpha \vee \sim \beta))$ premise
3. $\alpha \vee \sim \beta \vee (\alpha \vee \sim \beta)$ (T_6), 2, and $(MP)^*$
4. $\alpha \vee \sim \beta$ (T_1), 3 and $(MP)^*$
5. α (A_{13}), 1, 4 and $(MP)^*$.

The sequence of formulas

1. $\sim (\alpha \wedge_d \sim \alpha)$ (A_9)
2. $\sim \alpha \vee \sim\sim \alpha$ (A_{15}), 1 and $(MP)^*$
3. $\sim \alpha \vee \alpha$ (A_{16}), 2 and $(MP)^*$
4. $\alpha \vee \sim \alpha$ (T_2), 3 and $(MP)^*$.

constitutes a proof of (T_7).

(R_1) - (R_3) are obvious due to (A_6), (A_5), (A_4), (A_7), (A_8) and $(MP)^*$. ∎

COROLLARY 6. *The axiom schemata* (A_1) - (A_{17}) *become theses of the classical propositional calculus,* CPC, *after replacing in each of the schemata all the discursive connectives with their classical counterparts, i.e.* $\to_d / \to$ *and* $\wedge_d / \wedge$. *The rule* $(MP)^*$ *becomes an admissible rule of* CPC *after replacing* $\to_d$ *with* $\to$.

Let $(D_2) = \{\alpha : \vdash_{(D2)} \alpha\}$ denote the system described in Corollary 6 and $CPC = \{\alpha : \vdash_{CPC} \alpha\}$.

COROLLARY 7. $(D_2) \subset CPC$.

4   Soundness and Completeness

THEOREM 8 (Soundness). $\vdash_{D2} \alpha \Rightarrow \models \alpha$.

Proof. By induction. ∎

THEOREM 9 (Completeness). $\models \alpha \Rightarrow \vdash_{D2} \alpha$.

The initial idea of the proof traces back to [Reichbach, 1957]. Also, the observation that D_2^+ coincides with CPC^+ will be of great help.[6]

---

[6] D_2^+ (CPC^+) is an abbreviation for the positive part of D_2 (CPC).

Proof. Assume that $\nvdash_{D2} \alpha$ and $\models \alpha$. Define a sequence of all the formulas of D_2 as follows:

$\Gamma = \gamma_1, \gamma_2, \gamma_3, \ldots$ \qquad where $\gamma_1 = \alpha$

Next define a family of (finite) subsequences of Γ:

$\Delta_1 = \delta_1$ \qquad where $\delta_1 = \gamma_1 = \alpha$

$\Delta_2 = \delta_1, \delta_2$ \qquad where $\delta_1 = \gamma_1 = \alpha$ and $\delta_2 = \gamma_2$ if $\nvdash_{D2} \delta_1 \vee \delta_2$; otherwise $\delta_2 \neq \gamma_2$ and $\Delta_2 = \Delta_1$

$\vdots$

$\Delta_i = \delta_1, \delta_2, \delta_3$ \qquad for $i \geq 3$, where $\delta_1 = \alpha$, $\delta_2 = \gamma_k$ and $\delta_3 = \gamma_{k+1}$ if $\nvdash_{D2} \delta_1 \vee \delta_2 \vee \delta_3$; otherwise $\delta_3 \neq \gamma_{k+1}$ and $\Delta_i = \Delta_2$

$\vdots$

$\Delta_n = \delta_1, \delta_2, \delta_3, \ldots, \delta_n$

$\vdots$

Finally define:

$\nabla_1 = \Delta_1, \Delta_2, \Delta_3, \ldots$
$\nabla_2 = \Delta_2, \Delta_3, \Delta_4, \ldots$
$\nabla_3 = \Delta_3, \Delta_4, \Delta_5, \ldots$
$\vdots$
$\nabla_n = \Delta_n, \Delta_{n+1}, \Delta_{n+2}, \ldots$
$\vdots$

From now on we use ∇_i, where $i \in \{1, 2, 3, \ldots\}$, to denote both the i-sequence and the set of formulas which contains all the elements of the i-sequence. Let ∇ stand for $\{\nabla_1, \nabla_2, \ldots, \nabla_i, \ldots, \nabla_n, \ldots\}$. ∎

LEMMA 10.

(i) $\nvdash_{D2} \delta_1 \vee \ldots \vee \delta_n$, for any $n \in N$.

(ii) if $\beta \notin \nabla_i$, then $\vdash_{D2} \delta_1 \vee \ldots \vee \delta_k \vee \beta$, for some $k \in N$.

Proof. (i)–(ii) apply the definition of ∇_i, where $i \in \{1, 2, 3, \ldots\}$. ∎

DEFINITION 11. $\nabla_i \mathbf{R} \nabla_k \Leftrightarrow (\nabla_i = \nabla_k)$, for every $\nabla_i, \nabla_k \in \nabla$.

LEMMA 12. $\mathbf{R}$ *is the equivalence relation on* ∇.

Proof. Immediately from Definition 11. ∎

In Section 2, we mentioned that the connectives of $\leftrightarrow_d$ and $\rightarrow_d$ were redundant. This fact simplifies a proof of the next lemma.

LEMMA 13. *For every $\beta, \gamma \in For_{D2}$, for any $\nabla_i, \nabla_k \in \nabla$:*

(i) $\beta \vee \gamma \in \nabla_i \;\Leftrightarrow\; \beta \in \nabla_i$ *and* $\gamma \in \nabla_i$

(ii) $\beta \wedge_d \gamma \in \nabla_i \;\Leftrightarrow\; \beta \in \nabla_i$ *or* $\forall_{\nabla_k \in \nabla} (\nabla_i R \nabla_k \Rightarrow \gamma \in \nabla_k)$.

Proof. (i)$\Rightarrow$. Assume $\beta \vee \gamma \in \nabla_i$ and $\{\beta \notin \nabla_i$ or $\gamma \notin \nabla_i\}$. Apply Lemma 10(ii) to $\beta \notin \nabla_i$ (or $\gamma \notin \nabla_i$ as the second possibility). Then use as many times as necessary (R_3), (T_1), (T_2), (T_3), (T_4) and $(MP)^*$ to finally obtain $\vdash_{D2} \delta_1 \vee \ldots \vee \delta_i \vee (\beta \vee \gamma)$, for some $i \in N$. But $\delta_1, \ldots \delta_i, (\beta \vee \gamma) \in \nabla_i$. This is in contradiction to Lemma 10(i).
(i)$\Leftarrow$. Suppose that $\beta \in \nabla_i$, $\gamma \in \nabla_i$ and $\beta \vee \gamma \notin \nabla_i$. Make use of Lemma 10(ii) to deduce $\vdash_{D2} \delta_1 \vee \ldots \vee \delta_k \vee (\beta \vee \gamma)$, for some $k \in N$, and next (T_2), (T_3), $(MP)^*$ to be in contradiction to Lemma 10(i).
(ii)$\Rightarrow$. Let $\beta \wedge_d \gamma \in \nabla_i$, $\beta \notin \nabla_i$ and $\exists_{\nabla_k \in \nabla} (\nabla_i R \nabla_k$ and $\gamma \notin \nabla_k)$. Apply Definition 11, Lemma 10(ii), (R_3), (T_1), (T_2), (T_3), (T_4), (R_1), (T_5) and $(MP)^*$ to obtain $\vdash_{D2} \delta_1 \vee \ldots \vee \delta_i \vee (\beta \wedge_d \gamma)$. Note that $\delta_1, \ldots \delta_i, (\beta \wedge_d \gamma) \in \nabla_i$, for some $i \in N$. This is in contradiction to Lemma 10(i).
(ii)$\Leftarrow$. Suppose $\{\beta \in \nabla_i$ or $\forall_{\nabla_k \in \nabla} (\nabla_i R \nabla_k \Rightarrow \gamma \in \nabla_k)\}$ and $\beta \wedge_d \gamma \notin \nabla_i$. If $\beta \in \nabla_i$ and $\beta \wedge_d \gamma \notin \nabla_i$, use Lemma 10(ii), (T_5), (R_2), $(MP)^*$ to be in contradiction to Lemma 10(i). If $\forall_{\nabla_k \in \nabla} (\nabla_i R \nabla_k \Rightarrow \gamma \in \nabla_k)$ and $\beta \wedge_d \gamma \notin \nabla_i$, make use of **R** and proceed analogously to the previous case.   ■

The proof of completeness additionally requires a definition and two lemmas.

Let ∇_i be a sequence and $i \in \{1, 2, 3, \ldots\}$. Then

$$\nabla_i^\star = \delta_1^\star, \ldots, \delta_i^\star, \ldots$$

where (a) $\delta_1^\star = \delta_1 = \gamma_1 = \alpha$
(b) $(\delta_n = \delta_k^\star)$ if $\nvdash_{D2} \sim (\sim \delta_1^\star \wedge_d \ldots \sim \delta_k^\star \wedge_d \sim (\delta_1^\star \vee \ldots \vee \delta_k^\star))$, for every $\delta_n \in \nabla_i$, $k, n \in N$ and $n \geq k$; otherwise $\delta_n \neq \delta_k^\star$.

DEFINITION 14. We call a formula β *discursive* if it contains at least one discursive connective. A formula β is a *discursive thesis* if it is a thesis and discursive.

LEMMA 15.

(i) $\nabla_i^\star \subseteq \nabla_i$, *for every* $i \in \{1, 2, 3, \ldots\}$

(ii) $\nvdash_{D2} \sim (\sim \delta_1^\star \wedge_d \ldots \wedge_d \sim \delta_n^\star \wedge_d \sim (\delta_1^\star \vee \ldots \vee \delta_n^\star))$, *for every* $n \in N$

(iii) if β is not a discursive thesis, $\beta \notin \nabla_i$, then $\vdash_{D2} \sim (\sim \delta_1^\star \wedge_d \ldots \wedge_d \sim \delta_k^\star \wedge_d \sim \beta \wedge_d \sim (\delta_1^\star \vee \ldots \vee \delta_k^\star \vee \beta))$, for some $k \in N$.

Proof. (i)–(ii) immediately from the definitions of $\nabla_i^\star$, where $i \in \{1, 2, 3, \ldots\}$ (iii) from the characteristic of ∇_i ($\nabla_i^\star$), the fact that $\nabla_i^\star \subseteq \nabla_i$, for every $i \in \{1, 2, \ldots\}$, (A_{10}), (A_{11}), (A_{12}) and $(MP)^\star$. ∎

LEMMA 16. *For every $\beta \in For_{D2}$, for any $\nabla_i \in \nabla$:*
(i) $\sim \beta \in \nabla_i \Leftrightarrow \beta \notin \nabla_i$.

Proof. (i)$\Rightarrow$. Let (1) $\sim \beta \in \nabla_i$ and (2) $\beta \in \nabla_i$. It implies, by Lemma 10(i), that $\nvdash_{D2} \beta \vee \sim \beta$. A contradiction to (T_7).
(i)$\Leftarrow$. Assume (1) $\beta \notin \nabla_i$ and (2) $\sim \beta \notin \nabla_i$. Apply Lemma 10(ii) to (1) and (2), to obtain (3) $\vdash_{D2} \delta_1 \vee \ldots \vee \delta_m \vee \sim \beta$, for some $m \in N$, (4) $\vdash_{D2} \delta_1 \vee \ldots \vee \delta_p \vee \beta$, for some $p \in N$. Evidently, $m \geq p$ or $m < p$. Suppose that $m \geq p$ (the case $m < p$ is similar to $m \geq p$). If $\beta \notin \nabla_i$, $\sim \beta \notin \nabla_i$ and $\nabla_i^\star \subseteq \nabla_i$, then (5) $\beta \notin \nabla_i^\star$, (6) $\sim \beta \notin \nabla_i^\star$.
 We need to show three subcases:

(a)   neither β nor $\sim \beta$ is a *discursive thesis*

(b) β is a *discursive thesis*, but $\sim \beta$ is not a *discursive thesis*

(c) $\sim \beta$ is a *discursive thesis*, but β is not a *discursive thesis*.

Observe that the fourth subcase (both β and $\sim \beta$ is a *discursive thesis*) is impossible due to *Soundness*.
 Subcase (a). Let $m = 1$. Then (7) $\vdash_{D2} \sim (\sim \delta_1^\star \wedge_d \sim \beta \wedge_d \sim (\delta_1^\star \vee \beta))$ and (8) $\vdash_{D2} \sim (\sim \delta_1^\star \wedge_d \sim\sim \beta \wedge_d \sim (\delta_1^\star \vee \sim \beta))$ by Lemma 15(iii), (2), (1). Next use (T_8) and $(MP)^\star$ to (7) and (8), to produce $\vdash_{D2} \delta_1^\star$. But $\delta_1^\star = \delta_1 = \gamma_1 = \alpha$. A contradiction.
 Let $m > 1$. Then (7)' $\vdash_{D2} \sim (\sim \delta_1^\star \wedge_d \ldots \sim \delta_k^\star \wedge_d \sim \beta \wedge_d \sim (\delta_1^\star \vee \ldots \vee \delta_k^\star \vee \beta))$, for some $k \in N$, by Lemma 15(iii) and (1). If $k < m$, apply (A_{11}), (A_{12}) and $(MP)^\star$. It yields (8)' $\vdash_{D2} \sim (\sim \delta_1^\star \wedge_d \ldots \sim \delta_m^\star \wedge_d \sim \beta \wedge_d \sim (\delta_1^\star \vee \ldots \vee \delta_m^\star \vee \beta))$, where $\delta_1^\star = \delta_1$, $\delta_2^\star = \delta_2$, ..., $\delta_m^\star = \delta_m$. Take (8)', (3), (A_{14}) and $(MP)^\star$, to obtain (9)' $\vdash_{D2} \delta_1 \vee \ldots \vee \delta_m$. But δ_1, ... , $\delta_m \in \nabla_i$. A contradiction due to Lemma 10(i).
 We prove the subcases (b) and (c) in a very similar way. However, we are not allowed to apply Lemma 15(iii) to $\beta \notin \nabla_i$, in Subcase (b), and to $\sim \beta \notin \nabla_i$, in Subcase (c).
Now consider a canonical model for $\vdash_{D2}$. Let $\langle \nabla, \mathbf{R}, v_c \rangle$ be such a model. The canonical valuation $v_c : For_{D2} \times \nabla \Rightarrow \{1, 0\}$ is defined:

$$v_c(\beta, \nabla_i) = \left\{ \begin{array}{ll} 1, & \text{if } \beta \notin \nabla_i \\ 0, & \text{if } \beta \in \nabla_i \end{array} \right.$$

We have to show that the conditions $(\vee)$, $(\wedge_d)$ and $(\sim)$ hold for v_c, which is obvious due to Lemmas 13 and 16.

Recall $\nvdash_{D2} \alpha$ and $\models \alpha$. Notice that the formula α is the first element of each i-sequence, where $i \in \{1, 2, 3, \ldots\}$. Since $\alpha \in \nabla_i$, then the formula α is not valid in $\langle \nabla, \mathbf{R}, v_c \rangle$ and $\nvDash \alpha$. A contradiction. ∎

Acknowledgements

I wish to express my gratitude to one of the anonymous Referees for useful suggestions and comments that were helpful to improve the presentation of the results in this paper.

BIBLIOGRAPHY

[Achtelik *et al.*, 1981] G. Achtelik, L. Dubikajtis, E. Dudek, J. Kanior, *On Independence of Axioms of Jaśkowski Discussive Propositional Calculus*, **Reports on Mathematical Logic**, 11, 1981, pp. 3–11.

[Ciuciura, 1999] J. Ciuciura, *History and Development of the Discursive Logic*, **Logica Trianguli**, 3, 1999, pp. 3–31.

[Ciuciura, 2003] J. Ciuciura, *Logika dyskusyjna*, **Principia**, 35–36, 2003, pp. 279–291.

[Ciuciura, 2004] J. Ciuciura, *Labelled Tableaux for D_2*, **Bulletin of the Section of Logic**, 33(4), 2004, pp. 223–236.

[Ciuciura, 2005] J. Ciuciura, *On the da Costa, Dubikajtis and Kotas' system of the discursive logic, D_2^**, **Logic and Logical Philosophy**, 15, 2005, pp. 99–116.

[Da Costa and Dubikajtis, 1977] N.C.A. da Costa, L. Dubikajtis, *New Axiomatization for the Discussive Propositional Calculus*, in: A.I. Arruda, N.C.A. da Costa, R. Chuaqui (ed): **Non Classical Logics, Model Theory and Computability**, North-Holland Publishing, Amsterdam 1977, pp. 45–55.

[Jaśkowski, 1948] S. Jaśkowski, *Rachunek zdań dla systemów dedukcyjnych sprzecznych*, **Societatis Scientiarum Torunensis**, Sect. A, I, 5, 1948, pp. 57–77. First English translation: *Propositional Calculus for Contradictory Deductive Systems*, **Studia Logica** 24, 1969, pp. 143–157. Second one: *A Propositional Calculus for Inconsistent Deductive Systems*, **Logic and Logical Philosophy**, 7(1), 1999, pp. 35–56.

[Jaśkowski, 1949] S. Jaśkowski, *O koniunkcji dyskusyjnej w rachuneku zdań dla systemów dedukcyjnych sprzecznych*, **Societatis Scientiarum Torunensis**, Sect. A, I, 8, 1949, pp. 171–172. English translation: *On the Discussive Conjuntion in the Propositional Calculus for Inconsistent Deductive Systems*, **Logic and Logical Philosophy**, 7(1), 1999, pp. 57–59.

[Kotas, 1975] J. Kotas, *Discussive Sentential Calculus of Jaśkowski*, **Studia Logica**, 34(2), 1975, pp. 149–168.

[Kotas and da Costa, 1977] J. Kotas, N.C.A. da Costa, *On Some Modal Logical Systems Defined in Connexion with Jaśkowski's Problem*, in: A.I. Arruda, N.C.A. da Costa, R. Chuaqui (ed): **Non Classical Logics, Model Theory and Computability**, North-Holland Publishing, Amsterdam 1977, pp. 57–73.

[Reichbach, 1957] J. Reichbach, *On the first-order functional calculus and the truncation of models*, **Studia Logica**, 7, 1957, pp. 181–220.

Part V

Philosophical Aspects of Paraconsistent Logic

Reasoning and Modelling: Two Views of Inconsistency Handling

Paul Wong

ABSTRACT. Proponents of paraconsistent logics usually motivate *paraconsistency* by appealing to the need to reason with inconsistent theories in a non-explosive manner. The aim of this paper is to provide an alternative motivation for paraconsistency. Along the way we'll examine some of the methodological and foundational assumptions behind the use of paraconsistent logics for handling inconsistent information. Our interest here is partly methodological and partly conceptual – we want to distinguish between different senses of *using a logic* and different senses of *inconsistent information*. From the logic end, we argue that logic is and should be concerned with more than just inferences. From the inconsistencies handling end, we argue that paraconsistency is and should be concerned with more than non-explosiveness.

1 Introduction

The study of logic usually begins with one of two approaches. According to what Priest [Priest, 2001] calls the *canonical* approach, the aim of logic is to establish a standard for evaluating arguments – a standard by which we judge whether a conclusion can be legitimately inferred from a body of assumptions.[1] The legitimacy of an inference turns on the notion of a consequence relation which can be defined proof theoretically in terms of deduction or semantically in terms of class containment of models. Legitimate or valid inferences are those that are sanctioned by our consequence relation specified in standard proof theoretic or model theoretic ways. The usual completeness theorem of a logic is in turn an assurance that the proof theory and the semantics capture one and the same consequence relation.

According to the *representational* approach however, logic is understood as the study of the relationship between a formal language and its associated domains. It addresses issues concerning *how* to express and *what* can

---

[1]We shall set aside issues concerning formal vs informal logic. By a 'logic', we mean a logic with some underlying formal language.

be expressed in a formal language. Although the two approaches set out different aims, they are clearly related. Amongst the sort of things we want a formal language to be able to express are *declarative* sentences (in contrast to *imperative* sentences, e.g. `goto s` in some programming languages). The content of these declarative sentences is fixed by their *truth conditions* which in turn inform us that entities or states in the domain are in one way but not another. The availability of sentences bearing truth conditions allows us to be in the business of reasoning and inference again. If a body of declarative sentences truthfully represents the domain, we can infer further truthful sentences about the domain.[2]

Given these two approaches to logic, it is not surprising that paraconsistent logics are typically motivated in one of two ways. According to the epistemic account declarative sentences in a formal language can be used to represent states of a domain, in particular they can be used to represent states of the actual world. We may think of these representations as logical descriptions with empirical contents. Although we are the masters of our own language, infallibility is not a given. We make mistakes and some of them turn up as inconsistencies in our data and theories.[3] According to this view certain inconsistencies are just misrepresentations. Since logic is about consequence, it is the logician's business to sort out what can be deduced from these misrepresentations. Classical logic is of no help here since it does not distinguish between different sorts of mistakes and hence all sorts of mistakes can be inferred. Adopting alternative logics is one way we can continue to draw inferences under the threat of erroneous data.

According to the ontological account however, not all inconsistent descriptions are infected with errors or misrepresentations. Instead an alternative hypothesis is that certain inconsistent descriptions are just *correct* descriptions of entities or states that are inconsistent in and of themselves. As Priest would say, some inconsistent information or theories are *true* [Priest, 2002]. According to the dialetheic thesis the recurrence of certain paradoxical statements is not to be explained away in terms of mistakes on our part or defects in our language. Instead *the best explanation* of the persistence of these paradoxes is that they truly describe an ontology populated with

---

[2]We use the more neutral *declarative sentence* instead of *proposition*. Some may insist that it is *propositions* that are being expressed and that a consequence relation is defined over *propositions*. These are complicated issues which we cannot address fully here. But if it is of any comfort, we may tentatively identify *propositions* as declarative sentences with truth conditions.

[3]As Wheeler pointed out in [Wheeler, 2000], some of these mistakes, e.g. measurement errors, are so fundamental to the way we interact with the world that any attempt to eliminate them is a practical impossibility. Our scientific theories and thus scientific reasoning must face up to the force of inconsistencies.

inconsistent entities. The implication of the ontological or dialetheic account is that just as we need alternative logics to reason with potentially erroneous data or theories, we need alternative logics to reason with inconsistent entities and states. Once again, classical logic is of no help here since it provides no provision to deal with inconsistent entities or states.

While dialetheism, the hypothesis that there are true inconsistent theories, is a contentious claim, the aim of this paper is neither to defend nor to criticise this view. We take it as given that the epistemic account is a plausible motivation for paraconsistency. Like the epistemic account, we are interested in languages whose domains are constrained by the physical world. But we would like to motivate paraconsistency from a more practical standpoint. In particular, we would like to consider cases where the source of inconsistencies is neither rooted in errors nor in inconsistent ontologies. In drawing out these cases, our emphasis is on taking a logic as a formalism for representation. Seen in this light, inference is one of many functions a logic can deliver. A logic can also be extremely useful in *modelling* practical problems as well as structuring concepts and thoughts. The lesson to be drawn is that there is more to being a paraconsistent logic than being a non-explosive inference engine.

2   Reasoning with Inconsistencies as Reasoning with Misrepresentation

We'll begin with an imaginary scenario. Consider a situation in which a stationary physical object O may be located in one out of nine distinct possible locations. Information about the location of O is encoded in a simple language with p's and q's representing the coordinates of the object (see figure (1)). Complex expressions are generated using the usual connectives, $\{\neg, \wedge, \vee\}$ with their usual truth conditions. We are interested in locating O and information is gathered from various sources about the location of O.[4]

	q_1	q_2	q_3
p_1	×		×
p_2			
p_3			

Figure 1. A simple logical representation of an object's location.

---

[4]This is known as the information fusion problem within the AI community. For a highly non-trivial information fusion problem involving movement analysis, see The DARPA High-Performance Knowledge Bases Project described in [Cohen *et al.*, 1998].

Suppose we receive two messages:

$$A : p_1 \qquad\qquad B : \neg q_2$$

From the received messages we conclude that the possible location of O is:

$$C : (p_1 \wedge q_1) \vee (p_1 \wedge q_3)$$

But now suppose that we receive a further message:

$$D : p_3$$

Our example highlights several important methodological points. The first is the obvious point that information about the state of the world can be encapsulated in a formal language. The practical corollary of this is that more expressive formal languages are required for more demanding representational tasks. But more importantly, since a more expressive language may involve a greater computational cost, the choice of language should be gauged in terms of the representational task at issue. In our example it is clear that a simple propositional language suffices for the representational task.

The second methodological point is that contextual information is often crucial to a reasoning task. In our example, the background information is that the object O is located in exactly one and no more than one location, and that there are exactly nine possible locations of O. It is only in the context of this background information that we can deduce C from A and B. Some background information goes even further, it may include a theory which helps us interpret and fix the content of data or observations. That's why sometimes we insist that data or observations are theory laden. Without these background theories, we would not be able to register the noise in the air or ink mark on the page as genuine data.

Thirdly, our example illustrates how the process of reasoning can be viewed as exploration in the space of possibilities – eliminating some and further exploring others. Our symbolic representations A and B impose certain restrictions on the space of possibilities. The truth conditions of these expressions provide the basis for us to eliminate certain possibilities and thereby infer the possible location of O. Note how we have tied reasoning, information and possibilities in one conceptual swoop. If reasoning is the process of elimination and exploration in the space of possibilities, then information is just the instructions for what to do and where to go in this space. Logical deduction in this account is the regimentation and codification of this process of reasoning. The conclusion C is simply what is possible relative to the restrictions imposed by A and B together with the background information.

Finally and most importantly for us, our example illustrates and explains why inconsistencies (sometimes) interact badly with reasoning. Since our background information is that the nine locations are distinct and no physical object can be at different locations at the same time, A and D are inconsistent. While the information given by A tells us to eliminate p_2 and p_3, D tells us to eliminate p_1 and p_2, so we end up eliminating *all* possibilities. Inconsistencies in this case allow us to do too much – they wipe out the entire space of possibilities.

Our example is set up in such a way that the inconsistency between A and D is understood as a result of misrepresentation. The implicit assumption here is that the entities or states in the represented domain are not inconsistent. So no further explanation is required to account for the inconsistency – our data simply misrepresents the domain. Under normal circumstances the correct action is to identify the error and revise our data. But this is not always feasible. There may be insufficient information to identify the error or there may be hidden costs involved in revision. In some cases, the cost may simply outweigh the gain. Thus the use of paraconsistent logic is sometimes justified on grounds of instrumental rationality. In the absence of further information to guide revision and in circumstances in which we need to continue to reason without being undermined by inconsistencies, we need a *fault tolerant* logic – it must be able to operate under the explicit assumption that any given piece of information may be erroneous. Note however that this sort of instrumental justification of a paraconsistent logic carries weight only if its (expected) benefit is demonstrably greater than its (expected) cost. In cases where safety is a critical concern, there may be no option but to revise.

Assuming then that there are circumstances in which the option to revise is instrumentally unsound, there remains the question why paraconsistent logics should be adopted as the underlying inference engine from the epistemic point of view – i.e. from the point of view of minimizing errors. The usual move is to appeal to the fact that paraconsistent logics are nonexplosive. Classical logic in contrast is explosive – anything can be deduced from any inconsistent collection of assumptions. So paraconsistent logics trump classical logic on the grounds that they don't make things worse. While this sort of justification does give paraconsistent logics *some* credentials and cash value, it does not address our question fully. An adequate cost-benefit analysis of a logic cannot be given solely in terms of a property that a logic *fails* to have. Consider our reaction to a doctor's recommendation to consume plenty of water on the grounds that water is *not poisonous*.

Many things are not poisonous but they hardly count as remedies. The fact that paraconsistent logics are non-explosive may put them one up on classical logic, but this does not provide a sufficient explanation as to why paraconsistent logics are beneficial to handling inconsistencies *qua* errors and misrepresentation. To put the matter differently, if inconsistencies are understood as errors, then in the absence of further information to root out the errors or to revise, any inference, classical or not, is risky. What is needed then is a standard to assess this risk.

It is not our aim to challenge the claim that paraconsistent logics are useful fault-tolerant logics. But we do think that there is a problem lurking in the background. Our diagnosis of the problem lies in the standard characterization of paraconsistency as non-explosiveness. The notion of non-explosive inference seems too negative and too weak to do real theoretical work. In particular, we are confounded with a dilemma. On the one hand there is a proliferation of new paraconsistent logics, all of which claim to be inconsistency-tolerant and hence fault-tolerant.[5] On the other hand many standard and familiar logics get reinterpreted and become paraconsistent in the name of non-explosiveness. As Restall [Restall, 2003] pointed out recently, intuitionistic logic, orthologic and quantum logic are non-explosive with respect to a certain class of classical inconsistencies. So these logics are all inconsistency-tolerant in some respects. Béziau goes even further in [Béziau, 2002] to point out that even the modal logic $S5$ and standard first order logic are in a sense paraconsistent. For some this is good news – paraconsistency is (unexpectedly) everywhere. Classicists and intuitionists have always been paraconsistentists unknowingly. But for others with a genuine need for a decent logic for working with error-infested data, they are left with a large catalog of logics, all of which come with some form of guarantee for non-explosiveness and thus claim to be superior over classical logic, but none of which offers an explanation why it is best suited for its purpose.

3   Modeling with Inconsistencies

If our assessment seems too negative and dismissive, it is not because we are unsympathetic to the cause of paraconsistency. Far from it, we would like to see paraconsistent logics gain wider acceptance. But our vision of what a paraconsistent logic should do is rather different from the epistemologically driven concern for fault-tolerant deduction. Instead we agree with the dialetheist that not every case of inconsistent description is a case of misrepresentation. In this section we want to consider cases where logically

---

[5]A quick search of recently published papers in conferences or journals relating to paraconsistent reasoning can confirm that.

inconsistent descriptions can be deployed in a way that can be said to be correctly representing certain sorts of problem space. Unlike the epistemic account however, the source of inconsistencies in these cases do not depend on errors or mistakes being made. And unlike the ontological account, insisting that our inconsistent descriptions are *true* adds nothing by way of explanation or understanding of these cases (hence we can't appeal to 'inference to the best explanation' to justify an inconsistent ontology).

3.1 Strategic Interactions

Strategic interactions such as group decision making, voting and negotiation are rich sources of conflicts where the preferences and goals of individuals often diverge. Depending on the type of interaction, conflict resolution may not be an achievable end. In a business transaction where the buyer and the seller cannot establish an agreed price all deals are off. These situations provide a plausible motivation for the use of logically inconsistent descriptions to model the underlying strategic conflicts. Consider the following imaginary scenario:

Larry, Harry and Barry want to share a pizza but cannot agree on the selection of toppings and the local pizza maker will not add a topping to only a part of a pizza, but to the entire pizza. So they put the matter to a *blind vote*. They each write on a separate piece of paper (without revealing their identities) their own preference for toppings. Moreover their preferences are 'flat', i.e. there is no ordering imposed on the toppings. They either prefer a particular topping to be on the pizza or prefer it not to be on the pizza. The resulting votes are as follows:

A:	pineapple	anchovies	pickled jalapeño
B:	no pineapple	anchovies	no pickled jalapeño
C:	pineapple	no anchovies	no pickled jalapeño

Furthermore let suppose that Larry, Harry and Barry all agree that pickled jalapeño goes well with both pineapple and anchovies, i.e. they all agree that a pizza should have pickled jalpeño if and only if it also has both pineapple and anchovies.

The preferences of these individuals can be encapsulated in a very simple propositional language using p, q and r to represent various 'pizza topping states' – 'p' for 'pineapple', 'q' for 'anchovies' and 'r' for 'pickled jalapeño'. The three sentences corresponding to their votes above are:

A':	$p \land q \land r$
B':	$\neg p \land q \land \neg r$
C':	$p \land \neg q \land \neg r$

The sentence which corresponds to their agreement is:

$$D' : \ r \leftrightarrow (p \wedge q)$$

We take it as uncontroversial that our sentences A', B', C' and D' do capture the salient features of the strategic interaction.[6] For instance, everyone disagrees with the others on 2 choices of toppings and there is no unanimous consensus on any one topping. Strategic interactions like the pizza voting scenario are in fact well known to social choice theorists. Within legal contexts, it is known as the doctrinal paradox, and more generally it is known as the discursive dilemma.[7] So called because the standard majority rule *fails* to resolve the inconsistencies of A', B', C' and D' – we have a majority for p, a majority for q, a majority of $\neg r$ and an unanimous agreement on $r \leftrightarrow (p \wedge q)$. In fact, majority rule is not the only procedure that would fail. Under some general conditions on how individual preferences or judgements may be *aggregated* into collective choice, List and Pettit [List and Pettit, 2002; List and Pettit, 2004] show that no aggregation function satisfying these general conditions would yield consistent collective choice in certain situations. More specifically let $\mathcal{V}$ denote the class of all classical truth assignments over propositional language generated by propositional atoms under the usual truth functional connectives and let $N = \{1, \ldots, n\}$ denote a set of $n \ (\geq 2)$ decision makers. An aggregation function is a map $\mathcal{A} : \mathcal{V}^n \longrightarrow \mathcal{V}$. A *choice method* is a function $\mathcal{C} : \ \{1, 0\}^n \longrightarrow \{1, 0\}$. Consider an arbitrary but fixed set of propositional formulae Γ and consider the following properties on an aggregation function $\mathcal{A}$:

Universal Domain : $\mathcal{A}$ has the universal domain property if $dom(\mathcal{A}) = \mathcal{V}^n$, i.e. $\mathcal{A}$ is a total function well defined on every element of $\mathcal{V}^n$.

Anonymity : For any permutation $f : N \longrightarrow N$, any element $\langle v_1, \ldots, v_n \rangle \in \mathcal{V}^n$, and any formula $A \in \Gamma$, $\mathcal{A}(\langle v_1, \ldots, v_n \rangle)(A) = \mathcal{A}(\langle v_{f(1)} \ldots v_{f(n)} \rangle)(A)$, i.e. $\mathcal{A}$ is invariant under all permutations of N for each $A \in \Gamma$ and

---

[6] As an aside, there are *substructural* logics (e.g. relevant and linear logics) in which the fundamental consequence relations are defined over *multisets* rather then sets. In multisets, the ordering of sentences does not matter, but the number of occurrences does, e.g. $[A, B, A]$ is the same multiset as $[A, A, B]$ but not $[A, B]$. This way of bookkeeping the number of sentence tokens (of the same type) clearly makes sense if we think of sentence tokens as *votes*.

[7] In the standard legal scenario, the puzzle involves a three-judge count deciding whether (1) a defendant is liable r, (2) whether a valid legal contract is in order p, and (3) whether the defendant has breached the contract q. All three judges agree with the legal doctrine that a person is liable if and only if there is a valid contract in order and the person has breached the contract. It is clear that our pizza voting scenario is exactly analogous to the standard legal scenario.

thus the identity of an individual plays no role the determination of the collective choice.

Systematicity : There exists a choice method $\mathcal{C}$ such that for any element $\langle v_1, \ldots, v_n \rangle \in \mathcal{V}^n$ and any $A \in \Gamma$, $\mathcal{A}(\langle v_1, \ldots, v_n \rangle)(A) = \mathcal{C}(\langle v_1(A), \ldots v_n(A) \rangle)$, i.e. $\mathcal{A}$ should determine the collective choice on A solely on the basis of the pattern of votes on A; moreover, if A, $B \in \Gamma$ are such that the pattern of votes on A and B are exactly the same, then $\mathcal{A}$ should aggregate their votes in exactly the way.

THEOREM 1. *(List and Pettit [List and Pettit, 2002]) If* $\{p, q, p \wedge q, \neg(p \wedge q)\} \subseteq \Gamma$, *then there is no aggregation function satisfying universal domain, anonymity, and systematicity.*

Theorem (1) naturally invites comparison with Arrow's impossibility theorem. In [List and Pettit, 2004], List and Pettit show that the general theoretical framework of preference orderings deployed to derive Arrow's impossibility theorem can be mapped into the framework of aggregation of judgements.[8] However, whether there is a reverse mapping remains an open question. Other important open questions involve finding escape routes to these impossibilities and finding alternative conditions on aggregation functions. Both universal domain and anonymity for instance are relatively uncontroversial. Systematicity however has been challenged by Chapman in [Chapman, 2002]. In a forthcoming paper [Pauly and van Hees, 2004], Pauly and van Hees make a number of suggestions to weaken systematicity. For our purposes here however, we would like to return the issue of using logic as a *representational language* to model a problem domain.

Our sentences A', B', C' and D' are simple but efficient encodings of the structural features of the pizza voting situation. We can literally read information directly from our representations. So the first important point is that, though inconsistent collectively, our representations do not involve any misrepresentation or error. In fact their being inconsistent is a reason for their being a correct representation of the underlying strategic conflict. However, the notion of correctness invoked here is *not* an epistemic one, i.e. these sentences are correct collectively not because they are *true* collectively (though they do have truth conditions). The notion of correctness is a pragmatic one – they are correct representations in so far as they allow

---

[8]We note that according to List and Pettit [List and Pettit, 2002], the crucial difference between aggregation of preference orderings and aggregation of judgements is that 'aggregating preference orderings arises with issues that are treated as *independent* ... [whereas] aggregating sets of judgements arsises with a number of issues that are treated as *interconnected* rather than *independent*, ... '.

us to encapsulate salient features that are of interest to us. The important point here is that the deployment of a representational language as a
modeling tool involves *abstraction* in the sense that while certain features
of the problem domain are brought into focus other irrelevant features are
ignored. Abstraction is a standard technique used in engineering disciplines
for modeling. A capacitor for instance can be modeled by a set of mathematical equations. These equations constitute an idealized and simplified
model of the capacitor which ignores details such as the colour or shape of
the capacity. From the point of view of a circuit designer these are irrelevant details. Our point here is that modern logic, like mathematics, is an
invaluable intellectual tool of abstraction – it provides a unifying theoretical framework to describe and model diverse situations, phenomena, and
systems. We note that in the pizza voting scenario the votes are casted
blindly, certain *indexical information* are not captured by our representational scheme. For instance, we make no claim that A' represents Larry's
preference. Implicit in our assumption of the blind vote is that the class of
aggregation functions under consideration are those that respect anonymity.
Indexical information of the sort that reveal the identity of an individual are
irrelevant to the determination of collective choice under the constraint of
anonymity. In other situations, information about identities of individuals
are crucial to the determination of collective choice – for instance in situations in which certain individuals have veto power. But note that these
are cases in which anonymity cannot hold. The underlying structures of
these strategic interactions are sufficiently different from the pizza voting
scenario that a different representational scheme is warranted. So although
our logical representations omit certain (irrelevant) details of the situation,
they do not constitute a case of misrepresentation.[9]

Given what is said, the second important point is that the temptation
to infer that our sentences collectively constitute a true inconsistent theory
should be resisted here. Our sentences achieve their representational task
by being structurally similar to the domain of preferences under the broad
constraint of anonymity, – sentences A', B', C' collectively represent the
pattern and distribution of votes casted blindly by our voters. The pattern
and distribution of votes in turn has an internal structure which points to
a certain kind of impossibility – the impossibility to use a majority rule to
resolve the conflict. The collective inconsistency of our sentences allow us to
capture precisely this kind of impossibility inherent in the given situation.

---

[9]Nor should the omittance be construed as an invitation to equivocate between individual preferences and the joint satisfaction of their conflicting preferences. The distinction
is clear: each one of A', B' and C' represents a single vote by an individual and they
collectively represent all votes.

The underlying structure (and inconsistencies) exemplified by our sentences collectively is of course a combinatorial product of the truth conditions of our sentences individually. Formal semantics indeed fix the 'meaning' of our symbols, but the representation fit between our symbols and the domain represented is mainly a pragmatic issue. Which combinatorial configurations instantiated by our sentences are best suited to represent a domain depends on which features of the domain are of interest to us. As to the ontological question about the existence of the inconsistent imaginary pizza, ordered collectively by Larry, Harry and Barry, coming with both anchovies and no anchovies, we see no explanatory advantage in postulating its existence here. Strategic conflicts of the sort exemplified by the pizza voting scenario do not require the postulation of the existence of inconsistent entities.

3.2   Optimization and Decision Problems

Another rich source of conflicts involving matters of practical concern are certain classes of computational problems called optimization problems. All optimization problems have a similar form: they ask for a 'solution' (in a solution space) under some specified conditions or constraints where certain values are either maximized or minimized. Corresponding to each (class of) optimization problem is a decision version of the problem – can a universal turing machine answer a yes-no type question concerning the existence of a solution to the problem. Obviously if we can produce an algorithm to generate the solution or search through the entire solution space, then we have an answer to the decision problem. The interest in the decision problem is not necessarily to find a solution but to sort out the computational resources (time and space) required to produce an answer. The focus of decision problems is on providing the theoretical basis for classifying problems according to their complexity – a measurement by which we judge the difficulty of a problem class.

Instances of optimization problems can be found everywhere – optimal route plotting, goods packing, meeting scheduling etc. For simplicity, we'll focus on one particular class of optimization problems known as finite constraint satisfaction problems (**FCSP**). Nothing substantive hinges on the use of **FCSP** here. Any appropriately formulated combinatorial optimization problem can serve to illustrate the same point. A **FCSP** problem involves

1. a set of variables, $X_1, \ldots, X_n$

2. associated with each variable, X_i, is a finite domain, D_i of values

3. a set of constraints, $C_1, \ldots, C_m$, each is defined on a subset of variables and is a subset of the Cartesian product of the associated domains,

i.e.

$$C_i(X_{i_1}, \ldots, X_{i_k}) \subseteq (D_{i_1} \times \ldots \times D_{i_k})$$

A *solution* to a **FCSP** is simply an assignment of values to variables such that all constraints are satisfied. What is interesting for us is that many 'real world' problems, represented as instances of **FCSP**, are *over-constrained* – every assignment of values to variables fails to satisfy at least one constraint. An over-constrained **FCSP** does not have a solution. For instance,

EXAMPLE 2. Let X, Y, Z be variables whose domain is $\{1, 2, 3\}$. Let the constraints be: $X < Y$, $Y < Z$ and $Z < X$ where $<$ is understood as the usual ordering on natural numbers

Clearly, this is over-constrained since no natural numbers can satisfy all three constraints. What is more interesting is that there is a close relationship between **FCSP**s and logic (see [Bibel, 1988; Mackworth, 1992]). Any instance of **FCSP** can be represented as an equivalent logic problem in a variety of settings. In the *model checking* approach for instance, a **FCSP** is taken to have a solution iff a certain set of propositional sentences Γ is satisfiable. In fact, the solutions are just the set of models of Γ. In this scheme, the set Γ is constructed as a set of formulae in conjunctive normal form such that

1. Each possible combination of values for variables is represented by a set of propositional variables, $p_{d_1}^{x_1}, \ldots, p_{d_n}^{x_m}, \ldots$, where intuitively, $p_{d_j}^{x_i}$ is the proposition which says that the variable x_i is instantiated by the value d_j. For instance, the sentence $(p_{d_j}^{x_i} \vee p_{d_k}^{x_i})$ says that the variable x_i is instantiated by at least one of values d_j and d_k.

2. A constraint is stated *negatively* in terms of values that are forbidden, e.g. the sentence $\neg p_{d_j}^{x_i}$ says that x_i are never instantiated to value d_j, the sentence $\neg p_{d_k}^{x_i} \vee \neg p_{d_k}^{x_j}$ says that x_i and x_j are never instantiated to the same value d_k. The set of all constraints is represented by a set of propositional formulae in the variables, $p_{d_1}^{x_1}, \ldots, p_{d_n}^{x_m}, \ldots$.

Hence every instance of over-constrained **FCSP** can be associated with a corresponding inconsistent set of sentences in CNF. Once again inconsistencies find their way back. But given what we have said in the last section, inconsistencies are to be expected here. They represent constraints residing in a certain problem space that is populated with conflicts.

In terms of using a formal language to model 'real world' problems, there is a subtle question of whether inconsistency in the strictly logical sense is the right formalism for modeling conflicts. Conceivably we can deploy a very different formalism e.g. graphs or linear equations etc., to represent these problems so that the resulting representation is no longer inconsistent in the strictly logical sense. In fact mathematicians do use these formalisms. But to dismiss the representational power of logic on such grounds is to miss an important point. What makes an over-constrained scheduling problem interesting and difficult is precisely that the real world cannot meet its demands. We cannot schedule 4 consecutive eight-hour shifts in a day because there are only 24 hours in a day on Earth. A change in formalism may allow us to find hidden structures of the problem more easily or to perform computation over the representation more efficiently, but this by itself would not resolve the underlying conflict. There is a genuine sense in which the salient features of practical conflict is captured in terms of logical inconsistency.

4   Paraconsistency Again

How does all this bear on the issue of paraconsistency? Our examples, simple as they may be, are supposed to highlight the fact that logic is as much about *representation* as it is about *consequence*. In viewing logic as a language for modelling practical and abstract problems, the emphasis is on the discriminatory power of our language. The main issue in our examples is not fault tolerant deduction per se or the reality of an inconsistent ontology. Rather, it is the analysis of the structure and the underlying combinatorial properties of our logical representation which in turn inform us about the nature of the situation or problem under consideration. Of course to provide such an analysis, our logical description must capture the salient features of the problem *at some appropriate level of abstraction*. But this is very much a question about the representational efficacy of the formal language *and* the representational fit between the formal language and the problem domain. In saying this, we do not intend to suggest that representational issues have *nothing* to do with deduction. Far from it, deduction is related to meaning. As is well known, the meaning of a logical connective can be specified by the use of introduction and elimination rules. So deduction can be used to ground and fix the meaning of a logical representation.[10] But note that this way of bringing deduction back into the picture requires no tacit assumption about epistemic error or inconsistent ontology. Our examples serve to demonstrate this very point – there are cases in which a problem domain is best modelled by logically inconsistent descriptions involving neither epistemic error nor ontological assertion about the existence of inconsistent entities. Of course the representational fit between a

---

[10]A note of caution however is in order given the remarks of Prior [Prior, 1967] and Belnap [Belnap, 1967]. Giving the meaning of a logical connective via deducibility relation requires proving at least consistency (existence), and in some case also uniqueness.

logical description and a problem domain must be evaluated in the context of a machinery for specifying the meaning of the description. But for our purpose here, we need not be committed to a particular way to accomplish the task. There is no harm in being a methodological pluralist. Whether one opts for a model theoretic or proof theoretic machinery, incompleteness and unsoundness are genuine possibilities. There is no guarantee that a model theoretic specification must have a corresponding proof theoretic specification or vice versa.

The usefulness of paraconsistent logics as a way to ground the meaning of formal languages is perhaps analogous to the usefulness of a scientific instrument. Ancient astronomers carved out the constellations with their bare eyes, charting the night sky into distinct heavenly bodies and regions. They did what they could given what was available at the time. Astronomers in the Renaissance were bestowed with the gift of the telescope. They could now chart the night sky with finer precision and distinctions that were not seen by ancient astronomers. Modern astronomers go one step further by tapping into the unexplored territory of radio frequencies. Formal languages are the symbolic constellations for the modern logician. The history of modern logics too is punctuated with remarkable changes in the discriminatory and expressive power of logics. Propositional logic delivers the calculus of *propositions*. But to formalize 'Every natural number has a successor', we have to await for the advent of quantification theory. To formalize 'A relation is well-ordered if every non-empty subset has a least element', we have to go second-order or employ set theory of some form.[11] At each turn of this refinement, more can be said and more can be discerned. But at the same time it is also surprisingly conservative. In classical logics, sentences are sorted into three distinct classes – those that are tautologous (true in all models), inconsistent (false in all models), and contingent (true in some and false in some models).[12] Even in a simple classical propositional logic with countably many propositional atoms, there are at least countably many distinct non-equivalent classes of contingent sentences – one for each distinct atom. But oddly, there can only be *one* equivalence class of tautologous sentences and only *one* equivalence class of inconsistent sentences. Within the classical scheme a contradiction, $(p \wedge \neg p)$, is indistinguishable from a denial of the excluded middle, $\neg(q \vee \neg q)$ – no classical model, no classical proof will separate them. But note that we do distinguish these sentences

---

[11]We dare not say 'the set theory' here. As we all know ZF is to be distinguished from ZFC (with choice axiom), from ZFA (with anti-foundation axiom) from NBG (NBG for von Neumman, Bernays and Godel, not to be confused with the epithet 'No Bloody Good'), from (Quine's) NF. Set theories, like logics, come in many varieties.

[12]Given soundness and completeness of classical logics, the reader may use the appropriate proof-theoretic substitutes for 'true in all models' etc.

meta-logically. They are not merely distinct syntactic tokens of distinct types – while one can be used to assert a contradiction the other can be used to reject the law of excluded middle. The distinction also carries a certain semantic weight.

Our take home message then is this: formal languages which express inconsistencies are rich in structure and expressive power. Our complaint against classical consequence and classical semantics is that they do not make room for the discrimination between different inconsistencies within a formal language. Under the classical scheme, all inconsistencies are proof-theoretically and model-theoretically *equivalent*. But recall that the study of formal languages, standard first order model theory in particular, is very much concerned with the discriminatory power of languages and models. To make distinctions, we must be able to partition a language into distinct equivalence classes. In fact, in a very general sense all formal languages are concerned with equivalence classes – namely classes that are organized under the 'sameness-in-meaning' relation. Indeed this is one of the main goals (and advantages) of the study of formal languages – given any two expressions in a formal language we want to provide a systematic and rigorous method to determine if the two have the same meaning. Logic provides a paradigmatic method to do this – in fact it provides two methods, one via proofs the other via models.[13] So the inability of the classical scheme to discern different inconsistencies is a failing on its part to do its job.

Viewing the matter in this light gives us the satisfaction of putting a positive spin on paraconsistency and turning the tables on the classicist. It is often said that paraconsistent logics are simply too weak to do any real work – they give up too many classically acceptable rules of inference. Our rejoinder is that sometimes weakness is also its strength. Recall that in the study of modalities for standard modal logics, strong modal logics such as $S4$ and $S5$ have finitely many modalities.[14] More precisely, $S4$ has 14 distinct (non-equivalent) modalities while $S5$ has only 6 distinct modalities. So in terms of the discriminatory power of these logics, we can only express 14 and 6 distinct types of modal statements. These logics are strong, but they don't necessarily give us more distinctions. The comparison between classical and

---

[13]We take this to be at least a necessary condition for such a semantic specification. However, it is debatable whether it is also sufficient. Some may insist that the equivalence relation induced by the underlying logic must also be a *congruence* relation. This amounts to the requirement that intersubstitutivity of provable equivalents preserves equivalence. As is well known many paraconsistent logics e.g. Priest's LP and da Costa's C-systems, do not have such a property. We do not wish to settle the issue here. But we do think that it is a research direction worth further investigation.

[14]A *modality* is just any finite sequence of $\neg, \Box, \Diamond$ prefixing a well formed formula. See Chellas [Chellas, 1990] for more details.

paraconsistent logics is analogous to the comparison between strong and weak modal logics. Strength in deducibility is not tantamount to strength in discriminatory power. Paraconsistent logics and semantics are not merely non-explosive, they also allow us to preserve important distinctions. Not all inconsistencies are equal and they should not be. Paraconsistent logics are endowed with the power to discriminate between different inconsistencies. This, we maintain, is another way to 'go beyond consistency'.

BIBLIOGRAPHY

[Belnap, 1967] N. D. Belnap. Tonk, Plonk and Plink. In P. F. Strawson, editor, *Philosophical Logic*, pages 132–137. Oxford University Press, 1967. First appeared in *Analysis* Vol. 22 pp. 130–134 1962.

[Béziau, 2002] J. Y. Béziau. *S*5 is a Paraconsistent Logic and so is First-Order Classical Logic. *Logical Journal Studies*, 9:301–309, 2002.

[Bibel, 1988] W. Bibel. Constraint Satisfaction from a Deductive Viewpoint. *Artificial Intelligence*, 35:401–413, 1988.

[Chapman, 2002] B. Chapman. Rational Aggregation. *Politics, Philosophy and Economics*, 1 (3):337–354, 2002.

[Chellas, 1990] B. F. Chellas. *Modal Logic: An Introduction*. Cambridge University Press, 1990.

[Cohen *et al.*, 1998] P. Cohen, R. Schrag, E. Jones, A. Pease, A. Lin, B. Starr, D. Gunning, and M. Burke. The DARPA High-Performance Knowldge Bases Project. *AI Magazine*, 19 (4):25–49, 1998.

[List and Pettit, 2002] C. List and P. Pettit. Aggregating Sets of Judgments: An Impossibility Result. *Economics and Philosophy*, 18 (1):89–110, 2002.

[List and Pettit, 2004] C. List and P. Pettit. Aggregating Sets of Judgments: Two Impossibility Results Compared. *Synthese*, 140:207–235, 2004.

[Mackworth, 1992] A. K. Mackworth. The Logic of Constraint Satisfaction. *Artificial Intelligence*, 58:3–20, 1992.

[Pauly and van Hees, 2004] M. Pauly and M. van Hees. Logical Constraints on Judgement Aggregation. forthcoming in *Journal of Philosophical Logic*, 2004.

[Priest, 2001] G. Priest. Logic: One or Many. In J. Wood and B. Brown, editors, *Logical Consequence Rival Approaches: Proceedings of the 1999 Conference of the Society of Exact Philosophy, Volume One*. Hermes Science Press, 2001.

[Priest, 2002] G. Priest. Paraconsistent Logic. In D. M. Gabbay, editor, *Handbook of Philosophical Logic Volume 6*. Kluwer Academic Pub., second edition, 2002.

[Prior, 1967] A. N. Prior. The Runabout Inference Ticket. In P. F. Strawson, editor, *Philosophical Logic*, pages 129–131. Oxford University Press, 1967. First appeared in *Analysis* Vol. 21 pp. 38–39 1960.

[Restall, 2003] G. Restall. Paraconsistency Everywhere. *Notre Dame Journal of Formal Logic*, 43 (3):147–156, 2003.

[Wheeler, 2000] G. R. Wheeler. Kinds of Inconsistency. In *Paraconsistency: The Logical Way to the Inconsistent, Proceedings of the Second World Congress on Paraconsistency (WCP 2000)*, pages 511–522. Marcel Dekker, Inc., 2000.

Paul Wong
Automated Reasoning Group, Computer Science Laboratory, Research School of Information Sciences and Engineering, Australian National University, ACT 0200, Australia.
wongas@mail.rsise.anu.edu.au

Dialetheias are Mental Confusions

HARTLEY SLATER

ABSTRACT. I show in this paper that the 'true contradictions' favoured by certain paraconsistent logicians are better seen as contradictory thoughts. I start by reminding readers that a similar transformation of an extensional logic into an intensional logic is equally appropriate in the case of Intuitionism. The matter involves a close examination of the nature of negation. For, finding themselves short of a word to translate their various 'negations' into, many Intuitionist, and Paraconsistent mathematicians have perforce used 'not'. Lax attention to natural language has encouraged this. But strict attention shows it is not possible. Intuitionistic '¬α' merely entails 'not α' since that '¬' is a contrary-forming operator, as is now, quite widely, well known. But not even that can be said for paraconsistent '¬', as we shall see, since paraconsistent '¬α' does not oppose 'α'; that '¬' is merely a subcontrary-forming operator. In both cases the 'negation' involves a modal notion, and so should be symbolised using an intensional operator.

1

Discussions of the philosophical import of Intuitionistic Logic have been mostly concerned with Brouwer's school's historic reasoning, and what can be said either for or against it. But historical studies are not confined to discussing authors just in terms of the concepts they were conscious of, since further researches may well point to external facts which are still highly relevant. This is illustrated in the many epistemic notions Brouwer used, unsymbolised, in his work, about undecided properties of decimal expansions. If the postulated difference between 'not-not-p' and 'p' is properly one between 'not demonstrated that not-p' and 'p' then it should be symbolised differently.

Gödel's well known translation of Intuitionistic Logic into S4, McKinsey and Tarski's different one, and the semantics Kripke provided for Intuitionistic Logic in terms of stages of knowledge, are even more decisive external facts of the same kind, since they show explicitly how most, or all of Intuitionistic Logic can be seen as not an alternative *propositional* logic, but

instead a *modal, or intensional* logic. Certainly no-one wants to assert 'p v
L¬p', but that does not show in any way that one should not assert 'p v ¬p'.
Moreover, if the former is represented as 'p v ¬p' then the '¬' should not be
read simply as 'not'. We now turn to the detail of how a similar situation
arises in the dual, paraconsistent case. I shall talk mainly about Graham
Priest's work: it is mainly Priest who has insisted that there are not only
contradictory thoughts but also contradictory truths, i.e. dialetheias. It
is the latter extension of other, more valid ideas within the paraconsistent
tradition which is argued against here.

2

Graham Priest agrees that Intuitionistic Logic is not a rival to classical
logic, but still wants to maintain that Paraconsistent Logic is a rival. He
says [8, 110]:

> In the case of Intuitionism, where truth and falsity are not ex-
> haustive, I argued that intuitionist negation is not a contradictory-
> forming operator, and that we can define a genuine such operator
> by the condition: ¬α is true iff α fails to be true. It is natural to
> suppose that a similar objection can be made here. Dialetheic
> negation is merely a subcontrary-forming operator. The same
> clause still defines the genuine contrary-forming operator.

The objection that dialetheic negation is merely a subcontrary-forming operator was considered by Priest before, and rejected on other grounds (see
[10]). Subsequent debate has taken off from Slater's defense of the subcon-
tariety point, and here Priest tries to make out that dialetheic negation is
still a contradictory-forming operator, first because it captures so much of
the classical account, and second because any argument against it 'can rest
solely on the fact that the truth of ¬α does not rule out that of α'. Indeed
it does rest solely on this, so how does Priest try to show that, nevertheless,
dialetheic negation is a contradictory-forming operator?

He does not consider the proof of the subcontrary point, offered in [10],
that his 'A is true' and 'A is false' are equivalent to 'V(A) $\geq$ 0' and 'V(A)
$\leq$ 0', whereas they should be equivalent to, say 'V(A)$\geq$ 0' and 'V(A) $<$
0' (or their equivalents), if they are to be contradictories. Instead he tries
to show that Boolean negation likewise involves an operator for which the
truth of ¬α does not rule out that of α. But, even if this was true, it
would merely show that Boolean negation was not a contradiction-forming
operator, it would not show that, somehow, dialetheic negation was one. So
the argument is a red herring.

How is it that Priest does not show that the truth of a Boolean negative does not rule out the truth of the correlative Boolean positive? His argument runs like this [8, 110]:

> Suppose we define an operator, $\neg$, such that $\neg\alpha$ is true iff α is not true, and let us say, false otherwise...The behaviour of $\neg$ requires...careful examination... In particular, why should one suppose that we can never have both α and $\neg\alpha$? The natural argument is simply that if α and $\neg\alpha$ are true then α is both true and not true, and this cannot arise. But we cannot argue this way without seriously begging the question. If, as the dialetheist claims, some statements and their negations are both true, maybe α can be both true and not true...The Boolean properties of 'Boolean negation' may therefore be an illusion.

Boolean negation, i.e. contradiction, however, was not traditionally characterised as Priest supposes, but instead: contradictories cannot be true together, or false together. On that basis Priest's argument for the possibility that contradictories might be both true collapses immediately: he has himself begged the question, by presuming that contradictories are defined otherwise. But even if they were defined otherwise, what sort of a case would Priest have got? He says at the same place: 'Indeed, if a dialetheic solution to the semantic paradoxes is correct, and α is 'α is not true' then α *is* both true and not true'. But, first, the fact is that one never can get the supposed identity

$\alpha = $ 'α is not true',

if it is meant to be of the form

'p' $=$ '''p'' is not true',

since that would require a whole to be a proper part of itself. And if what is involved instead is a referring phrase 's' to a sentence such that, for instance,

s $=$'s is not true',

then the T-scheme is required to generate any paradox, which could quite straightforwardly lead to the denial of that scheme, by *Reductio*. It is well known that there are lines of thought about the Liar which deny that sentences are true or false, but instead insist that propositions are [5]. That means shifting from taking 'is true' to be a predicate of sentences, to taking it to be a predicate of 'that' clauses, as in the locutions 'that p is true', 'it is true that p'. There is then no trouble with self-referential sentences, even of the strengthened sort: if b='b is neither true nor false' then what is true is that b is neither true nor false, but that is not to say that what is true is b, i.e. 'b is neither true nor false', so there is no contradiction. Of course there are no self-referential propositions comparable to self-referential sentences, since "p'='that p is false" is ruled out by mereology - something like '5=5+0' is possible, but not anything like "5'='5+0", since the right hand expression is longer than the left hand one.

Priest has pursued his contrary view of the matter, however, trying to show, on the same basis as above, that the traditional law *ex contradictione quodlibet*, (ECQ), does not hold for Boolean negation. But his argument is incorrect just because of this issue of what is true: he thinks a formula, i.e. a mentioned sentence, might be what is true, but instead it is the referent of a 'that' clause which includes that formula as a used sentence. He goes on [8, 110]:

> We can now deal with another law of negation: *ex contradictione quodlibet* (ECQ): $\alpha.\neg\alpha \vdash \beta$...Given the present discussion, it can clearly be seen to fail. For we can take an α which is both true and false, and a β that is not true. This instance of the inference is not truth preserving, and hence the inference is not valid (truth preservation being at least a necessary condition for validity). For good measure, the equally contentious inference of Antecedent Falsity (AF) $\neg\alpha \vdash \alpha \rightarrow \beta$, must also be invalid, for exactly the same reason (*modus ponens* holding); as again, and more contentiously, must be the disjunctive syllogism (DS): $\alpha, \neg\alpha \vee \beta \vdash \beta$.

But clearly, by using 'α', and '$\neg\alpha$', so that such may be true together, none of this goes against the classical rules ECQ, AF and DS, since they deal with *contradictories*, i.e. things which cannot be true together. Disjunctive Syllogism, for instance, isn't just *the formula*

'$\alpha, \neg\alpha \vee \beta \vdash \beta$',

otherwise there would be no chance that it was necessary. Disjunctive Syllogism is this formula interpreted a certain way, in particular so that the truth of $\neg\alpha$ does rule out that of α. As a result, while, for instance, the formula

'$\alpha.\neg\alpha \vdash \beta$',

fails on a dialetheic interpretation, the explanation is that the '$\neg$' then is just a subcontrary-forming operator. Priest wants to say it fails on a Boolean interpretation also, but that would again only show that his definition of Boolean '$\neg$' did not make it a full contradictory-forming operator. If one says that the conjunction of a formula and its negation entails everything, then what one then says, i.e. the proposition one then expresses, is necessarily true, if one is using 'negation' to mean Boolean negation.

3

It is only one aspect of the matter, however, that Priest's logic (LP) concerns itself with subcontraries. For in [7, 4] (c.f. [8, 102]) he said:

> How could a contradiction be true? After all, orthodox logic assures us that for every statement, α, only one of α and $\neg\alpha$ is true. The simple answer is that orthodox logic, however well entrenched, is just a *theory* of how the logical particles, like negation work; and there is no a priori guarantee that it is correct.

Orthodox logic is inevitably correct about what the traditional logical connectives, such as Boolean negation, consists in, since that is just a matter of definition. And what kind of 'theory' is Boolean negation? Boolean negation is merely one relation amongst others which propositions may fall into. As a result, if one chooses propositions which fall into this relation, and use the standard conjunction and disjunction, all the above classical laws will hold. Moreover the relation of Boolean negation was what was traditionally called 'contradiction', so if '$\neg$' symbolises that relation there is no way that a contradiction can be true. Priest is more likely using the term 'contradiction' not for this relation, however, but just for the sign '$\neg$', since certainly 'p.$\neg$p' might be given a non-traditional interpretation on which it was not true. But that, of course, would in no way conflict with the necessary fact expressed before.

If Boolean negation comes into some theory about 'the logical particles', then that theory must be about how one does or should use words and symbols like 'not', and '$\neg$': for instance, that we do or should use 'not' or '$\neg$' to symbolise what was classically called 'contradiction' rather than some other relation (c.f. [6, 225]). But that is quite a different matter; it is a conventional, or legislative matter about how we do or should use words, not directly a logical issue. Some people cannot say 'no', it is well known, or if they do say this they do not mean it, as their children might say. Such people are not really denying anything when they say 'no', and might be said to have a theory about how to use the word - so that it does not oppose, or contradict. But it is just a matter of choice to use the negative particle in this, or some other way; for instance, as normally, in the traditional sense involving denial. So this is quite a separate question from the question whether contradictories may be both true: that question is about a certain propositional relation, and its connection with truth.

And that question, as was pre-figured at the start, is also to be crucially distinguished from the question of whether contradictories may be *thought to be both true*. Priest says [8, 114]:

In explaining their views, people often assert contradictions, un-
wittingly. In this way, they discover - or someone else points
out to them - that their views are inconsistent. In virtue of this,
they may wish to revise their views. But in asserting $\neg\alpha$ in this
context, they are not expressing a refusal to accept α, i.e. denying it. It is precisely the fact that they accept *both* α and $\neg\alpha$
that tends to promote belief revision. It may even be rational
sometimes - as a number of classical logicians have suggested -
to hang on to both beliefs and continue to assert them: consider,
for example, the Paradox of the Preface. If to assert $\neg\alpha$ is to
deny α mooting this possibility would not even make sense.

One point arising from this is that Priest has had difficulty saying what
denial is, if it is not asserting the reverse [8, 114] - he is like the above weak
father with 'no'. A further point which needs to be emphasised is that the
modalities in the above are not the usual ones, and quite deliberately so.
Normally, we expect people to be consistent, and the criticism that someone
is being illogical by contradicting themselves involves a considerable censure.
Priest is trying to make us more tolerant, it might be said: it's not that
people *ought to* revise their views, if they are inconsistent, it's that people
'may wish' to, he says. And Priest's view is that such inconsistencies 'tend
to' promote belief revision, whereas a traditional logician does not see the
matter in these terms, as if it was some natural, maybe causal process. The
traditional logician would insist that the muddled get out of their muddle
and become consistent, taking that *action* to be a choice the muddled simply
have not made. Certainly it is quite natural for people to be muddled, and
to stay muddled - but only until they meet a traditional logician. For,
unlike Priest, who does not want to be intolerant, the traditional logician
will not be so easy going. He will impose upon his pupils the requirement
of determinateness of sense, since only in that way can anyone have clear
thoughts.

But such facts about the social value of being consistent, and the forces
which might insist on this, or contrariwise might be more slack, like the
question over the use of 'not', are not themselves matters of a priori logic.
They are matters to do with getting the rigour of a priori logic at least to
some extent into people's lives. How is it then that they enter the discussion
of paraconsistent logic? They enter it, surely, because, as with the give-away,
but unacknowledged mention of what is demonstrated with Brouwer, they
show what Priest *has in mind*. He has in mind the undoubted possibility of
mental confusions. In place of 'p.$\neg$p', in his Paraconsistent Logic, what is
really meant is something like 'Tp.T$\neg$p' - where 'T' is not 'it is true that'
but 'it is thought that'. Thus formulated Paraconsistent Logic is about the

possible contents of people's minds, and in particular the above passage is reminding us of the possible contents of people's minds before they meet a strict teacher.

Priest also mentions 'belief revision' in the above passage, but only one way of looking at this is 'classical', in the sense that the belief set has to be consistent, and deductively closed. This is the line taken in the standard AGM account [2]. Priest has essayed, instead, a non-classical, paraconsistent theory of belief [9]; but it stays within the 'belief revision' tradition, from the present point of view, because it focuses on the propositions in a belief set, without symbolising them *as beliefs*. It thus represents the matter in terms of an alternative extensional logic: the logic which supposedly operates on the propositions in the believer's mind. What is wanted, by contrast, is an intensional logic, which makes explicit the fact that the propositions concerned are in someone's mind through the introduction of an intensional operator which expresses this. If we produce the logic of that operator, we produce a logic which is different from classical propositional logic, certainly, but merely because it supplements it, instead of being a rival to it on its own ground (c.f. [3, Ch 1]).

One such intensional logic, in fact, is quite well studied already, in the standard, Bayesian semantics for belief. It was Ramsey, of course, who linked beliefs to subjective probabilities; later writers have constructed more complete Bayesian theories, using conditionalisation as the guide to belief revision. The formal consequences of treating belief in terms of probability without that dynamic aspect are detailed in [4]. It is worth reminding ourselves what laws hold, in that case.

Within this kind of logic we do not get '*ex contradictione quodlibet*' in the form 'Bp.B¬p ⊢ Bq', or 'Disjunctive Syllogism' in the form 'Bp.B(¬p ∨ q) ⊢ Bq', when 'Bp' means, at least, that $pr(p) \geq 1/2$. And while we do still get 'Antecedent Falsity', in the form 'B¬p ⊢ B(p ⊃ q)', we do not in the form 'B¬p ⊢ Bp ⊃ Bq', and 'Modus Ponens' fails in this way, i.e. we do not have 'Bp.B(p ⊃ q) ⊢ Bq', which means we do not necessarily have deductive closure of the belief set. The Paradox of the Preface, which Priest mentioned above, is then resolved immediately, since 'Bp.Bq.Br...B(¬p ∨ ¬q ∨ ¬r...)', is quite consistent. And going with that, 'Adjunction' does not hold in the relevant sense, i.e. we do not have 'Bp.Bq.Br... ⊢ B(p.q.r...)', which shows dramatically how it can be that people 'do not put 2 and 2 together'. More fundamentally, since it is evidently quite possible that 'Bp.B¬p', it is this which most closely represents a 'true contradiction', although we now see it is just a matter of mental confusion - whether allowable or not.

The Bayesian treatment, however, does not parallel Priest's LP very closely. Certainly if 'Bp' is true if and only if $pr(p) \geq n$, for some threshold n, that allows both Bp and B¬p to be true for certain values of 'n'. But LP only had three values [6, 226], whereas the probabilistic one is continuously valued. This asymmetry, however, can be removed if we treat Belief as also three valued, say, Belief (V(p) = 1), Wonder (V(p) = 0), and Disbelief (V(p) = -1). We can construct the matter simply using the operator 'it is thought that' as before, with the three categories then being 'Tp.¬T¬p', 'Tp.T¬p', and 'T¬p.¬Tp' - where Tp if and only if $V(p) \geq 0$. In Priest's logic the same semantic relation holds with 'T' read in terms of truth, and he accepts 'A is true iff A' [6, 238]. It follows that there is an interpretation of LP in terms of mental confusions. Q.E.D.

4

But while this reformulation of LP may be acceptable to many paraconsistent logicians, and even provable, Priest is definitely committed to denying it. Priest has expressly developed his Dialetheism as a doctrine about contradictions being in *things* rather than in *thoughts*, i.e. in the extensional world, and not an intensional one. This is not just because of a resistance to moving away from extensional logic, with the consequent resistance to symbolising human confusion not extentsionally, but with an intensional operator. The logical paradoxes centrally have troubled Priest, as we saw, in part, with the previous discussion of propositional self-reference. The fact that a paraconsistent view of the Liar involves adherence to the disquotation-scheme, however, shows that the contradictions Priest thinks inevitable are in fact avoidable. And the first thing needed to show that any contradictions are in a person's mind and not reality is just that: that there is an alternative reading of the facts which does not find them paradoxical. We can see this crucial matter again in Priest's treatment of Heterologicality.

On this matter, Priest significantly does not mention an assumption which Copi, for instance, makes explicit. In place of Copi's formalisation of 'x is not self-applicable', which, in Priest's terms, is equivalent to

$\exists Y(xRY.(Z)(xRZ \supset Z=Y).\neg Yx)$,

and contains a uniqueness clause [1, 301], Priest merely writes [7, 163-4]:

$\exists Y(xRY.\neg Yx)$,

for the relevant predicate. Abbreviating this to 'Fx', he goes on to argue that, given 'F'RF, from ¬F'F' one can obtain $\exists Y('F'RY.\neg Y'F')$, i.e. F'F'. He also argues that from F'F' one can get ¬F'F', just because F'F' entails $\exists Y('F'RY.\neg Y'F')$. But even given 'F'RF, ¬F'F' does not follow from $\exists Y('F'RY. \neg Y'F')$, i.e. the fact that some Y which 'F'Rs does not apply to 'F', since one also needs a proof of the uniqueness clause before one can say that *the only* Y which 'F'Rs is F, and therefore ¬Y'F' for that Y. No contradiction is obtainable without a uniqueness proof, which would exclude non-univocality in 'F'. Priest is therefore confused if he thinks that the contradiction follows as a matter of demonstrated fact. The necessary truth of a contradiction is not established by Priest's argument, simply because there is a hidden premise, which it is open to anyone to deny, to free their minds of the contradiction. Priest chooses not to deny this premise, putting the contradiction in his mind simply by an act of his own will.

And it must stay in his mind, because of a further feature of the case: the fact that the relevant predicate 'is not self-applicable' is undoubtedly ambiguous. Certainly the predicate 'is not self-applicable' is not like 'is a bank' in allowing semantic disambiguation into 'is a money bank' and 'is a river bank', but it still is dependent on the context for its sense, to pick up the appropriate referent for 'self', so its meaning is not constant. In the sense, if not the syntax of

x is not self-applicable if and only if x is not applicable to x

there are *four* occurrences of 'x' and not just *three*, and so the predicate on the left is not replaceable by any constant one. Here is an independent reason for the conclusion that there is ambiguity in connection with Heterologicality, and it shows that a defence of the uniqueness assumption explicit in Copi, and hidden or ignored in Priest, just is not obtainable. Another illustration arises in Quine's famous example: 'yields a falsehood when used as a predicate appended to its own quotation' yields a falsehood when used as a predicate appended to its own quotation. There is a use (as well as a mention) of 'its own' in this case, and so it is like the 'itself' in: 'is not applicable to itself' is not applicable to itself.

There is also a further thought which hinders seeing this, amongst the learned: the common reading of lambda abstraction expressions as referring to concepts. Isn't 'y is not self-applicable', equivalent to 'λx(x is not applicable to x)y', and so doesn't the lambda term, in which the 'x' is bound, denote the required constant concept of Heterologicality? The lambda abstraction merely isolates the syntactic form of a sentence, however, and clearly whether a single concept is associated with the predicate in that sentence is not deducible from that - as 'is not applicable to x', and 'yields a falsehood when used as a predicate appended to x's quotation' again illustrate.

BIBLIOGRAPHY

[1]   I. Copi. *Symbolic Logic*. Macmillan, New York, 4 edition, 1973.
[2]   P. Gärdenfors. *Knowledge in Flux*. M.I.T. Press, Cambridge MA, 1988.
[3]   S. Haack. *Deviant Logic, Fuzzy Logic*. University of Chicago Press, Chicago, 1996.
[4]   J. Hawthorne and L. Bovens. The preface, the lottery, and the logic of belief. *Mind*, 108.429:241–64, 1999.
[5]   W. Kneale. Propositions and truth in natural lanugages. *Mind*, 81:225–43, 1972.
[6]   G.G. Priest. The logic of paradox. *Journal of Philosophical Logic*, 8:219–41, 1979.

[7]   G.G. Priest. *Beyond the Limits of Thought*. C.U.P., Cambridge, 1995.
[8]   G.G. Priest. What not? a defence of dialetheic theory of negation. In D. Gabbay and
      H. Wansing, editors, *What is Negation?*, pages 101–120. Kluwer, Dordrecht, 1999.
[9]   G.G. Priest. Paraconsistent belief revision. *Theoria*, LXVII:214–28, 2001.
[10] B.H. Slater. Paraconsistent logics? *Journal of Philosophical Logic*, 24:451–4, 1995.

Hartley Slater

Philosophy, University Western Australia, 35 Stirling Highway, Crawley
W.A. 6009WA, Australia.

slaterbh@cyllene.uwa.edu.au

Reply to Slater

GRAHAM PRIEST

In his 'Dialetheias are Mental Confusions' (200a), Slater argues that dialetheism may be dismissed very swiftly. People's beliefs can be inconsistent; the truth cannot. The negation symbol of a paraconsistent logic does express real negation. For such negation, A and $\neg A$ cannot be true together—by definition. End of story. Slater also essays a consistent account of the semantic paradoxes. In this note I will explain why I am unpersuaded about both matters.

1 Contradictory or Sub-Contrary?

Slater and I agree that negation, whatever it is, is a contradictory-forming operator (cfo). It is the relation that obtains between pairs such as 'Socrates is mortal', 'Socrates is not mortal' and 'Some person is mortal', 'No person is mortal'. The crucial question, then, is what exactly, this relationship amounts to.

Traditional logic—by which I mean logic in the Aristotelian tradition—characterises the relation in a familiar way. A and B are contradictories if you must have one or the other, but you can't have both. That is, $\Box(A \vee B)$ and $\neg\Diamond(A \wedge B)$. Hence, A and $\neg A$ are contradictories if we have $\Box(A \vee \neg A)$ and $\neg\Diamond(A \wedge \neg A)$, that is, $\Box\neg(A \wedge \neg A)$. Consider any propositional logic, the modal extension of which satisfies Necessitation (if $\vdash A$ then $\vdash \Box A$), as it should. Then if it is such that both:

1. $A \vee \neg A$

2. $\neg(A \wedge \neg A)$

are logical truths, $\neg$ is a cfo. Since LP satisfies these conditions, its negation symbol is a cfo. Note that this is not true of either intuitionistic logic or some paraconsistent logics, such as the da Costa C-systems. In the first of these, 1 is not a logical truth; in the second, 2 is not. This is why the charge that the negation operator in those logics is not really a cfo gets its bite.

Naturally, in a paraconsistent context 1 and 2 do not stop $A \wedge \neg A$ holding as well. But this does not show that $\neg$ is not a contradictory-forming operator. It just shows that there is *more* to it than one might have thought. Let us call this more, for want of a better phrase, its *surplus content*.[1]

In the paper where Slater first levelled his charge that the negation symbol of LP is not a cfo, he observed, correctly, that in the semantics for LP there can be interpretations, ρ, and sentences, A, such that $A\rho1$ and $\neg A\,\rho1$, i.e., $A\rho0$. That is, sentences are both true and false in some interpretations.[2] The relevance of this observation is, however, less than transparent. The theory of interpretations is a twentieth (or late nineteenth) century construction aimed at giving an account of validity—an account that has somewhat tenuous links to Aristotle's own notion of syllogistic validity. Now, in the model theory of classical logic—by which I mean logical theory in the tradition of Frege and Russell—A cannot be both true-in-an-interpretation and false-in-an-interpretation. This delivers the logical truth of both 1 and 2, and so the fact that negation is a cfo. The semantics of LP do likewise, but also make room for negation to have surplus content.[3] They also show why the inference of Explosion:

$$A \wedge \neg A \vdash B$$

is invalid—which it is, incidentally, in syllogistic. (Contradictory premises do not suffice to make a syllogistic inference valid.)[4]

2   Definition

One can, of course, contest the view that negation has a surplus content. For that matter, one can contest the view that a cfo really satisfies conditions 1 and 2, as do intuitionists.[5] The relation of being contradictories is one of a whole bunch of notions that play a central role in logic. Others include implication, modal status, generality. These are all things that are

---

[1] And if it be retored that $\neg$ cannot have surplus content, the reply is 'Of course: $\neg\Diamond(A \wedge \neg A)$'!

[2] Slater (1995). I take semantic evaluations to be relations between formulas and truth values. Slater formulates LP as a three-valued logic, and takes evaluations, V, to be functions from formulas to the values $\{1, 0, -1\}$. Hence, what I write as $A\rho1$, he writes as $V(A) \geq 0$. Nothing turns on this, in this context, I think.

[3] In section 3 of his paper, Slater points out that one may interpret the semantics of LP in intentional terms. This, however, shows nothing. One can interpret the semantics of classical logic in terms of switching-circuits. That does not show that classical logic is about electronics. All semantics are subject to multiple interpretations. And in the intended interpretation of LP semantics, $\neg$ is a purely extensional operator which simply operates on the truth(-in-an-interpretation) values of its inputs.

[4] For a discussion, see Priest (200b), 2.1.

[5] Slater claims (p. 2) that I do not think that intuitionist logic is a rival to classical logic. It certainly is. It disagrees with classical logic as to how a cfo should behave—as well as many other things.

contentious—indeed, in the history of Western philosophy, logicians have frequently contended about them.[6] In logic, just as much as physics, people put forward theories of how the relevant notions behave (and we have to judge the theories by the usual criteria of theoretical adequacy).[7] The inferential behaviour of a cfo cannot, therefore, be settled by definition, as Slater thinks it can.

But maybe one can define some operator, let us call this $, which is a cfo and has no surplus content. If one can, then, in some ways, the behaviour of $\neg$ is beside the point: a classical logician can simply concede it to the paraconsistent logician, and make their point in terms of $. If we interpret Slater's remarks about definition in this way, they have a point. The crucial question then becomes: how is $ to be defined? Slater does not say exactly what sort of definition he has in mind, nor what, exactly, he takes the definition to be. Clearly, an explicit definition, of the form '*dialetheia* means *true contradiction*', is not going to get us very far. Such definitions are eliminable without loss—or if they are not, they are creative, and so objectionable. We must appeal therefore to some notion of implicit definition.

There are two plausible strategies here.[8] The first is proof-theoretic. One simply takes $ to be a connective that satisfies all the rules of inference governing negation in classical logic. In this case, we will have Explosion, which effectively rules out surplus content—on pain of any surplus turning into the total content of everything.[9]

The problem with this strategy is that there is no guarantee that this specification determines a notion with any sense. As Prior pointed out,[10] an arbitrary set of rules may well not succeed in capturing any meaningful notion. Prior's example was a connective, $*$ (tonk), taken to be governed by the rules $A \vdash A * B$, $A * B \vdash B$. Clearly, if tonk were a legitimate notion, we could prove everything. But if $ were a legitimate notion we could, similarly, prove everything, given only that we have the T-schema and some way of forming self-referential truth-bearers. (We simply formulate a liar

---

[6] And unless one is a nominalist, these disputes are not simply about the way that words are used.

[7] The theories, of course, have to answer to certain data. Thus, theories of consequence have to answer to the particular inferences that strike us as valid or invalid. But as in all theoretical enterprises, the data is defeasible.

[8] These are discussed further in Priest (1990).

[9] Note that simply having a connective satisfying Explosion is not sufficient to damage dialetheism. In a logic appropriate for dialetheism one may have a logical constant, $\bot$, such that $\bot \vdash B$, for all B. (See Priest (1987), 8.5.) If one defines $-A$ as $A \to \bot$, then one has $A, -A \vdash B$.

[10] Prior (1960).

sentence, L, of the form $\$T\langle L\rangle$, and establish $L \wedge \$L$ in the usual way.)[11]

The other strategy is to characterise $\$$ model-theoretically. The natural thought here is to specify a connective whose truth-in-an-interpretation conditions are:

0. $\$A\rho 1$ iff it is not the case that $A\rho 1$

(and, if you like: $\$A\rho 0$ iff $A\rho 1$). One might contest the claim that these truth conditions determine a meaningful connective. But let us grant that they do.[12]

Given that conjunction and disjunction behave in the usual fashion, it is straightforward to establish that, for all ρ, $(A \vee \$A)\rho 1$, and:

1. for no ρ, $(A \wedge \$A)\rho 1$

so, for all ρ, $\$(A \vee \$A)\rho 1$. Hence, given that modal operators behave in a natural way, we can establish that $\$$ is a cfo. But does $\$$ satisfy Explosion? According to the model-theoretic account of validity, $A \wedge \neg A \models B$ iff:

2. for all ρ, **if** $(A \wedge \neg A)\rho 1$, $B\rho 1$

Keep your eye on the boldfaced **if**. To establish 2, we have to infer it from 1. The inference is a quantified version of the inference:

3. it is not the case that C; so **if** C, D

If **if** satisfies *modus ponens*, if it is, say, the conditional of a relevant logic, then the inference 3 is not valid. If, on the other hand, **if**, does not satisfy *modus ponens*, say it is the material conditional, then 3 may well be valid. But now the validity of Explosion does not rule $\$$ out from having surplus content. One cannot get from $(A \wedge \$A)\rho 1$ and $A \wedge \$A \models B$ to $B\rho 1$, since this would use an instance of the Disjunctive Syllogism, invalid in all paraconsistent logics:

$(A \wedge \$A)\rho 1$ and (it is not the case that $(A \wedge \$A)\rho 1$ or $B\rho 1$); so $B\rho 1$

Either way, then, $\$$ fails to perform as required. The dialetheist can accept the implicit definition involved in 0. They just have no reason to suppose that the connective, so defined, rules out surplus content.

---

[11]For other arguments to the effect that an explosive $\$$ is, indeed, meaningless, see Priest (200a), ch. 5.

[12]There are at least semantics for relevant logics where a connective is given such truth conditions. See Meyer and Routley (1973).

All this assumes that the 'not' in 0 behaves as paraconsistent logic says that it does. It might be thought that we would fare better if we take it to be $ itself. But in fact, in that case, we get nowhere. Given that the truth conditions for $ themselves employ $, we have to know what inferences govern $ before we can infer anything about how sentences containing $ behave. This cannot be taken for granted: the whole point was precisely to justify various inferences concerning $. It might be suggested that we can simply *assume* that $ is a connective that satisfies the principles of classical negation; but the problem with this is obvious enough. We were supposed to be in the process of showing that $, as characterised by its truth conditions, satisfies Explosion (and so rules out surplus content). So such an assumption is clearly question-begging.

Though the route has been a long one, the point is simple. The phrasing of an implicit definition, on its own, does not get us very far. One needs to show that this succeeds in characterising a notion. One then has to show that the notion characterised has the properties that one claims it does. To establish this, one needs to make inferences from the defining conditions. What we have just seen is that one can make the case that $, characterised semantically, rules out surplus content only by begging the question in one way or another.[13]

3   Self-Reference

Of course, none of these considerations show that negation does have surplus content. That is the onus of arguments for dialetheism—which brings us to one of these, the paradoxes of self-reference. Slater finds flaws in the arguments involved in the liar paradox and the heterological paradox.

For the liar paradox, he claims (p. 4) that truth is not a property of sentences, but of propositions, and that once one sees this, it is impossible to come up with the sort of self-referential proposition required for the liar. Now, for myself, I see nothing wrong with taking sentences to be truth bearers, provided that we are talking about *interpreted* sentences here (and not simply grammatical strings), and that the sentences do not contain index-ical phrases, such as 'I' and 'now'. But let us grant that it is propositions which are to be regarded, strictly speaking, as truth bearers. The claim that one cannot come up with the appropriate self-referential propositions

---

[13]Before we leave the notion of negation, a quick comment on two other points. Slater claims (p. 6) that dialetheists have a problem expressing denial, and that one can have inconsistent beliefs only by being muddled. Both claims are false. Denial is a speach act that is not the same as the assertion of a negation; and the "preface paradox" shows that rational people can have inconsistent beliefs. On both of these matters see, e.g., Priest (1993), and (200a), ch. 6.

is false. One can do this with appropriate demonstratives, as in 'this is false', where 'this' refers to the proposition the sentence expresses.

Slater claims (p. 4) that there cannot be such propositions, since they would have to be part of themselves. But propositions are abstract objects, and it is not at all clear that they cannot contain themselves as parts. Indeed, using representations of propositions employing non-well-founded set theory, one can show that there are self-referential propositions of exactly the kind employed in the liar paradox—as Barwise and Etchemendy have demonstrated.[14]

Another way to obtain an appropriate self-referential sentence is as follows. Those who take sentences to express propositions make use of those very words. But now we can employ these words to formulate the sentence: the proposition expressed by this sentence is false. Reasoning as usual leads to contradiction.[15] Putting matters this way invites the objection that the description 'the proposition expressed by this sentence' fails to refer. This may be because the sentence expresses no proposition or because it expresses more than one proposition. Of course, merely to moot this possibility is not to solve the paradox (more of this in a moment): one needs to give reasons to suppose the claim to be true. Using the Barwise-Etchemendy account of propositions it is demonstrably false.

In any case, the move is of no avail, since we can formulate an extended version of the paradox:

(*) either this sentence expresses no unique proposition or it expresses a
      false one.

If it expresses a unique proposition, we have the usual contradiction. And now if we claim that (*) does not express a unique proposition, since, presumably, we must be taking ourselves to be expressing a true proposition in saying this, (*) would appear to express a true disjunctive proposition.

Slater's account of the heterological paradox is somewhat different. He diagnoses a suppressed premise, concerning ambiguity, in the argument. The argument can then be taken as a *reductio* of that premise. Actually, this is a quite general strategy for attempting to solve the self-referential paradoxes. Take some premise, A, in the argument involved, claim that this is true only on condition that B, and then deny B. We have, in fact, just looked at a version of this strategy in connection with the liar paradox (where B was a claim to the effect that a certain sentence does not express a unique proposition). But just because this strategy is available to virtually *any* supposed solution to the paradoxes, it is worth very little unless there

---

[14]Barwise and Etchemendy (1987), esp. chs. 3, 4.
[15]See Asher andr Kamp (1986).

is independent reason for supposing that B fails—and even then, as we have already seen, this strategy may still fail, due to extended paradoxes.

In the case of the Heterological paradox, the suppressed premise that Slater claims to find is that the heterological predicate, H—that is, in the notation he uses, $\exists Y(xRY \wedge \neg Yx)$—is univocal. Now the predicate H certainly does not appear to be ambiguous, as Slater seems to concede (p. 8). He offers an argument to the effect that it is; but I must confess that I am at a loss to make anything sensible of it. The thought appears to be that the denotation of 'self' is context-dependent. Hence, the meaning of 'self-applicable' depends on the context. Whether or not 'self' is generally context-dependent, in 'x is self-applicable', 'x is self-pitying', 'x is self-aggrandizing', etc., the 'self' is just an anaphoric back reference to x. No context is needed to determine this.

Worse, even if H is ambiguous, we can just disambiguate in an appropriate fashion and run the argument for the relevant sense. Note that if H is ambiguous, this is presumably due to an ambiguity in the predicate R ('refers to'). But if this is ambiguous, so is Slater's claim that H is ambiguous: $\exists X \exists Y(\langle H \rangle RX) \wedge \langle H \rangle RY \wedge X \neq Y$. He, I assume, had a particular sense of 'refer' in mind.

Finally, if one is prepared to talk of properties, one can formulate the Heterological paradox simply in terms of them. H, now, is the property of not applying to itself, $\lambda P(\neg PP)$. We then have:

$$HH \leftrightarrow \lambda P(\neg PP)H \leftrightarrow \neg HH$$

This invites a reply (which Slater, in effect, makes) to the effect that the lambda term fails to pick out a unique property. Let us call such a term 'undefined', and write Ux for 'x is undefined'.

We now have to face the issue of the truth conditions of sentences of the form $\lambda P(A)Q$ when $\lambda P(A)$ is undefined. There would seem to be (at least) three options here, in accord with standard policies of defective reference:

$\lambda P(A)Q$ is false

$\lambda P(A)Q$ is true if *some* denotation of '$\lambda P(A)$' applies to some (or every) denotation of 'Q' (and false otherwise)

$\lambda P(A)Q$ is true if *every* denotation of '$\lambda P(A)$' applies to some (or every) denotation of 'Q' (and false otherwise)

Each would seem to make perfectly good sense, giving rise to a different notion of predication. We will operate in terms of the first (in line with Russell's theory of definite descriptions).

Having got this straight, we can now formulate an extended version of the paradox. By standard fixed-point constructions, we can generate a predicate, H^*, of the form $\lambda P(\neg PP \vee U \langle H^* \rangle)$ (the property of not applying to itself or of this specification being defined). We then have:

$$H^*H^* \leftrightarrow \lambda P(\neg PP \vee U \langle H^* \rangle)H^* \leftrightarrow (\neg H^*H^* \vee U \langle H^* \rangle)$$

If H^* is defined, we have the usual contradiction. If not, then the right hand side is true. So, then, is H^*H^*. So H^* is defined. We are back with contradiction.

BIBLIOGRAPHY

[1] N. Asher and J. Kamp (1986), 'The Knower's Paradox and Representational Theories of Attitudes', pp. 131-47 of J. Y. Halpern (ed.), *Theoretical Aspects of Reasoning about Knowledge (Proceedings of the 1986 Conference)*, Los Altos, CA: Morgan Kaufmann.

[2] J. Barwise and J. Etchemendy (1987), *The Liar: an Essay on Truth and Circularity*, Oxford: Oxford University Press.

[3] R. K. Meyer and R. Routley (1973), 'Classical Relevant Logics, I', *Studia Logica* 32, 51-68.

[4] G. Priest (1987), *In Contradiction*, Dordrecht: Kluwer Academic Publishers; second edition, Oxford: Oxford University Press, forthcoming.

[5] G. Priest (1990), 'Boolean Negation and All That', *Journal of Philosophical Logic* 19, 201-15; a revised version appears as ch. 5 of Priest (200a).

[6] G. Priest (1993), 'Can Contradictions be True? II', *Proceedings of the Aristotelian Society, Supplementary Volume* 67, 35-54.

[7] G. Priest (200a), *Doubt Truth to be a Liar*, Oxford: Oxford University Press, forthcoming.

[8] G. Priest (200b), 'Paraconsistency and Dialetheism', Ch. 3, Vol. 8 of D. Gabbay and J. Woods (eds.), *Handbook of the History of Logic*, Amsterdam: Elsevier, to appear.

[9] A. Prior (1960), 'The Runabout Inference-Ticket', *Analysis* 21, 38-9; reprinted as pp. 129-31 of P. Strawson (ed.), *Philosophical Logic*, Oxford: Oxford University Press, 1967.

[10] B. H. Slater (1995), 'Paraconsistent Logic?', *Journal of Philosophical Logic* 24, 451-54.

[11] B. H. Slater (200a), 'Dialetheias are Mental Confusions', this volume.

Graham Priest
Philosophy, University of Melbourne, Australia.
g.priest@unimelb.edu.au

Response to Priest

HARTLEY SLATER

There is no difficulty in seeing where Priest has gone wrong in each of his many arguments against 'Dialetheias are Mental Confusions'.

In section 1 Priest misses, yet again, the elementary point that if valuations are functions from formulas to the values 1, 0, -1, then some value's being greater or equal to 0 is subcontrary, and not contradictory to its being less than or equal to 0. What is contradictory to the value being greater or equal to 0 is its being -1. In section 2 Priest finds difficulties with several ways in which a 'classical logician' might define a contradictory forming operator, but his objection to the first of these is that 'we could ... prove everything, given only that we have the T-schema and some way of forming self-referential truth-bearers'. But I have denied that the latter is possible in the propositional case.

In section 3 there are several further misunderstandings. First Priest represents my argument against self-referential propositions as being 'they would have to be part of themselves', objecting that 'propositions are abstract objects, and it is not at all clear that they cannot contain themselves as parts'. But my argument was about the impossibility, for any 'p', of 'it is not true that p' being itself 'p', i.e. it was about the mereology of certain concrete objects, sentences. There follows a clear *non-sequitur* in Priest's analysis of the meaning of the sentence 'either this sentence expresses no unique proposition or it expresses a false one'. He says 'If it expresses a unique proposition we have the usual contradiction. And now if we claim that [it] does not express a unique proposition... [it] would appear to express a true disjunctive proposition'. So there is still a problem? But there is no problem if the sentence does not express a unique proposition, since the disjunctive proposition expressed is then only one of the propositions expressed.

Priest argues against ambiguity in such cases by saying 'Of course, merely to moot [the possibility that a sentence expresses more than one proposition] is not to solve the paradox...one needs to give reasons to suppose the claim to be true', and later, 'Actually, this is a quite general strategy for attempting to solve the self-referential paradoxes. Take some premise, A, in the argument involved, claim that this is true only on condition B, and then deny B...But [this strategy] is worth very little unless there is independent reason for supposing that B fails'. The boot, however, is on the other foot, if anyone wants to maintain there is provably a paradox in such cases. If A leads to paradox, and A is true only if B, then to prove there is actually a paradox one must show that B definitely is true. The onus of proof is on Priest, therefore, to demonstrate conclusively that there is no ambiguity in such a case as that above.

And likewise in the case of Heterologicality. First, in my paper, I gave two formulations of 'x is not self-applicable', one by Copi involving an explicit uniqueness clause, and one by Priest, lacking such a clause. I pointed out that, as a consequence, there was an undischarged assumption, and specifically an unproved supposition of the above kind, in Priest's (and also Copi's) derivation of a contradiction. In response Priest considers only his own formalisation saying my claim was that his formal expression might be ambiguous. But no: again, the claim was merely that Priest's formal expression lacked an explicit uniqueness clause, and that proof of the uniqueness expressed by that clause was needed before a contradiction could be conclusively obtained. Then, finally, Priest misses my point about the context sensitivity of reflexive pronouns proving ambiguity in the case of 'is not self-applicable'. He had excluded, shortly before, indexicals such as 'I' and 'now', but thinks the same caution does not arise with 'self'. He says 'in 'x is self-applicable'...the 'self' is just an anaphoric back reference to x. No context is needed to determine this'. Still, the value of 'x' is needed to make determinate the property expressed by 'is x-applicable'. As it stands this predicate expresses a functional property, i.e. a property which varies with the value of 'x'. So it cannot be formalised using the constant lambda term which Priest goes on to provide. The error, of course, is a big one, and goes back to Frege. For more on this, as well as propositional liars, and the location of the ambiguity in the sentential T-scheme see my 'A Poor Concept Script' in the (electronic) *Australasian Journal of Logic*, 2004, and 'Choice and Logic' in the *Journal of Philosophical Logic*, 2005.

Hartley Slater
Philosophy, University Western Australia, 35 Stirling Highway, Crawley W.A. 6009WA, Australia.
slaterbh@cyllene.uwa.edu.auM207:

Contradiction: The Real Philosophical Challenge for Paraconsistent Logic

CATARINA DUTILH NOVAES

1 Introduction

"The negation is one of the most important topics of (philosophical) discussion concerning paraconsistent logic."[1] This statement is not nearly as trivial as it may seem at first sight. Within the paraconsistent framework, a pair of propositions that, according to classical logic, would form a contradiction may in some circumstances belong to the same set of propositions, without provoking the 'explosion' of this set, as classical logic would have it. Clearly, the main concept here at stake is that of contradiction. But since the concepts of negation and contradiction have been conflated in the logical developments of roughly the last century, it is often taken for granted that a reformulation of the notion of negation is the most pressing task for paraconsistency.

In what follows, I will try to disentangle the misunderstanding that consists in identifying contradiction with negation; a historical analysis will show that, in the development of logic, various (legitimate) conceptions of negation were not directly associated with that of contradiction. Insofar as negation is a syntactical notion, whereas contradiction is essentially a semantic notion, they are independent concepts. Important philosophical conclusions for paraconsistency can be drawn from this analysis: I argue that paraconsistent systems are not constrained by certain specific requirements in order to feature 'the real negation', but that they still have to give an account of the notion of contradiction.

The profound revision of classical logic proposed by paraconsistent logicians bears primarily upon the classical concept of contradiction, and exactly for this reason it is mistaken to simply assume that a paraconsistent

---

[1] Throughout this text, I will use the term 'paraconsistent logic' in the singular, even though there exist various different paraconsistent systems. With 'paraconsistent logic' and 'paraconsistency' I will be referring to the conceptual base that all those systems have in common, assuming that this simplification will not be harmful in the present context.

account of this concept is not needed. In fact, I will argue that a total dismissal of the concept of contradiction, understood as pairs of propositions which cannot be true together (and cannot be false together), is a move just as trivializing as the dismissal of all inconsistent sets of propositions as explosive. Paraconsistency must give an account of the notion of contradiction within a paraconsistent system; if this is successfully done, then an enrichment of the very framework of paraconsistency is likely to be achieved.

This paper is composed of three parts. In part one, a historical analysis of the notion of negation is sketched, yielding the conclusion that negation is not the greatest challenge for paraconsistency. In part two, the need for a paraconsistent account of the notion of contradiction is argued for, and some possible answers are presented. In part three, I draw concluding remarks.

2   Negation

2.1   The real negation

Some of the attacks against paraconsistent logic were based on the argument that the paraconsistent negation (in its different versions, according to the system in question) is not the 'real negation'. This kind of argument has been put forward even by proponents of paraconsistency, referring to other paraconsistent systems:

> That an account of negation violates the law of non-contradiction therefore provides prima facie evidence that the account is wrong. This is the second piece of evidence that da Costa negation is not negation. [PR89][p.164/5]

In [Sla95], Slater correctly notes that, in any system where the negation is defined as usual, but where A is a proposition of the system and A and $\neg A$ can both be true, '$\neg$' is not a contradiction-forming functor, but rather a sub-contrariety forming functor, given the very definitions of contradiction and sub-contrariety. Indeed, the crucial concept underlying paraconsistent logic is precisely that, for some proposition A, A and $\neg A$ can both be true. Hence, thus defined, the negation of any paraconsistent system[2], not only da Costa's, is bound to be a sub-contrariety forming functor. From this fact, Slater concludes that no paraconsistent negation can be the real negation, and therefore that there is no paraconsistent logic [Sla95][p.451].

It is obvious that, according to this view, the so-called 'real negation' is propositional negation, in particular insofar as it is a contradiction-forming functor. Such attacks are neutralized if it is shown that the conflation

---

[2]But notice that paraconsistent systems often feature two or more kinds of negation.

between contradiction and negation is not legitimate; thus, there would be nothing intrinsically wrong with the negation having properties other than contradiction formation in a given logical system.

Another traditional way of defining the 'real negation' is to postulate compliance with the law of the excluded middle (LEM) - written $A \vee \neg A$ - and to the law of non-contradiction (LNC) - written $\neg(A \wedge \neg A)$. But again, this approach simply begs the question, for the use of the '$\neg A$' notation already presupposes the contradiction-forming negation. What the observance of LEM and LNC defines is the concept of contradiction, and that can be done without using the negation: A and B are contradictory propositions iff $A \vee B$ holds and $A \wedge B$ does not hold, regardless of the form of A and B.

A quick look at the history of logic shows that the (syntactical) notion of negation and the (semantic) notion of contradiction are in fact not only conceptually but also historically independent, even though there is an obvious connection between them.

2.2   Historical remarks: in search of the real negation

The history of the concept of negation in logic is not a new topic[3], and an extensive analysis of it falls out of the scope of the present text, so my aim with the historical sketch below is a modest one.[4] Yet, I hope to be able to convince the reader that there is more in the world of negation than contradiction-forming propositional negation.

In fact, the association between negation and contradiction is, historically, anything but predominant; indeed, the longest-lasting notion of negation is the notion of term or copula negation. This fact is related to the predominance of the Aristotelian term-based paradigm in logic, which was dominant up to Kant (with a few honorable exceptions such as Stoic logic). It only became entirely surpassed by the propositional paradigm in the 19th / 20th century, in particular as a result of the influence of Frege's works. Hence, the change occurred with respect to the notion of negation appears simply to have accompanied the more general change of paradigm in logic, from term logic to proposition-based logic.

Aristotle

First of all, it must be noticed that the notion of contradiction in traditional Aristotelian logic does not have a straightforward syntactical counterpart. In *De Interpretatione* 19b5 - 20b12 [Ari28], Aristotle analyzes the different effects obtained by the addition of the particle 'not' in different

---

[3]Horn's A *Natural History of Negation* [Hor89] is an authoritative study on this topic.

[4]I will only approach the so-called Western (European) tradition in logic, which is the relevant one for this discussion anyway.

positions of the subject-copula-predicate schema, indicating thus that there
is no uniform treatment of contradiction in terms of the negation. In this
passage, the famous Square of Opposition is defined. Worth noting is the
fact that, as depicted in the Square, contradiction is only one of the relations of opposition between propositions related to the negation, the others
being contrariety and sub-contrariety. Moreover, according to Aristotle in
this passage, there exists the possibility of iteration of negation particles,
generating unlikely propositions such as *Not man is not not-just* (19b36).

Contradiction is indeed a strictly semantic notion for Aristotle: 'the most
indisputable of all beliefs is that contradictory statements are not at the
same time true' (*Met.* 1011b12-13). Propositions such as 'Socrates is alive'
and 'Socrates is dead' are contradictory in virtue of the meaning of their
terms (they cannot be both true at the same time and cannot be both false
at the same time[5]), and not in virtue of the presence or absence of negating
particles.

Moreover, nothing remotely similar to the concept of contradiction-forming
propositional negation can be found in Aristotle's writings. In fact, the first
propositional treatment of the negation occurred only in Stoic logic - the
Stoics were the first logicians of the Western tradition to take the unana-
lyzed proposition as the basic unit of their logical system. But the Stoics
distinguished three varieties of negation, so even in their system propositional negation did not reign absolute (Cf. [Hor89][p.21-23]).

(Later) medieval logic

Later medieval logic is, as is well known, extremely marked by the influence
of Aristotelian texts. Yet, some interesting developments in this period concerning the negation deserve our attention. In particular, it is worth noting
that, while non-propositional negation remained predominant, some 14th
century logicians were familiar with the concept of propositional negation
(probably an indirect influence from Stoic logic, although the path of influence has not yet been established with certainty by historians of logic - cf.
[Hor89][p.26]).

One of them was John Buridan. Buridan's works, such as his *Summula
de Suppositionibus* [Bur98], show a concern with the syntactical aspects of
logical analysis that often goes beyond most of his contemporaries' mild
interest in syntax (which was rather the domain of grammarians, not of
logicians). In this text, he contrasts the concepts of 'negating negation'
[*negatio negans*] and 'infinitizing negation' [*negatio infinitans*]: the former
affects (ranges over) both terms of the proposition, while the latter affects
(ranges over) only a given part of the proposition (cf. [Bur98] [p.57-59]).

---

[5] Disregarding possible concerns with vagueness.

Although not exactly identical to our concept of propositional negation, Buridan's negating negation is obviously a deviation from the traditional Aristotelian term negation.

However, once more the relation between negation and contradiction is not at all as tight as it could be expected, not even in the case of negating negation. In fact, the semantic effect of the negation is rather explained in terms of the 'distribution' of the terms, which has no direct connection with the concept of contradiction.[6]

The establishment of negation as a contradiction-forming functor

The emergence of contradiction-forming negation happened, as to be expected, simultaneously with the emergence of proposition-based logic. G. Frege is usually considered to be its pioneer.[7]

Frege's definition of the negation as a function that maps a truth-value into the opposite truth-value was certainly one of the main motors behind the dissemination of the propositional concept of negation in the 20th century. Here is his characterization of the negation function (notice that contradiction is not explicitly mentioned):

> The next simplest function, we may say, is the one whose value is the False for just those arguments for which the value of $—x$ is the True, and, conversely, is the True for the arguments for which the value of $—x$ is the false. I symbolize it thus
>
>
>
> and here I call the little vertical stroke the stroke of negation. [Fre80][p.34/5]

Frege's rather unhandy notation did not survive, but the concept of negation as a function (more precisely called a functor later on) mapping truth into falsehood and vice-versa is, as we all know, alive and kicking. Frege is indeed the main source of our current notion of negation, since Whitehead and Russell virtually copied Frege's definition of it in *Principia*, only introducing symbolic simplifications (and, arguably, conceptual impoverishment); now, the logical system presented in *Principia* is roughly propositional and predicate logic as we know it.

---

[6]For more on the concept of distribution, see [Hod98].

[7]The role of Frege's predecessors in preparing the terrain for his work and the elements that he borrowed from them are often neglected. It is evident however that he did not make it all up by himself, and there is interesting work being done on the connections between Frege and his predecessors. However, he certainly produced the first comprehensive system of proposition-based logic.

The Contradictory Function with argument p, where p is any
proposition, is the proposition which is the contradictory of p,
that is, the proposition asserting that p is not true. This is
denoted by $\sim p$. Thus $\sim p$ is the contradictory function with p as
argument and means the negation of the proposition p. It will
also be referred to as the proposition *not-p*. Thus $\sim p$ means
not-p, which means the negation of p. [WR10][p.6]

Frege's and Whitehead/Russell's notion of the negation as the syntactical counterpart of contradiction have left such deep marks in the logical
panorama since then[8] that historians of logic must, from time to time, re-
mind logicians that the contradiction-forming negation is by far not the
only and not even the most primitive concept of negation to be traced in
the development of logic.

2.3   Conclusion: there is no real negation

Hence, paraconsistent negation is in principle as real a negation as any
other.

It is high time that the widespread conflation between negation and contradiction be challenged; negation and contradiction are two separate notions, which, due to a narrow understanding of the negation, were assimilated to one another in modern logic. Negation has become the syntactical
counterpart of contradiction, and that is, in itself, a most welcome development: it is certainly convenient to have a simple syntactical device to
express so fundamental a notion such as that of contradiction. However, if
one takes a glimpse at the history of negation, as we have just done, it is
evident that propositional negation is only one of the many sorts of negation
that have been in operation throughout the history of logic, and most of
these negations are not contradiction-forming functors.

The various kinds of negation are also widely known to linguists and
semanticists, who have been working with the idea of different sorts of
negation in natural languages for many years - interestingly, many of them
argue that propositional negation simply isn't the most convenient and natural concept of negation when it comes to studying natural languages (Cf.
[Hor89]).

In sum, it seems that paraconsistent logicians have a considerable (but
not unlimited) degree of freedom to temper with the concept of negation
in the systems they develop, since the history of logic and the analysis of

---

[8]'Deviations aside, it is indisputable that the Fregean model has carried the day.
The syntax of negation in the first-order predicate calculus is simply $\sim p$, where p is
any proposition. The semantics is equally straightforward, at least if presuppositional
phenomena are ignored: $\sim p$ is true if and only if p is false.' (Horn 1989, 43).

natural languages show that there exists a plethora of different concepts of negation, none of them being prima facie more legitimate than the others.

3   Contradiction

3.1   Contradiction is the real challenge for paraconsistency

However, that negation is not as serious a challenge for paraconsistency as it might be expected only means that the paraconsistent logician is left with another, perhaps more complicated problem. The liberty concerning the notion of negation is not extended to the concept of contradiction; an account of this concept within the framework of paraconsistency must be formulated. Unlike negation, the concept of contradiction did not undergo historical variations, so the plausibility of alternative formulations of this concept is harder to argue for.

Traditionally, contradiction is the property of a pair of propositions which cannot both be true and cannot both be false at the same time; well, the core idea of paraconsistency is precisely that in some circumstances (in belief contexts, in certain scientific theories), two propositions that are contradictory according to classical logic - henceforth, C-contradictions - can in some sense be held true at the same time. But if this is the case, then the very definition of contradiction is violated and, therefore, is no longer applicable. So paraconsistent logicians must give an account of what contradiction amounts to within a paraconsistent system. Which are, if any, the pairs of propositions that, within paraconsistent logic, should not be both held to be true/false at the same time, that is, that are P-contradictions?

3.2   Constradition + logical consequence = explosion

Traditional notion of contradiction

I have already defined contradictory propositions as those which cannot be true and cannot be false together. But the concept of contradiction deserves a closer inspection. 'Etymology, of course, only tells us what words do not mean any longer.'[9] So etymology should be used with caution when it comes to understanding a concept, yet it can no doubt be useful.

The term 'contradiction' has its origin from the Latin *contradictio*, the corresponding verb being *contradicere*, to contradict. The particle *contra* expresses opposition, and *dicere* means 'to say'. Thus, it simply means 'to say the opposite'. The Greek term, *antiphasis*, means exactly the same, the particle *anti* expresses opposition and *phasis* is the noun corresponding to the verb *phanai*, which also means 'to say', 'to speak'.

Thus, properly speaking, the term 'contradiction' applies only to entities

---

[9] J. Hoyrup 2003, 'Bronze Age Formal Science?' University of Roskilde, pre-print, p.1.

belonging to the linguistic realm: propositions, sentences, statements can
be contradictory to each other. On the ontological realm, contradictions
would correspond approximately to impossibilities, that is, to situations
that cannot obtain. Derivatively, one often says that facts are contradictory when they are not compossible, although properly speaking only the
propositions/sentences describing them can be said to form a contradiction.[10]   Properly speaking, contradiction is the linguistic counterpart of
ontological impossibility.

In model-theoretic terms, a pair of contradictory propositions is such
that there is no model that satisfies them both, and there is no model that
falsifies them both. Similarly, in possible-world semantics terms, a pair of
contradictory propositions is such that there is no possible world in which
both are true and there is no possible world in which both are false. Treating
models and possible worlds as roughly equivalent notions[11], we have:

- A and B are contradictory propositions

- $\Leftrightarrow$ There is no model/possible world $\aleph$ such that $\aleph \models A$ and $\aleph \models B$
  and there is no model/possible world $\aleph'$ such that $\aleph' \nvDash A$ and $\aleph' \nvDash B$.

Traditional notion of logical consequence

It is in fact inaccurate to talk about **the** traditional notion of logical consequence, as much as it is inaccurate to talk about **the** real negation. In
the history of logic as much as in current developments, there are a variety
of notions of logical consequence being put to use. But one can said with
reasonable certainty that the model-theoretic notion of logical consequence
has dominated the logical landscape for the last five or six decades, under the influence of Tarski's achievements. But the debate concerning the
'correct' notion of logical consequence still goes on, involving partisans of
modal, epistemic and semantic notions of logical consequence.[12]

My aim in this section is to show that the *ex falso* rule, responsible for
the phenomenon of explosion of inconsistent sets of propositions, is not
a unnatural rule - a logicians' trick -, as some seem to think, but rather
a very natural corollary of the combination of the traditional notion of

---

[10]'Getting back to the Latin etymology, there is not much sense in saying that facts
'contradict' each other (in Latin *contradicere*, where *dicere* means to tell, to speak, to
say), since facts do not 'say' anything' [...] (Bobenrieth 1998, 29).

[11]Which they obviously are not, but for the present purposes, this assimilation is
harmless.

[12]The debate on logical consequence was re-ignited by J. Echemendy's 1990 book *The
Concept of Logical Consequence*. A list of the main conceptions of logical consequence
can be found in S. Shapiro, 'Logical Consequence: Models and Modality', in M. Schirn
(ed), *The Philosophy of Mathematics Today*, Clarendon, Oxford, 1999 (p. 132).

contradiction and (some of) the traditional notion(s) of logical consequence. The notions of logical consequence that are particularly vulnerable to the *ex falso* paradox are the modal and the model-theoretic ones, so I will concentrate on these. (I disregard thus the epistemic and semantic notions).

The traditional modal definition of logical consequence, possibly to be found in Aristotle, but certainly stated explicitly in some medieval treatises, goes as follows:

- A set of propositions Δ implies a proposition P iff it is impossible for Δ to be true and P false.

Given the notion of contradiction sketched above, that is, as the linguistic counterpart of impossibility, we have:

- If Δ is contradictory, it cannot be true, then a fortiori it cannot be true while any proposition P is false.

A simplified formulation of the model-theoretic definition of logical consequence would be:

- A set of propositions Δ implies a proposition P iff all models that satisfy Δ also satisfy P.

A corollary of this formulation and of the model-theoretic notion of contradiction is:

- If Δ is contradictory, it is not satisfied by any model, then a fortiori all models that satisfy Δ (namely, none) also satisfy any proposition P.

Explosion

From both corollaries to the *ex falso* rule is just a step. If a set of propositions Δ cannot be true, it cannot be true while any proposition P is false. Therefore, according to the modal definition of logical consequence, any proposition follows from a contradictory set Δ.

Similarly, if a set of propositions Δ is satisfied by no model at all, then the class of models satisfying Δ (the empty class) satisfies any proposition P. Therefore, according to the model-theoretic definition of logical consequence, any proposition follows from a contradictory set Δ . Thus, contradictory premises validate any conclusion - *ex falso sequitur quodlibet*. Representing logical consequence by '⊢', we have:

- If Δ is contradictory, then, for all propositions P, $\Delta \vdash P$.

- ⇔ If Δ is contradictory, Δ 'explodes'.

Therefore, if one wants to avoid this phenomenon, one has to reformulate at least one of the two notions that are its cause, namely the notions of contradiction and of logical consequence. Relevance logicians investigate the effect of reformulating the notion of logical consequence; paraconsistent logicians choose the path of reformulating the notion of contradiction.

3.3   Paraconsistent objection: 'Not every' or 'No'?

In its early days, the birth of paraconsistent logic was motivated by the realization that many theories (in particular in the natural sciences), which had been used with great success for centuries, turned out to be inconsistent when axiomatized. But their very success (in particular with respect to predictions) was the proof that they were not trivial, that is, that it was not the case that any proposition could be derived from them (since some propositions could clearly **not** be derived from these theories; for example, that an apple would float in the air if thrown upwards cannot be derived from classical mechanics). In other words, at least some inconsistent theories did not explode, as classical logic would have it (as a consequence of *ex falso*).

Thus, originally, paraconsistency was not a challenge to the very notions of contradiction, consistency and explosion; it was rather the realization that contradiction was perhaps not the best criterion for identifying the explosion and trivialization of a theory.

Paraconsistency thus challenged the statement:

- (1) Every inconsistent set of propositions explodes.

As any person minimally familiarized with the principles of logic knows, the contradictory of a universal statement is an existential statement. Thus, challenging (1) simply amounts to claiming:

- (2) Not every inconsistent set of propositions explodes.⇔
- (2') Some inconsistent set of propositions does not explode.

Notice that (2) is perfectly compatible with:

- (3) Some inconsistent set of propositions explodes.

But it seems that, at least for many people (both among the critics as well as among the supporters of paraconsistent logic), the paraconsistent dismissal of (1) amounted to endorsing a negative universal statement:

- (4) No inconsistent set of propositions explodes.

Endorsing (4) seems to be a more radical form of commitment to paraconsistency, whereas endorsing (2) would be a milder form of this commitment.
Moreover, these two positions obviously have different answers to the issue
of contradictions within paraconsistency. I will call the position that adheres to (4) **Answer 1**, and the position that adheres to (2) (and thus to
(2') and possibly (3)) **Answer 2**. I will also argue that the position defined
by (4) - Answer 1 - is not a more radical version of paraconsistency, but
rather a position that comes dangerously close to trivialism.

Either way, a revision of the notion of contradiction is very much needed,
as the following argument shows:

- At least some contradictory set of proposition does not explode.

- $\Leftrightarrow$ Not every proposition can be inferred from any inconsistent set of
  propositions.

- $\Leftrightarrow$ There is an inconsistent set Δ and a proposition P such that $\Delta \nvdash P$

- $\Leftrightarrow$ (Def. of consequence) There is a model/possible world $\aleph$ such that
  $\aleph \vDash \Delta$ and $\aleph \nvDash P$.

- $\Leftrightarrow$ (A fortiori) There is an inconsistent set of propositions Δ and a
  model/possible world $\aleph$ such that $\aleph \vDash \Delta$.[13]

- $\Leftrightarrow$ There is a pair of contradictory propositions A and B such that
  $\aleph \vDash A$ and $\aleph \vDash B$.

Thus, paraconsistency clearly violates the very definition of classical contradiction, C-contradiction. It must therefore develop its own notion of contradiction, P-contradiction. In particular, the classical notion of contradiction does not hold anymore in terms of its extension, that is, the set of pairs
of C-contradictions is not equivalent to the set of pairs of P-contradictions.
Answer 1 corresponds to holding that the set of P-contradictions is empty,
whereas Answer 2 sets up the task of defining the members of the set of
P-contradictions.

Answer 1

One answer to the issue of contradictions within paraconsistency is the
contention that:

- No inconsistent set explodes.

---

[13]Of course, paraconsistency must also give an account of the counterintuitive idea of
'true contradictions' (which is the field of dialethists) and of the putative existence of
'impossible situations', impossible worlds (on that, see the issue of *Notre Dame Journal
of Formal Logic* dedicated to the topic of impossible worlds, 38(4) / 1997).

That is, just as much as, for classical logic, all inconsistent sets of proposition are equally 'bad', for a paraconsistent logician adhering to Answer 1, all inconsistent sets of propositions are equally 'good'. The task of the paraconsistent logician would be to provide technical tools to cope with C-contradictions, but not to develop ways to differentiate one contradiction from another. According to this approach, to the eyes of paraconsistent logic, all contradictions are the same.

Therefore, the notion of contradiction would be an idle notion for paraconsistency, and the set of P-contradictions would be empty. Any pair of propositions can be considered P-consistent. This position also assumes that paraconsistent systems are immune to paradox, that is, to the derivation of two different propositions that cannot reasonably be held to be true at the same time. If this is the case, then the comparison between two competing paraconsistent systems cannot be resolved on the basis of paradox-based criteria.

This approach may be fruitful, especially if paraconsistent logic is primarily thought to have a meta-function with respect to inconsistent theories, in the sense of defining the inferential mechanisms used by the proponents of these theories in their reasoning (since classical logic cannot provide a rationale for non-trivial inference-making from inconsistent sets).

From a philosophical perspective however, this development is somewhat disappointing. The fundamental insight of paraconsistency is that the concept of contradiction is not sufficiently fine-grained to discriminate between 'good' and 'bad' theories. But in the present account, paraconsistent logic is even less able to discriminate between good and bad theories; if classical logic at least has a criterion, albeit a crude one (consistency-contradiction), the paraconsistent perspective seems to be left with none, if it cannot define which pairs of propositions should definitely not be both derivable from a theory. It is as if, from the paraconsistent perspective, all inconsistent theories were equally good, and that is just as trivializing as the classical view according to which all inconsistent theories are equally bad.

Answer 2

By contrast, Answer 2 seems to be more in the spirit of paraconsistency, that is, the general refinement of the notion of rationality beyond the coarse criterion of consistency. Answer 2 takes the crucial paraconsistent contention to be an existential claim, that is, that some inconsistent sets of propositions do not explode, and therefore accepts (the possibility) that some inconsistent sets of propositions do explode, even within a paraconsistent framework.

- There are some pairs of propositions that, within a paraconsistent system, cannot and should not be true at the same time.

The recently introduced concept of hyper-contradiction (Cf. [Bre]) can be seen as the realization that some C-contradictions can be harmful even to paraconsistent systems, that is, that P-contradictions can rise in a paraconsistent system, and that they may cause the explosion (trivialization) of the system all the same. The system being considered on the object-level, these hyper-contradictions should be avoided, since they are taken to be a sign that something is very wrong with the system in question.

In sum, the results related to the notion of hyper-contradiction show that paraconsistent systems are not immune to paradox, and therefore that paraconsistent logicians should address the question of which pairs of propositions they do not want to see derived from their systems, namely those that would cause the trivialization even of a paraconsistent system.

Changing the perspective, and considering the case of a paraconsistent system being used as a meta-logic to analyze a scientific theory or the like, it is also the task of paraconsistent logic to define P-contradictions, that is, contradictions that are so threatening to a theory that they really compromise rational inference-making within it. Of course, many theories that, when formalized, turn out to be inconsistent are not trivial, but possibly many theories that are inconsistent are simply so badly inconsistent that they are in fact trivial, even if treated with the paraconsistent apparatus. This 'bad' kind of inconsistency can be quantitative (too many C-contradictions may be a sign that even paraconsistency cannot save the theory) or qualitative - that is, the C-contradiction in question is so strong (for example, the fact that the core statements of the theory can be proved both true and false) that it is also a P-contradiction, a contradiction that even a paraconsistent logician cannot accept.

Therefore, within a paraconsistent system, the C-contradictory of a proposition that is not also a P-contradictory of this proposition can be represented by a negation symbol which is indeed a a sub-contrariety forming functor, as argued by Slater. But, in the same system, it can be expected that the behavior of pairs of P-contradictions would follow the usual behavior of Boolean, contradictory forming negation. In other words, it would appear that the semantic difference between C-contradictions and P-contradictions in a paraconsistent system amounts, on the syntactical level, to the presence of (at least) two kinds of negations, and that the negation corresponding to P-contradictions (which do cause explosion) would be a simple Boolean negation. In effect, many paraconsistent systems hitherto proposed feature two or more negations; in many cases, one of these negations is precisely a Boolean negation, which suggests that such systems are underlined by the (presumably sound) idea that not all contradictions are of the same kind.

This argument implies the assumption that the set of P-contradictions is a subset of the set of C-contradictions, and that is indeed a rather intuitive idea. But I cannot think of any conclusive argument against the existence of a P-contradiction that is not a C-contradiction.

For example, S. Shapiro has argued recently that, while a paraconsistent arithmetic containing its own semantics can be formulated in a recursively axiomatized formal system PA* (which seems to be good news for the dialetheist), it turns out that

> There are purely arithmetic (indeed, Π_0) sentences that are both provable and refutable in PA*. [...] For the thorough dialetheist, it follows that ordinary PA and even Robinson arithmetic are themselves inconsistent theories. I argue that this is a bitter pill for the dialetheist to swallow. [Sha02][p.817]

Contradictions formed by purely arithmetic sentences and their contradictories seem to be a case of contradictions that even the radical paraconsistent logician would not want to hold as 'true contradictions'. In other words, such contradictions would be good candidates for the list of P-contradictions.

In sum, the determination of the set of P-contradictions seems to me to be one of the most pressing tasks in the current stage of development of paraconsistent logic. It may turn out that this set is empty (as a defendant of Answer 1 might claim), but this claim must be argued for and not simply assumed as a corollary of the basic intuitions of paraconsistency, as I hope to have shown.

4   Conclusion

Since the very early stages of philosophy, consistency and contradiction have always been taken to be reliable criteria for rationality or the lack thereof. The Socratic strategy of refutation, described in Plato's dialogues, consists of forcing somebody to grant two contradictory propositions that are both derived from the position he/she is trying to defend, since this is taken to be a sign that the position in question is defective and therefore should be dismissed. Consistency and rationality have always gone hand in hand, as much as their duals, contradiction and irrationality.

There were, of course, dissident voices in the history of philosophy, claiming that an understanding of (ir)rationality based only on the notions of consistency and contradiction was too crude, or even simply dead wrong (Hegel, for example). But these voices never managed to convince the majority.

Hence, on the philosophical level, consistency has always been a fundamental component of the concept of rationality. But on the pre-theoretical level, on the level of how we actually perform non-trivial reasoning, it is arguable that inconsistency has never been so serious a threat as assumed by the theoretical accounts of rationality.

As already said, paraconsistent logic was in some sense born from the realization that consistency, in its classical sense, was not sufficient to discriminate between good and bad theories, exactly because our actual reasoning is, it seems, much more capable of coping with inconsistent premises than classical logic. Indeed, it has become a motto in non-classical circles that classical logic simply is not an accurate model of human rationality.

Therefore, a corollary of the paraconsistent position is that more fine-grained criteria for rational inference-drawing are needed. However, surprisingly, the paraconsistent position is often taken to be a sort of trivialism / relativism, a position according to which there are no criteria for rationality, whereas the original objection of paraconsistency was simply that the familiar criteria were not good enough. A natural development from this objection is the search for new, more accurate criteria, going beyond the crude dichotomy consistency / contradiction. It seems to me that this is (still) a crucial task for the paraconsistent enterprise.

One promising line of research, which has already begun to be explored in some paraconsistent circles, is the development of technical means to differentiate one C-contradictory set from another. Arguably, there are degrees of contradiction; some contradictions are just 'a bit' contradictory, whereas others are so hopelessly contradictory that, if these hopeless cases are derivable from a theory, not even a paraconsistent logician would be ready to jump into the sea to rescue the theory in question. These are precisely the contradictions that even a paraconsistent logician would not be willing to accept as true, the P-contradictions.

In any case, an important step towards the definition of P-contradictions is the dissociation of the notions of negation and contradiction, otherwise the debate is likely to take place only on the syntactical level and would probably not go very far (since the problem in question really is essentially semantic).

This paper was an attempt to convince the paraconsistent audience of the need and fruitfulness of defining the concept of P-contradiction, as well as its extension. Of course, as is often the case, the task of the philosopher is to outline problems to be solved, and then leave the hard work for the logician - in this case, that would amount to formulating precise, technical accounts of what I called P-contradictions. Such developments, it seems to me, could represent a significant enrichment of the framework of paraconsistency as a better model of human rationality.

Faculty of Philosophy - University of Leiden
Postbus 9515 2300 RA Leiden
The Netherlands
C.Dutilh-Novaes@let.leidenuniv.nl

BIBLIOGRAPHY

[Ari28] Aristotle, *The Works of Aristotle vol. 1.* Oxford, Oxford University Press, 1928.

[Bob98] A. Bobenrieth, 'Five Philosophical Problems Related to Paraconsistent Logic'. *Logique et Analyse* 41, 1998.

[Bre]   M. Bremer, 'Hyper-Contradictions'. In this volume.

[Bur98] J. Buridan, *Summula de Suppositionibus.* Nijmegen, Ingenium, 1998.

[Fre80] G. Frege, 'Function and Concept'. In *Translations from the Philosophical Writings of Gottlob Frege*, eds. P. Geach and M. Black, Oxford, Blackwell, 1980.

[Hod98] W. Hodges 'The laws of distribution for syllogisms'. *Notre Dame Journal of Formal Logic*, 39(2), 1998.

[Hor89] L. Horn, *A Natural History of Negation.* Chicago, University of Chicago, 1989.

[PR89]   G. Priest and R. Routley, 'Systems of Paraconsistent Logic'. In *Paraconsistent Logic - Essays on the Inconsistent*, eds. G. Priest, R. Routley and J. Norman, München, Philosophia, 1989.

[Sha02] S.Shapiro, 'Incompleteness and Inconsistency'. *Mind* 111, 2002.

[Sla95] B. H. Slater, 'Paraconsistent Logics?'. *Journal of Philosophical Logic* 24, 1995.

[WR10]   A. Whitehead and B. Russell, *Principia Mathematica.* Cambridge, Cambridge University Press, 1910.

Catarina Dutilh Nóvaes

Philosophy Faculty, University of Leiden, The Netherlands.

C.Dutilh-Novaes@let.leidenuniv.nl

Paraconsistency and the Consistency or Inconsistency of the World

ANDRÉS BOBENRIETH M.

ABSTRACT.   The purpose of this paper is to analyse the claim
that the world or reality is consistent, as well as the contrary thesis, i.e. that the world is inconsistent, in order to see how they can
be understood. Firstly, I will review the main definitions of consis-
tency/inconsistency with the purpose of showing several difficulties
that emerge if one wants to apply them to the non-linguistic components of the world. Secondly, I will present a classificatory schema of
the main philosophical views on the subject. Thirdly, several positions that have been proposed within paraconsistency that are particularly relevant to the subject will be briefly reviewed, followed by
some comments on Priest's view. Then, in sec. 5, I will argue that an
alternative view can unfold from radically assuming that the world or
mind-independent reality is neither consistent nor inconsistent, thus
postulating that inconsistency is not an ontological issue but a problem of rationality. In support of that I will claim that the main
notions behind inconsistency are negation and incompatibility, and
then point out that the traditional approach to the problem is based
on very strong realistic assumptions about them. Then, I will suggest that rather than remaining trapped there, one can adopt a form
of anti-realism about inconsistencies, which would allow us to deal
with several problems related with inconsistency in several fructifer-
ous ways. For instance, it may pave the way to give an e0planation
of both, the frequent appearance of inconsistency among our explanations and theories about the world, and our persistent search for
consistency (which I tried to do in Bobenrieth 2003: chap. VII).

1   Can consistency/inconsistency be applied to the world?

Several definitions of consistency can be found in reference books of philosophy and/or logic. They have great similarities, yet there are important

differences, particularly concerning the terms in which they are articulated
and to which logical notions they appeal.[1] One can find them articulated
in terms of a set of statements, sentences, propositions and formulae (or a
combination of them); also an important part of them appeal to syntactical
notions, mainly "derivability", while others appeal to semantic notions such
as: "logical implication" (in the sense of semantic validity), "true under an
interpretation" and "to have a model". Consequently, if one wants to artic-
ulate a definition of inconsistency based on them, the same variations will
be present.

Considering the question of whether the world (or mind-independent reality) is consistent or inconsistent, these definitions are not useful. The
main reason for this is that, in order to talk about a mind-independent
reality, the straightforward option would be to talk about objects, events,
state of affairs and/or some similar notions, which are not present in the
standard definitions. Moreover, these definitions are articulated in terms of
truth-bearers, while objects, events and state of affairs, normally, are not
considered as truth bearers. Thus, despite the differences about which is
the preferred option in each definition about truth-bearers, they cannot be
applied –as they are– to non-truth-bearers. To overcome this problem, one
could try to change the clause "group of statements (sentences, propositions,
formulae)" to something like "group of elements", thus allowing that these
definitions could be applicable to other elements, such as objects, events
and/or state of affairs. But then a mayor problem will emerge. As we
saw, these definitions employ syntactic and semantic notions that cannot
be applied to these other elements without falling in some kind of category
mistake: it would be wrong to say that an object is derivable from another
object(s), or that an object logically implies another object, or that a set of

---

[1] With the aim of grouping them up, one can present the following definitions:

1) A set of statements* is consistent if for no statement* it is the case that the conjunction of that statement* and its negation is derivable from that set of statements*
using the accepted rules of derivability.

2) A set of statements* is consistent if for no statement* it is the case that both that
statement* and its negation are logically implied by the set of statements*.

The first one is usually known as the syntactic sense of consistency and the second as
the semantic sense (cf. Detlefsen / McCarty / Bacon 1999: p. 25). Yet there are other
notions that have been traditionally considered as semantic definitions of consistency (cf.
Audi (ed.) 1999: p. 177), among which the most prominent could be presented thus:

3) A set of statements* is consistent if they can all be true under at least one interpretation.

And in a more specific model-theoretic vein we found this:

4) "A set of formulae X is called semantically consistent if X has a model." (Marciszewski 1981: p. 276).

* In these definitions, 'statements' may be substituted (or followed) by 'sentences',
'propositions' or 'formulae'.

events is truth under some interpretation, and so on. Consequently, neither the syntactic definitions of consistency/inconsistency, nor the semantic definitions can be used to talk about the world (or mind-independent reality), and it would not only be a matter of expanding the domain of application of the definitions, since it is mainly due to the concepts upon which each definition is structured.

The definitions of consistency/inconsistency can also be extended to include thoughts, beliefs and theories (as in Mautner 1997: p. 101), but still they will not be talking directly about the world (in the sense of mind-independent), or at least not about of a non-Platonic world.

One way of overcoming this kind of problems has been suggested by Priest, thus:

> "Of course, the world as such is not the *kind* of thing that can be consistent or inconsistent. Consistency is a property of sentences (statements, or whatever), not tables, chairs, stars and people. However, it might be suggested, to say that the world is consistent is to say that any true purely descriptive sentence about the world is consistent." (Priest 1987: p. 200)

Yet he immediately admits that "What we are to make of the notion of a purely descriptive statement is a moot point." (Ibid. p. 200). Taking that suggestion forward, we could assume that the claim 'the world is inconsistent' would be understood as 'some true purely descriptive sentences about the world are inconsistent'[2]. But if we focus on the consistent/inconsistent part, several problems emerge.

Batens' approach to the subject is quite judicious (cf. Batens 1999: pp. 266f.); indeed, he has recently said:

> "Consistency refers to negation, and that occurs in sentences, not in facts, events, and processes that make up the world. So, the claim [the world is consistent] refers to a description of the world, and hence presupposes a language L and a relation R that ties the language to the world." (Batens 2002: pp. 131.)

After analysing several problems, he comes to some conclusions, being the most relevant, for my concerns here, the following: First, no one has got even close to prove that "whatever the world looks like, the true description as determined by L and R is consistent." (Batens 2002: p. 131; cf. Batens 1999: p. 266f.). Second, even if that was accepted for the sake of the

---

[2]Note that I am not interpreting it as '*any* true purely descriptive sentence about the world is inconsistent' which would be a position that Priest has characterized as "*trivialist*" (cf. Priest 2000: p. 3).

discussion, "there is no warrant at all that humans will ever be able to handle L or to sufficiently get a grasp on R in such a way that our knowledge of the world will be consistent." (Batens 1999: p. 267f.). Third, he adds that "no one demonstrated that a specific possible structure of the world *cannot* be consistently described with respect to some L and R, and that no one demonstrated that our future knowledge is necessarily inconsistent" (ibid. p. 268). So, under this understanding of the claim that the world is consistent, neither the positive nor the negative have been proved. Batens, in these and other works, has given several reasons why it is plausible to think that inconsistencies will often emerge within our knowledge, which justifies the need for paraconsistent logic, independently of ontological claims about the consistency or inconsistency of the world.

In Bobenrieth 2003 (particularly in pp. 236-240) I have raised some questions about the possible interpretation of the claim that 'a true purely descriptive sentence about the world is consistent/inconsistent'. For me the main problem is how consistency/inconsistency is to be understood here. It is useful to focus on the following question: with what would be inconsistent this pure descriptive sentence. Three main options emerge: 1) with other sentences, 2) with the world, and 3) with itself. Each option has to be carefully analysed, but let me highlight some problems. If we take the first option, then the in/consistency of the world would be reduced to a relation among sentences. If we take the second option, then our standard definitions of in/consistency are not applicable, as long as we aim to deal with a world that contains some non-linguistic constituents. The third option seems to me the most revealing, because in analysing it we get to two issues that I think are the main ones in relation with in/consistency: negation and the relations of compatibility/incompatibility. An important part of Bobenrieth 2003 is devoted to study these issues, yet we will come back to them in the last section of this paper.

At this point, I would like to raise a question that I did not present in that text. As it is well known, Priest's preferred semantic is a relational one, where a "true contradiction" gets both truth values: {true, false}. Now, assuming that the claim that the world is inconsistent is to be understood as 'some true purely descriptive sentences about the world are inconsistent', then, being those sentences "rightly" inconsistent, both truth values would be assigned to them, hence we will have that our 'true purely descriptive sentence' is false (yet also true). Considering this, one may conclude that it is not a "correct", "right" or "adequate" description of the world, simply because it is stated as false. It could be said that this is question begging because, accordingly to that view, the fact that the world is inconsistent is tantamount to accept that some of its "correct" descriptive sentences are true and also false, which is fair enough. But in order to make sense of the situation I can see two options: first, a distinction would have to be provided between the meaning of 'true' within the phrase 'true purely descriptive sentence' and 'true' as a semantic evaluation, which would be particularly troublesome for a view close to a disquotational conception of truth. A second option would be to incorporate 'false' within the description of the sentence, so it would end up being something like 'to say that the world is inconsistent is to say that some true and also false purely descriptive sentences about the world are inconsistent', which –I think– would be clear, but not clarifying at all.

In this section, I have tried to show that the claim that 'the world is consistent' leads to several problems of interpretation: first, it is patent (even obvious) that the usual definitions of consistency used in logic and philosophy are not applicable without substantial amendments; second, the way these modifications are to be carried out is anything but a straightforward matter; third, other possible interpretations of that claim, such as Priest's, lead us to further problems and do not help much in clarifying the issue. Finally, let me emphasize that those problems also apply to the opposite claim, i.e. 'the world is inconsistent'.

2   General positions

Considering that just an elementary analysis leads us to see those problems, for me it is somehow bewildering that there has been a long-standing philosophical debate articulated in terms of the consistency or inconsistency of the world. Moreover, there are three questions than can help us in order to schematise the main philosophical views about consistency/inconsistency. First, is the world or reality consistent or inconsistent? Second, is consistency an absolute requirement of rationality?[3] And, third, what is the relationship between these two issues? These questions can be interpreted in different ways, and, in general, there have been different answers put forward to them throughout the history of philosophy. My aim here is not to review them, but, rather, to present a schematic view of possible answers.

The most common option, which could be labelled as the classical position, would answer affirmatively the first two questions: the world is consistent and consistency is an absolute requirement of rationality. Concerning the third question, there has been more diversity among the people who defend the two previous answers, but the common core would be that there is a close link between the two issues, and the variance would be about

---

[3]This question does not assume a particular view on rationality, rather, it points to rationality as a generic issue or problem within philosophical thought. Carlos Verdugo made me see that this clarification was needed.

which is the most determinant, and how one affects the other. If we follow ukasiewicz's (cf. ukasiewicz [1910] 1971: p. 488) formulation of the principle of non-contradiction, we can distinguish two main possible options within the classical position: one stressing the ontological formulation of the principle of non-contradiction and the other stressing the logical formulation. They constitute different ways of answering the question about the relationship between the two issues. The first option would emphasise that it is the constitution of the world what establishes the requirement of non-contradictoriness, which would be an Aristotelian position; while the second option would emphasise non-contradictoriness as an internal requirement of our knowledge about the world or reality, which can be seen as a Kantian position.[4] A third position would be not to emphasise any of the two instances, but simply to use one formulation or the other depending on the context, without engaging in an attempt to answer the third question. I think this is quite a common attitude among the people that have accepted the classical view without needing to go deeper into its foundations. Certainly, other options may be identified within the classical conception, but these three seem to be the main ones.

The opposite flank would claim that the world or reality is inconsistent, and that consistency is not an absolute requirement of rationality. Again, both assertions can be interpreted in different ways, but whoever accepts both of them, formulated in these or in some other general terms, would position him/herself in clear antagonism to the classical position. I will call it the anti-classical position. With respect to the third question, and in parallel to the situation within the classical position, there are several options, but similarly the main two would be: one stressing the role of inconsistency within rationality and the other stressing the inconsistency within reality. Hegel and Marx may be considered as paradigmatic figures in each option, and the famous inversion of the Hegelian dialectics by Marx may be read in these terms.

Having these classical and anti-classical positions as parameters, we can find several other possibilities somewhere in between. The first one would be to claim that the world or reality may be inconsistent but still consistency has to be a prerequisite of rationality. Thus, the fact that the world may be irrational, as it were, would not necessarily undermine consistency as a rational requirement. We may aim for consistency in our discourse about the world even if it may be itself inconsistent. The straightforward way to support this view would be answering the third question by saying that they are independent issues and each one has its own grounds. Note that to say that the world may be inconsistent does not mean that it is inconsistent in

---

[4]See, for example, Kant [1800] 1988: p. 57.

all cases; just one case of inconsistency in the world would be sufficient.

The converse position would be to say that the world or reality is consistent but consistency does not have to be a requirement of rationality. Again, the straightforward option would be to claim that consistency of reality is one thing and another is whether any structure has to be consistent in order to be considered as rational; but then further explanations would have to be given in order to justify the exclusion of the requirement of consistency from rationality. As in the previous position, only one case where keeping an inconsistency would appear as the rational thing to do, would –in principle– be sufficient, but considering that consistency is regarded as such a basic requisite of rationality, it seems that a stronger case would have to be built to defend this position.

These are the four options in which either an affirmative or negative answer to the first two questions can be given; yet, the third question allows several variations within the range of these four main options. However, there is also the option of assuming an agnostic position with respect of the first two questions, which opens up more possibilities: it is possible to be agnostic, in the sense of suspending the judgment, with respect to the consistency of the world or reality, but still hold a position about the second question, either postulating that consistency is a necessary requirement of rationality or denying that it has to be accepted unrestrictedly. Likewise, it is possible to be agnostic, or not to assume a position, about consistency as a rational requirement but still hold a position about the consistency or inconsistency of the world or reality. Finally, it would be possible to be agnostic about both issues, which would constitute a fully-fledged scepticism about the whole matter. Among these agnostic positions, the third question does not seem to be so relevant, because being agnostic about any of the two parts of the relation will affect the need for an explanation about their relationship.

Despite the diversity of these positions, they share the assumption that the first question is sound, i.e. that it makes sense to ask whether the world or reality is consistent or inconsistent. So, I am understanding the agnostic attitude not as denying that there is a fact of the matter in saying that the world is consistent or inconsistent, but as avoiding committing oneself in one way or the other. That is, as a suspension of judgment, but not as a rejection of the prospect of the judgment in itself.

3   Paraconsistent positions

Among the researchers in paraconsistent logic one can find different answers to those three questions, yet not necessarily addressing them directly or in the terms that I have proposed, but I think this schema is useful in order to set together the different views. In general, one can say that all the paraconsistent positions answer affirmatively the second question, otherwise the whole enterprise would be a form of irrationalism, and I do not know of any paraconsistent logician that is prepared to accept that. Regarding the first and third question, there are several positions among the people working on paraconsistency, even though, there are many researches that have not addressed them in their publications, and some of them seem to be quite happy with that.

Let us concentrate on the first question. Considering the consistency or inconsistency of the world there is a whole range of possible positions within paraconsistency. At one extreme, would be the view that the world is consistent, but due to diverse reasons we have to deal with inconsistencies so we have to have some sort of paraconsistent logic. At the other extreme, would be the view that the world is inconsistent and consequently the correct logic to deal with reality has to be one that allows to handle inconsistencies. Several ways of categorizing the positions that tend to one or to the other end have been proposed, yet, informally, they are frequently called: 'weak paraconsistency' and 'strong paraconsistency', respectively. I think this is a misleading terminology, mainly because one can be a "strong supporter of paraconsistency", without supporting the claim that the world is inconsistent.[5]

Let me briefly review the positions that, for our purposes, are the most relevant.[6] Considering the "weak" side, it is difficult to find a "fully-fledged" defence for the consistency of the world made by someone working on paraconsistency. However, in the last years some authors, most of them living in Canada, have orientated their work in an opposite way to what they present as the "radical" option within paraconsistency (i.e. dialetheism) (cf. Brown 1999, Brown / Schotch 1999: p. 269) and even have explicitly defended some kind of 'weak paraconsistency' (cf. Woods 2003: p. 108ff). Furthermore, even though from a different perspective, Batens has defended for years the idea that "Even if the world is consistent, our knowledge often requires that we apply a paraconsistent logic rather than a Classical Logic." (Batens 1999: p. 268). Though, it has to be noticed that this is a

---

[5] As it has been said many times, one can be a supporter of paraconsistency without supporting dialetheism (which is the position normally associated with 'strong paraconsistency'); moreover, comparing the option of supporting *only* paraconsistency and supporting paraconsistency and dialetheism, the latter does not add anything to paraconsistency as such, while it suggests that paraconsistency "is not enough", so it seems that the first option is a form of "stronger" paraconsistency (paraconsistency and nothing more than paraconsistency). Anyhow, there is no much point in quarrelling about who is the "strongest paraconsistentist".

[6] For a more extended presentation see Bobenrieth 1996: pp. 332-351; Bobenrieth 2003: pp. 215-243.

conditional statement because –as we have seen– he has pointed out several problems related to the statement 'the world is consistent'.

A large majority of the people working on paraconsistency could be placed in an intermediate position, where there is no much commitment about the consistency or inconsistency of the world. Newton da Costa has defended such a position in several places, being da Costa 1980 the most comprehensive presentation. There he says: "What one can say is that *a priori*, especially appealing to logic, contradictions can be neither justified nor excluded. The existence, or not, of real contradictions will only be settled down *a posteriori* by science." (da Costa 1980: p. 208 [tran.]). Even so, he is inclined to believe that it is more plausible to prove the existence of real contradictions, given that only one case would be needed to show that they do exist, while to refute it a finite number of cases —no matter how big this number may be— would not be enough. One of his overall conclusions is that: "Knowledge is possible even if the universe is inconsistent." (Ibid. p. 222 [tran.]).

Later on, da Costa has developed his ideas, especially in works with Béziau and Bueno. Among them special attention should be given to da Costa / Bueno 1996, where they start proposing a "division of labour", between research on foundations of paraconsistency and the philosophy of paraconsistency, thus making a direct analogy to the difference that has been made between the foundations of mathematics and the philosophy of mathematics. Within this framework, da Costa and Bueno propose an agnosticism in two senses: agnosticism with regard to the truth of paraconsistency, and agnosticism in connection with the existence of true contradictions (cf. da Costa / Bueno 1996: p. 57). Thus, they are affirming that it is not necessary to assume a commitment with a unique "paraconsistent vision" that is right, and, therefore, to exclude other perspectives about contradictions. They consider that notions like 'pragmatic truth', illustrate how heuristic and pragmatic considerations are enough to provide sense to investigations of formal systems that accept some type of contradictions. It is then, according to the authors, a question of articulating a fallibilist and pluralistic proposal, and not a relativist one. (cf. ibid. p. 59).

This double-headed agnosticism also points out that to postulate the existence of contradictions at the abstract level of formal sciences, such as Russell's set, is different from defending that reality is actually contradictory (cf. ibid. p. 58). Then, although they assume that in certain ways "true" contradictions can "exist", this does not affect the possibility of being agnostic with regard to the existence of real contradictions (unless a certain type of Platonic world is postulated). So, considering the three questions that we have been reviewing, the position of da Costa and Bueno would be agnosticism with respect to the first question, but not with respect to the second because they think that there are several contexts and/or matters in which it is rational to accept some contradictions.

There are other proposals made by different authors that are quite relevant to the topic. For me, the ideas put forward by Carnielli, Lima Marques and Fariñas del Cerro[7] deserve especial attention. In some papers these authors assume that contradictions are linguistic entities that are given in certain conceptual system linguistically structured; therefore, to accept something as a contradiction requires that in the system there is a negation operator that allows such articulation. They illustrate this point stating that they do not worry whether a simultaneously round and square object exists or not, they only worry about how this situation is expressed in the formal system, thus accepting it as a contradiction only if within the system it is established that something that is round is *not* square. This is a position that does not have an ontological commitment with contradictions, but a "linguistic" commitment, and they link the concrete intuitive concept of contradiction with negation as a linguistic reality.

Moving forward to the other side, i.e. authors for whom the answer to the first question is that the world is inconsistent, we find proposals such as Peña's view[8] on the "contradictoriness of the real" grounded on his "gradualistic" conception of reality; yet, without any doubt, dialetheism is by far the most prominent. It was defined by Priest and Routley as the view that there are true contradictions (dialetheias). In their first systematic presentation, after giving several arguments to support dialetheism, they concluded that "What mainly prevents the acceptance of this view is the ideology of consistency: the deepseated and irrational view that the world is consistent." (Priest / Routley 1989: p. 528). In Priest 1987 there is a major defence of dialetheism, which was followed by many articles discussing it and some related issues. These are so well known among people working on the philosophical aspect of paraconsistency that there is no much point in reviewing them here.

4   Some comments on Priest

Here I just want to highlight some points and make some comments about Priest's view. According to him, to accept that dialetheism is true is to accept that there are some sentences that are both true and false (cf. Priest 1987: p. 84), being the liar sentences the most paradigmatic case. However,

---

[7]See Carnielli / Lima Marques 1991 and 1992, Carnielli / Lima Marques / Fariñas del Cerro 1992.

[8]See Peña 1991: pp. 90 ff; 1993: pp. 259 passim. For a contrast with dialetheism see Peña 1996.

besides the case of the paradoxes of self-reference, the following are the main examples of dialetheias presented by Priest: 1) transition states, 2) some of Zeno's paradoxes, 3) certain kinds of moral and legal dilemmas, 4) borderline cases of vague predicates, 5) certain quantum mechanical states, 6) multi-criterial terms (cf. Priest 1998, see also Priest 2002: p. 291). Now, one may ask: does considering them as *true* contradictions necessarily yield to accept them as *real* contradictions? For example, to grant that the liar sentence is both true and false, does it make it a *real* contradiction? Of course, the main point here is how 'real' will be understood; for instance, it can be said that language is part of reality, so sentences are "real". But, in that case, it does not seem to be much opposition against the "ideology of consistency", because many of the people that assume that 'the world is consistent' will also be prepared to accept that natural languages have some sort of inconsistency; hence, for them, that part of "reality" would be inconsistent. Even if someone might accept that the best semantic theory for natural languages has to be inconsistent, it does not follow that s/he is accepting that 'the world is inconsistent', having in mind an extra-linguistic world. The other examples of dialetheias would require careful examination, but I cannot go deeper into that now; however it is worth noticing that examples 4 and 6 also inherently involve language.

At this point, it is important to recall what we have seen (sec. 1) about Priest's proposal to understand the claim that 'the world is consistent'. Recently, he has revisited the issue in relation with empirical sciences, and he has presented the following conclusion:

> "Thus, if we are realists, we will let our best theory, provided that it is not ruled out as a candidate for truth on other grounds, inform us as to what reality is like; and if our theory is inconsistent there is no reason to suppose that the theory does not get it right: reality itself is inconsistent. In other words, inconsistencies of *this* kind in science do not mandate that the acceptance of the theory or theories in question be provisional in any special way." (Priest 2002a: p. 126)

Meanwhile, in Priest 1999a after saying that is possible to perceive that something is not the case, he argues in favor of the thesis that if there were contradictions in our perceivable world, then we will be able to perceive them. After that, his conclusion is this:

> "Consider the observable world, i.e., all that is observably the case. If there were inconsistencies in this, it would follow from the above that we would perceive them. But apart from the odd visual illusions, we do not: our perceptions of the world are

entirely consistent. Hence, the observable world is consistent."
(Priest 1999a: p. 444)

So, our observable world is said to be consistent (yet, this is presented as an *a posteriori* and fallible thesis) and the world in general would be inconsistent if our best theories tell us so. But considering that our current 'best theories' tend to be pretty much consistent (and if they appear to be inconsistent, their proponents tend to make good efforts to solve the problems or at least to hide them), then we are not left with much of an 'inconsistent world'. There, the cases of inconsistency, leaving language peculiarities aside, will be mainly events that we cannot perceive because they happen too fast or because they belong to the quantum level. Although it is right to say that just one case would be enough to refute the thesis that the world is consistent, much more is needed to defeat the "ideology of consistency", particularly if it works so well in *our* world: the world that we perceive and talk about.

5   An anti-realist view about inconsistencies

At the end of sec. 2, I have pointed out that those general philosophical positions share the assumption that it makes sense to inquire whether the world is consistent or inconsistent. Furthermore, according to what we have seen in sec. 3, it is also the case for several paraconsistent views: whether claiming that they do not need to go against the classical position, or postulating that the world is inconsistent, or even saying that it is not known which is the case, and thus being agnostic about the point, they are accepting that there is a fact of the matter on whether the world is consistent or inconsistent. My proposal is to challenge this common view by radically assuming that the world is neither consistent nor inconsistent and exploring the consequences that follow from that, aiming to disentangle the problems around consistency/inconsistency from the ontological stance to which it is usually related.

It could be said that what I am saying is quite obvious, because *strictly* speaking the world is neither consistent nor inconsistent, as far as we are considering the non-linguistic components of the world or reality, and I will be happy with that. The main problem lies in that people, although accepting that "rigorist" statement, tend to assume that there are some ways in which to talk about the in/consistency of the world makes sense. But then, those "analogical" or "extended" interpretations of in/consistency need to be analysed in order to find out on which ground they may be based. I have undertake this task with several of them in Bobenrieth 2003 (particularly chap. I-IV), which allowed me to indicate some strong ontological presuppositions and/or requirements that would be needed to sustain such interpretations. Nevertheless, I cannot prove the needed features (or elements) of reality do not hold or exist, but I can argue that we can do fine, or better, without assuming or accepting them. So, it is clearly an anti-realist move.[9]

Due to limits of space, I just want to touch now on perhaps the most common of those interpretations. Someone can say: you are right that the world is not consistent or inconsistent as such, but what is meant by those claims is that there are contradictions in the world, or there are not. Then, a study of the definitions of 'contradiction' is required (see Bobenrieth 2003: chap. I); yet, at the very beginning we face a similar problem with the case of 'consistent' and 'inconsistent': the usual definitions of 'contradiction' are also articulated in terms of linguistic items. Again, we may try to "extend" the definitions in order to include other elements, but that is not the only problem, because these definitions are grounded on notions that are, in principle, only applicable to linguistic items. I have pointed out (Bobenrieth 1998: p. 24) that there are two main tendencies among the definitions of 'contradiction': to refer to the situation where some statement is the negation of another, or to the situation where two statements cannot be true, neither can be both false. Classically these definitions are taken as closely interlinked, but paraconsistent logic has shown that this involves several assumptions. Thus, taking them separately, and with the aim of "extending" these definitions in order to apply them to facts, objects, events, state of affairs, etc., we have to ask questions like: what could it mean to say that one fact is the negation of another? What could it mean the claim that two states of affairs cannot be both true neither can be both false? A plausible step would be to make further modifications such as to relate negation with properties, and, for the second case, not to talk about being true or false, talking instead about two states of affairs that cannot hold together, but one of them has to hold. But then we would have to find two states of affairs that are not only mutually incompatible but also jointly exhaustive, which would normally require the use of negation. From there several other problems emerge, but for the moment it is important to realize that these "extensions" have taken us quite far from those original definitions of 'contradiction'.

Another route to understand what does it mean that the world has con-

---

[9]Interestingly enough, Priest, in his "An anti-realist account of mathematical truth", has said:

"I do not wish to give a critique of realism here. Its problems are well known. However, the sheer *prima facie* implausibility of these supra-sensible objects (to which it is easy to become blind) means that, other things being equal, an account of mathematical truth which avoids invoking real mathematical objects is preferable. It is my claim that such an account can be given. [...]" Priest [1983] 2002: p. 119.

tradictions or not, could be to use some formulation of the principle of non contradiction: for example: "a single thing cannot have contrary properties in the same part of itself, at the same time and in the same respect." (Detlefsen / McCarty / Bacon 1999: p. 60). Considering that it is also possible to talk about contradictory properties (not only mutually exclusive but also jointly exhaustive[10]), I have proposed[11] to use the term 'incompatible' in order to embrace both situations. Consequently, a better formulation would be: 'a single thing cannot have incompatible properties...'. There a main question would emerge: What determines that a property is incompatible with another? That may point to the long standing discussion on the nature of the properties, with its traditional positions: realism, conceptualism and nominalism, but the key issue now is not so much the nature of the properties themselves, but the nature of the relationships of incompatibility among them (although the differences in the positions about the former problem would have consequences in the latter).

Considering that my main interest here is how we can understand the claim that the world is consistent, let us focus on how this can be expressed in terms of properties. It would have to be something like: there are properties that are incompatible and in the world (or reality) there is not any instance in which these incompatible properties obtain at the same time and in the same respect. Several problems may arise here, but let me concentrate on the following point: What happens if a case emerges where two properties, that had been considered as incompatible, seem to obtain at the same time and in the same sense with respect of the same entity? One option would be to say that there is an error concerning our appreciation that those are the properties that are actually obtaining; another would be to say that there is an error regarding their simultaneity, or that they are obtaining in the same respect or related to the same entity. But if none of these errors seem to be present, then there is always the possibility of saying that that by means of that particular case it has become evident that these properties were not incompatible, since there is at least one case where they are in fact compatible.

A good example of this is the discovery of the platypus in Australia.

---

[10]I am using the terminology that Brody (1967: p. 61) applies to terms.

[11]In Bobenrieth 2003: chap. II and III, I have proposed to use 'incompatible' as a generic term that would include incompatibility among statements (sentences, propositions and formulae) and also incompatibility among other elements such as properties and terms. For the first case of incompatibility I have suggested that a more specific name can be 'inconsistency', which would include both the contradictory opposition among statement (sentences,...), as well as the contrary opposition. It has to be noticed that this is only a terminological proposal, that comes after, and as a result of, analysing different situations that one may find described as "inconsistency".

Before that event it was assumed, in Western science, that being a mammal was incompatible with being oviparous (i.e. to lay eggs). The platypus had all the other characteristics of mammals, but it reproduced itself by means of eggs in a similar way to reptiles and birds. Originally it was thought that there ought to be a mistake, but later it was confirmed that the information was correct: the same animal has almost all the other characteristics of mammals (hair, feeding their offspring with milk by means of mammary glands, a large brain, and a complete diaphragm) but it lays eggs. It was also discovered that, like birds and reptiles, it has cloaca. Then the solution was to establish a new order (monotremes) within the class of mammals. Later on, it was discovered that two other species of animals from Australia and New Guinea, called echidnas, share the same characteristics.

Thus, it was found that two properties, that were considered incompatible, actually obtained at the same time and in the same respect in a group of individuals. Then, the solution was to modify the judgment of incompatibility that so far had been made regarding mammals and laying eggs. According to the previous taxonomical conception it would constitute an inconsistency, but instead of taking it as "an inconsistent reality" the option that was taken was to say that the properties of being a mammal and being oviparous were actually not incompatible. In short, it was *not* accepted as a case in which 'the world is inconsistent'.

In the light of cases like this, the principle of non-contradiction, and the related statement that 'the world is consistent,' amount to stating a triviality, something like: what is incompatible has to be actually incompatible because if it ends up being compatible then it is (was) not incompatible. Furthermore, the statement 'the world is inconsistent' would amount to say: there are supposedly incompatible properties that are compatible. Neither of these options helps much to make sense of the respective positions.

Perhaps a better way of presenting the matter could be to claim that the world is constituted by objects, events, state of affairs, facts, etc., and they have properties which are, also, in the world, or are real, in the sense of being mind-independent. On top of that, there would be relations among properties, incompatibility being one of them, which would also be, somehow, in the world, in the sense of mind-independent. Then, our judgments of incompatibility would have to correspond to those "real incompatibilities" among those properties in reality. Consequently, every time that two properties that were assumed as incompatible, actually appear in the same entity at the same time, the conclusion would have to be that we were wrong about them, that is, our judgment did not correspond to or reflect those real incompatibilities among properties.

Such a view would constitute a strong type of realism, and as such it could be submitted to the nowadays customary criticisms made to realist proposals. But in this case it would not only have to face the problems that confront any realist proposal about properties, but also would have to defend the mind-independent existence of relations among properties. Having done that, it would then have to show how compatibility/incompatibility would exist in reality. Paraphrasing the known expression, it would amount not only to try to carve reality in the right joints, but also to expect to get from reality the perfect-fitting packages for the pieces.

Negation is the other big issue related with contradictions, particularly the bearing of negation in reality. This is a subject that could take us to the long-standing philosophical discussion about "negative facts",[12] or to many other logical and philosophical discussions. Due to limitations of space, I will limit myself to summarize my position about negation.

The main idea is that there is nothing in the mind-independent reality that corresponds to, or is the correlate of, the mental operation of negation. This seems to me as a quite straightforward thesis, which was already present in the traditional classification of negative particles, like 'not', as syncategorematic expressions; as Batens (1999: p. 270) has said, 'not' belongs to the category of "non-referring words". At the basis of my view lies the rejection of any kind of "negative perception" or perception of negative facts, so I am in the line of something that has been said many times, but particularly by the two forerunners of paraconsistent logic ukasiewicz (cf. [1910] 1970: p. 507f.) and Vasiliev (cf. Vasiliev's text in Arruda 1990: p. 44), and it stands against opposite views, such as Priest's (cf. Priest 1999a: p. 444; 2002a: 120f). The other main base of my position is that negation has to do only with meanings, or, as Heinemann (1944: p. 147) said, it is a relation of meaning and not of being. Only where there are meanings it makes sense to talk about negation, understanding meaning as mind-dependent. Negation does not reflect or represent something in reality but something that we do with reality.

Furthermore, negation may be involved in different forms of opposition but it must not be identified or restricted to any of them. There are different types of negative expressions, as Horn has convincingly argued for (see Horn 1989); in different contexts, or applied to different expressions, they can have different meanings. Our natural languages contain several negative particles which can be used in different ways and their contribution to the meaning of the expression in which they are used depends on several factors. They generally aim to express some form of opposition, but there are different types of opposition, ranging from a radical opposition that would

---

[12]It is worth noticing that it is an issue very much discussed in the "classical Indian philosophy" (see Gillon 1998).

aim to establish a total differentiation between two or more elements, to a minimal differentiation between very similar elements. Both ends of this range are ideal limits since, in principle, at one end it always seems possible to find some common features between two very different entities, and, at the other, minute differences may not be enough to establish a distinction. Yet, the central goal is the establishment of dichotomies through simple "all-purpose" linguistic-conceptual devices, but that can be done in different levels and in different ways, so we must be aware of the differences in the articulation and use of such dichotomies. Thus, we have different forms of negation that can express different types of opposition (without having a one to one correlation). Although, there is a common core that makes them recognizable as negation. That common core allows us to distinguish negation from other notions; in particular, I agree with Priest's distinction between denial and negation (cf. Priest 1999: 113f.).

In sum, I think that negation is a mental operation that we use in order to understand reality and ourselves. It is a cognitive tool as well as an expressive tool; we use it to establish distinctions about what we experience and what we think. It is not given to us by reality, but it is one of the most basic ways in which we deal with reality. It is at the basis of our conceptual apparatus, yet it is used in several different ways. This position can be seen as an anti-realistic stance about negation.

Coming back to our main problem, if the notion of contradiction is taken as based on negation, then, due to the separation between negation and mind-independent reality that my position sustains, it would also yield to separate contradiction from mind-independent reality. If we need to have negations in order to have contradictions, then we cannot have contradictions in mind-independent reality. Contradictions would be made of something that does not belong to the world.

This view about negation is closely related with the thesis that the world is neither consistent nor inconsistent. It can constitute a main component of an anti-realist view about contradictions. The next step would be to relate negation with the relations of incompatibility in order to cover the two main senses in which 'contradiction' is understood (I have attempted to do that in Bobenrieth 2003: chap. VII). Now, if we link all of this with the difficulties that presents the attempt to apply 'consistency' or 'inconsistency' to the world (to its non-linguistic components), then we have the basic setting for an anti-realist view about inconsistencies.

6   Conclusion

I have tried to show that an alternative view to the main views about inconsistency ought to be considered. Regarding the three main questions proposed, such a view will dismiss the question of whether the world is consistent or inconsistent, and will apply itself to fully address the second question: consistency as a requirement of rationality. Consequently, the third question, i.e. about the relation between both issues, would be answered by saying that the problem of inconsistency is not an ontological problem but a problem of rationality, and only as such it can be properly addressed.

This alternative view may contribute to change the discussion about inconsistency; particularly, it would escape from the opposition between the classical and anti-classical positions about contradictions/inconsistencies by arguing that they are both wrong in assuming that the world is either consistent or inconsistent[13]. It would not aim to refute the traditional Aristotelian view, but rather to show that the framework in which it is placed is inadequate and misleading. The goal would be to shift from the usual ontological arguments to arguments regarding rationality.

This is just a starting point, since if we assume that inconsistency is not a problem about reality, a whole range of questions emerge. For instance, one of the most important is to understand why consistency is so widely considered as a corner-stone of rationality. I think it is not enough to say that people have been indoctrinated in "the ideology of consistency"; there is much more in the drive towards consistency that a "received attitude". In Bobenrieth 2003 (chap. VII) I have tried to give an explanation, but the most important thing here is that an anti-realist view about inconsistencies may open up the discussion, allowing alternative ways of dealing with this and other relevant issues.

One last comment: it is clear that the polemic about the liar sentence is an issue where a paraconsistent view has much to contribute, but to my mind it has been somehow overestimated. There are many other issues in which different views about inconsistencies are much needed. For instance,

---

[13] PS.: Priest commenting this paper has suggested that my position could fit into what Mares calls 'semantic dialetheism'. According to Mares, semantic paraconsistency holds that "there are no inconsistent things but inconsistencies arise (or may arise) because of the relation between language and the world" (Mares 2004: p. 265). I fully agree with the second part of that claim, but I disagree with what is implicit in the first part since although I would accept that there are no *inconsistent* things, yet I will have to add that neither there are *consistent* things. If we look at how Mares characterizes the 'semantic theories' on vagueness we find that they "try to retain traditional metaphysical intuitions –such as the view that real things and properties have sharp boundaries" (Mares 2004: p. 264). Concerning semantic paraconsistency, he says "it does not force us to relinquish the intuition that the world is consistent" (ibid. p. 274), and semantic dialetheism would be a particular version of semantic paraconsistency. I do challenge the traditional metaphysical intuition that the world is consistent, but also the opposite intuition, i.e. that the world is inconsistent, so I do not think that my proposal fits into Mares' classification, which is also the case for the agnostic position, as Mares himself accepts (cf. ibid. p. 266).

to get deeply involved in the discussion about rationality would be a step forward for paraconsistency; many exciting problems will emerge if we focus not so much on the paradoxes of language and more on the "paradoxes of human mind".

BIBLIOGRAPHY

[1] ARRUDA, Ayda I.: 1990 *N. A. Vasiliev e a Lógica Paraconsistente*. Campinas: Centro de Lógica, Epistemologia e História da Ciência-UNICAMP, 1990.

[2] AUDI, Robert (ed.): 1999 *The Cambridge Dictionary of Philosophy*. Cambridge: Cambridge University Press, 1st edn. 1995, 2nd 1999.

[3] BATENS, Diderik: 1999 "Paraconsistency and its relations to world views", *Foundations of Science* 3, 1999, pp. 259-283.

[4] BATENS, Diderik: 2002 "In defence of a programme for handling inconsistencies", in Meheus (ed.) 2002: pp. 129-150.

[5] BOBENRIETH M., Andrés: 1996 *Inconsistencias ¿Por qué no? Un estudio filosófico sobre la lógica paraconsistente*. Bogotá: Colcultura, 1996.

[6] BOBENRIETH M., Andrés: 1998 "Five philosophical problem related with paraconsistent logic", *Logique et Analyse* 161-162-163, 1998, pp. 21-30.

[7] BOBENRIETH M., Andrés: 2003 *Inconsistencies and their rational tracks*. University of Leeds, PhD Thesis, March 2003.

[8] BRODY, Boruch A.: 1967 "Glossary of Logical Terms", in Edwards (ed.): *The Encyclopedia of Philosophy*. New York: The Macmillan Company & the Free Press / London: Collier-Macmillan Limited, 1967: vol. 5, pp. 58-77.

[9] BROWN, Bryson: 1999 "Yes, Virginia, there Really are Paraconsistent Logics", *Journal of Philosophical Logic* 28, 1999, pp. 489-500.

[10] BROWN, Bryson / SCHOTCH, Peter: 1999 "Logic and Aggregation", *Journal of Philosophical Logic* 28, 1999, pp. 265-287.

[11] CARNIELLI, Walter / LIMA MARQUES, Mamede: 1991 "Razão e irracionalidade na representación do conhecimiento", *Trans/Form/Ação* (São Paulo) 14, 1991, pp. 165-177.

[12] CARNIELLI, Walter / LIMA MARQUES, Mamede:1992 "Reasoning under inconsistent knowledge", *Journal of Applied Non-Classical Logic* 2 (1), 1992, pp. 49-79.

[13] CARNIELLI, Walter / FARIÑAS DEL CERRO, Luis / LIMA MARQUES, Mamede: 1991 "Contextual negations and reasoning with contradictions", *Knowledge Representation: Reasoning with Inconsistency* (Proceedings of the Twelfth International Conference on Artificial Intelligence [IJCAI-91]. Sydney, Australia: 24-30 August 1991) pp. 532-537.

[14] DA COSTA, Newton C.A. 1980 *Ensaio sobre os Fundamentos da Lógica*. São Paulo: Hucitec y Editora da Universidade de São Paulo, 1980. [French translation : *Logiques classiques et non classiques*. Paris: Masson, 1997].

[15] DA COSTA, Newton C. A. / BUENO, Otávio S. A.: 1996 "Consistency, paraconsistency and truth (logic, the whole logic and nothing but logic)", *Ideas y Valores* 100, 1996, pp. 48-60.

[16] DETLEFSEN, Michael / MCCARTY, David Charles / BACON, John B.: 1999 *Logic from A to Z*. London / New York: Routledge. [Consist of the separated publication of the "Glossary of logical and mathematical terms" of the *Routledge Encyclopedia of Philosophy* (CRAIG / FLORIDI 1998)].

[17] GILLON, Brendan S.: 1998 "Negative facts in classical Indian philosophy" in *Routledge Encyclopedia of Philosophy* (CRAIG / FLORIDI ed.) London: Routledge, 1998.

[18] KANT, Imanuel. [1800] 1988 "*Logic*. New York: Dover Publications, 1988. (1st. edn. In German 1800).

[19] ŁUKASIEWICZ, Jan: [1910]1971 "On The Principle of Contradiction in Aristotle", *Review of Metaphysics* 24, 1971, pp. 485-509. [1st edn. in German 1910].

[20] HEINEMANN, F. H.: 1944 "The Meaning of Negation", *Proceedings of the Aristotelian Society* 44, 1944, pp. 127-152.

[21] HORN , Laurence R.: 1989 *A Natural History of Negation*. Chicago: The University of Chicago Press, 1989.

[22] MARCISZEWSKI, Witold (ed.): 1981 *Dictionary of Logic as Applied in the Study of Language*. The Hague: Martinus Nijhoff Publishers, 1981.

[23] MARES, Edwin D.: 2004 "Semantic dialetheism" in Priest, G. / Beall, JC / Armour-Grab, B.: *The Law of Non-Contradiction; new philosophical essays*. Oxford: Clarendon Press, 2004.

[24] MAUTNER, Thomas (ed.) 1997 *Dictionary of Philosophy*. London: Penguin books, 1997 (1st edn. Oxford: Blackwell, 1996).

[25] MEHEUS, Joke (ed.) 2002 *Inconsistency in Science*. Dordrecht / Boston / London: Kluwer Academic Publishers, 2002.

[26] PEÑA, Lorenzo: 1991 *Rudimentos de lógica matemática*. Madrid: Consejo Superior de Investigaciones Científicas, 1991.

[27] PEÑA, Lorenzo: 1993 *Introducción a las lógicas no clásicas*. México: Universidad Nacional Autónoma de México, 1993.

[28] PEÑA, Lorenzo: 1996 "Graham Priest's "Dialetheism"– is it altogether true?" (sic.), *Sorites* 7, 1996, pp. 28-56.

[29] PRIEST, Graham: [1983]2002 "An anti-realist account of mathematical truth", in Jacquette, D. (ed.): *Philosophy of Mathematics* (Oxford: Blackwell, 2002): pp. 119-127. [Originally published in *Synthese* 57 (1983), pp. 49-65.]

[30] PRIEST, Graham: 1987 *In Contradiction. A study in transconsistent*. Dordrecht / Boston / Lancaster: Marttinus Nijhoff Publishers, 1987.

[31] PRIEST,   Graham: 1998   "Dialetheism"   in   Zalta,   E. (ed.)   *The   Stanford   Encyclopedia   of   Philosophy*   (1998   edn.) httt://plato.stanford.edu/archives/win1998/entries/dialetheism.

[32] PRIEST, Graham: 1999 "What not? A defence of dialectical theory of negation", in Gabbay / Wansig (eds.) *What is negation*. Dordrecht: Kluwer Academic Publishers, 1999: pp. 101-120.

[33] PRIEST, Graham: 1999a "Perceiving contradictions", *Australasian Journal of Philosophy* 77(4), 1999, pp. 439-446.

[34] PRIEST, Graham: 2000 "Paraconsistency and Dialetheism", text for a "Research Seminar" in The University of St. Andrews, October-December 2000.

[35] PRIEST, Graham: 2002 "Paraconsistency", in Gabbay, D. M. / Guenthner, F. (eds.): *Handbook of Philosophical Logic*, vol. 6. Dordrecht: Kluwer Academic Publishers, 2002, pp. 287-394.

[36] PRIEST, Graham: 2002a "Inconsistency and the empirical Sciences", in Meheus (ed.) 2002: pp. 119-128.

[37] PRIEST, Graham / ROUTLEY, Richard: 1989 "The philosophical significance and inevitability of paraconsistency", in Priest / Routley / Norman (eds.) *Paraconsistent Logic, Essays on the Inconsistent*. München, Hamden, Wien: Philosophia Verlag, 1989, pp. 483-539.

[38] WOODS, John: 2003 *Paradox and Paraconsistency. Conflict Resolution in the Abstract Sciences*. Cambridge: Cambridge University Press, 2003.

Andrés Bobenrieth M.
Universidad de Valparaíso – Chile.
andres.bobenrieth@uv.cl

www.ingramcontent.com/pod-product-compliance
Lightning Source LLC
LaVergne TN
LVHW012326060326
832902LV00011B/1737